D0368991

143587

MODERN METHODS FOR
TRACE ELEMENT ANALYSIS

MAURICE PINTA

Research Director
Office de la Recherche Scientifique
et Technique Outre-Mer
93140 — Bondy — France

Ronald M. Scott, Professor, and **Krishnaswamy Rengan**,
Associate Professor, Department of Chemistry, Eastern
Michigan University, Ypsilanti, Technical Advisors

Translated from the French by

STS, Incorporated
Ann Arbor, Michigan

54 3
P659

ANN ARBOR SCIENCE
PUBLISHERS INC
P.O. BOX 1425 • ANN ARBOR, MICH. 48106

Alverno College
Library Media Center
Milwaukee, Wisconsin

Second Printing, 1978

Copyright © 1978 by Ann Arbor Science Publishers, Inc.
230 Collingwood, P.O. Box 1425, Ann Arbor, Michigan 48106

Library of Congress Catalog Card No. 76-050988
ISBN 0-250-40152-5

Manufactured in the United States of America
All Rights Reserved

Alverno College
Library Media Center
Milwaukee, Wisconsin

REMOVED FROM THE
ALVERNO COLLEGE LIBRARY

INTRODUCTION

While it was customary in the past to define a "trace" as any element having a concentration insufficient to permit its quantitative determination, this is no longer true. Instrumental methods have made trace analysis one of the most important fields in analytical chemistry. The term "trace" today implies a very small but measurable concentration. Our knowledge of trace elements has advanced very broadly in the past ten years because of the spectacular development of instrumental analytical techniques. Several years ago, the detection limits of the classical instrumental methods, such as colorimetry, atomic and molecular absorption spectrometry, emission spectrometry and polarography ranged between a few ppm (10^{-6}) and a few dozen ppm. Today, methods such as neutron activation, spark mass spectrometry or flameless atomic absorption spectrometry allow detections of a few parts per billion (ppb or 10^{-9}) in complex media.

The variety of physicochemical and instrumental methods of analysis is extensive today. The following are methods that are either common today or likely to be common within the next few years:

- colorimetry and absorption molecular spectrometry
- fluorometry or molecular fluorometry
- anodic stripping voltametry
- differential pulse polarography
- ionometry or specific electrodes
- atomic absorption spectrometry (flame and nonflame)
- atomic fluorescence spectrometry
- atomic emission spectrometry
- X-ray fluorescence spectrometry
- electron microprobe analysis
- ion microprobe analysis
- electron spectroscopy for chemical analysis (ESCA)
- Auger electron spectroscopy
- spark source mass spectroscopy
- gas chromatography
- activation analysis

Modern Methods for Trace Element Analysis should permit the reader to choose the most suitable method and conditions to be adapted to each particular problem. Since the objective is of a practical nature, we have included only the methods which have been proved or which are complementary to classical techniques. In this context, some explanations concerning the choice of the methods described as well as those omitted are included. Descriptions of the 17 techniques listed above would justify several books. However, for a variety of reasons, they do not meet our general objective. We will rapidly review these techniques:

Colorimetry, or more generally molecular absorption spectrometry, was the method of choice 20 years ago. It has now been superseded by atomic absorption spectrometry, which is simpler and more specific.

Fluorometry or molecular fluorometry represents a great advance when judged by the impressive number of publications, even though these deal more with the field of organic chemistry. We considered it useful to indicate the possible application of fluorometry in trace analysis in the first chapter. We consider this technique to be complementary, particularly to atomic absorption.

The *electrical methods, i.e.,* anodic stripping voltametry and differential pulse polarography, are certainly among the most sensitive methods, permitting detection limits of up to 0.01 ppb to be obtained under certain conditions. Ionometry, the use of ion specific electrodes, is another new technique. These methods are still in the stage of instrumental development and elaboration of specific methods. The operating conditions are highly dependent on the matrix composition and the pH of the polarograph solution. Because of this, these methods will not be discussed.

Atomic absorption, whether by flame or flameless techniques, will have an important place in the book. There is no doubt that they can be placed at the top of the list by their importance in trace analysis. Flameless atomic absorption spectrometry allows the development of trace analysis in the field of microchemistry.

Atomic fluorescence spectrometry cannot be neglected, although the real place of this technique cannot yet be estimated. Numerous authors consider it to be a necessary complement to the above-mentioned techniques. In particular, it lends itself to multielemental analysis.

Atomic emission spectrometry has also been included because of recent developments of new excitation sources, such as the laser, hollow cathode and, especially, the plasma (particularly a coupled inductive plasma). Emission spectrometry thus remains at the top of the list of multielemental analysis methods. It is a technique of choice because of its speed, especially in semiquantitative analysis.

X-Ray fluorescence spectrometry has advanced greatly in importance in trace analysis particularly because of the development of nondispersive analysis and the use of radioactive isotopes as excitation sources. X-Ray fluorescence has numerous advantages: it is nondestructive, it permits multielemental analysis and it displays both speed and accuracy.

Electron microprobe and *ion microprobe analysis, electron spectroscopy for chemical analysis* (ESCA) and *Auger electron spectroscopy* detect trace elements with very high sensitivity. Actually, these are methods of surface analysis applicable to the *in situ* analysis of samples in a thin and homogeneous film. They still remain in the realm of highly specialized laboratories.

Spark mass spectrometry does not belong in this book. It requires specifically qualified personnel and very costly equipment. The analysis deals with microsamples, generally in the solid state. It is a microanalytical advance, but at the limit, the representativeness of the sample is often debatable. It is difficult to apply to quantitative analysis.

Gas chromatography has been investigated for the determination of trace elements. Elements forming volatile and thermally stable complexes can be detected and measured. Numerous metallic fluoroacetylacetonates are known which can be separated by gas chromatography and measured with an electron capture detector. The useful quantities are smaller than 10^{-13} g. Actual applications of gas chromatography to trace analysis are still rare. The analysis of selenium in seawater separated after complexing with 5-nitropiaselenol and measured by gas chromatography and electron capture is a typical example of possible applications thus far.

Activation methods—neutron activation, charged particle activation and *photon activation*—are used only by a very limited number of laboratories because of the equipment necessary for irradiation of the sample as well as for the chemical and physical operations resulting in a quantitative determination. However, in view of the innumerable possibilities offered, the activation methods cannot be neglected. We wanted to present this technique in parallel with the others in order to permit the reader to evaluate the principle and deduce its possible applications.

The chapter on "Activation Analysis" is directed essentially at users of other techniques and not at activation specialists. Our intent was to show the contribution made by activation analysis to the classical analytical chemist. It is a general multielemental technique which can very often be used as a reference method, particularly for trace analysis. It is undoubtedly the most sensitive analytical method. It is appropriate that the reader will find at least two if not three methods applicable to his own problems in this book. In fact, it is very important for the validity of an analytical result that it be verified by another technique.

We have made no attempt to solve all analytical problems, particularly with regard to the investigated medium. Stress is placed on the analytical aspect of trace elements in the environment considered in the broadest possible context and including the mineral, plant and animal realms. In contrast, we do not deal with trace elements in metallurgy, since we believe that this is a special technique for which numerous procedures exist. However, after solubilization of the sample, nothing stands in the way of the methods described in connection with different media.

The techniques described in this book are based on original and care-fully selected notes. However, the applications presented should not be considered an exhaustive review of the scientific literature.

Maurice Pinta

CONTENTS

CHAPTER 1

FLUOROMETRY

1. INTRODUCTION

1.1 Fluorescence Spectra

Numerous inorganic and organic compounds are photoluminescent; that is, they emit radiative energy after having absorbed a radiation of given frequency. If the phenomenon has a measurable time constant of more than 10^{-9} sec between absorption and reemission, the process is termed phosphorescence. If the time constant is too small to be measurable, the phenomenon is fluorescence. When the energy originates from a chemical reaction, it is called chemiluminescence. If this energy is of biological origin, it is bioluminescence. Generally, the emitted radiation has a wavelength greater than or equal to the absorbed radiation. The fluorescence is called resonance.

When an atomic population in the ground state absorbs the resonance radiation of this atom (in the visible and ultraviolet region of the spectrum), it can reemit its resonance radiations in all directions. This is atomic fluorescence and will be studied in another chapter. In the X-ray region, atoms excited by X-radiation of suitable wavelength can also reemit an X-ray fluorescence spectrum under certain conditions. This is X-ray fluorescence spectrometry, which is likewise the subject of a special chapter.

The present chapter is limited to visible and UV radiations emitted by solutions of certain metal complexes (chelates). These are generally in solution when they absorb suitable radiation, although solids have been studied. The resulting emission spectrum occurs in a specific wavelength region.

Figure 1 is a diagram of the phenomenon that occurs in the molecule during fluorescence. The absorption of energy brings the molecules

1

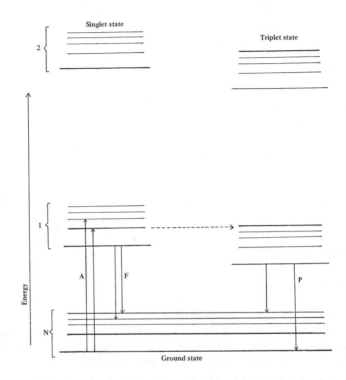

Figure 1. Fluorescence and absorption transition in a molecule.
N = vibration levels; 1 = first excitation state;
2 = second excitation state; A = absorption transitions;
F = fluorescence transition; P = phosphorescence transition.

to one of numerous vibration levels (singlets) from their lowest excited electron state. Any energy in excess of that of the lower level of the first excited state is rapidly lost in the form of collisions and energy transfer to other modes of molecular rotation and vibration. The molecule then returns to one of the excited vibratory levels of its ground electron state. In Figure 1, the vertical bars represent the energy differences between states. The energy corresponding to the fluorescence transition F is lower than that of the absorption transition A.

In certain molecules, a second series of excited states exists, the triplet levels, the energy values of which are similar to those of the singlet levels. These molecules can change directly from the single first excited state to one of the triplets in a radiationless process. In Figure 1, this transition is indicated by dots. Transitions with energy emission or absorption between singlet and triplet states are forbidden. The phosphorescence transition P between the ground vibratory state of the triplet level to one of those of the single-line level has a duration ranging from 12^{-2} sec to several seconds.

Each electronic state of a molecule consists of numerous vibration levels. Consequently, the absorption as well as the fluorescence and phosphorescence spectrum has the form of broad bands.

1.2 Fluorometry and Chemical Analysis

Fluorescence is the radiation emitted by a substance returning to its ground state after having absorbed a certain amount of energy when excited by a given kind of radiation. In principle, the excitation source and the emitted energy can each be any wavelength of the visible or invisible spectrum, but fluorometry is usually understood to mean the measurement of visible spectral energy emitted by a substance excited by ultraviolet light.

Classical fluorometry is a simple technique, in which the only apparatus required is a spectrophotometer and an ultraviolet excitation source. It is applicable to both solid samples and samples in solution. The intensity of a fluorescent radiation is given by the equation:

$$F = A \, I_o \cdot (1\text{-}10^{-kcl})$$

where F is the intensity of the fluorescent radiation, I_o is the intensity of the incident radiation, A is the fraction of the radiation which has been absorbed, k is constant for a given material, solvent, and wavelength, c is the concentration of the fluorescent substance, and l is the optical density of the solution.

If the product kcl is small (less than 0.01), the above expression may be written as

$$F = 2.3 \, AI_o kcl$$

It will be seen from the formula that F is proportional to c. As a matter of fact, at high concentrations the relationship is no longer valid and the curve $F = f(c)$ tends to become concave towards the concentration axis. In other cases, the curve may display a maximum, i.e., the fluorescence may decrease when c is increased beyond a certain value.

Figure 2 shows the variation of the fluorescence F as a function of concentration. It will be noted that the representative curve passes through a maximum M. In practice, the useful analytical range is defined by point A, beyond which the curve has too strong a deflection, which consequently reduces the fluorescence sensitivity.

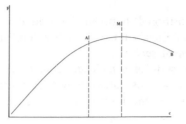

Figure 2. Variation of fluorescence as a function of concentration.

1.3   Apparatus

The fluorometric apparatus is similar to a classical absorption spec-trophotometer. The continuous radiation source is replaced by a source of ultraviolet radiation (Wood's lamp, xenon lamp or mercury vapor lamp) that projects a vertical on a lateral beam on the solution of the sample studied. The fluorescent radiation is spectrally analyzed in a certain wave-length interval either by using a filter instrument or by scanning the wave-lengths with an instrument equipped with a monochromator (Figure 3).

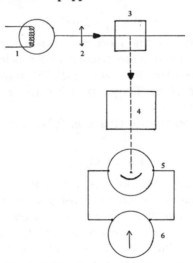

Figure 3. Fluorometer.

The intensity of the fluorescent radiation at a given wavelength is the difference between the intensity of the radiation emitted by the solution of the element and that emitted by the solvent or any other •

reference solution whose composition is identical with that of the sample solution except that the fluorescent element is absent. The emission is measured at 90° to the excitation direction to minimize scattering effects.

2. ANALYTICAL METHOD

2.1 Characteristics of the Method

For a given concentration, the sensitivity is determined by the strength of the fluorescence signal equal to at least two times the electron background accompanying it. When fluorescence is measured with a single-beam technique, the absolute radiation quantity is determined. The background depends on the gain of the recording receiver (detector/amplifier). It can be controlled with excitation sources of different intensities, with the minimum electron gain necessary for analysis, and finally with the use of detectors of different sensitivities.

Fluorometry is a specific method permitting the analysis of complex mixtures. When two complexes are fluorescent, they are not always excited at the same wavelength and the fluorescence wavelengths are also different. It is often possible with a particular chelating complex to involve a choice of the excitation radiation and the wavelength of the emitted radiation that eliminates interference by the fluorescence effect produced by other metals or elements present in the sample. This results in a better specificity than in atomic spectrometry. A relation analogous to that of the Beer-Lambert law in molecular absorption spectrometry does not exist between the fluorescence intensity and the concentration of the element or its complex. Finally, while fluorometry is applied very widely, it can not be considered a universal method for trace element analysis. While some highly sensitive practical techniques exist, they are limited in number.

The methods described below are often preferred to techniques such as atomic absorption, emission spectroscopy and mass spectroscopy because of their simplicity in operation and apparatus. (White *et al.*, 1970; Winefordner *et al.*, 1972).

2.2 Applications

The applications listed below should not be considered an exhaustive review of the practical possibilities offered by fluorometry. We have indicated only the elements having an intrinsic fluorescence (such as uranium) or those forming chelates, of which the fluorescence is particularly sensitive and the analysis of which is of true practical interest.

Within this context and that of the introduction, a choice was made among published studies. Naturally, the criteria of this selection are (1) the validity of the method and (2) its practical interest. The methods can therefore be considered examples, which can be easily adapted to any other similar medium. The methods are listed element by element in alphabetical order.

In the description of the procedures, the principles to be observed in trace analysis in particular and the precautions to be taken to reduce or at least control the sources of contamination were assumed to be known: quality of reagents and particularly of the deionized water, the glassware, the equipment, and the laboratory atmosphere (Lytle, 1970). A blank test should accompany all series of analyses. The necessary care taken when preparing standard solutions will not be repeated each time a new method is described. The basic rules concerning classical procedures in analytical chemistry, which are extensively involved here, are assumed: procedures of precipitation, filtration, evaporation, extraction of metals in the form of organic complexes in a suitable solvent, separation and washing of phases, fractionation on ion exchanger and elution. The precision and accuracy of the results depend on these indispensable precautions.

3. PRACTICAL METHOD

3.1 Aluminum

This element, among the first to have been analyzed by fluorometry, forms strongly fluorescent complexes with numerous reagents. The specificity of fluorescence permits the analysis of diverse media. A study of this subject was published by Lytle (1970). In addition to the reagents listed in Tables I and II, a few others of practical interest can be cited. The most classical method is based on measurement of the fluorescence of the Al-hydroxyquinolate complex in chloroform. The sensitivity and specificity are better than with the method of absorption of the complex.

The procedure of the practical method is as follows : an aliquot of the sample solution containing between 2 and 16 μg Al is diluted to 50 ml with 1:17 sulfuric acid, followed by addition of 2 ml 2% 8-quinolinol solution in acetic acid and 2 ml ammonium acetate buffer (dissolve 200 g ammonium acetate and 70 ml sodium hydroxide in concentrated solution in 1 liter of water). Adjust the pH to 8 ± 1.5 with 1:2 ammonium hydroxide, extract the Al-8-hydroxyquinolate complex with two portions of 1.5 ml chloroform, filter the organic solution, and dilute to a known volume. Read the fluorescence at 520 nm with excitation at 365 nm.

Table I. Classical Fluorometric Methods in Trace Analysis

Elements	Reagents	Chemicals Conditions	Excitation (nm)	Fluorescence Emission (nm)	Detection (μg/ml)	Interferences
Al	Alizarin Garnet R	pH 4.6	470	500	0.01	Be,Co,Cr,Cu,Fe,Ni,Zr
	Morin	pH 3.3	430	500	0.05	Ag,As,Be,Cr,Fe,Ga,R.E.
	2,3-Hydroxynaphtoic acid	pH 5.8	370	460	0.002	Be,Mo,W
	Salicylidene-o-aminophenol	pH 5.6	410	520	0.0002	Cr,Sc,Th
B ($B_4O_7^{2-}$)	Benzoin	EtOH, pH 12.8	370	480	0.04	Be,Sb
Be	2,3-Hydroxynaphtoic acid	pH 7.5	380	460	0.0002	Cr,Bi,Sn,Ti,Fe
	8-Hydroxyquinaldine	$CHCl_3$	-	-	0.001	Al,Bi,Cd,Cr,Cu,Fe,Sn,Ti,Zn
Cu (Cu II)	Tetrachlorotetraiodo fluorescein o-phenanthroline	pH 8.0, pH 7	500	570	0.001	-
F (F^-)	Al complex of alizarin Garnet R	pH 4.6 (quenching) (CHCl₃)	470	590	0.001	Be,Co,Cr,Cu,Fe,Ni,Zr
Ga	8-Hydroxyquinaldine	pH 3.9	-	492	0.02	Cu
In	Salicylidene-o-aminophenol	pH 4.0 ($CHCl_3$)	420	520	0.007	Al,Be,Cu,Fe,Zr
	8-Hydroxyqunoline	pH 5.1	-	-	0.04	
Mg	Bissalicylidene diaminobenzofuran	pH 10.5	475	545	0.002	Mn
Mo	Carminic acid	pH 5.2	560	590	0.1	-
Sc	Salicylaldehyde-semicarbazone	pH 6.0 (toluene)	370	455	0.002	-
Se	Diaminobenzidine	pH 7.4	420	550-600	0.002	-
Sn (Sn IV)	Flavanol	0.1 M H_2SO_4	400	470	0.1	P,Zr
Tl (Tl I)	HCl - KCl (Saturated solution)	3.3 M HCl	250	430	0.01	Au,Bi,Pt,Sb
	Rhodamine B	0.8 M KCl	-	640	0.03	
U	Fused NaF + LiF	-	-	550	0.0001	
W (W VI)	Carminic acid	pH 4.6	515	585	0.04	-

Table II. Recent Fluorometric Methods

Elements	Reagents	Detection ($\mu g/ml$)	References
Ag	1,10-Phenanthroline	0.1	Lisitsyna *et al.*, 1970
Al	Salicylaldehyde	0.1	Ben–Dor *et al.*, 1970
	2-Hydroxy-1-naphtaldehyde	0.1	
	Lumogallion	0.1	Shigematsu *et al.*, 1970
B	Hydroxy-2-methoxy-4-chloro-4'-benzophenone in 90% H_2SO_4	0.001	Monnier *et al.*, 1972a; Monnier *et al.*, 1972b
	Rhodamine 6G (C_6H_6)	–	Vasilevskaya *et al.*, 1971
Be	Tetracycline + 5,5-diethyl-2-thiobarbituric acid	0.1	Naito *et al.*, 1969
	3,5,7-Trihydroxyfavone	–	Hayashi *et al.*, 1970
	Acetylacetonate-trylon	0.07	Morgen *et al.*, 1970
	o-(salicylidene amine) Phenyl arsonic acid	0.001	Talipov *et al.*, 1972a; Talipov *et al.*, 1973
Br	Fluorescein	0.002	Axelrod *et al.*, 1970
Ca	Chlorotetracycline	–	Hallett *et al.*, 1972
Cd	Eosine	0.01	Matveets, 1971
	8-*p*-Tosylaminoquinoline, pH 11	–	Vesiene *et al.*, 1972
Cu	Cysteine	–	Anglin *et al.*, 1971
F	Zr-calcein	0.002	Har *et al.*, 1971
Ga	Lumogallion (isoamylalcohol)	0.01	Zweidinger *et al.*, 1973
Ge	2,2',4'-Trihydroxy-3-arseno-5-chloroazobenzene	0.02	Shcherbov *et al.*, 1970
Hg (HgII)	Oxidation of thiamine pH 9.5	0.1	Holzbecher *et al.*, 1973
Li	8-Hydroxyquinoline	–	Schulman *et al.*, 1971
N(NO_2^-)	2,3-Diaminonaphtalene	0.007	Sawicki, 1971
N(NO_3^-)	Fluorescein in 95% H_2SO_4	0.01	Axelrod *et al.*, 1970
Pb	CaO-Pb	–	Vasil'ev *et al.*, 1973
	Zn S film	10^{-11}g	Gurvich *et al.*, 1970
S(H_2S)	Fluorescein-tetramercurate (quenching)	0.001	Bozhevol'nov *et al.*, 1971; Gruenert *et al.*, 1971; Wronski, 1971
Se	2,3-Diaminonaphtalene	0.001	Nazarenko *et al.*, 1970; Passwater, 1973
Sb	Morin	–	Shcherbov *et al.*, 1973
Sn	Morin	–	Shcherbov *et al.*, 1973
Tl	Tellin 3,3 M, HCl-0.8 M, KCl	0.01	Konig *et al.*, 1971
V	Benzoic acid + Zn amalgam	0.001	Koch *et al.*, 1971
Zn	Benzimidazole carboxaldehyde, 8-quinolylhydrazone	0.001	Ryan *et al.*, 1972
Zr	Salicylidene-4-aminoantipyrine	0.008	Talipov *et al.*, 1972b

Shigematsu (1970) uses lumogallion for the analysis of seawater and shows that the fluorides and phosphates do not interfere. Morin, one of the oldest reagents used for aluminum analysis, is still employed by Baker *et al.* (1970) for the analysis of atmospheric dust particles. The morin solution is evaporated on spots of filter paper in the ring-oven technique. This reagent is also used by Eggert (1970) to determine aluminum in plant tissue. Puech *et al.* (1972) determine aluminum in microgram quantities in biological media by fluorometry of the complex formed with Pontochrome Blue Black indicator. The specificity is excellent.

3.2 Nitrogen

Under particular conditions, certain nitrogen compounds—nitrites, nitrates, oxides—can form fluorescent chelates, which have a specificity and sensitivity permitting a trace analysis. We should note, however, that the truly tested practical methods are still limited.

3.2.1 Nitrate and Nitrite Determination

A particularly sensitive method for nitrate analysis is based on the reduction of nitrates into nitrites by hydrazine sulfate, followed by reading of the fluorescence of the compound formed with 2,3-diaminonaphthalene. The detection limit is 0.01 μg NO_3 (Sawicki, 1971).

In concentrated sulfuric acid medium, nitrates inhibit the fluorescence of fluorescein. A measurement of the fluorescein extinction permits the detection of 0.01 μg NO_3^-/ml of the final solution (Axelrod *et al.*, 1970). The practical method is as follows: a suitable volume of nitrate solution is added to a given volume of concentrated sulfuric acid (95%) containing 5×10^{-7} M of disodium fluorescein so that the final sulfuric acid strength is not less than 87%. The mixture is allowed to stand for 45 min. Fluorescence is then read at 485 nm with excitation at 435 nm. The operating range is $0.5\text{-}2.0 \times 10^{-6}$ M NO_3^- in 90% sulfuric acid. The sulfuric acid strength is important because the sensitivity decreases with the acid strength, particularly below 90% of acid. Chlorides interfere: a quantity of chlorides equal to 100 times the nitrates completely suppresses the reaction with the formation of nitrosyl chloride. This interference is eliminated in the presence of antimony(III). Iron Fe(II) also interferes, but not Fe(III); therefore, it is advantageous to oxidize it in advance.

3.2.2 Determination of Nitrites in Biological Media

Nitrites in biological materials and food products can also be determined at levels of a few nanograms by diazotizing chloraniline followed by coupling with 2,6-diaminopyridine (Dombrowski *et al*, 1972). The azo compound formed is extracted with benzene and, after elimination of the solvent, is treated with ammoniacal cupric sulfate with the final formation of a triazole compound. This, in a medium of moderate acidity, is strongly fluorescent at 430 nm with excitation at 360 nm. The detection limit is 20 ng in 10 ml of final solution, and the range of determination extends from 20 to 200 ng.

The practical method is as follows: the sample (containing 20-200 ng of NO_2^- ions) is treated with 0.1 ml 6 N HCl, 0.3 ml of 0.06% *p*-chloroaniline and placed into an ice bath for 15 min. Then 0.3 ml of ammonium sulfamate is added, followed 5 min later by 2 ml of 2,6-diaminopyridine (0.025% solution buffered to pH 5 with ammonium acetate). The mixture is allowed to stand for 30 min in an ice bath and then placed into a decanting flask with 6 ml benzene and agitated for 30 sec. The organic phase is separated and then first washed with sodium acetate solution (pH 5) and then with water. The benzene is evaporated, the residue is taken up in 2 ml water and 0.4 ml ammoniacal cupric sulfate solution (1 g $CuSO_4$, 3 ml of 28% ammonia and 7 ml water), and heated on a boiling water bath for 30 min in a stoppered flask. The solution is then cooled, acidified with 0.45 ml 6 M HCl and finally diluted to 10 ml in a volumetric flask. This solution is ready for the fluorometric reading. A blank is prepared under the same conditions with 2 ml distilled water.

3.2.3 Determination of Ammonia

A method for the determination of ammonia has been proposed by Sardesai *et al.* (1969) with application to blood analysis. Ammonia, separated by the action of sulfuric acid, reacts with 3,5-diacetyl-1,4-dihydro-lutidine, forming a fluorescent compound.

3.2.4 Determination of Nitrogen Dioxide

Nitrogen dioxide in air is determined from a concentration of 10^{-6} vol % by reading the chemiluminescence formed with siloxene (Agranov *et al.,* 1970).

3.3 Beryllium

The classical method (Morgen *et al.*, 1970), is based on a fluorometric reading at 380 nm of the Be-8-hydroxyquinaldine complex extracted into chloroform at pH 8 ± 0.2; the analysis requires between 7
and 70 ng Be/ml.

Recently other reagents have been proposed (Table II). In particular, the method published recently by Talipov *et al.* (1972-1973) makes
use of beryllium complexing with *o*-(salicylideneamino) phenylarsonic acid
at pH 6 and seems to be one of the most sensitive. It permits the detection of 0.001 µg Be/ml and at the same time is one of the most specific
techniques. Thus far, the authors do not seem to have reported practical
applications.

3.4 Boron

Several reagents producing a specific fluorescence have been proposed
for boron analysis. Boric acid forms a fluorescent complex with dibenzo-
phenone in concentrated sulfuric acid solution. We should also note boron
analysis in water and plants with 2-hydroxy-4-methoxy-4'-chlorobenzene,
which permits the determination of a range of 0.074-5 ppm. The method
(Liebich *et al.*, 1970) is described below.

3.4.1 Boron Analysis in Natural Water

The sample can be directly diluted in sulfuric acid under such conditions that it does not exceed 0.2 ml in 3.8 ml concentrated sulfuric
acid. The fluorescent solution should contain between 0.001 and 1 µg
B/ml.

When the quantity of boron in the sample is too small or when the
sample contains organic materials, mineralization becomes necessary
(Monnier *et al.*, 1966, 1969). A suitable volume is evaporated to dryness
in a quartz dish in the presence of 0.3 ml saturated $Ca(OH)_2$ solution.
In the presence of organic substances, the residue is treated with 3.7 ml
concentrated sulfuric acid of analytical grade and 1 ml 30% hydrogen
peroxide, heated to 100° and then to 190° for 50 min. The solution
obtained is treated with 0.2 ml HMCB solution (1.05 mg hydroxy-2-
methoxy-4-chloro-4'-benzephenone in 10 ml concentrated sulfuric acid),
heated to 70° for 40 min and cooled for 30 min. The fluorescence is
read at 490 nm with excitation at 365 nm. A blank as well as standards
(5-50 ng B/ml) must be prepared under the same mineralization conditions
as used for the samples.

The following sample quantities are used: 0.6-1 ml for B contents of 10 ppb, 0.2-0.5 ml for 100-1000 ppb of B. For sea water (5000 ppm), the sample should first be diluted: 1 ml sea water in 100 ml sulfuric acid, followed by simple dilution of volumes of 0.3-0.7 ml of the prepared solution. For plant analysis, the sample is mineralized in sulfuric acid containing a hydrogen peroxide addition.

Vasilevskaya (1971) also proposed a method for determining boron in water, rocks, plants and coals. It is based on a luminescence reading of a complex formed by boric acid with rhodamine 6 Zh extracted into benzene. Excitation is produced at 366 nm and luminescence is read at 546 nm. The method is sensitive up to 0.01 μg B/ml.

3.5 Bromine

A sensitive fluorometric method permits a bromine analysis based on the quenching of the fluorescein fluorescence in the presence of bromine (formation of tetrabromofluorescein). The solution is excited to 440 nm and fluorescence is read at 470 nm. The detection limit is 0.002 μg Br/ml. An application is described below.

3.5.1 Bromine Analysis in the Atmosphere

A simple method has been proposed by Axelrod et al. (1971) for determining bromine in aerosols and dust particles of the atmosphere. A suitable volume of air is filtered on a fiberglass filter. An aliquot of the filter is washed with a specific quantity of glacial acetic acid. A 0.5-ml sample is treated with 9.5 ml of a solution prepared from 9 ml 1.1×10^{-4} M fluorescein in glacial acetic acid and 0.5 ml 30% hydrogen peroxide. The solution is allowed to stand for 45 min at room temperature. Fluorescence quenching by bromine is read at 470 nm with excitation at 440 nm.

The precision of the method is a function of the acetic acid concentration of the test solution. It must be greater than 90% since the sensitivity decreases with lower concentrations. The hydrogen peroxide concentration in the test solution must be at least 1%. Accuracy is good for bromine concentrations up to 2 ng/ml; at a level of 20 ng/ml, the standard deviation is 2%.

No interference is produced by the following ions: SO_3^{2-}, NO_3^-, Cl^-, K^+, NH_4^+, Mg^{2+}, Ca^{2+}, Fe^{2+}, Cu^{2+}, Pb^{2+} (100 times more than bromine). Iodine (100 x Br) causes an error by deficit of 20%. A Cl/Br ratio of 10,000 does not interfere.

3.6 Calcium

The fluorescence of alkaline earth chelates has received little attention because it involves metals that are easier to determine analytically with other techniques such as atomic absorption spectrometry. However, in recent years numerous authors have proposed interesting applications of fluorometry to calcium analysis.

With a mixture of phenyltrifluoroacetone and rhodamine S, the Ca, Sr and Mg ions form complexes that are extractable with benzene. These have absorption maxima at 540 nm for Ca and Sr and at 544 for Mg, and fluorescence maxima at 565 nm for Ca and Sr and 570 for Mg. Tetracycline is also a fluorescence indicator for calcium.

3.6.1 Determination of Calcium in Biological Media

Fingerhut *et al.* (1969) proposed a method for the analysis of calcium in blood serum by reading the fluorescence of the complex formed with calceine. The sample is dialyzed to eliminate organic substances that can produce fluorescence. Proteins are separated by treatment with 0.2 N hydrochloric acid. Calceine is added in alkaline solution (2 N NaOH) in an amount of 0.03 mg/l. Where magnesium does not react with calceine, the pH of the test solution should actually be higher than 13.6. The method proposed by the authors is intended to be employed with the fluorometric autoanalyzer.

Hallett *et al.* (1972) used chlorotetracycline for the determination and study of the *in situ* biological variation of calcium in nerve tissue during nerve excitation. In fact, the fluorescence of tetracycline is enhanced when it is chelated with bivalent metals. The reagent is introduced into the analyzed tissue by injection or perfusion. It is then excited to 400 nm by a quasi-monochromatic filtered light. The fluorescence emission is recorded between 400 and 700 nm while the nerve tissue is treated, particularly during excitation.

3.7 Copper

Few applications of fluorometry exist for trace analysis of copper. We may note fluorescence of the compound formed with cysteine (Anglin *et al.*, 1971) for the analysis of biological tissue and particularly of skin extracts.

Ritchie *et al.* (1969) studied the biochemistry of 1, 1, 3-tricyano-2-amino-1-propene and showed that this compound reacted with Cu(II), producing a fluorescence in the presence of an excess of acid. Applied to biological media, the method is simple, sensitive and particularly specific.

3.7.1 Determination of Copper in Biological Media

The sample (Ritchie *et al.*, 1969) is mineralized by acid attack, and the solution obtained is adjusted to pH 8.5 with sodium hydroxide. A volume of 1 ml of this solution is placed into a test tube with 0.3 ml 0.001 M 1,1,3-tricyano-2-amino-1-propene* and 0.5 ml 0.002 M imidazole (13.60 mg of reagent in 10 ml water). The mixture is held at 37-40° for 15 min with periodic agitation.

A series of standard solutions containing 0-0.5 μg Cu/ml is prepared with the addition of the above reagents under the same conditions. Each solution (sample and standards) is treated with 6 ml water and 1 drop 5 N HCl. After agitation, fluorescence is read under the following conditions: excitation at 365 nm, emission at 510 nm. Ca, Mn, Na, Zn, Fe, Mg—elements that are normally present in biological media—practically do not interfere.

3.8   Gallium

The Lumogallion (2,2',4-trihydroxy-5-chloro-1,1'-azobenzene-3-sulfonic acid) is among the most sensitive fluorometric reagents for gallium with a detection limit of 1 ng/ml (Onishi, 1955). It is also quite specific. The method reported below is one application of it.

Gallium, particularly in the presence of aluminum, can also be determined by precipitation with ammonium hydroxide, complexing with rhodamine B, and extraction of the gallium complex into benzene. The fluorescence excitation wavelength is near 530 nm, and the limit of detection is 0.1 μg Ga/ml in the extract.

3.8.1 Determination of Gallium in Biological Media

This method (Zweidinger *et al.*, 1973) has a detection limit of 15 ng Ga/g of material. The sample (1 g) is mineralized by sulfuric-nitric acid and perchloric acid attack until there is complete destruction of the organic materials. The residue is redissolved in 10 ml 6 N HCl. This solution is loaded on a column of 2.5 mm x 7 cm Dowex 1 x 8 cation exchange resin (50-100 mesh). After washing the column with 20 ml of 6 N HCl, gallium is eluted with 10 ml N HCl. Then 0.1 ml 1% ascorbic acid is added to the eluate, followed by 0.1 ml 0.25% $Na_2S_2O_3$ solution. The pH is adjusted to 2.25 ± 0.05 with 1.5 M Na_2CO_3

---

* Dissolve 13.2 mg reagent in 100 ml water, which is stored in darkness at 4° for not more than two weeks (Carboni *et al.*, 1958).

solution, and 0.8 ml Lumogallion (0.01% solution in 0.06 M HCl) is added. The final solution is agitated and allowed to stand for 60 min, followed by extraction with 3 ml isoamyl alcohol. Fluorescence is read at 570 nm with excitation at 490 nm. The standard solutions and blank are prepared in the same way.   Between 0.005 and 5 μg Ga must be present in the final solution of 3 ml isoamyl alcohol.

Al, Fe and Cu essentially do not interfere in the concentrations at which they are present in biological media. For the gallium contents involved, it is indispensable to check the purity of the reagents, particularly that of sodium carbonate.

3.9   Iodine

3.9.1 Iodine Determination in Rocks

In an acid medium and in the presence of bromine ion (Br⁻), iodine (I⁻) forms a complex ion with Butylrhodamine C, which can be extracted with benzene. In the presence of methylketone, it produces an intense fluorescence at 590 nm that is stable for 6 h (Podberezskaya *et al.*, 1971, 1973).

A sample of 5 g of rock is molten at $700° \pm 5°$ in a mixture of sodium and potassium double carbonate and zinc oxide. The iodine is separated by oxidation with ammonium peroxydisulfate [$(NH_4)_2 S_2 O_8$] in 5 N sulfuric acid and extraction from benzene. The extract is agitated with 11 N sulfuric acid containing 1% potassium bromide and 0.001% Butylrhodamine.

The fluorescence of 0.02-1 μg of iodine can thus be determined. The error for iodine contents $\leqslant$ 0.2 ppm is $<$ 24% and the accuracy is less than 32%.

3.10 Oxygen

3.10.1 Determination of Ozone in the Atmosphere

Several fluorometric methods have been proposed recently for the determination of ozone. Amos (1970) suggested a specific method for the detection of 0.02 ppm in the atmosphere. Ozone is collected in a chloroform solution of 1,2-di(4-pyridyl)ethylene, where it forms pyridene-4-aldehyde, which reacts with 2-(diphenylacetyl)-1,3-indandione 1-hydrazone to form a strongly fluorescent compound.

The method based on chemiluminescence of a compound formed by ozone with Rhodamine B adsorbed on silica gel can also be used. Fluorescence of the pigment in the red region permits the detection of 0.001 ppm ozone in the atmosphere.

In addition, an automatic analyzer has been proposed by Hodgeson *et al.* (1970). The recommended adsorbent consists of silica gel plates for thin-layer chromatography (ECS-6061-Eastman). The support is dried for 15-30 min in an oven at 110°, immersed in a 1:6 solution of 60% SR-82 resin in benzene for 1 min, dried in the oven at 110° for 24-48 hr, and immediately immersed for 1 min in a solution of 0.1 g/l of Rhodamine B in acetone. This is followed finally by drying in a nitrogen stream and storage in the absence of air and moisture. Any trace of moisture considerably reduces the chemiluminescence of the compound formed. This method is highly specific. In fact, NO, NO_2, SO_2, H_2S, NH_3, Cl_2 essentially do not interfere. This method permits the determination of traces ranging between 0.001 and 1 ppm (Bersis *et al.*, 1966; Hodgeson *et al.*, 1970).

3.11   Phosphorus

With Rhodamine B, the phosphomolybdenum compound forms a fluorescent complex that can be extracted with a chloroform-butanol system (4/1 v/v). Excitation is produced at 350 nm.

3.11.1 Trace Analysis of Orthophosphates

A sample (1-2 ml) containing between 0 and 0.6 μg phosphorus in the form of orthophosphate is placed into a decanting flask with 5 ml molybdenum reagent*, 0.5 ml concentrated hydrochloric acid, and 5 ml Rhodamine B (0.115 g Rhodamine B in 1 liter of distilled water). The mixture is agitated for 5 min. The excess of reagent is extracted three times from 5 ml chloroform, after which the complex is extracted with 10 ml of chloroform-butanol (4/1 v/v).

The organic phase is separated and read by fluorometry at 575 nm with excitation at 350 nm. A blank should be prepared and the fluorescence subtracted from that of the sample. Standards are prepared from monopotassium phosphate in water. The technique is specific since the majority of elements normally present in the natural medium (Ca, Al, Fe, Si) do not interfere, with the exception of arsenic As(V).

This method (Kirkbright *et al.*, 1971) is applicable to plants and biological media as well as to water.

---

*Dissolve 30 g ammonium molybdate in 1 liter H_2O. Add HCl so that the solution diluted to 1 liter is molar in acid. Purify the solution by extraction with four successive portions of 100 ml chloroform.

3.12 Selenium

Selenium with a valence of 4 (selenite) forms strongly fluorescent "piazselenols" with 2,3-diaminonaphthalene and 3,3'-diaminobenzidine. In particular, the "piazselenol" of diaminonaphthalene extracted from cyclohexane produces sensitive fluorescence with 0.02 µg. The fluorometric analysis of selenium is among the most sensitive techniques, and no other classical and sensitive method exists for this element. Fluorometry is applicable to trace analysis of selenium in the majority of chemical and natural media, in industrial products and water. One of the principal difficulties of analysis results from the losses of selenium that may occur when the sample is dissolved.

3.12.1 Determination of Selenium in Soils and Sediments

The sample is dissolved by digestion with nitric-perchloric acid. Selenium is reduced to the state of selenite, complexed with diaminonaphthalene, extracted with cyclohexane, and analyzed by fluorometry.

The following practical method is used. The sample (1 g) is digested with a mixture of nitric-perchloric acid (4/1) in two portions of 10 ml. After 1.30 hr, the cooled residue is taken up with 10 ml of 6 M HCl, heated to boiling, cooled and filtered. The digestion vessel and filter are washed with 10 and then with 5 ml of 6 M HCl. These washings are combined with the filtrate. The latter is treated with 5 ml arsenic-reducing solution (see the preparation below) and 10 ml 50% hypophosphorus acid.

The solution is heated to complete precipitation of the arsenic. The precipitate, which also contains the selenium, is filtered and taken up in 2 ml HNO_3 and a small amount of distilled water. Solubilization is obtained after 2-3 min of heating. After cooling, 2.5 ml formic acid and 1.0 ml 0.5 M EDTA are added. At this stage of sample preparation, a series of standards is also prepared to cover a range of 0-4 µg of selenium in the form of selenious acid.* The pH of the solutions is adjusted to 1.5 ± 0.1 with NH_4OH or HCl.

The solutions are diluted to 40 ml and extracted with 5 ml diaminonaphthalene**. The mixture is agitated, allowed to stand for 1 hr

---

* Preparation of the selenium stock solution: dissolve selenous anhydride in 4 N hydrochloric acid and dilute the solution to obtain 0.2 µg Se/ml.

** To be prepared daily: dissolve 0.1 g 2,3-diaminonaphthalene in 100 ml 0.1 N HCl and extract this solution twice with 40 ml cyclohexane.

in darkness and then extracted with exactly 10 ml cyclohexane. The organic phase is separated and washed with 15 ml 0.1 M HCl. The solution is ready for the fluorometric reading.

This method permits the determination of selenium contents ranging between 0.5 and 3 ppm of dry product. An important precaution to be taken is to use analytical-grade reagents and products. Preparation of the arsenic-reducing solution (Wiersma et al., 1971; Nazarenko et al., 1971b; Van Duuren et al., 1971):

(a) Dissolve 0.315 g As_2O_3 and 10 tablets NaOH in 50 ml H_2O
(b) Mix 75 ml concentrated HCl and 25 ml 50% H_3PO_2

Displace the traces of selenium by adding 0.8 ml solution (a) by bringing to boiling for 5 min and filtering the precipitate (As + Se). The filtrate represents the arsenic-reducing solution.

3.12.2 Determination of Selenium in Plant and Biological Media

The described method (Lamand, 1969) permits the determination of contents between 0.005 and 0.1 ppm. The principle is similar to that of the preceding method.

The sample (1 g) is digested in a volumetric flask under a condenser with 5 ml nitric acid for 16 hr in the cold, brought to boiling, and with 2 ml of perchloric acid and brought to boiling for 30 min. The condenser is removed and the flask is heated up to the development of white fumes and maintained at this temperature for 15 min.

Selenium is present in this solution with a valence of 6. It is reduced to a valence of 4 by addition of 1.5 ml concentrated HCl and is brought to boiling for 10 min. After cooling, 6 ml H_2O are added. At the same time, a series of standards is prepared (0.05-0.5 μg Se) in 6 ml H_2O and 2 ml 10% HCl (see preparation of the stock selenium solution in the preceding method). Each solution (sample and standard) is treated with 2 ml 0.04 M EDTA and 1 drop of indicator (aqueous solution of 0.02% o-cresolsulfonphthalein). The pH is adjusted to 1 with ammonia and the solution is brought to 50 ml with 0.1 M HCl containing an addition of 5 ml diaminonaphthalene (0.1% solution in 0.1 M HCl). This procedure must be carried out in the absence of light.

The solutions are placed in a water bath at 50° for 20 min (in darkness). After cooling, they are extracted with 10 ml cyclohexane and the organic phase is washed twice with 25 ml 0.1 M HCl. The cyclohexane solution is centrifuged and the fluorescence is read at 525 nm with excitation at 366 nm. (Schroeder et al., 1970; Olson, 1969; Watkinson, 1966; Patrias et al., 1969).

Note: Contaminations are undesirable; the acids and reagents must be purified.

3.13 Sulfur

There are few sensitive methods for the determination of sulfates and sulfites. The principle of the method of Axelrod *et al.* (1970) might be suggested: sulfur dioxide as well as the sulfate ions are bound by $HgCl_4{}^{2-}$ ions and react with formaldehyde HCHO to form a complex $HSO_3{}^{2-}$-HCHO. This, when added to 5-aminofluorescein, inhibits its fluorescence by forming a nonfluorescent complex. This reaction permits the determination of 0.02 μg SO_2/ml in the $HgCl_4{}^{2-}$ solution

We should also note a sulfate analysis in natural water by inhibition of the fluorescence of the morin complex. While the method is sensitive, it is nevertheless subject to interference due to phosphates, fluorides and various cations.

A larger number of methods exists for the determination of sulfides. Hydrogen sulfide inhibits the fluorescence of fluorescein-mercuric acetate and tetramercuric fluorescein complexes. Two methods based on this principle are described below to determine hydrogen sulfide in water and air.

3.13.1 Determination of Hydrogen Sulfide in Water

Hydrogen sulfide is separated from the medium in the form of triethyllead sulfide $(C_2H_5)_3Pb\text{-}S\text{-}(C_2H_5)_3$, extracted into hexanol, and then analyzed by fluorometry with tetramercuric fluorescein (Wronski, 1971).

The method permits the determination of traces of 0.01 μg H_2S/ml of water. The sample (500 ml) is agitated with 0.5 ml of a solution of 2.5 N NaOH and 0.5 N EDTA and 10 ml of 0.01% triethyllead chloride in hexanol. To 3.5 ml of the extract obtained, 0.5 ml of solution A (see below) is added, and the mixture obtained is titrated with an alcoholic solution of tetramethylfluorescein (TMF) (solution B, see below) by recording the fluorescence at an excitation wavelength of 520 nm. The necessary quantity of TMF to obtain fluorescence is a function of the quantity of hydrogen sulfide present.

In practice, the curve on which the fluorescence intensity is plotted on the ordinate and the volume of tetramercuric fluorescein solution on the abscissa is constructed. This shows a change of slope at the titration end point; the abscissa of this point is a function of the quantity of hydrogen sulfide determined. Preparation of reagents:

> Solution A: Mix 20 ml saturated methylphenobarbitone solution in 96% ethanol with 1 ml triethylamine, and bring to 100 ml with 96% ethanol.

Solution B:(tetramercuric fluorescein):   Mix 4 ml aqueous
5×10^{-4} N tetramercuric fluorescein with 16 ml H_2O, 20 ml
saturated methylphenobarbitone solution in 96% ethanol, and
0.5 ml triethylamine and dilute to 100 ml with 96% ethanol.

3.13.2 Determination of Hydrogen Sulfide in Air

The method proposed by Hardwick *et al.*(1970) is based on fluorescence quenching of the fluorescein-mercuric acetate complex.   A suitable quantity of air sample is bubbled into an alkaline solution of fluorescein-mercuric acetate, FMA $[C_{20}H_{10}O_2Hg(COOCH_3)_2]$ of suitable titer.
The fluorescein is read before and after the bubbling step.

Starting with 25 ml 0.00075% FMA titration solution in 0.1 N
NaOH, fluorescence quenching permits a determination of hydrogen sulfide in quantities of 2-7 μg.   Moreover, from 25 ml 0.00025% FMA in
0.1 N NaOH, we can determine 0.3-2 μg H_2S, and with 25 ml 0.006%
FMA, 15-60 μg H_2S can be found.   Thus, with an air volume of 4.5
liters and an 0.00025% FMA solution in 0.1 N NaOH, 0.05 ppm (v/v)
H_2S/air can be detected.

The fluorescence quenching is not proportional to the quantity of
hydrogen sulfide present, and this limits the analysis range for a given
quantity of FMA between two values, beyond which the method loses all
accuracy (Figure 4).   In particular, the decay of fluorescence is absent
above a certain quantity of hydrogen sulfide.

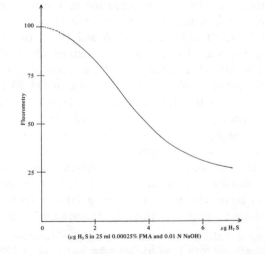

Figure 4. Fluorescence intensity of the hydrogen sulfide solution
in 25 ml 0.00025% FMA and 0.01 N NaOH

The FMA reagent is stable for 14 days if stored at low temperature and in the absence of light. The standard curve is obtained from a suitable quantity of sodium sulfide introduced into the FMA solution with a micropipette. The authors verified that this curve was practically identical to that obtained from standard mixtures of hydrogen sulfide in pure air.

Finally, the trace analysis of hydrogen sulfide is not sensitive to interference by gaseous CO_2 and SO_2 if the sum of these does not exceed 500 ppm in an air sample of 4.5 liters. On the other hand, compounds such as the mercaptans and disulfide also quench fluorescence and interfere with the analysis.

3.14 Uranium

Fluorometry is one of the most suitable methods for the analysis of uranium present as traces in the majority of media: water, rocks and minerals, plant and biological products.

The practical techniques are based on (1) a solid-state determination of the fluorescence of uranium in a sodium fluoride fusion or in a mixture of sodium carbonate and fluoride (detection of 10^{-9} g) or (2) the use of a sulfuric acid-phosphoric acid solution (Danielson et al., 1973) in which the fluorescence of complexes formed with Rhodamine B (Leung et al., 1972) or with thenoyltrifluoroacetone in butyl alcohol medium is read (Dobrolyubskaya, 1971). Both of these methods require a chemical separation to enrich the end product, which is subjected to fluorescence as well as to separate interfering substances, particularly organic products.

Finally, the most common method remains the fluorescence of the fusion product with sodium carbonate and fluoride. However, the fluorescence reading is strongly influenced by the sample preparation conditions: composition of the flux, melting temperature and time, and storage time of the sample.

3.14.1 Determination of Uranium in Minerals

The sample (1 g) is calcined at 500° and placed into a Teflon* dish. At the same time, a blank test and a uranium standard containing 10 μg U are prepared in two other similar dishes. In each dish, 10 ml of concentrated HNO_3 and 10 ml HF are added. The dishes are kept on a water bath for 16 hr. This attack is continued at elevated temperature up to complete elimination of silica. The residue is taken up with

---

*Registered trademark of E. I. duPont de Nemours & Company, Inc., Wilmington, Delaware.

5 ml HNO_3 and 25 ml water. The solution is filtered and the filtrate evaporated to 10 ml. The residue on the filter is calcined in a platinum crucible at 550°, melted with a Meker burner with 0.5 g Na_2CO_3, taken up in 5 ml water and 2-3 drops ethanol, and combined with the preceding filtrate. The crucible is rinsed with 1 ml nitric acid and 4 ml water. The solution is evaporated to 10 ml with the addition of 1 drop of 30% hydrogen peroxide and heated to decoloration of the titanium complex.

The uranium is then extracted with ethyl acetate. The preceding solutions are placed into decanting flasks and agitated with 13 ml of aluminum nitrate solution* and 20 ml ethyl acetate. The organic phase is separated and evaporated to less than 10 ml at room temperature (16 hr) and then at slightly elevated temperature. After cooling, the solution is brought to 10 ml.

Volumes of 0.1 ml of the latter solution (for two or three duplicate determinations) are evaporated and the residues are treated with 0.6 g of a mixture of 98% NaF and 2% LiF. They are then fused on a burner suitably adjusted to obtain melting in 90 sec and held in the molten stage for 120 sec with agitation and cooled. After 15 to 60 min, the fluorescence is read in the yellow-green region of the spectrum with excitation at 365 nm. Generally, the blank test has a fluorescence that must be taken into account (Dobrolyubskaya, 1971; Pakalns, 1970; Maxwell, 1968; and Smith et al., 1969).

3.14.2 Determination of Uranium in Plant Products

In this method (Huffman et al., 1970) a sample of 10 g is calcined at 550° for 2 hr, a quantity of 50 mg of ash is solubilized in 6 ml 85% nitric acid with an addition of 9.5 g aluminum nitrate. Then 10 ml ethyl acetate are added, the mixture is agitated and centrifuged. The organic phase, which contains uranium, is separated. The separated aqueous phase contains most elements capable of fluorescence quenching. The separated ethyl acetate solution is placed in a platinum dish, the solvent burned off, and the residue heated to red heat to eliminate remaining traces of nitric acid.

Two grams of flux (45.5 parts Na_2CO_3, 45.5 parts K_2CO_3 and 9 parts NaF) are added to this dish, which is heated gently with agitation up to melting. It is then held at this temperature for 1 min while continuing agitation so as to obtain a uniform melt. The dish is then cooled, and fluorescence is read directly on the sample in the dish with excitation at 365 nm.

---

*Dissolve 2.500 g aluminum nitrate in 2 liters water. Dilute 100 ml of this solution with 73 ml water and 4 ml concentrated HNO_3.

A series of standards is prepared from ash samples that contain increasing additions of uranium (0-0.05 μg U or 0-0.5 μg U). These are then treated as above.

The method is sensitive to 0.01 μg in 50 mg of initial ash, corresponding to about 1 ppm in the sample. At this concentration level, the accuracy error is in the order of 8%.

3.14.3 Determination of Uranium in Biological Media

After mineralization by calcining, biological tissue can be analyzed under the same conditions as plants. The method proposed by Korkisch (1973) is suitable for a urine analysis. Mineralization of the sample is unnecessary, and separation of uranium on a strong basic anion exchanger is used instead. A suitable volume of urine (containing 0.1-1 μg U) is acidified with concentrated hydrochloric acid and loaded on Dowex 1 x 8 anion resin. Under these conditions, most organic and inorganic compounds as well as phosphoric acid are displaced. The foreign ions that might be found on the resin at the same time as uranium, such as iron or certain pigments, are displaced by washing the resin with a mixture containing 50% tetrahydrofuran, 40% methyl glycol and 10% 6 N hydrochloric acid. Uranium is eluted with N hydrochloric acid. Fluorometric analysis of this eluate is interference-free; it is performed by the classical technique.

3.14.4 Determination of Uranium in Water

The method of Danielson et al. (1973) can be used. These authors bind uranium on ion exchanger in order to separate substances interfering with fluorescence and to enrich the medium. Fluorometry is performed directly on the eluate.

A sample volume of 200 ml is treated with 1 ml 2 M sulfuric acid, filtered on a 1.2 μg cellulose nitrate filter, and then passed through a resin column (8 mm x 8 mm) of Dowex 1 x 8 (100-200 mesh) in the sulfate form at a flow rate of 180 ml/hr. The column is washed with 120 ml 0.01 M sulfuric acid at a flow rate of 5 ml/hr. The eluate is treated with 3.0 ml 4 M phosphoric acid. A volume of 3 ml of the latter solution is sufficient to read and record the fluorescence between 470 and 510 nm, the maximum being at 493 nm.

The standard is prepared by adding a known quantity of uranium to 3 ml of the analytical solution, i.e., 0.3 ml of uranium solution containing 1.5 or 2 μg/ml. If the water sample contains organic matter, the eluate must be evaporated, calcined and treated with 1.5 ml of 30%

hydrogen peroxide. The residue is taken up at elevated temperature in 10 ml of a mixture of 0.8 M phosphoric acid and 0.01 M sulfuric acid. The solution is held at elevated temperature for 30 min, and the volume is finally adjusted to 6 ml to be read as before. The limit of detection of the method is 0.1 μg U/ml of water.

3.14.5 Determination of Uranium in Sea Water

The method proposed by Kim *et al.* (1971) and Leung *et al.* (1972) involves the use of a uranium separation by precipitation and flotation, followed by complexing with Rhodamine B, extraction of the complex with ethyl acetate and reading of the fluorescence.

The following procedure is used. A 500-ml sample containing between 1 and 6 μg U is treated with 2 ml 0.1 M thorium nitrate solution, and the pH is adjusted to 5.7 ± 0.1 with M hydrochloric acid. The solution is placed into the flotation cell with 3 ml 0.05% (w/v) dodecanoate solution in 95% ethanol. An air stream is passed through the solution at a flow rate of 10 ml ± 2 ml/min for 5 min. The uranium is collected in the flotation product. It is recovered by aspiration and placed into an Erlenmeyer flask. It is dissolved in 3-4 ml of a 4:1 mixture of 12 N HCl-16 M HNO$_3$. The solution is evaporated to a few ml and then to dryness, and the residue is taken up in 10 ml of a saline complexing solution (550 g Ca(NO$_3$)$_2$ x 4 H$_2$O, 7.5 g Na$_2$ EDTA in 125 ml of water). After solubilization at elevated temperature and cooling, the uranium is extracted by agitating with 6 ml ethyl acetate. After phase separation by centrifuging, 5 ml of the organic solution are evaporated. The residue is taken up in a few drops of dilute nitric acid, evaporated and finally taken up in 1 ml of pH 7.5 buffer (solution of 0.05 M ammonium hydroxide and 0.9 M ammonium nitrate) and 2 ml Rhodamine B (reagent prepared by saturation of benzene rhodamine dried on sodium containing 1% benzoic acid). The fluorescence of the uranium-Rhodamine B complex is measured in this solution at 575 nm.

The standard is prepared by the method of additions to sea water samples (500 ml with additions of 2, 4, and 6 μg U and treated as above). For additional information, consult Florence *et al.*, 1969, and Viswanathan *et al.*, 1968.

REFERENCES

Agranov, K. I., A. S. Klimentov, L. V. Reiman and V. L. Shcheglov. *J. Anal. Chem.* (USSR) 25:1127 (1970).

Amos, D. *Anal. Chem.* 42:842 (1970).

Anglin, J. H., Jr., W. H. Batten, A. I. Raz and R. M. Sayre. *Photochem. Photobiol.* 13:279 (1971).

Axelrod, H. D., J. E. Bonelli and J. P. Lodge. *Anal. Chim. Acta* 51(1):21-24 (1970).

Axelrod, H. D., J. E. Bonelli and J. P. Lodge. *J. Environ. Sci. Technol.* 5:421 (1971).

Baker, J. T. *Chemical News Leaflet* (Phillipsburg, New Jersey: Spring, 1970.

Ben-Dor, L., and E. Jungreis. *Isr. J. Chem.* 8:951 (1970).

Bersis, D., and E. Vassiliou. *Analyst* 91:499 (1966).

Bozhevol'nov, E. A., S. U. Kreingol'd, and L. I. Sosenkova. *Metody Anal. Khim. Reaktiv. Prep.* 19:25 (1971).

Carboni, R. A., D. D. Coffman, and E. G. Howard. *J. Amer. Chem. Soc.* 80:2838 (1958).

Dianelson, A., B. Roennhalm, L. E. Kjeilstroem, and F. Ingman. *Talanta* 20(2):185 (1973).

Dobrolyubskaya, T. S. *Zh. Anal. Khim.* 26(9):1835 (1971).

Dombrowski, L. J., and E. J. Pratt. *Anal. Chem.* 44(14):2268-2272 (1972).

Eggert, D. A. *Stain Technol.* 45:301 (1970).

Fingerhut, B., A. Poock, and H. Miller. *Clin. Chem.* 15:870 (1969).

Florence, T. M., D. A. Johnson, and Y. J. Farrar. *Anal. Chem.* 41:1652 (1969).

Gruenert, A., and G. Toelg. *Talanta* 18:881 (1971).

Gurvich, A. M., A. P. Nikiforova, and M. I. Tombak. *J. Anal. Chem.* (USSR) 25:1319 (1970).

Hallett, M., A. S. Schneider, and E. Carbone. *J. Membrane Biol.* 10(1):31 (1972).

Har, L. T., and T. S. West. *Anal. Chem.* 43(1):136-139 (1971).

Hardwick, B. A., D. K. B. Thistlethwayte, and R. J. Fowler. *Atmos. Environ.* 4:379 (1970).

Hayashi, T., K. Hara, S. Kawai, and T. Ohno. *Chem. Pharm. Bull.* 18:1112 (1970).

Hodgeson, J. A., K. J. Krost, A. E. O'Feeffe, and R. K. Stevens. *Anal. Chem.* 42:1795 (1970).

Holzbecher, J., and D. E. Ryan. *Anal. Chim. Acta* 64(3):333 (1973).

Huffman, C., and L. B. Riley. *U. S. Geol. Surv., Prof. Paper* 700-B:181 (1970).

Kim, Y. S., and H. Zeitlin. *Anal. Chem.* 43(11):1390-1393 (1971).

Kirkbright, G. F., R. Narayanaswamy, and T. S. West. *Anal. Chem.* 43(11):1434-1438 (1971).

Koch, K. J., and D. E. Ryan. *Anal. Chim. Acta* 57(2):295-300 (1971).

Konig, F. H. P., G. Den Boef, and H. Poppe. *Fresenius'Z. Anal. Chem.* 256:270 (1971).

Korkisch, J., and I. Steffan. *Mikrochim. Acta* 2:273 (1973).

Lamand, M. *Ann. Fals. Exp. Chim.*, pp. 4-12.
Leung, G., Y. S. Kim, and H. Zeitlin. *Anal. Chim Acta* 60(1):229-233 (1972).
Liebich, B., D. Monnier, and M. Marcantonatos. *Anal. Chim. Acta* 52(2):305-312 (1970).
Lisitsyna, D. N., and D. P. Shcherbov. *J. Anal. Chem.* (USSR) 25:1986 (1970).
Lytle, F. E. *Appl. Spectrosc.* 24(3):319-326 (1970).
Matveets, M. A., and D. P. Shcherbov. *J. Anal. Chem.* (USSR) 26 (4, Pt. 2):723 (1971).
Maxwell, J. A. *Rock and Mineral Analysis* (New York: Intersciences Publishers, 1968), p. 475.
Monnier, D., and M. Marcantonatos. *Anal. Chim. Acta* 36:360 (1966).
Monnier, D., B. Liebich, and M. Marcantonatos. *Z. Anal. Chem.* 247:188 (1969).
Monnier, D., and M. Marcantonatos. *Mitt. Geb. Lebensmittelunters Hyg.* 63(2):212 (1972a).
Monnier, D., C. A. Menzinger, and M. Marcantonatos. *Anal. Chim. Acta* 60(1):233-238 (1972b).
Morgen, E. A., E. A. Vlasov, and A. Z. Serykh. *Zh. Pribl. Khim.* (Lenograd) 43:2744 (1970).
Naito, T., H. Nagano, and T. Yasui. *Japan Anal.* 18:1068 (1969).
Nazarenko, I. I., A. M. Koslov, and I. V. Koslova. *Zh. Anal. Khim.* 25:1135 (1970).
Nazarenko, I. I., and I. V. Koslova. *Zavod Lab.* 37:414 (1971a).
Nazarenko, I. I., A. M. Koslov, and I. V. Koslova. *Metody Anal. Khim. Reaktiv. Prep.* 19:32 (1971b).
Onishi, H. *Anal. Chem.* 27:832 (1955).
Olson, O. E. *J. Assoc. Offic. Anal. Chem.* 52(3):627-634 (1969).
Pakalns, P. *Aust. At. Energy Comm.*, AAEC/TM (Rep.), AAEC/TM-552(6):1 (1970).
Passwater, R. A. *Fluorescence News* 7(2):11 (1973).
Patrias, G., and O. E. Olson. *Feedstuffs* 41(43):32 (1969).
Puech, A., G. Kister and J. Chanal. *Zentralbl. Pharm., Pharmakother. Laboratoriumsdiagn.* 111(1):7 (1972).
Podberezskaya, N. K., V. A. Sushkova, and E. A. Shilenkov. *Issled. O Obl. Khim. Fiz. Metodov. Anal. Miner. Syr'ya* 87 (1971).
Podberezskaya, N. K., and V. A. Sushkova. *Zavod. Lab.* 39(7):774 (1973).
Ritchie, K., and J. Harris. *Anal. Chem.* 41(1):163-166 (1969).
Ryan, D. E., F. Snape, and M. Winpe. *Anal. Chim. Acta* 58(1):101 (1972).
Sardesai, V. M., and H. S. Provido. *Mikrochem. J.* 14:550 (1969).
Sawicki, C. R. *Anal. Letter.* 4:761 (1971).
Schroeder, H. A., D. V. Frost, and J. J. Balass. *J. Chronic Dis.* 23:227 (1970).
Schulman, S. G., and M. S. Rietta. *J. Pharm. Sci.* 60(11):1962 (1971).
Shcherbov, D. P., R. N. Plotnikova, and I. N. Astaf'eva. *Zavod. Lab.* 36:528 (1970).
Shcherbov, D. P., I. N. Astaf'eva, and R. N. Plotnikova. *Zavod. Lab.* 39(5):546 (1973).

Shigematsu, T., Y. Nishikawa, K. Hiraki, and N. Nagano. *Bunseki Kagaku* 19:551 (1970).

Smith, A. Y., and J. J. Lynch. *Can. Dep. Energy, Mines Resources, Geol. Surv. Can. Paper* 69-40 (1969).

Talipov, S. T., A. T. Tashkhodzhaev, L. E. Zel'tser, and K. Khikmatov. *Dokl. Akad. Nauk Uzb. SSR* 29(5):34 (1972a).

Talipov, S. T., A. T. Tashkhodzhaev, L. E. Zel'tser, and K. Khikmatov. *Nauch. Tr. Tashkent. Univ.* 419:89 (1972b).

Talipov, S. T., A. T. Tashkhodzhaev, L. E. Zel'tser and K. Khikmatov. *Zh. Anal. Khim.* 28(4):807 (1973).

Tashkhodzhaev, A. T., L. E. Zel'tser, and K. Khikmatov. *Uzb. Khim. Zh.* 16(3):22 (1972).

Van Duuren, B. L., and T. L. Chan. *Advan. Anal. Chem. Instrum.* 9:387 (1971).

Vasil'ev, E. N., O. A. Fakeeva, E. A. Solov'ev, and E. A. Kozhevol'nov. *Zh. Anal. Khim.* 28(4):688 (1973).

Vasilevskaya, A. E. *Nauch. Tr., Vses. Inst. Miner. Resur.* (Russ) USSR 5:22-31 (1971).

Vesiene, T., and B. Raieinskyte. *Liet. TSR Mokslu Akad. Dard., Ser. B.* 6:115 (1972).

Viswanathan, R., S. M. Ramaswami, and C. K. Anni. *Bull. Nat. Inst. Sci. India* 38 (Pt. 1):284 (1968).

Watkinson, J. H. *Anal. Chem.* 38 (1):92-97 (1966).

White, C. E., and R. J. Argauer. *Fluorescence Analysis - A Practical Approach.* (New York: Marcel Dekker, 1970), p. 251.

Wiersma. J. H., and G. F. Lee. *Environ. Sci. Technol.* 5(12):1203 (1971).

Winefordner, J. D., S. G. Schulman, and T. O'Haver. *Luminescence Spectrometry in Analytical Chemistry* (New York: Wiley-Interscience, 1972).

Wronski, M. *Anal. Chem.* 43:606 (1971).

Zweidinger, R. A., L. Barnett, and C. G. Pitt. *Anal. Chem.* 45(8):1563-1564 (1973).

CHAPTER 2

EMISSION SPECTROSCOPY

1. ELEMENTARY THEORETICAL CONCEPTS*

Every spectral radiation produces a sinusoidal electromagnetic disturbance of the medium in which it propagates. This oscillation is defined by its period T or its frequency $\nu = 1/T$ and by its amplitude. The wavelength of a radiation *in vacuo* is:

$$\lambda_o = c\,T = \frac{c}{\nu}$$

where $c \cong 3 \cdot 10^{10}$ cm/sec is the velocity of light. The wave number is the reciprocal of the wavelength:

$$\nu' = \frac{1}{\lambda_o}\ \text{cm}^{-1} = \frac{\nu}{c}$$

Wavelengths are expressed in angstroms (Å) or in nm:

$$1\ \text{Å} = 10^{-1}\ \text{nm}$$

Bohr showed that the atom consists of a nucleus surrounded by electrons, distributed in different shells or orbits. Under excitation, the electron passes from one stable orbit to another, and its energy changes by an amount dW. On returning to its initial orbit, the electron emits this energy as electromagnetic radiation in accordance with the relationship established by Planck in 1900:

$$dW = h\nu_o$$

where h is Planck's constant, $6.55 \cdot 10^{-27}$ ergs/sec, and ν_o is the frequency of the radiation in vacuum.

The energy of the electron in its stable state has been calculated to be:

---

*Reproduced from "Recherche et Dosage des Elements Traces," by M. Pinta, 1961, with permission of Dunod Ed, Paris.

$$W = - \frac{2\pi^2 me^4}{h^2} \times \frac{1}{n^2}$$

where   m is the mass of the electron,
     e is the charge of the electron,
     h is Planck's constant, and
     n is a positive integer.

Letting

$$R = \frac{2\pi^2 me^4}{h^3 c} \qquad \text{(where R is Rydberg's constant)}$$

we have:

$$W = - \frac{Rhc}{n^2}$$

In general, radiation is emitted when the energy of an electron changes from the value W_2 (n_2 shell) to the value W_1, which corresponds to the ground state (n_1 shell).

When electrons originally in orbit n_1 enter, while under excitation, orbits n_2, n_3 . . . n, and then return to the orbit n_1; the energy that they liberate is manifested by the emission of a series of spectral lines. The frequency ν of each of these lines is given by the expression:

$$\nu = \frac{W_2 - W_1}{h}$$

$$\nu = Rc \left(\frac{1}{n_1^2} - \frac{1}{n^2} \right)$$

the corresponding wave number ($\nu' = \nu/c$) is

$$\nu' = R \left(\frac{1}{n_1^2} - \frac{1}{n^2} \right)$$

This formula was experimentally established by Balmer in 1885 for the hydrogen atom: when $n_1 = 2$ and n = 3, 4, 5 . . . we obtain the series of lines discovered by Balmer in the visible spectrum:

$$\nu' = R \left(\frac{1}{2^2} - \frac{1}{n^2} \right) \qquad (n > 2)$$

where $R = 109,677 \cdot 6 \ cm^{-1}$ (Rydberg's constant).

Subsequently, other series were discovered: the Lyman series in the ultraviolet:

$$\nu' = R \left(\frac{1}{1^2} - \frac{1}{n^2} \right) \qquad (n > 1)$$

and the Paschen series in the infrared:

$$\nu' = R \left(\frac{1}{3^2} - \frac{1}{n^2} \right) \qquad (n > 3)$$

In general, the frequencies of all the radiations emitted by hydrogen can be obtained by the difference between two values of Rc/n^2, called spectral terms:

$$\nu = \Delta \frac{Rc}{n^2}$$

where n is any integral number.

For other elements, Ritz and Rydberg proposed analogous formulas:

$$\nu' = A - \frac{R}{(n + \mu)^2} \qquad \text{(wave number)}$$

$R = 109,677 \cdot 6 \text{ cm}^{-1}$, or:

$$\nu' = A - \frac{R}{(n + \mu + \delta/n)^2}$$

where μ and δ are correction factors that vary with the particular element and are always less than unity. A is a constant representing the convergence limit of ν' as n increases infinitely. These formulas can be used to obtain the series of spectral lines of the elements.

Conversely, if an atom at the energy level W_1 receives radiation consisting of rays of frequency ν, it can absorb a quantum $h\nu$ of energy from these rays, and pass to the energy state W_2:

$$W_2 = W_1 + h\nu$$

Kirchhoff summarized the phenomena of atomic absorption in the following law: "a body subjected to certain excitation conditions can only emit radiations that it is able to absorb under the same conditions."

In practice, the only lines that can be absorbed are those which during emission result in the lowest energy level (resonance lines). A method of spectral analysis is based on these principles.

2. PRINCIPLES OF SPECTRAL ANALYSIS

The energy that must be supplied to an atom for it to emit a line of frequency $\nu = (W_2 - W_1)/h$ is called the excitation potential, and is expressed in electron-volts. Thus, the appearance of a spectral line essentially depends on the source of excitation. If the energy of excitation of the source is sufficient to expel the electron completely from the atom, the atom becomes ionized; the ionization potential is an expression of the energy required to ionize the atom. Any energy in excess of the ionization potential causes displacement of other electrons, and the ionic spectrum is formed.

The spectrum of the neutral atom is relatively easy to produce: burner flames have sufficient energy to excite the spectra of many atoms, such as the alkali metals, alkaline earths, chromium, and manganese. The electric arc or spark can excite the spectra of all the neutral atoms and of some singly, doubly or multiply ionized atoms. The lines of neutral atoms are designated by the symbol of the element followed by the number I, *e.g.*, Na I; the lines of singly ionized atoms which have lost one electron) are designated by the symbol and the number II, *e.g.*, Na II; the lines of doubly ionized atoms (which have lost two electrons) are designated as Na III, and so on. In fact, it is often difficult to distinguish between flame, arc, or spark spectra because some lines appear in all these spectra.

In qualitative spectral analysis the elements are identified by the spectra of their neutral or ionized atoms. The intensity of a given line is a function of the number of atoms undergoing excitation; if I is the intensity of the line and N is the number of atoms involved, N will be proportional to I only under special conditions, which are often difficult to realize. Then

$$I = K \cdot N \text{ or } \log I = \log K + \log N$$

where K is a constant.

This relationship is obeyed in particular in the case of very small traces of the element, when there is no danger of interference by reabsorption (or self-absorption) of the lines by neutral, nonexcited atoms. The experimentally established relationship is

$$I = KN^m$$

or

$$\log I = \log K + m \log N$$

where K and m are constants that depend on the line itself, or, in other words, on the excited atom or ion and on the conditions of excitation: m is generally less than unity ($m < 1$).

This relationship is the basis of quantitative spectrographic analysis. Under given conditions of excitation, the intensities of the lines of the elements constituting a complex medium are dependent only on the concentration of the elements; in general a small variation in the concentration of an element in the emission source does not produce any variation in the intensities of the lines of the other elements. On the other hand, the ratio of the intensities of two lines emitted by the same element or by two different elements can vary considerably with the mode of excitation (flame, arc, or spark). Reproducible excitation conditions must be used as far as possible: in flame spectrography, the rate of introduction of the sample into the flame and the nature and the rate of flow of the gases must be rigorously controlled. In arc and spark spectrography, the nature and resistance of the electrodes, the current, and the potential difference between the electrodes must be checked and kept constant. In quantitative analysis, the material to be studied is compared with a series of synthetic samples of the same base composition.

Best results are obtained if the errors due to variations in the source are eliminated by comparing two lines, one of the element to be determined, and the other of a suitable element present in a known quantity in the source, and acting as an internal standard.

If I_X and I_E are the respective intensities of a characteristic line of the element X to be studied, and of the internal standard element E, and C_X and C_E are their respective concentrations, we may write:

$$C_X = C_E \cdot f\left(\frac{I_X}{I_E}\right)$$

The determination of the concentration C_X thus amounts to the measurement of the ratio between the intensities of spectral lines. The element used as internal standard and the lines used for the determination must be suitably chosen. This is the principle of the method of internal standards (Gerlach and Schweitzer, 1930).

The *persistent lines* (raies ultimes) defined by Gramont (1920) are the most sensitive lines of the various elements, *i.e.,* those that persist in the spectrum when the concentration of the element in the source is very low. The identity of these lines depends on the source of excitation, and the lines themselves are not always the most intense ones in the spectrum of the element. The raies ultimes are those generally used for the analysis of traces of elements. Two tables of raies ultimes are given in appendix 3 and 4.

In general, a line can be used in quantitative analysis if the ratio dI/dC—that is, the variation in the intensity of the line as a function of the concentration—is sufficiently large. Gerlach and Schweitzer showed that the lines corresponding to a fundamental level are easily reabsorbed when the concentration of the element increases. In fact, as the number of emitter atoms increases, there is a simultaneous reabsorption of the lines by neutral atoms. Thus, for the raies ultimes the proportionality between I and C is obtained only at very low concentrations. To obtain quantitative results it is often necessary that the determination be carried out on trace quantities of the element in the source. This is achieved by suitable dilution, and the determination of major elements thus often becomes an analysis of trace elements.

The phenomenon of reabsorption or self-absorption, while interfering with emission spectrographic analysis, is the basis of a new analytical method called atomic absorption (see Chapter 4).

In addition to the method of internal standards, which is most often employed at present, we may note the comparison method, in which a suitable line of the element being determined is compared under the same spectroscopic conditions with the same line of the same element in a standard. This technique, which is due to Grammont, is an excellent semiquantitative method, particularly well-suited to routine analysis. Since the fluctuations of the source interfere, the method is unsuitable for quantitative analysis by arc or spark spectrography, but is generally used in flame spectrography or spectrophotometry.

3.   METHOD OF INTERNAL STANDARDS

This method, the principle of which has just been discussed, is the most widely used spectrographic method, particularly in trace element analysis. Let I_X and I_E be the intensities of the characteristic lines of the element X to be determined, and of the internal standard E, whose concentrations in the excitation source are C_X and C_E, respectively; then

$$I_X = K_X \cdot C_X^{m_X} \text{ and } I_E = K_E \cdot C_E^{m_E}$$

(K_X, K_E, m_X and m_E being the constants defined above); hence:

$$\log \frac{I_X}{I_E} = \log \frac{K_X}{K_E} + \log C_X^{m_X} - \log C_E^{m_E}$$

Since $C_E^{m_E}$ is practically constant, the equation becomes:

$$\log \frac{I_X}{I_E} = \log K + m_X \log C_X$$

The curve representing $\log I_X/I_E$ as a function of $\log C_X$ reflects the accuracy of the determination. If it is a straight line, the accuracy of the determination is greater, the steeper the slope. Thus, the sensitivity and accuracy of the method of internal standards are a function of the conditions of excitation, of the nature of the element taken as the internal standard, and of the lines chosen.

A number of factors must be considered when choosing the internal standard:

1. The internal standard is a metal, an oxide, or salt. Its concentration in the excitation source should usually be small.

2. The rates of volatilization of the internal standard and of the element to be determined must be similar.

3. The lines of the element to be determined and of the internal standard must have similar excitation potentials.

4. The line of the internal standard must have as small a self-absorption (reabsorption of the line by the atoms in their fundamental state) as possible.

5. The element introduced into the sample to act as internal standard must be sufficiently pure.

6. When the internal standard is added in very small amounts to the material to be studied, the analyst must make sure that the sample does not already contain traces of this element, or at least that these traces are negligible in comparison with the quantity of the standard element added.

7. The lines of the element to be determined and of the internal standard must be sufficiently close to each other in the spectrum.

Tables I and II (Ahrens, 1950) give some useful information for the choice of internal standards: the boiling points of the elements and their oxides and the order of volatilization of some elements as elements, sulfides, oxides, sulfates, carbonates, silicates and phosphates.

4. APPARATUS

4.1 Excitation Sources

The following paragraphs should not be considered as an inventory of the apparatus and instruments used in spectroscopy. Only some of the apparatus currently used in the analysis of trace elements—in qualitative, semiquantitative, and quantitative analysis—will be described. A spectrographic set-up consists of:

1. an excitation source: flame, arc, spark, plasma, or laser
2. a dispersing unit: quartz prism spectrograph
3. a line reading unit: comparator or microphotometer, direct reading
4. accessories: electrode cutter, logarithmic sector, optical accessories, and photographic materials.

4.1.1 Flames

From the point of view of spectroscopy, flames are purely thermal sources giving a temperature of 1500 to 3000°, according to their nature. The excitation energy is relatively low, and only low-potential elements, such as the alkali and alkaline earth metals, can be excited to any great extent.

Table I. Boiling Points of Some Elements and Oxides (°C)

Ag	1950	Cr$_2$O$_3$	mp:1990	MnO	mp:1650	Si	3025
Ag$_2$O	d:300	Cs	480	Mn$_3$O$_4$	mp:1705	SiO$_2$	2230-2590
Al	2360	Cs$_2$O	d:360-400	Mg	1000	Sr	1250
Al$_2$O$_3$	2250	Cu	2465	MgO	mp:2500-2800	SrO	mp:2430
As	s:510	CuO	d:1030	Mo	4540	SnO$_2$	d: 1130
As$_2$O$_3$	s:190	Fe	2775	MoO$_3$	s:	Ta	5130
Au	2700	FeO	mp:1420	Na	780	Ta$_2$O$_5$	d: 1470
Ba	1520	Fe$_2$O$_3$	mp:1565	Na$_2$O	s: 1275	Te	4500
BaO	cv: 2000	Ga	2300	Nb	4300	TeO$_2$	s: 450
Be	2380	Ga$_2$O$_3$	mp:1900	Nb$_2$O$_5$	mp:1520	Tl	1330
BeO	cv: 3900	Ge	2620	Ni	2785	Tl$_2$O	1080
Bi	1470	GeO$_2$	mp:1100	NiO	3380	Th	4650
Bi$_2$O$_3$	1890(?)	Hf	4400	Os	4900	ThO$_2$	4400
B	3575	HfO$_2$	5400	Pb	1610	Ti	3200
B$_2$O$_3$	1230	Hg	250	PbO	mp: 890	TiO$_2$	d: 1640
C	4200	HgO	d: 500	Pd	2950	V	3300
Ca	1375	In	1900	Pt	3750	V$_2$O$_5$	d: 1750
CaO	2850	In$_2$O$_3$	d: 850	P	280	W	5400
Cd	767	Ir	4400-5300	P$_2$O$_5$	s: 350	WO$_3$	s:
CdO	d:950-1000	K	760	Rb	600	Y	3200
Ce	3365	La	3230	Rb$_2$O	d: 400	Y$_2$O$_3$	mp:2400
CeO$_2$	mp:1950	La$_2$O$_3$	4200	Rh	3625	Zn	810
Co	2800	Li	1220	Ru	4000	ZnO	s: 1800
CoO	mp:1935	Li$_2$O	mp:1700	Sc	2900	Zr	4370
Cr	2560	Mn	1900	Sc$_2$O$_3$	4450	ZrO$_2$	5000

d: decomposed; mp: melting point; s: sublimed; cv: calculated value.

Table II. Order of Volatilization of Elements in the Electric Arc

In the elemental form	$Hg > As > Cd > Zn > Sb \geqslant Bi > Tl > Mn > Ag, Sn, Cu >$ In, Ga, Ge $> Au > Fe, Co, Ni, \gg Pt \gg Zr, Mo, Re, Ta, W$
As sulfides	As, $Hg > Sn, Ge \geqslant Cd > Sb, Pb \geqslant Bi > Zn, Tl > In >$ $Cu > Fe, Co, Ni, Mn, Ag \gg Mo, Re$
As oxides, sulfates, carbonates, silicates, or phosphates	As, $Hg > Cd > Pb, Bi, Tl > In, Ag, Zn > |Cu, Ga > Sn >$ Li, Na, K, Rb, Cs $> |Mn > Cr, Mo?, W?, Si, Fe, Co, Ni >$ $|Mg > Al, Ca, Ba, Sr, V > Ti > |Be, Ta, Nb > Se, La, Y$ and most of the rare earths $> |Zr, Hf$

The sign | divides the elements into three groups: volatile elements, moderately volatile elements, and nonvolatile elements. These groups are subdivided into subgroups by the sign |. $\gg$ means: much more than.

Table III (Mavrodineanu and Boiteux, 1965) gives the flame temperatures usually employed. In analysis of trace elements hydrogen or acetylene flames (nitrous oxide acetylene, 2900°) are commonly used. The oxyhydrogen (2600°) and oxyacetylene (3100°) flames are frequently used since they are the hottest. However, the oxyacetylene flame has a considerable spectral background, which reduces the sensitivity of the analysis and thus limits its possible applications. The air-acetylene flame is less sensitive, but is nevertheless often used, as it is easy to produce and control.

The sample is solubilized in an acid and is usually atomized by the carrier gas in an atomizing chamber designed to homogenize the spray formed. The finest spray particles are fed into the burner, while the larger droplets are evacuated from the atomization chamber, recovered, and atomized again (Figures 1 and 2). With a suitably adjusted apparatus it is possible to take pictures of spectra for 10 to 15 min starting with 10 ml of solution.

The sample is usually introduced into the source in aqueous solution, but an organic solvent such as acetone, benzene, toluene, or chloroform can also be used. Flames are a sufficiently stable source for direct use in quantitative analysis without any need to introduce an internal standard element into the sample solution.

The use of a flame is specially recommended in the spectral analysis of trace elements. Lithium, sodium, potassium, rubidium, calcium and strontium are analyzed in acetylene-air flame, and Al, Cr, Dy, Er, Eu, Fe, Gd, In, La, Mn, Ni, Pr, Sn, Tb, V, Y, Yb, in nitrous oxide acetylene flame (see Table IV).

Table III. Flame Temperatures

Fuel	Oxidant	Theoretical Reactions (Stoichiometry)	Limits of Inflammability at $20°$, Fuel (%)	Maximum Temperatures ($°C$) Experimental
Hydrogen	Oxygen	$H_2 + \frac{1}{2}O_2 \rightarrow H_2O + 57{,}800$ cal	4-94	2560
	Air	$H_2 + \frac{1}{2}O_2 + 2N_2 \rightarrow H_2O + 2N_2 + 58{,}000$ cal	4-75	1950
Propane	Oxygen	$C_3H_8 + 5O_2 \rightarrow 3CO_2 + 4H_2O + 530{,}570$ cal		2750
	Air	$C_3H_8 + 5O_2 + 20N_2 \rightarrow 3CO_2 + 4H_2O + 20N_2 + 530{,}570$ cal	31-9.4	1820
Butane	Oxygen	$C_4H_{10} + 6.5O_2 \rightarrow 4CO_2 + 5H_2O + 687{,}940$ cal		2800
	Air	$C_4H_{10} + 6.5O_2 + 26N_2 \rightarrow 4CO_2 + 5H_2O + 26N_2 + 687{,}940$ cal	1.9-8.4	1790
Acetylene	Oxygen	$C_2H_2 + O_2 \rightarrow 2CO + H_2 + 300{,}100$ cal	2-95	2785
	Air	$C_2H_2 + O_2 + 4N_2 \rightarrow 2CO + H_2 + 4N_2 + 300{,}100$ cal	2.5-80	2125
	N_2O	$C_2H_2 + 5/2\ N_2O \rightarrow 2CO_2 + H_2O + 5/2\ N_2 + 408{,}100$ cal	2.2-67	2700
Town gas	Oxygen	Town gas $+ 0.98\ O_2 \rightarrow CO_2 + H_2O + 108{,}790$ cal	10-73.6	2640
	Air	Town gas $+ 0.98\ O_2 + 3.9\ N_2 \rightarrow CO_2 + H_2O + 3.9\ N_2 + 108{,}790$ cal	9.8-24.8	1600
Cyanogen	Oxygen	$(CN)_2 + O_2 \rightarrow 2CO + N_2 + 126{,}700$ cal		4640

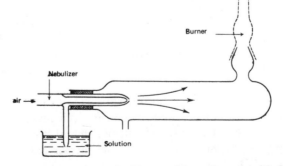

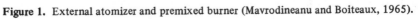

Figure 1. External atomizer and premixed burner (Mavrodineanu and Boiteaux, 1965).

Figure 2. Direct injection burner.

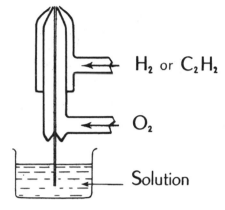

Table IV. Limits of Detection for Some Elements in Different Flames (μg/ml)

	Air-Acetylene Flame	Oxygen-Acetylene Flame	Nitrous Oxide-Acetylene Flame
Al	–		0.05
Ca	1	0.2	0.001
Cr	–	0.5	0.004
Dy	–	–	0.05
Er	–		0.07
Eu	–	–	0.001
Fe	–	1	0.01
Gd	–	–	0.02
In	–	–	0.07
K	0.1	0.02	0.03
La	–	–	–
Li	0.1	0.01	0.01
Mn	–	0.1	–
Na	0.1	0.01	0.01
Ni	–	–	–
Pr	–	–	0.02
Rb	1	0.2	0.07
Sn	–	–	–
Sr	1	0.2	0.2
Tb	–	–	0.01
Tl	–	–	0.02
V	–	5	0.1
Y	–	–	0.03
Yb	–	–	0.006

4.1.2 Electric Arcs

The arc is produced by passing a current of 2 to 30 A between two electrodes at a potential of several dozen or several hundred volts. This is the source of thermal excitation, and its temperature varies from 2000 to 8000° according to its characteristics, its nature, the form of the electrodes and the location of the discharge. The arc is a much more powerful spectral source than the flame. Many elements, *e.g.*, most of the metals and some metalloids, are excited by the arc with a sensitivity sufficient for trace analysis.

Table V gives the ionization potentials of some elements in the electric arc, and the working temperatures used. The types of arc most often used are:

1. low voltage direct current arc
2. high voltage alternating current arc
3. interrupted direct current arc

4.1.2.1 Direct Current Arc. The direct current is produced by a rectifier or a generator. A diagram of the circuit, given in Figure 3, includes a resistance R_1 of 110 ohms (15 A) and a resistance R_2 (190 Ω, 2.8 A) in parallel with R_1. This combination of resistances can be used for accurate adjustment of the current in the arc. The inductance L is optional. It helps to reduce fluctuations in the arc. The dc arc between carbon or graphite electrodes is the source most frequently used in the analysis of trace elements. It is particularly sensitive and uses only milligram amounts of the sample.

The sample to be studied is placed either in the negative electrode (cathodic arc method) or in the positive electrode (anodic arc method). In either case, the discharge between the electrodes is spectrographed. The most commonly used electrodes are shown in Figure 4.

Mannkopff and Peters (1931) showed that the region of the most intense emission in the carbon arc lies in the part of the discharge close to the cathode. This apparent increase in the intensity is probably due to the high concentration of the ions and the atoms there. This method of excitation at the "cathode layer" is widely employed in the analysis of trace elements. The sample is introduced into an appropriate electrode (Figure 4, A, E, F) that acts as the cathode of an arc 10 to 12 mm long. The region near the cathode (2 mm) is spectrographed by means of a suitable optical apparatus.

The dc arc (cathodic or anodic) is widely used in qualitative and semi-quantitative analysis. Due to its instability, quantitative analysis is rather difficult. This problem can be remedied by constant surveillance of the conditions of excitation such as current intensity, distance between the electrodes, and the optical centering of the source.

Table V. Ionization and Excitation Potentials of Elements

Element	Ionization Potential (volts)	Excitation Potential (volts)
Ag	7.6	3.6
Al	6.0	3.1
As	10.5	6.5
Au	9.2	4.6
B	8.3	4.9
Be	9.3	5.4
C	11.3	7.7
Ca	6.1	2.9
Ce	6.6	
Co	7.9	3.6
Cr	6.8	2.9
Cs	3.9	1.4
Cu	7.7	3.8
Fe	7.9	3.3
Hg	10.4	4.9
K	4.3	1.6
La	5.6	2.2
Li	5.4	1.8
Mg	7.7	4.3
Mn	7.4	3.1
Mo	7.4	3.2
Na	5.1	2.1
P	11.0	7.1
Pb	7.4	4.3
Pt	9.0	4.0
Sb	8.6	5.3
Si	8.2	4.9
Sn	7.3	4.3
Ti	6.8	3.3
W	8.0	3.2
Y	6.6	2.7
Zn	9.4	6.6
Zr	7.0	3.4

Excitation potential—energy in volts to excite the resonance line of atom.
Ionization potential—energy in volts necessary to remove an electron from the neutral atom.

The nature of the electrodes is another important factor. Graphite or carbon is generally used, although spectrographers often tend to confuse these two forms. Even though chemically identical, their physical structures are very different, since graphite is crystalline while carbon is amorphous. In the electric arc these two forms behave very differently. Carbon is the poorer

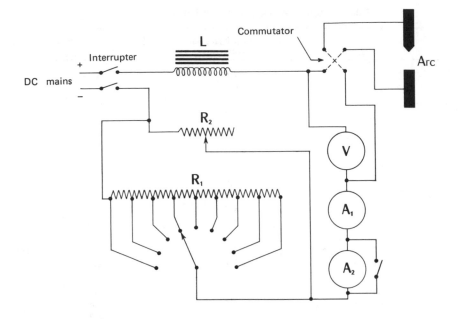

Figure 3. Circuit of dc arc (Mitchell, 1948).

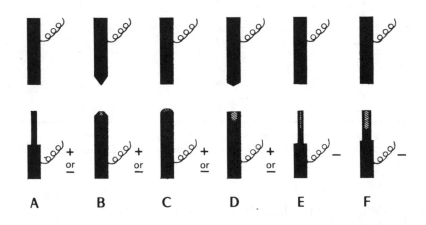

Figure 4. Electrodes used in arc spectrography.

conductor of both heat and electricity, and is also more rapidly used up by the heat. The electrodes B, C, and D should be made of graphite, while the electrodes A, E, and F should be made of carbon. Due to its form, the electrode must be consumed at the same rate as the substance placed inside it. Thus it is necessary to have the electrodes in constant motion while the arc is burning in order to maintain a constant interelectrode gap. Carbon is mainly used for cathodic and cathodic layer excitation.

The stability of combustion of the arc during the spectrographic determination is an important factor, which largely depends on the experience of the operator. Combustion in the arc is most often effected in the atmosphere, so draughts must be avoided. If several elements are determined in the same spectrum, the selectivity of the volatilization of the elements in the arc is an important source of error. An interesting apparatus has been proposed by Hoens and Smit (1957) to stabilize the vaporization rate of the substance filling the electrode in cathodic layer excitation. The upper part of the electrode, rotated at 15 rps, is concentrically cooled by water during the combustion. The device is shown in Figure 5.

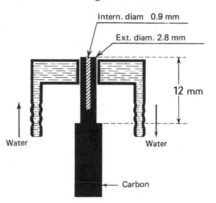

Figure 5. Water-cooled electrode.

Direct current arc combustion in an atmosphere of a noble gas is used to attenuate the cyanogen band spectrum and to stabilize the volatilization of the elements (Figure 6).

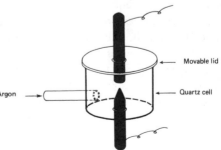

Figure 6. Arc in controlled atmosphere.

4.1.2.2 High-Voltage Alternating Current Arc. Duffendack and Thompson (1936) suggested an ac arc of 2000 to 4000 V with a current intensity of 2-4 A and a charging voltage of 50 to 100 V. The circuit, which is diagrammatically shown in Figure 7, consists of a transformer T, 110/5000 V, a choke L, and a resistor R in the primary circuit. This source is much more stable than the dc arc. Its sensitivity is lower, the electrodes do not heat up as much, and the spectral background is weaker because the electrodes cool off every time the current passes through zero. This arc has been successfully used for the analysis of solutions.

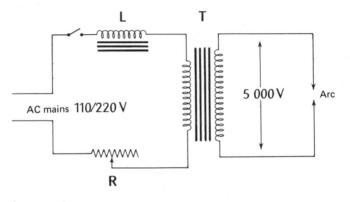

Figure 7. An ac arc circuit.

4.1.2.3 Interrupted Arc. To avoid heating the electrodes, the arc may be periodically interrupted with automatic reignition. The current is broken by a rotatory interrupter and the reignition effected by a high-frequency spark between the electrodes. The Pfeilsticker circuit is diagrammatically shown in Figure 8. It consists of an oscillatory circuit $C_1 L_1$ fed by the transformer T, producing a high-frequency current that passes through the circuit $C_2 L_2$ of the arc through the Tesla transformer $L_1 L_2$. A high-frequency spark is thus maintained between the electrodes E_1, igniting the arc fed by the mains each time the rotating interrupter closes the circuit. Under these conditions, the arc is struck intermittently and the frequency of ignition (1 to 10/sec) as well as the duration of each elementary arc (20 to 200 msec) can be adjusted. The arc, which can be fed by ac or dc, is highly luminous and its stability is satisfactory. It is possible to feed high-voltage alternating current arcs and interrupted arcs with solutions. Two devices are shown in Figures 9 and 10.

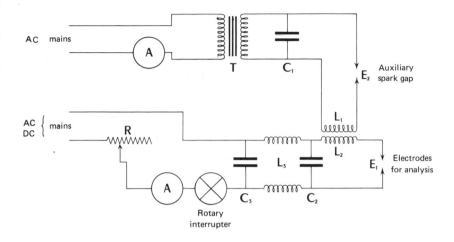

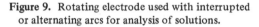

Figure 8. Interrupted arc circuit.

Figure 9. Rotating electrode used with interrupted or alternating arcs for analysis of solutions.

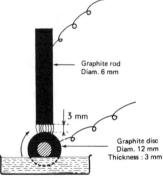

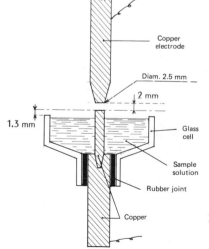

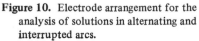

Figure 10. Electrode arrangement for the analysis of solutions in alternating and interrupted arcs.

4.1.3 The Spark

The spark is a disruptive discharge between two electrodes under high voltage. It causes ionization of the atoms constituting the electrodes. The high potential difference produces ionized atoms, and the spectrum obtained consists of the spectra of the ions formed superposed on one another. Because the electrodes are not hot, the sample to be studied can be placed in the electrode as an aqueous or organic solution.

An electric spark is usually produced by the discharge of a condenser C in the electrode circuit, which also includes an inductance L and a resistance R (Figure 11). This is the principle of the condensed

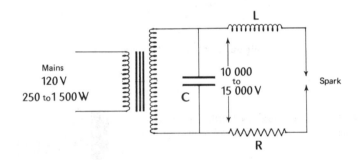

Figure 11. Condensed spark circuit.

spark. The discharge current is an oscillating current, and is given by the formula

$$I = V \sqrt{C/L}$$

where I is the current intensity (in amperes), V the voltage (in volts) of the discharge across the terminals of the condenser of capacity C (in farads), and L is the inductance (in henrys).

The value of I can reach several hundreds of amperes, which means that the source has a high ionization energy. Thus, for example, we may cite the working parameters of the Durr generator:

V = 14,000 to 21,000 V

C = 0.001 to 0.12 μF

L = 0.05 to 1 mH

R = a few dozen ohms

In quantitative analysis, the voltage V and the number of discharges per second must be stabilized. The disruptive voltage in the circuit just described is a function of the state of the surface of the electrodes and of the ionization of the air.

Feussner's circuit (Figure 12) is designed to stabilize the discharges. The controlled spark circuit includes in its discharge circuit a synchronized rotary interrupter that permits the passage of an auxiliary spark when the mobile electrode of the spark gap is situated opposite the fixed electrode. The rotor of the interrupter is so adjusted that the electrodes face one another when the charge on the condenser is at the maximum.

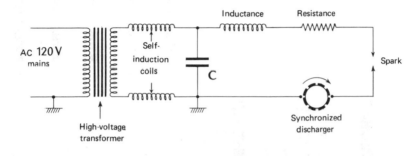

Figure 12. Feussner spark generator.

In some modern generators, the discharge is triggered by a high voltage spark controlled by a synchronized interrupter placed in a low power auxiliary circuit. The circuit is shown diagrammatically in Figure 13.

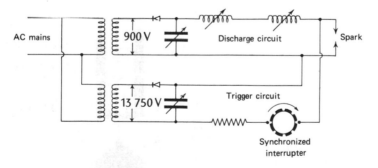

Figure 13. Multisource spark generator.

Several types of electrodes are used in spark spectroscopy. The powders are placed into the well of an electrode similar to that used with the electric arc or are bound with graphite powder in the form of an electrode.

Samples in solution can be evaporated at the tip of an electrode (Figure 14) or analyzed with either the concentric cup electrode (Figure 15), the porous electrode (Figure 16), in which the sample is supplied to the spark through the base of the upper electrode by its porosity, or the rotary electrode (Figure 17).

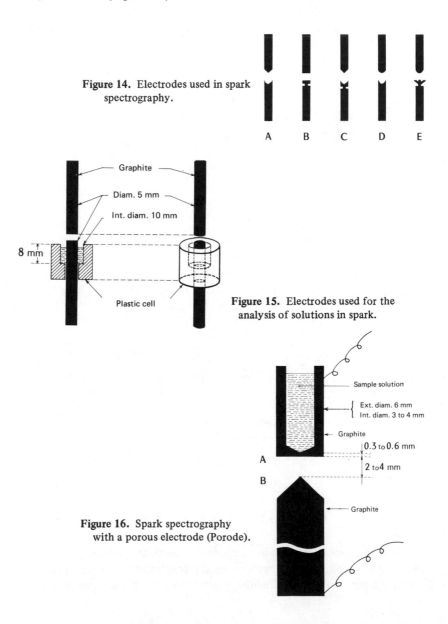

Figure 14. Electrodes used in spark spectrography.

A   B   C   D   E

Figure 15. Electrodes used for the analysis of solutions in spark.

Graphite

Diam. 5 mm

Int. diam. 10 mm

8 mm

Plastic cell

Sample solution

Ext. diam. 6 mm
Int. diam. 3 to 4 mm

Graphite

0.3 to 0.6 mm

2 to 4 mm

A

B

Graphite

Figure 16. Spark spectrography with a porous electrode (Porode).

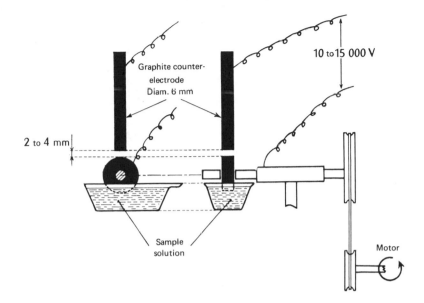

Figure 17. Spark spectrography with the rotating electrode (Rotrode).

4.1.4 Glow Discharge

If a voltage (300-1500 V) is applied between two plane and parallel electrodes placed in a glass tube filled with argon at 0.1-20 torr, a discharge is produced with dark and bright zones with an electric current of between 50 and 400 mA. The cathode is distinguished from the anode by the Aston dark space, the cathode envelope, the Hittorf dark space, negative light, the Faraday dark space, the positive column and finally the anode dark space. The negative glow consists of the emission of the gas (argon) and the cathode material.

The positive column is eliminated with the use of an annular anode that penetrates just into the cathode space (Figure 18). Thus, the negative glow will fill the anode-cathode space. It is precisely this zone that is analyzed by spectrography. The sample is a constituent of the cathode.

Before considering the analytical applications, we should note the characteristics of the glow discharge. Since the pressure in the tube is low, the lines are free of Stark, Lorentz and Holsmark broadening. The Doppler effect, which is a function of the cathode temperature, controls only the emitted line width. Mechanistically the field between the electrodes causes an ion, when ejected from the plasma by thermal motion, to be accelerated and to impinge on the cathode, causing evaporation of

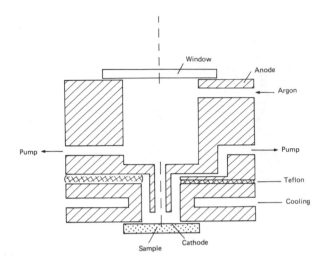

Figure 18. Glow discharge lamp.

the cathodic material. The atoms constituting the cathode thus enter and are excited in the plasma. The emitted spectrum has the following characteristics: fine lines, excellent stability and absence of a spectral background. The sensitivity will depend on the degree of sputtering, which is a function of the nature of the analyzed material, the discharge gas pressure, and the energy dissipated in the light source.

The glow discharge has been widely used for the analysis of metals (cast iron and steel) in which the absence of a matrix effect and particularly of a structure effect has been noted. The precision is excellent (Baudin, 1975). Glow discharge excitation has been applied for the analysis of nonconductive media such as geological specimens. In this case, the powdered specimen (100 μ) only needs to be pelletized with a metal powder (copper or silver), or better with graphite powder in a ratio of 1:3.

Ropert (1971) determined the major rock elements (Si, Ca, Al, Fe, K, Mg, Ti) with concentrations generally greater than 1%. However, it cannot be ruled out that the level of trace elements can also be attained.

4.1.4.1   Studies to be Consulted on the Analytical Applications of the Glow Discharge

Analysis of rocks and minerals: Ropert (1971); El Alfy *et al.* (1973).
Analysis of impurities in semiconductors: Greene *et al.* (1973).
Determination of Se, F, Cl in uranium oxide: Czarkow (1972).

Determination of traces (Bi, Cd, Co, Cu, In, Ni, Pb, Zn) in organic
acids: Krasilshchik *et al.* (1971).
Determination of Al, Fe and Cu in silica: Kuzovlev *et al.* (1973).

4.1.5 *Plasmas*

In high-intensity electromagnetic fields, the gases become conduc-
tors, and complex electrical charge transfer phenomena occur, which are
called gas discharges. Thus the result of a gas discharge is the production
of an ionized gas containing N_e electrons, N_i positive ions and N neutral
molecules per unit of volume. It is said that a plasma has been produced.
Generally, the gas is macroscopically neutral, $N_e = N_i$.

The state of an ionized gas in equilibrium can be characterized by
the following three properties: α (degree of ionization), T (temperature)
and N (number of neutral molecules).

$$\alpha = \frac{N_i}{N_i + N}$$

Colloquially, any more or less ionized gas has been called a plasma. In
practice, α varies between 10^{-10} and 1.

Plasmas tend to be used increasingly as a spectral emission source.
Generally, the sample is introduced in the form of a solution that is
atomized by the carrier gas by either a pneumatic or ultrasonic process.
Several types of plasmas are used as spectroscopic excitation sources.

4.1.5.1 Arc Plasma (Plasma Jet). The first important studies concerning the use of a plasma jet in spectroscopy were published by Margoshes
and Scribner in 1959. They used helium or argon as the plasma-generating
gas together with pneumatic atomization. They studied the emission spectra of elements introduced into the plasma: Fe, Ni, Cr, V, Cu, Cd, Ca,
Co, Mg, Mn, Zn. Quantitative determinations were made with Fe, Ni and
Cr, with vanadium as the internal standard. The precision was good and
the coefficient of variation was 2%.

Subsequently, Owen (1961) improved the Margoshes plasma technique by adding a stabilizing electrode at the tip of the discharge (Figure
19). This device reduces the coefficient of variation to 0.5%. Later,
Greenfield *et al.* (1965) published results obtained with a Margoshes-type
plasma jet and compared them with those obtained with a HF-induced
plasma. They concluded that the HF-plasma was a better tool than the
jet plasma.

Other important studies on jet plasmas were published by Goto and
Atsuya (1966-1967) in Japan. The authors measured temperatures of
2700-5000°K for power values of 0.5-3 kW with plasma-generating gas

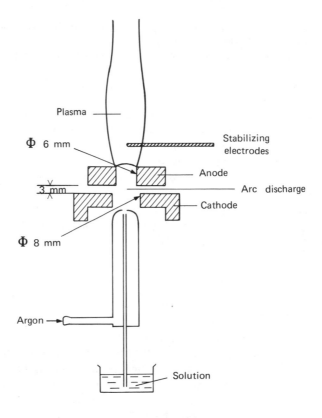

Figure 19. Arc plasma jet.

velocities approaching the subsonic velocity of Mach 20. The principal disadvantage of this technique remains the presence of electrodes.

A refined version of the plasma jet was proposed by Kranz (1967). The arc jumps between thorium-tungsten electrodes in the interior of a cooled copper cylinder. An opening at the top of the device allows injection of the sample aerosol and the radial discharge of the plasma jet. This plasma is remarkably stable and its temperature is favorable for applications in analytical spectroscopy—6500°K in the jet axis and 5000°K at a distance of 2.2 mm with nitrogen as the plasma-generating gas. This plasma can operate continuously for several dozen hours. Malinek and Massmann (1970) adopted a version of the Kranz model to determine boron by atomic absorption. They observed no interference from the seven investigated elements (Al, Ba, Fe, Li, K, Mn, Zn) and no difference of absorption signals as a function of the anions employed (Cl⁻, SO_4^{--} or NO_3^-).

The spectral excitation in the plasma jet was studied by Valente and Schrenk (1970), who compared the detection limits (concentration producing a signal two times the standard deviation of the background) in a plasma jet designed by them (low carrier gas consumption) with other plasmas and atomic absorption (Table VI).

Table VI. Comparison of Detection Limits of a Few Elements (μg/ml)[a]

	Plasma Jet (Valente)	Plasma Jet (Classical)	Plasma H.F.	Atomic Absorption
Ca	0.002	0.003	–	0.001
Cd	0.03	0.4	0.03	0.001
Cr	0.003	0.05	0.001	0.002
Fe	0.005	0.14	0.005	0.002
La	0.07	–	0.003	2
Li	0.001	0.0008	–	0.005
Ni	0.003	1	0.006	0.005
Pb	0.03	4	0.008	0.004
U	0.5	–	0.03	12
Y	0.008	–	0.0002	0.2
Zn	0.01	0.3	0.009	0.0005

[a]According to Valente and Schrenk, 1970.

For rock analysis, Yudelvich et al. (1972) introduced the powder sample directly into the plasma, improving the detection limit 10- to 100-fold. Chapman et al. (1973) recently described a plasma jet with an atomizing chamber in which the analysis solution is atomized not directly in the plasma but in a chamber similar to that used in flame emission spectrometry. The emission stability in the plasma as well as the detection limits are thus improved. Table VII lists some of the detection limits compared to those obtained with a nitrous oxide-acetylene flame (concentration producing a signal twice the standard deviation of the background).

4.1.5.2 Inductively Coupled Plasma.   In an induced plasma, a magnetic field is produced by high-frequency alternating current flowing through a solenoid inductance surrounding the induction furnace. The alternating field accelerates the electrons in the plasma with transfer of a fraction of the kinetic energy to the plasma atoms by collisions. According to the available energy, the atom can be excited or ionized. Ionization produces electrons that permit regeneration of the plasma. The induced current in the plasma heats it by the Joule effect, a process analogous to

Table VII. Detection Limits for a Few Elements in Plasma Jet[a]
and in the Nitrous Oxide-Acetylene Flame ($\mu g/ml^{-1}$)

	Plasma Jet	$N_2O\text{-}C_2H_2$ Flame (Emission)
Aluminum	0.3	0.06
Barium	0.2	0.03
Boron	0.05	—
Calcium	0.008	0.002
Chromium	0.4	0.01
Copper	0.2	0.2
Iron	0.2	0.3
Lithium	0.08	0.002
Magnesium	0.02	0.06
Manganese	0.04	0.04
Molybdenum	0.1	0.6
Nickel	0.3	0.4
Tin	0.6	—
Titanium	0.08	0.2
Zirconium	0.3	6

[a]Chapman et al., 1973.

induction heating of a metallic object. The energy absorbed by the plasma is restored in the form of heat and light when the dissociated molecules recombine.

When the gas is not ionized, the plasma cannot form and an electron flux must be generated to start it. The most common method of initiating the HF-plasma consists of induction-heating the sample in the solenoid. The heated material emits thermal electrons that ionize a fraction of the surrounding gas, forming a plasma that is subsequently self-sustained by the high-frequency field. The plasma has the following characteristics (Figure 20): frequency 4-50 MHz, power of 2-5 kW, and gas temperature of 9000-10,000°K.

The first investigations concerning the analytical possibilities of inductively coupled plasmas were made by Greenfield (1971) in Great Britain and, independently, by Wendt and Fassel (1965) in the U.S. These two groups described their equipment, the characteristics and appearance of the plasma, and the detection limits observed for a few elements in solution. Subsequently, Hoare and Mostyn (1976) and Veillon and Margoshes (1968) reported the detection limits observed for trace elements in solution.

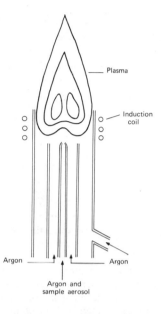

Figure 20. Inductively coupled plasma.

Hoare and Mostyn (1976) used a pure argon plasma and a flame identical to that of Greenfield for the analysis of boron and zirconium in nickel alloys. Ultrasonic atomization of the liquid sample is completed by a sampling device with multiple rotary cells. A few detection limits are reported, but only few elements produce better values than the best ones of Greenfield. With high concentrations in the sample solutions, the Zr and B interferences are negligible. Hoare and Mostyn also analyzed atomized lithium and aluminum salts in a plasma. These are the first reported applications of powder analysis.

Veillon and Margoshes (1968) reviewed the experimental study of Wendt and Fassel. Pneumatic atomization with desolvatation of samples was employed. Contrary to the results of the preceding studies, significant interelement effects were observed. Instead of a depressing effect of the emission of calcium, and of PO_4^{3-} and Al^{3+} ions, the authors observed an enhancement, which they were unable to explain. For their part, Mermet and Robin (1975) recently demonstrated that the chemical interferences were negligible; on the other hand, they noted ionization interactions.

Recently, Fassel and Kniseley (1974), Boumans and DeBoer (1972), and Souilliart and Robin (1972) discussed the analytical possibilities of inductively coupled plasmas. The detection limits found for several elements, compared with the flame emission and atomic absorption techniques (FES, AAS), are given in Table VIII. Furthermore, experiments were made with the direct analysis of powders by atomic emission by injecting the solid sample into the plasma.

Smith (1971) introduced the direct injection of powder into the plasma. In his study he concentrated on powders containing boron and beryllium, elements that are difficult to analyze by flame emission and classical atomic absorption.

Studies to be consulted for the analytical applications of inductively coupled plasmas are:

> Greenfield (1971): General review of applications.
> Dagnall et al. (1971): Determination of impurities in powders.
> Greenfield et al. (1972): Determination of trace elements from μl of organic and biological specimens.
> Savinova et al. (1972): Rock analysis.
> Boumans et al. (1972): Simultaneous multielemental analysis.
> Souilliart et al. (1972): Determination of rare earths—Y, La, Nd, Dy, Lu—in iron and zirconium matrices and in hafnium.
> Atsuya et al. (1973): Determination of rare earths.
> Scott et al. (1974): Determination of the detection limits of elements in solution (Al, Ba, Ca, Co, Cr, Ni, Pb, Se, Ti, V, Y Zn).
> Scott and Strasheim (1975): Determination of trace elements Fe, Mn, Cu, Al, B, Zn in plants.
> Scott and Kokot (1975): Determination of trace elements (Cu, Zn, Ni, Co, Pb) in soils; the results are in agreement with atomic absorption.

4.1.5.3 Microwave Plasma (Radiofrequency Plasma). The microwaves are coupled with a gas (argon) in a resonant cavity where standing waves are produced. This cavity is in capacitive or inductive resonance with the coupled magnetron circuit. This type of plasma has been developed since 1950 when magnetrons appeared on the market (Figure 21). It has the following characteristics: power 250 W, frequency 2450 MHz, maximum temperature 5000°K. The production of microwave or electrical plasmas was developed by Laroche (1954), Mavrodineanu and Hugues (1963). The latter compared the capacitive HF-plasma with the microwave plasma as sources for analytical spectrometry. Studies to be consulted on the analytical applications of the microwave plasma are:

> Kleinmann (1972): Determination of trace elements in solution.
> Layman et al. (1973): Determination of traces with atomization in a microarc and excitation in the microwave plasma.
> Lichte and Skogerboe (1973): Study of the detection limits of several elements.

Table VIII. Comparison of Experimentally Determined Detection Limits ($\mu g/ml$)

Element	Plasma			Flame	
	Fassel	Boumans	Souilliart	AAS	FES
Ag	0.004	–	0.03	0.005	0.008
Al	0.002	0.002	–	0.03	0.005
Au	0.04	–	0.04	0.02	4
B	0.005	0.08	0.03	6	30
Ba	0.0001	0.00002	–	0.05	0.002
Be	0.0005	0.0004	–	0.002	0.1
Bi	0.05	–	–	0.05	2
Ca	0.00007	0.00002	–	0.001	0.0001
Cd	0.002	0.003	–	0.001	0.8
Co	0.003	–	–	0.005	0.03
Cr	0.001	0.0003	–	0.003	0.004
Cu	0.001	0.0001	–	0.002	0.01
Fe	0.005	0.0003	–	0.005	0.03
Ga	0.014	0.0006	–	0.07	0.01
Ge	0.15	0.004	–	1	0.5
Hf	0.01	–	0.04	8	20
Hg	0.2	0.001	–	0.5	40
In	0.03	–	–	0.05	0.003
La	0.003	0.0004	0.006	2	2
Mg	0.0007	0.00005	0.00005	0.0001	0.005
Mn	0.0007	0.00006	–	0.002	0.005
Mo	0.005	0.0002	–	0.03	0.1
Na	0.0002	0.0003	–	0.002	0.0001
Ni	0.006	0.0004	–	0.005	0.02
Pb	0.008	0.002	–	0.01	0.1
Pt	0.08	–	–	0.1	2
Rh	0.003	–	–	0.03	0.02
Se	0.03	–	–	0.1	100
Si	0.01	–	–	0.1	5
Sn	0.3	0.03	–	0.02	0.3
Sr	0.00002	–	–	0.01	0.0002
Ta	0.07	–	0.03	5	20
Ti	0.003	0.0002	0.001	0.09	0.2
Tl	0.2	–	–	0.03	0.02
U	0.03	–	–	–	10
V	0.006	0.0002	0.001	0.02	0.01
W	0.002	0.001	0.1	3	0.5
Zn	0.002	0.016	0.05	0.002	50
Zr	0.005	0.0004	0.005	5	10

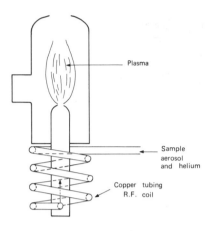

Figure 21. Ratio frequency plasma.

VanSandwijk *et al.* (1973): Determination of Cd, Tl, In, Pb and Hg
in nanogram quantities with excitation in a sealed tube in a microwave cavity.
Watling (1975): Mercury determination in water (10^{-12} g/l).

4.1.5.4 The Future of Plasma Sources. In recent years, a significant technological development of plasma as a spectral emission source
has been seen. Several devices are now available on the market (arc
plasma and inductive-coupled plasma) and can be used in spectrophotometry as well as in direct-reading spectrometry. Thus the plasmas are
now competing with the classical arc or spark sources. In any case, they
seem to have been designed for this purpose by the manufacturers, but
it is probable that in future years we will see these sources as equipment
of flame emission and atomic absorption spectrometers. While practical
methods still need to be developed in most cases, it seems that a certain
number of advantages may be expected, particularly concerning the
following: (1) analytical accuracy, because interactions seem to be much
less frequent than in a classical flame, (2) sensitivity, which is comparable,
(3) stability and therefore precision, and (4) the possibility of application
in simultaneous multielemental analysis.

4.1.6 Lasers

For several years, an incident laser beam has been used to volatilize
solid samples subjected to spectral analysis. The vapor produced contains
excited and even ionized atoms, the spectrum of which can be recorded

(Debras-Guedon, 1967), although it is also possible to excite the atoms produced in the vapor with an electrical spark or arc.

4.1.6.1 Direct Excitation. The first technique of direct excitation is used by Runge *et al.* (1964) of the U.S., by Debras-Guedon *et al.* (1963) of France, and by Karyakin *et al.* (1965) of the USSR, but it seems that the second has been most extensively developed in recent years with equipment constructed and marketed by Jarrell Ash in the U.S. and by Zeiss of Jena in Germany.

The principle of the method with direct excitation is shown schematically in Figure 22. The impact of the laser beam represents a few

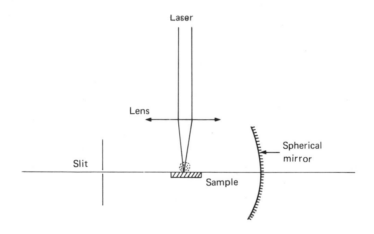

Figure 22. Laser excitation source.

hundredths to a few tenths of a millimeter in diameter. The laser-target contact time is on the order of 100 μsec and the total energy delivered in this very brief period is about 100 J, representing a power of 1 MW. The temperature attains 5,000-10,000°K. Utilization of this excitation technique for spectral analysis still has disadvantages, particularly because the emitted luminance has only a very short duration, reducing the sensitivity of analysis. The object is to obtain the maximum yield, *i.e.*, starting with a certain light emission, the most intense spectrum as possible will have to be recorded.

The most important factors involved to this end are the laser power and wavelength, the respective position of laser-target-slit of the spectrograph, focusing of the laser, focusing of the source images of the spectrograph slit, position of the reflecting mirror, luminosity of the spectrograph, and finally, the characteristics of the photographic emulsions.

Generally, a series of successive laser pulses may be necessary to determine low concentrations (a few dozen or hundred ppm). If spot analysis is not the primary objective, it is preferable to focus the laser on different but close points. The method has been utilized for the analysis of steels and nonconductive powders (Debras-Guedon, 1967).

The lines obtained generally are those of neutral atoms but also of ionized atoms (the excitation potentials attain up to 17 eV). Molecular bands (AlO, CN) are observed, and certain lines are self-absorbed. Finally, the sensitivity is better at long wavelengths. It is difficult to formulate a correlation between the sensitivity and excitation potential. Some elements that are not very sensitive in an electrical arc (Ca, Ti, Zr) produce much more intense spectra with this excitation technique. Inversely, other elements such as Pb and B are detected with difficulty or not at all in a concentration of several percent.

Compared to classical arc or spark spectroscopy, Morton *et al.* (1973) demonstrated that matrix effects are significantly reduced with laser excitation and an auxiliary spark. Finally, we should note the possible application in quantitative analysis if the use of an internal standard is possible. It should be pointed out that compared to arc spectra, the spectrum obtained is devoid of a background due to cyanogen bands. Finally, the following properties of this spectral excitation source may be considered established: spot analysis (homogeneity with concentrations of a few μ—0.1-0.1 μm^2) and surface analysis. The sensitivity is mediocre and trace analysis is difficult.

4.1.6.2 Excitation by Laser-Spark Coupling. This technique consists of exciting the vapor emitted by laser incidence by a spark and considerably improves the sensitivity and thus the detection limit. The apparatus is shown schematically in Figure 23. The sample (S) has a plane surface that is to be analyzed. This surface is parallel to and about 1 mm below the optical axis of the spectrograph. The spark electrodes (electrode gap of about 1 mm) are located in and perpendicular to the optical axis. The laser beam (perpendicular to the optical axis of the spectrograph) is focused on the sample with a microscope. This microscope also permits a visual observation of the sample zone to be analyzed.

At the moment of laser incidence, the sample is volatilized over a surface of 10-50 μ^2. The vapor contains excited atoms and ions that expand in the electrode gap under the voltage of a spark circuit. Ionization initiates the discharge, the energy of which contributes to excitation and ionization of the atoms in the vapor produced by the laser. The emitted spectrum is focused on the spectrograph and analyzed. This technique must also be considered a micromethod for surface analysis. Commercial instruments (Zeiss-Jena, Jarrell-Ash) utilize the described principle.

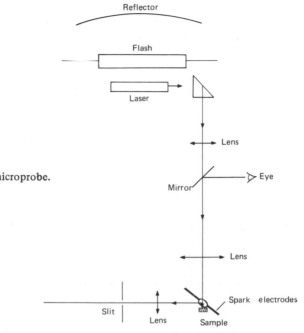

Figure 23. Laser-spark microprobe.

   The Zeiss laser microanalyzer consists of a neodymium-glass resonator with excitation by a xenon flash lamp under an elliptical reflector (Figure 23). The laser beam is focused on the sample with microscope optics, also allowing a very precise localization of the analyzed zone, which may have a diameter of 10-250 μm.

   A sample quantity of 1 μg (to initiate a flash) is vaporized between the electrodes, to which a potential difference of 1000-5000 V is applied. The spark is focused on the slit of a spectrograph of good luminosity (1/11-1/30) with average scattering (4 Å at 200 nm, 13.5 Å at 300 nm). The use of sensitive photographic emulsions is also recommended (astronomy emulsions). All useful information on chemical microanalysis with a laser can be found in the book by Moenke and Moenke (1973).

   Analytical applications are gradually developing. Published studies, still not very numerous, are defining the application range. This is a micromethod for surface analysis, but the quantity of sample involved must be indicated since it results from the dimensions of the crater produced by laser incidence. Depending on the energy, the diameter is 25-250 μ and the quantity of vaporized sample is 1 μg for a medium having a density of 10. These are only orders of magnitude. Numerous factors are involved, particularly the conductivity of the sample, its evaporation temperature and the surface condition.

Tables IX through XI list some detection limits in absolute value (Tables IX and X) as well as in concentration for a steel alloy (NBS 1100) (Table XI) for different quantities of vaporized samples (according to Jarrell Ash).

Table IX. Detection Limits in Quantities Essentially Corresponding to Absolute Volatilization of 1 μg of Sample

Element	ppm
Cr	250
Mn	500
Pb	450
Mg	300
Ni	1,000
Ti	400
Zn	5,500
Cu	500
Sn	3,000
Si	10,000

Table X. Detection Limits of Metals in Organic Matrix[a] (Treytl et al., 1972)

Element	Detection Limit μg
Li	2×10^{-7}
Mg	2×10^{-9}
Ca	1×10^{-8}
Fe	3×10^{-7}
Cu	2×10^{-9}
Zn	5×10^{-8}
Hg	3×10^{-7}
Pb	1×10^{-8}

[a]In optimal conditions and direct reading spectrometry.

For the moment at least, analytical applications are in the qualitative or semiquantitative area. While precision is good (errors smaller than 10%), the calibration conditions on which the accuracy depends are difficult to determine. The internal standard technique may improve precision as well as accuracy.

Analytical micromethods with laser excitation of the spectrum were applied originally to studies of the heterogeneity of metallic alloys. It is thus possible to analyze defects observed on a scale of a few dozen microns in a

Table XI. Detection Limits in Low Alloy Steel (NBS 1100)

| Crater Diameter | $135\,\mu$ | $80\,\mu$ | $50\,\mu$ |
| Weight Vaporized | 15 μg | 3 μg | 0.8 μg |
Element	ppm	ppm	ppm
Al	15	35	75
Cr	60	125	300
Cu	8	15	35
Mn	20	55	125
Mo	125	310	850
Ni	90	205	450
Si	35	95	225
Ag	10	25	50
Ti	8	20	45
V	25	80	200

plane surface. In this connection, laser "microanalysis" can be considered a method of chemical surface analysis intermediate between X-ray fluorescence spectrometry, which is applicable to surfaces of a few dozen mm², and electron microprobe analysis for a surface of a few μm². It is also a direct analytical method for the sample requiring practically no pretreatment. Only the surface condition has an influence in selecting the region to be analyzed with a microscope. Chemical analysis of minerals has been the subject of several publications. The method permits a study of the chemical composition of mineral elements forming rocks and this can be done directly on a rock microsection.

Biological applications include the study of point anomalies in tissue. Brech (unpublished) has cited various applications, particularly for point analysis of bone tissue. The author determined elements in the following concentration levels (%): 0.1-1 P, 0.01-1 Mg, 0.001-0.1 Al, 0.001-0.01 Cu, 0.001-0.01 Si, 0.001-0.1 Ti, 0.001-0.01 Zn. The application possibilities have been reviewed by Heyndryckx and Grangeon (1971). Webb and Webb (1971) defined the application conditions for semiquantitative analysis. Petkova and Petkov (1971) developed mineral analysis, with particular attention to the influence of the atmosphere on excitation (air, O_2, Ar and Ne). Saffir et al. (1972) analyzed biological media and Treytl et al. (1972) determined Ca, Fe and Zn. The book by Andersen (1973) also contains numerous geological and biological applications. Petkova et al. (1971) determined impurities in pyrite and galenite minerals.

4.2   Dispersing Apparatus

Optical systems for analyzing spectral radiations operate on the principle of prism dispersion or diffraction by a grating. A spectrograph (Figure 24) consists essentially of a slit F, which admits the incident light and the size of which can be adjusted from 1 to 1000 μ, an objective L_2 (collimator), a dispersion system (prism or grating), and an objective focusing system or camera L_3, which forms an image of the spectrum on a surface that holds a photographic plate. In spectrometers, the plate is replaced by a photoelectric recorder. The image of the excitation source is projected onto the slit of the spectrograph by a suitable lens. In the particular case of the spectrographic determination of trace elements in complex materials, the apparatus must have certain features.

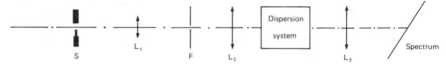

Figure 24. Diagram of the spectrograph.

4.2.1 Properties of Spectrographs

Without going into the theory of these instruments, it is useful to review briefly the main properties of the dispersing instruments.

4.2.1.1   Dispersion. Let us consider two rays of adjacent wavelengths, λ_1 and λ_2, differing by $d\lambda_1$, making an angle $d\theta$ as they emerge from the prism. The angular dispersion is then defined as $d\theta/d\lambda$. In the focal plane of the spectrum, the linear dispersion is defined as $dl/d\lambda$, where dl is the distance between two lines differing by $d\lambda$. The linear dispersion is a function of the focal length of the objective. In prism instruments it is a function of the refractive index of the refracting medium, and in grating instruments, of the number of lines per unit length and of the order of the spectrum. The dispersion varies inversely with the square of the wavelength in the case of the prism, and is independent of the wavelength in the case of the grating.

The reciprocal expression $(d\lambda/dl)$ is currently used to designate the dispersion of an instrument. Its dimension is thus Å/mm.

4.2.1.2 Separating Power; Resolving Power. Separating power is the capacity to separate two radiations of neighboring wavelengths to give two distinct images. The separating power at wavelength λ is defined as the smallest difference $\Delta\lambda$ between two wavelengths that can still be separated.

As every line on the photographic plate is an image of the slit of the spectrograph, it is clear that the separation will depend on the size of the slit used.

The resolving power is $R = \lambda/\Delta\lambda$. It is mainly a function of the dispersing element, and is larger for gratings than for prisms.

In a prism spectrograph, if e_1 and e_2 are the maximum and minimum paths of the beam in the prism, we have

$$R = \frac{\lambda}{\Delta\lambda} = (e_1 - e_2) \cdot \frac{dn}{d\lambda}$$

since $e_2 \cong 0$:

$$R = e \frac{dn}{d\lambda}$$

The resolving power can be increased either by a suitable choice of prism $(dn/d\lambda)$ or by using several prisms.

4.2.1.3 Luminosity; Transparency. For a given source, luminosity is defined as the intensity of the spectrum (the blackening of the photographic plate). It depends mainly on the aperture of the spectrograph, that is to say on the ratio between the diameter of the objective of the camera and its focal distance. Transparency is a function of the optical path, in particular of the layer thickness of the refractive substances traversed and the number of refracting or reflecting surfaces encountered.

4.2.2 Prism Spectrographs

The principle of the spectrograph is presented in Figure 25. The geometrical characteristics of the instrument will depend on the properties

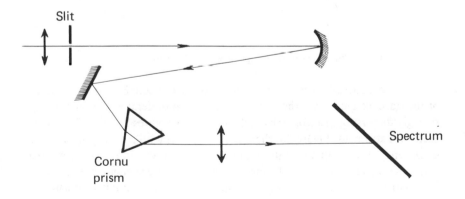

Figure 25. Cornu prism spectrograph.

desired. The angle of deviation θ of the light rays depends on the refractive index n and the angle A of the prism. It is at minimum at a certain incidence angle and is given by the formula:

$$\sin \frac{\theta + A}{2} = n \sin \frac{A}{2}$$

In practice, spectrographs are adjusted so that the ray of medium wavelength has a minimum deviation. The linear dispersion increases and the luminosity decreases with the focal length of the lens L_3. An instrument with a suitable dispersion and resolving power is necessary for spectrographic analysis of trace elements. Quartz, which is sufficiently transparent in the ultraviolet regions, must be used for the optical system, since the lines employed are situated in the ultraviolet between 2300 and 4500 Å.

Figures 25-27 show the principle of a few instrument models. The spectrograph shown in Figure 25 has a quartz prism of approximately 60° and a mean dispersion of 10 to 15 Å/mm at 3000 Å. The spectrum between 2300 and 4500 Å is 200 mm long.

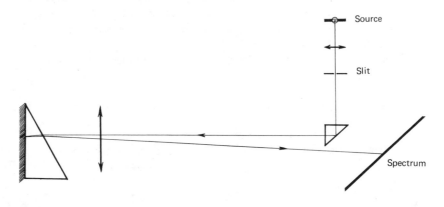

Figure 26. High-dispersion spectrograph, Littrow type.

Less strongly dispersing instruments are not suitable for the analysis of the trace elements. On the contrary, it is often preferable to use more strongly dispersing spectrographs such as those in Figures 26 and 27. In the Littrow instrument, Figure 26, there is a 30° prism with a reflecting base. A single lens acts as both collimator and camera objective. In this way, spectrographs with a large focal length, and thus with a high dispersing power, can be constructed. Figure 27 shows a two-prism high-dispersion instrument.

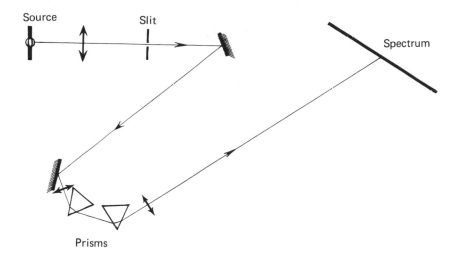

Source   Slit

Spectrum

Prisms

Figure 27. Dispersion spectrograph.

4.2.3 Grating Spectrographs

The diffraction grating consists of a large number of parallel lines ruled on a plane or on a concave surface, giving a succession of opaque and transparent portions (transmission grating) or opaque and reflecting portions (reflection grating). The gratings disperse the incident light into a certain number of spectra defined by their order K (K = 1, 2, 3, . . .). At any given diffraction angle, the diffracted beam is monochromatic. For example, a grating of 1000 lines per mm in conjunction with a focusing system of 2 m gives a linear dispersion of 5 Å per mm for the spectrum of the first order. The resolving power R = $\lambda/d\lambda$ is a function of the total number N of grating lines R = kN. For a 10-cm grating with 1000 lines per mm, the resolving power is R = 100,000 for the first order spectrum (K = 1), *i.e.*, two lines differing by dλ = λ/R = 0.04 Å can be separated at 4000 Å. These properties of the grating make it preferable to the prism. The luminosity is weaker in older grating instruments, but recent improvements in the design of the gratings make it possible to concentrate the light in a given order, thus obtaining a luminosity comparable to that of prism instruments.

Figures 28 and 29 give schematic representations of grating spectrographs. The most widely used grating spectrographs are made by A. R. L., Baird, Jarrel-Ash, and Hilger.

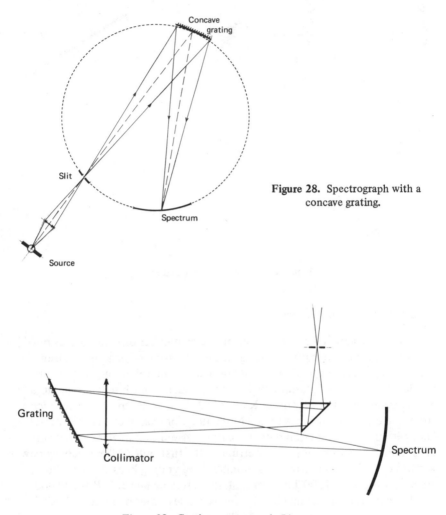

Figure 28. Spectrograph with a concave grating.

Figure 29. Grating spectrograph, Littrow type.

4.2.4 Projecting Optic

The projecting apparatus is usually mounted on an optical bench, affixed to the spectrograph. It consists of a series of suitable lenses designed to throw the image of the source onto the slit of the spectrograph. The focal length and position of the projection lens must be calculated so that the flux that passes through the spectrograph is used at maximum efficiency, that is to say, it should cover the entire area of the objective (collimator).

The apparatus shown in Figure 30 is the classic one for qualitative analysis. The slit is not evenly illuminated and the blackening of the lines on the plate is not uniform, so this technique is not suitable for quantitative analysis. The spectrograph is sometimes illuminated directly by placing the source a few centimeters away from the slit. This procedure is used in flame spectroscopy.

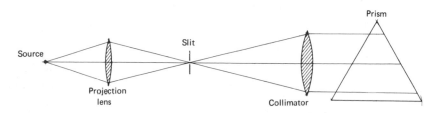

Figure 30. Apparatus in which the image of the source is focused on the slit of the spectrograph.

In quantitative analysis, the density of the lines on the photographic plate must be uniform along their entire length, which means that the slit of the spectrograph must be uniformly illuminated. This condition is fulfilled by projecting the source onto the collimator of prism. For this purpose, a short-focus lens (about half that of the collimator) is used. It is placed near the slit of the spectrograph, as shown in Figure 31. This technique has the disadvantage of producing a less intense spectrum because a large part of the light energy of the source does not enter the spectrograph.

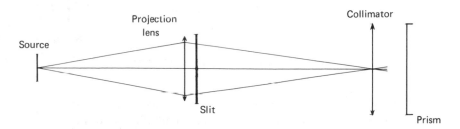

Figure 31. Apparatus in which the source is focused on the collimator.

Finally, the spectrum of only a part of the source may be sought. A typical case is cathodic layer excitation. The source is projected onto a screen with a hole a few millimeters in diameter, which is situated on the axis of the spectrograph and serves as a diaphragm. An image of the cathodic zone spectrum is formed on this diaphragm, which then serves as the spectral source. The apparatus is illustrated in Figure 32. The cathodic layer, which is situated

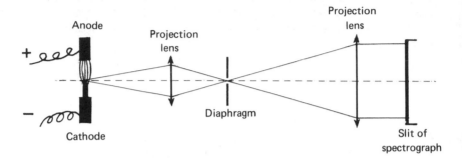

Figure 32. Cathodic layer arc spectrography.

in the optical axis, is projected directly onto the prism. The lower two-thirds of the height of the prism, which receives the central and anodic regions of the arc, is screened off (Figure 33).

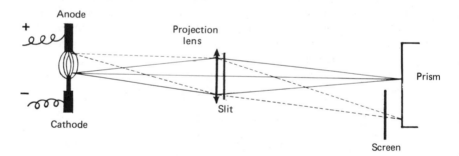

Figure 33. Cathodic layer arc spectrography.

4.2.4.1 The Step Sector. The step sector is disc-shaped, as shown in Figure 34, and revolves in front of the spectrograph slit so as to vary the exposure along the slit. The blackening of a line on the photographic plate then shows a number of steps corresponding to times of exposure: t, t/2, t/4, t/8, t/16, t/32 and t/64. Under these conditions the logarithm of the exposure varies very nearly 0.30 from one step to the next. It will be shown that this progression is particularly convenient for plotting the characteristic curves and for calculating the relative intensities of the spectral lines.

The step sector is indispensable in quantitative and semiquantitative spectrographic analysis because this device makes it possible to photograph

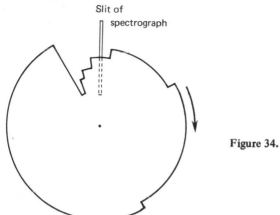

Figure 34. The step sector.

the spectrum of a sample at various exposures in a single determination, and the characteristic curve of the photographic emulsion can be plotted for each line. The step sector is placed in front of the spectrographic slit, which must be uniformly illuminated by the source of excitation. This is attained by the optical arrangement shown in Figure 35. The projection lens forms an image of the source on the collimator or prism of the spectrograph. The step sector is placed between this lens and the spectrograph.

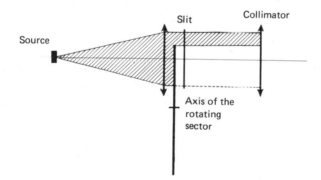

Figure 35. Position of the step sector.

The step sector is the most efficient means of producing a variable exposure over the length of the spectrographic slit. The variation is independent of the wavelength and is readily reproducible.

The use of the step sector is based on the assumption that the "reciprocity law" is obeyed for the photographic emulsion, and also that the "intermittency effect" is negligible. The exposure E is the product of the intensity of the spectral emission and the time t: E = It. The reciprocity law is obeyed if the optical density D of the image of the line on the plate is proportional to log E or log It, *i.e.*, D = log It.

Reciprocity between I and t is not always independent of the photographic emulsion used. The reciprocity conditions hold if the curves D = log It obtained at various exposure times have the same slope (are parallel) for any intensity I. If this is so, the emulsion can be used with the step sector.

The intermittency effect can sometimes cause trouble when the step sector is used. Experience has shown that an intermittent exposure does not affect the photographic emulsion in the same way as a continuous exposure lasting for an equal period of time. The blackening of the photographic plate resulting from an intermittent exposure is weaker and is a function of the emulsion and of the wavelength. This effect disappears if the intermittent frequency is higher than 10/sec. All spectra should be photographed for the same periods of time. Rates of rotation between 300 and 1000 revolutions per minute are usually suitable.

Kniseley and Fassel (1955) studied the use of the step sector with alternating and interrupted arcs. An error can result if the frequency of the light pulses happens to be synchronous with the frequency of the passage of the sector across the slit.

4.2.5 Instruments for Studying and Measuring the Lines

Accessories used in spectrographic analysis include the spectrum projector, the comparator, and the densitometer.

4.2.5.1 Spectrum Projector. This instrument, the principle of which is shown in Figure 36, is used to project the image of a part of the spectrum

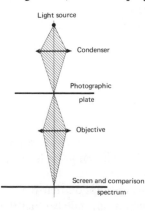

Light source

Condenser

Photographic plate

Objective

Screen and comparison spectrum

Figure 36. Diagram of the spectrum projector.

onto a plane with a known magnification (10 to 20). The spectrum projector is especially useful in qualitative analysis. The unknown spectrum is projected onto an enlargement of a comparison spectrum on which the persistent lines of the elements sought are marked. The comparison spectrum is obtained from a synthetic sample with a composition analogous to that of the sample being spectrographed. Identification of the lines is facilitated by photographing a spectrum of iron alongside each spectrum (the sample and the comparison spectrum).

4.2.5.2 **Comparator.** This instrument is used to locate and indicate the lines on the photographic plate. The sample and the comparison spectra, magnified to the same extent, are projected in juxtaposition and are compared visually as shown in Figure 37. A light source illuminates the sample and the comparison spectrum by way of two plane mirrors, M_1 and M_2. Two objectives, O_1 and O_2, in conjunction with pairs of plane mirrors—M'_1, M''_1 and M'_2, M''_2—project the two spectra side by side on the screen. These are then observed simultaneously through the eyepiece (in the Judd-Lewis instrument of Hilger), or are projected onto a ground-glass plate (in the Beaudoin instrument). In the latter case, the instrument also serves as spectrum projector. The comparator is most useful in qualitative and quantitative analysis.

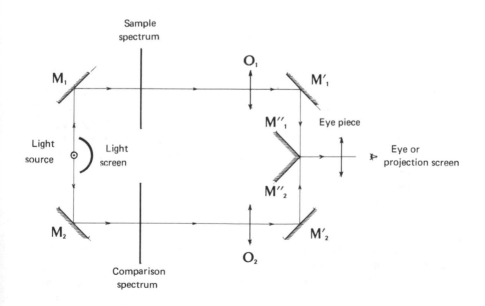

Figure 37. Diagram of the spectrum comparator.

4.2.5.3 Densitometer. The densitometer or microphotometer is an instrument used to determine the degree of blackening or optical density of the lines on the photographic plate. It measures the variation of a known light flux that has passed through the blackened layer. Figure 38 shows the operating principle of this instrument. A condenser C forms an image of the

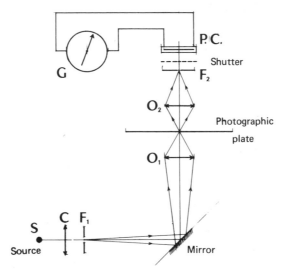

Figure 38. Diagram of the densitometer.

source S on the slit F_1, which admits a suitable part of the light. The light beam is then focused on the photographic plate by the objective O_1, is made to pass the plate, and is then projected onto the slit F_2 placed in front of the photoelectric cell P.C. The resulting electric current is measured by a galvanometer G. Two optical densities are measured: the flux i_0 penetrating the unmarked photographic plate and the flux i that strikes the cell when the line studied is interposed in the path of the beam.

The blackening of the line is given by the formula:

$$D = \log \frac{i_0}{i}$$

Determination of the optical density of the spectral background is carried out under the same conditions.

If the slit F_1 is removed, a larger part of the spectrum is projected onto the plane of the slit F_2, and the plate can easily be placed so that the line to be measured coincides as far as possible with the slit F_2. The size of the slit F_2 must be smaller than the image of the line to be measured, but the surface to be measured photometrically must not be too small or the error due to the grain of the photographic emulsion might be neglected. A slit 0.1 to 0.3 mm wide and 4 to 5 mm high is usually suitable. When the densitometer is used, it is essential that the light source employed in the measurements of i_o and i for a given blackening be stable.

4.2.5.4 Other Accessories. The following accessories are also necessary for spectrographic analysis.

>*Electrode cutter:* This may be an ordinary watchmaker's lathe.
>
>*Mixing mill:* This should be made of agate or tungsten carbide. A mechanical mill can be used to grind and homogenize the samples, and either a small mortar or a micromill with plastic balls to prepare the material to be filled into the electrodes.
>
>*Balances:* A 100-mg torsion balance is particularly useful for the preparation of spectrographic mixtures.
>
>*Photographic equipment:* Developing, fixing and washing must be carried out under rigorously controlled conditions. The choice of the plates depends on the nature of the work.

5. PHOTOMETRIC MEASUREMENT OF SPECTRAL LINES

5.1 Introduction

In quantitative spectrum analysis the concentration of an element present in a given excitation source is determined as a function of the relative intensity of a suitable line of the element, nonionized or ionized. It is important, therefore, that the intensity of the spectral line be measurable with sufficient accuracy. If the lines are emitted under constant conditions, the simplest and most accurate procedure is to always record the intensity of the line directly by a photocell. The method is not applicable to the spectrum analysis of trace elements because of their low concentrations and the complexity of the spectrographic medium.

Direct reading spectrum analysis, with arc or spark as the source of excitation, will be discussed. With the arc as well as with the spark, a convenient technique is to photograph the spectrum, and then measure the relative intensity of the blackening produced by the lines on the plate. Of the possible photometric methods, those usually employed are the comparison method and the method of internal standards.

In the first method, often used in semiquantitative analysis, the line of the element to be determined is measured and compared with the line of a few synthetic samples, the spectra of which are photographed on the same plate. The method of internal standards is the principle of quantitative analysis. The ratio of the intensities I_X/I_E of a line of the element X and of the internal standard element E is determined. We have already seen that the relationship between the intensity of a line and the concentration of the corresponding element is:

$$I = K \, C^m$$

where K and m are constants that depend on the particular line and on the conditions of excitation.

As has been shown:

$$I_X = K_X \cdot C^m{}_X X \qquad \text{(element being determined)}$$

and

$$I_E = K_E \cdot C^m{}_E E \qquad \text{(internal standard element)}$$

If the concentration of the internal standard is constant and is known, we have the standardization curve:

$$\log \frac{I_X}{I_E} = m_X \cdot \log C_X + \text{constant}$$

The accuracy of the determination is greater the steeper the slope of the line representing the last equation. This depends on m_X, and thus on the spectral line measured. The choice of the spectral lines to be measured is thus seen to be of considerable importance.

5.2 Characteristic Curve of an Emulsion

When a photographic emulsion is exposed to a light flux of relative value I, the resulting blackening, measured by its optical density D, is not proportional to E. In practice the curve D = f(log E) (Figure 39), called the

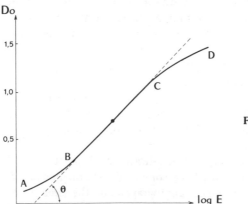

Figure 39. Characteristic curve of a photographic emulsion.

characteristic or blackening curve, is rectilinear between two values of log E, which determine the three portions of the curve: AB is the concave portion corresponding to the underexposure of the photographic emulsion; BC is the straight line, the region of normal exposure; and CD is the convex region of overexposure. The rectilinear portion is the most important in spectral analysis, since the measurements, which depend on the relative position of the characteristic curves, are most accurate in this region.

The contrast γ of an emulsion is the tangent of the angle θ formed by the rectilinear part of the characteristic curve with the abscissa axis: $\gamma = \tan \theta$. It will be seen that a comparison of several characteristic curves is only possible if γ is kept constant for each curve. Several factors may affect γ: the time of development of the plate, the type of developer, the type of emulsion, and the wavelength.

5.3 Quantitative Analysis Determination of the Intensity Ratio of Two Lines by the Methods of Internal Standards

5.3.1 Principle

The blackening curve is obtained from a line of relative intensity I by varying the exposure E with the aid of a step sector. The exposure is the product of the intensity I and the time t. A variation in t causes a variation in E, and thus also in the optical density D:

$$D = \log It$$

or

$$D = \log t + \text{constant}$$

If the times of exposure corresponding to each aperture in the sector are t, t/2, t/4, t/8, t/16, t/32 and t/64, the difference between the logarithms of the values of two adjacent exposures is equal to log 2. The logarithm of the exposure thus varies by 0.3 from one aperture to the next. The graph is plotted using an abscissa scale in which 1 cm corresponds to the antilogarithm of 0.1. The serial numbers of the steps are marked at 3 cm distance on the scale, with aperture No. 1 corresponding to a time of exposure t, and aperture No. 7 to time t/64 (see Figure 40).

Inversely, if in the equation

$$D = \log It$$

or

$$D = \log I + \log t$$

D is fixed, it is seen that log I and log t are inversely proportional. In other words, in Figure 40 the sequence of logarithms of the relative intensities, log I, is opposite to that of the logarithms of the times log t.

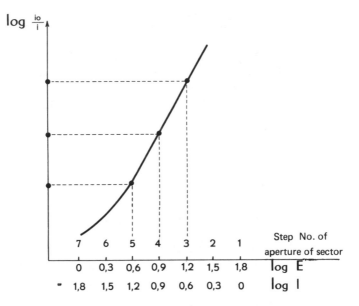

Figure 40. Determination of characteristic curve.

Table XII shows the relationship between the exposure and the relative intensity for each step of the sector. The characteristic curves of the respective intensities I_X and I_E of the element to be determined and the internal

Table XII. Variation in the Relative Exposures and Relative Intensities as a Function of the Time of Exposure

No. of Step	1	2	3	4	5	6	7
Time of exposure	t	t/2	t/4	t/8	t/16	t/32	t/64
K log relative exposure	1.8	1.5	1.2	0.9	0.6	0.3	0.0
K′ log relative intensity	0.0	0.3	0.6	0.9	1.2	1.5	1.8

standard are shown in Figure 41. The relative intensity of each line is determined by the relative position of each curve, which can be correlated by drawing a line parallel to the abscissa, with an ordinate corresponding to an optical density situated on the rectilinear portion of the characteristic curves, for example, with an ordinate D = 0.4. Under these conditions, we can write:

$$\log \frac{I_X}{I_E} = \log I_X - \log I_E = \overline{AC}$$

$$\log \frac{I_X}{I_E} = \overline{AC}$$

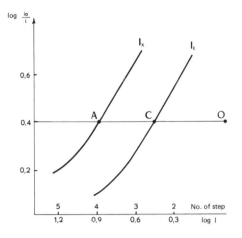

Figure 41. Characteristic curves of a pair of analytical lines.

By convention the distance $\overline{AC}$ is a positive logarithmic value if $I_X > I_E$ and negative if $I_X < I_E$. To plot the characteristic curve

$$D = \log E = \log It$$

the step sector is sometimes replaced by a step density filter, which can be used to vary I and thus log E.

5.3.2 Seidel's Method

The fact that the characteristic curves are not rectilinear makes the work difficult because four or five measurements of the optical density are necessary in order to plot each curve. Moreover, with some very weak lines it is not possible to plot the characteristic curve in the rectilinear region, and this limits the sensitivity of the method. To overcome these two problems, Seidel suggested that the optical density $\log (i_o/i)$ be replaced by the expression

$$\log \left(\frac{i_o}{i} - i \right)$$

which gives a practically linear characteristic curve, even at very low exposures. This method proved very useful, in particular in the determination of the intensity ratios.

The curves in Figure 41 are replaced by those in Figure 42 in which the number of the step of the step sector is plotted on the abscissa, and the Seidel density on the ordinate

$$\log \left(\frac{i_o}{i} - i \right)$$

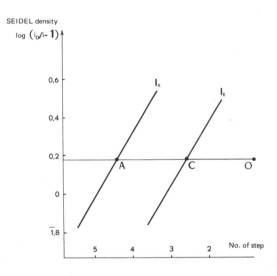

Figure 42. Characteristic curves determined by Seidel densities.

where i_o and i are the measurements, carried out with a densitometer, of the light flux passing through the nonblackened part of the photographic plate and through the line to be measured, respectively. In practice, two measurements are sufficient to plot each curve.

To avoid the labor of calculating log $(i_o/i$ - 1) as a function of i each time, the determinations can be carried out directly with a galvanometer graduated in Seidel densities, but this instrument is not readily available. However, a special table that gives values of the Seidel density as a function of i can be used. It is then advisable to give i_o a fixed value, for example, that corresponds to the maximum deflection of the galvanometer (see Mitchell, 1948).

5.3.3 Determination of the Intensity Ratio of Two Lines With Correction for the Spectral Background

In the preceding method, the value of log (I_X/I_E) represented by AC (Figures 41 and 42) does not take into account the spectral background that may be superimposed on the lines measured. In certain cases, the background is zero or negligible, and the length AC is a sufficiently accurate measure of log (I_X/I_E).

On the other hand, the background is often significant, and a correction is necessary. The blackening curves of the lines and the spectral background curves must then be measured and plotted. Generally the two spectral backgrounds are measurable and differ from one another.

The blackening curves plotted in Seidel densities are shown in Figure 43. The abscissa gives the logarithms of the relative intensities. If I_X, I_E, I_{fX} and I_{fE} are the relative intensities of the lines and the spectral backgrounds of the element to be analyzed and of the internal standard element, then:

$$\log (I_X + I_{fX}) - \log I_{fX} = \overline{BA}$$

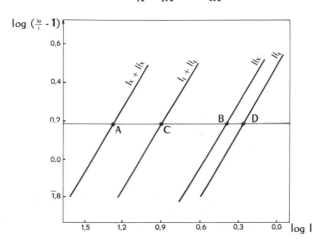
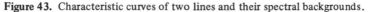

Figure 43. Characteristic curves of two lines and their spectral backgrounds.

It is possible to write:

$$\log I_X \cong \log (I_X + I_{fX}) - \log I_{fX}$$

and

$$\log I_X \cong \overline{BA}$$

and also

$$\log I_E \cong \overline{DC}$$

finally

$$\log \frac{I_X}{I_E} \cong \overline{BA} - \overline{DC}$$

In some particular cases the spectral backgrounds have the same value. The above expression becomes:

$$\log \frac{I_X}{I_E} \cong \overline{CA}$$

Generally, when it is necessary to take the spectral background into account, the calibration curve is plotted as $\log (I_X/I_E) = f (\log \text{ conc. } X)$,

Figure 44. This curve is readily obtained with the aid of a number of synthetic standards containing increasing concentrations of the element X, expressed in ppm, over a suitable range of concentrations.

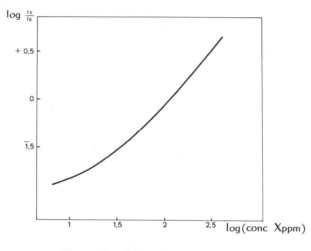

Figure 44. Calibration curve.

It is not usually necessary to determine the calibration curve experimentally each time. All the experimental conditions—composition of the material studied, excitation conditions, time of exposure, photographic emulsion, and development of the plate—must be reproduced with the greatest possible accuracy. Nevertheless, in order to obtain more accurate results, the spectra of two standards used to plot the calibration curve are photographed on the same plate as that of the sample to be studied.

5.4 Semiquantitative Analysis. Determination of the Relative Intensity of a Line in the Comparison Method

In the comparison method, the intensity of a line of the element to be determined, chosen from the spectrum of the sample, is compared with the intensity of the corresponding lines of synthetic samples of known concentrations obtained on the same plate. In this method, the intensity of a spectral line is defined by the position of the characteristic curve shown in Figure 40.

Two cases must be distinguished, according to whether the spectral background is negligible or not. In the absence of a significant spectral background for a line of intensity I_X, the position $\overline{OA}$ of the characteristic curve

(Figures 41 and 42) with respect to an arbitrarily chosen origin O is considered to be a function of the intensity I_X, and thus of the concentration of the element producing the line:

$$\overline{OA} = f(I_X) = f(\text{conc } X)$$

This equation represents the calibration curve and is determined experimentally by means of a series of synthetic samples with increasing concentrations of the trace element. In practical work, it is necessary to fix the position of this curve accurately for each photographic plate by means of two or three experimental points obtained by taking two or three calibration spectra. For this purpose, appropriate synthetic samples are chosen, preferably with trace element concentrations covering the range of their expected concentrations in the unknown samples.

The second case to consider occurs when the spectral background is significant. For a line of intensity I_X, we have the characteristic curves shown in Figure 43, that is $(I_X + I_{fX})$ and I_{fX} for the sum of the line and the spectral background, and for the background alone, respectively. The intensity of the line I_X is given by the length $\overline{BA}$ (Figure 43). The concentration of the element X is a function of the value of $\overline{BA}$. We can thus write $\overline{BA} = f(\text{conc } X)$, an equation representing the calibration curve, determined under the same conditions as above. Here too, the standardization must be carried out accurately for each photographic plate.

The relative line intensities determined by this method are affected by errors due to fluctuations in the excitation source. The method is typically semiquantitative.

6. DIRECT READING SPECTROMETRY

6.1 Apparatus

In direct reading spectrometry, the energy of a spectral emission is measured by an electron photomultiplier cell, which converts the energy into electric energy which is recorded, after amplification, by a galvanometer. The actual measurement is carried out by comparing the intensity of a line of the element being studied with that of a line of an internal standard element.

The conventional instruments use two types of photoelectric recording. The first has two photomultipliers: a fixed one that corresponds to the line of the internal standard and a mobile one that scans the whole spectrum. The second type has several photomultipliers: one is placed at the line of the internal standard and the others at the lines chosen for the elements being studied.

A direct reading spectrometer consists of three parts:

1. The generator, which feeds the excitation source, is a spark circuit that consists of a high voltage transformer, a condenser, and an inductance (Figure 12). The other sources of excitation often used in direct reading spectrometry are the interrupted arc and the ac arc.

2. The grating dispersion system consists of an adjustable entry slit to admit the light beam, a diffraction grating that reflects spectra of the first and second orders, and exit slits suitably arranged and placed on the Rowland circle of the spectroscope at the focal points of the radiations to be measured (Figure 28).

3. Photoelectric current integration, amplification, and measurement units.

The multichannel apparatus can have several electron photomultipliers, their number being limited by the resolution of the dispersive system and the mechanical limitations on the number of slits and photomultipliers. A modern instrument can have as many as 60 detector cells.

The scanning instruments consist of a first cell fixed on the line E of internal standardization, and a second cell that scans the spectrum and records successively the lines $X_1, X_2, X_3, \ldots$

In the optical system of the spectroscope itself, the grating, which has a better spectral resolution, tends to replace the prism.

6.2   Photoelectric Recording

Direct reading spectrometers use the principle of internal standardization with at least two radiation detectors, one for the element to be determined, X, and the other for internal standard element, E. The responses D_X and D_E of the detectors are proportional to the intensities of the lines, I_X and I_E:

$$D_X = k_1 \cdot I_X \text{ and } D_E = k_2 \cdot I_E$$

where k_1 and k_2 are instrument constants.

We have already seen that the ratio I_X/I_E is a function of the concentration C of the element X in the excitation source:

$$\log C_X = K + K' \log \frac{I_X}{I_E}$$

Hence:

$$\log C_X = K + K' \log \frac{D_X}{D_E}$$

This is the equation of the calibration curve for determining the element X, by comparison with the internal standard E. In practice, the two lines—that

of the element to be determined and that of the internal standard element—
are selected in the plane of the spectrum by two "precision" slits (Figure 45).

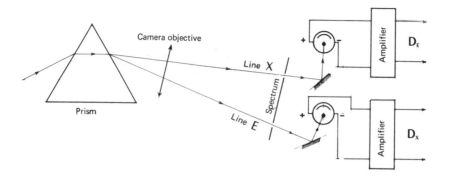

Figure 45. Measurement of the intensities of the lines in spectrometry.

6.3 Measurement of Lines

6.3.1 Spectrometer for Instantaneous Measurement

In these instruments, the photocell currents are measured separately,
after amplification, by microammeters. If D_X and D_E are the deflections obtained, the unknown concentration C_X can be found from the calibration
curve

$$C_X = f\left(\frac{D_X}{D_E}\right)$$

D_E is generally given a fixed value by adjustment of the excitation source or
the intensity of the flux entering the spectrometer. The ratio D_X/D_E is
most frequently evaluated directly with the aid of a potentiometer in which
the potentials of the two photomultipliers e_X and e_E ($e_X < e_E$) are opposed,
as shown in Figure 46. The intensity ratio of the lines is then equal to the
ratio e_X/e_E:

$$\frac{I_X}{I_E} = \frac{e_X}{e_E}$$

The current in the circuit (e_E, R) is:

$$i_1 = \frac{e_E}{R}$$

where R is the resistance of the potentiometer, and in the circuit (e_X, R_1)
it is:

$$i_2 = \frac{eX}{R_1}$$

The value of the resistance R_1 corresponds to the equilibrium of the potentiometer, that is to say to a zero current in the microammeter placed in the circuit. We then have:

$$\frac{eE}{R} = \frac{eX}{R_1}$$

or

$$\frac{eX}{eE} = \frac{R_1}{R} \text{ and } \frac{IX}{IE} = \frac{R_1}{R}$$

Hence:

$$\frac{IX}{IE} = k \cdot R_1$$

Thus, the value of R_1 given by the position of the potentiometer needle is proportional to the intensity ratio of the lines I_X/I_E. The measurement is generally carried out with a recording potentiometer.

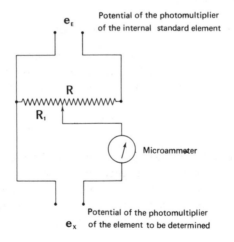

Figure 46. Measurement of the intensity ratio of two spectral lines.

6.3.2 Integrating Instruments

In these instruments, the lines whose intensities are I_X and I_E give the amplified photoelectric currents i_X and i_E ($i = K \cdot I$), which are made to charge condensers of capacity C. The charge is given by the formula:

$$Q = \int_0^t i \, dt,$$

The corresponding potential of the condenser is

$$V = \frac{Q}{C} = \frac{1}{C} \int_0^t i \, dt = \frac{1}{C} \int_0^t K \cdot I \cdot dt$$

and the expression

$$\int_0^t I \, dt$$

is the average illumination of the photomultiplier by the line of a given intensity during time t.

Let us consider, for example, the measurement of the ratio of the illumination of the lines X_1 of the element to be determined and E of the internal standard element. During the spectrometric exposure, the interruptors S and S_1 are open, the radiations E and X_1, detected by the photomultipliers P and P_1, give photoelectric currents that charge some condensers during a given time. The potentials of charging are, respectively:

$$V_{X_1} = k_1 \cdot X_1 \quad \text{and} \quad V_E = k \cdot E$$

But the ratio V_{X_1}/V_E represents a mean expression for the ratio of the intensities of the lines I_{X_1} and I_E, so that

$$\frac{I_{X_1}}{I_E} = \frac{V_{X_1}}{V_E}$$

The determination of the ratio of the charging potentials of the condensers gives the intensity ratio of the lines.

6.4 General Properties of Spectrometers

With regard to accuracy, the simplicity of the operation in comparison with spectrography helps to reduce the error. With a sufficiently stable source of emission, it is easy to reduce the standard error for a given determination to less than 1%. The mean error of a series of determinations is readily determined by a statistical method. Since the measurement can be repeated rapidly four or five times, accuracy is improved considerably.

Excellent spectral resolution can be obtained with concave grating instruments with a large focal length (1.5 to 3 m), which can be used for spectra of the first and second orders. It is thus possible to separate spectrometrically the lines of P and Cu at 2149.11 and 2148.97 Å, respectively, in the second order spectrum.

However, the sensitivity of a spectrometric analysis is often lower than when a photographic plate is used, despite the luminosity of modern grating spectrometers. Therefore, this technique is as yet far from being suitable for all laboratories carrying out analyses of trace elements.

Direct reading spectrometry is used chiefly in the metallurgical industry, where a permanent and rapid analytical control is necessary throughout the manufacturing process. The instrument is also suitable for geological, geochemical, and pedological prospecting analyses. It is, in fact, the future method for large scale routine analyses.

REFERENCES

Ahrens, L. H. *Spectrochemical Analysis* (London: E. Addison-Wesley Press, Inc., 1950).

Andersen, C. A. *Microprobe Analysis* (New York: Wiley, 1973).

Atsuya, I. and H. Goto. *Anal. Chim. Acta* 65: 303 (1973).

Baudin, G. *Bull. Inf. Sci. Tech.*, France, 200: 83-92 (1975).

Boumans, P. W. J. M. and F. J. De Boer. *Spectrochim. Acta* 27B: 391 (1972).

Brech, F. Reprint 2, *Jarrel Ash.* Unpublished.

Chapman, J. F., L. S. Dale and R. M. Whittem. *Analyst* 98: 529 (1973).

Czarkow, J. 7th Meeting Metallurgical Material Testing, Budapest (1972).

Dagnall, R. M., D. J. Smith and T. S. West. *Anal. Chim. Acta* 54: 397 (1971).

Debras-Guedon, J. and N. Liodec. *C. R. Acad. Sci.* (Paris) 257: 3336 (1963).

Debras-Guedon, J. *Chimie Anal.* 8: 419 (1967).

Duffendack, O. S. and K. B. Thompson. *Proc. Amer. Soc., Test. Material* 36: 301 (1936).

El Alfy, S., K. Laqua and H. Massmann. *Z. Anal. Chem.* 263: 1 (1973).

Fassel, V. A. and R. N. Kniseley. *Anal. Chem.* 46(3): 1110 A (1974).

Gerlach, W. and E. Schweitzer. *Die Chemische Emissions—Specktralanalyse* (Chemical Emission Spectral Analysis), Voss, Editor (Leipzig, 1931; English translation, London: A. Hilger).

Goto, H. and Z. Atsuya. *Z. Anal. Chem.* 38: 1763 (1966); 225, 121 (1967).

Grammont, A., de *C. R. Acad. Sci.* 171: 1106-1109 (1920).

Greenfield, S. I., L. I. Jones and C. I. Berry. *Analyst* 37: 920 (1965).

Greenfield, S. I. *Metron.* 3: 8 (1971).

Greenfield, S. I. and P. B. Smith. *Anal. Chim. Acta* 59: 34 (1972).

Greene, J. L. and J. M. Whelan. *J. Appl. Phys.* 44: 2509 (1973).

Heyndryckx, P. and R. Grangeon. *XVIe C.S.I.*, Heidelberg 1: 170 (1971).

Hoare, H. C. and R. A. Mostin. *Anal. Chem.* 39: 1153 (1967).

Hoens, M. F. A. and H. Smit. *Spectrochim. Acta* 192 (1957).

Karyakin, A. V., M. V. Achmanova and V. A. Kaigorodov. *XIIe C. S. I.* (Exeter: A. Hilger, 1965), p. 353.

Kawaguchi, H., M. Hasegawa and A. Mizuike. *Spectrochim. Acta* 27B: 205 (1972).

Kleinmann, I. *Spectrochim. Acta* 27B: 93 (1972).

Kniseley, R. N. and V. A. Fassel. *J. Opt. Soc. Amer.* 45: 1032-1034 (1955).

Kranz, E. *XIVe C.S.I.* (Debrecem: Hilger, 1967), Actes II, 697

Krasilshchik, V. Z., G. A. Steinberg and A. F. Yakovleva. *Zh. Anal. Khim.* 26: 1897, 1903 (1971).

Kuzovlev, I. A. *et al. Zav. Lab.* 9: 1071 (1971).

Laroche, M. J. "Chimie des hautes températures," *Coll. Nat. CNRS* 71 (1955).

Laymam, L. R. and G. M. Hieftje. *Pittsburg Conf. on Anal. Chem.*, Cleveland (March, 1973).

Lichte, F. E. and R. K. Skogerboe. *Anal. Chem.* 45:399 (1973).
Malinek, M. and H. Massmann. *Can. Spectros.* 15:1 (1970).
Mannkopff, R. and C. Peters. *Z. Physik.* 70:444 (1931).
Margoshes, M. and B. F. Scribner. *Spectrochim. Acta* 14:138 (1959).
Mavrodineanu, R. and R. C. Hughes. *Spectrochim. Acta* 19(8):1309 (1963).
Mavrodineanu, R. and H. Boiteux. *Flame Spectroscopy* (New York: Wiley, 1965), p. 621.
Mermet, J. M. and J. Robin. *Anal. Chim. Acta* 75:271 (1975).
Mitchell, R. L. *The Spectrographic Analysis of Soils, Plants and Related Materials* (London: Commonwealth Bureau of Soils Science, 1948).
Moenke, H. and L. Moenke. *Laser Microspectrochemical Analysis* (London: A. Hilger Ltd., 1973), p. 253.
Morton, K. L., J. D. Nohe and B. S. Madsen. *Appl. Spectros.* 27(2):109 (1973).
Owen, L. E. *Appl. Spectros.* 15:150 (1961).
Petkova, L. G. and A. P. Petkov. *XVIe C.S.I.* (Heidelberg, 1971), p. 197.
Petkova, L. G., A. P. Petkov, Tr. Dimitrov and A. Ivanova. *Bulg. Akad. Nauk.* 21:187 (1971).
Reed, T. B. *J. Appl. Phys.* 32:821 (1961).
Ropert, M. E. *Meth. Phys. Anal.* 7(3):239-244 (1971).
Runge, E. F., F. R. Bryan and R. W. Minck. *Can. Spectros.* 9:5 (1964).
Runge, E. F., S. Bonfiglio and F. R. Bryan. *Spectrochim. Acta* 22:1678 (1966).
Saffir, A. J., K. W. Marich, J. B. Orenberg and W. J. Treytl. *Appl. Spectros.* 26:469 (1972).
Savinova, E. N., A. V. Karyakin and T. P. Andreeva. *J. Anal. Chem. USSR* 27:682 (1972).
Scott, R. H., V. A. Fassel, R. N. Kniseley and D. E. Nixon. *Anal. Chem.* 46(1):75 (1974).
Scott, R. H. and M. L. Kokot. *Anal. Chim. Acta.* 75:257 (1975).
Scott, R. H. and A. Strasheim. *Anal. Chim. Acta* 76:71 (1975).
Smith, D. J. "Some Analytical Applications of High Frequency Plasma Torch," Thesis, Imperial College, London, 1971.
Souilliart, J. L. and J. P. Robin. *Analysis* 1:427 (1972).
Treytl, W. J., J. B. Orenberg, K. W. Marich, A. J. Saffir and D. Glick. *Anal. Chem.* 44:1903 (1972).
Valente, S. E. and W. G. Schrenk. *Appl. Spectros.* 24(2):197 (1970).
Veillon, C. and M. Margoshes. *Spectrochim. Acta* 23B:559 (1968).
Van Sandwijk, A., P. F. E. Van Montfort and J. Agterdenbos. *Talanta* 20:495 (1973).
Watling, R. J. *Anal. Chim. Acta* 75:281 (1975).
Webb, M. S. W. and R. J. Webb. *Anal. Chim. Acta* 55:67 (1971).
Wendt, R. H. and V. A. Fassel. *Anal. Chem.* 37:920 (1965).
Yudelvich, I. G., A. S. Cherevko and N. G. Skobelkina. *J. Anal. Chem. USSR* 27:1982 (1972).

CHAPTER 3

EMISSION SPECTROSCOPIC ANALYSIS—
METHODS AND APPLICATIONS

1. ANALYTICAL METHODS

1.1 Qualitative Analysis

Qualitative analysis can take two forms: (a) Search for the element or elements of interest by means of the characteristic lines of the element (see table in appendix); (b) Identification of the lines of a spectrum and determination of chemical elements; the complexity of the spectrum of certain elements makes the identification of all lines practically impossible, so it is advantageous to limit the search to persistent lines. In both cases, the analysis is a comparison with the spectrum of a standard.

In qualitative trace analysis, the sensitivity maximum is generally sought. The most sensitive excitation mode is an anodic arc: The sample (10-50 mg) is mixed with high-purity graphite powder in a ratio of 1:1-1:4 and placed into the well (2-3 mm diameter, 3-4 mm depth) of an electrode (Figure 4, Chapter 2), which serves as the anode in the dc arc. The discharge is focused on the spectrometer slit. The juxtaposition of the analyzed spectrum with a standard spectrum, such as iron spectrum, is made under the best experimental conditions with the use of a Hartmann diaphragm (Figure 1), which allows two or three spectra to be photographed without changing the position of the photographic plate. It uses a movable metal foil in front of the slit with three apertures of 1 mm allowing successive exposure of three contiguous slit zones. In practice, the reference spectrum is photographed with aperture 1 and the analyzed sample spectrum with apertures 2 and 3 with exposure times of 30 sec and 2 min. The standard spectrum is that of an artificial sample containing the analyzed trace elements in a concentration of 30 or 100 ppm in a matrix of major elements with a composition as close as possible to that of the analyzed samples.

91

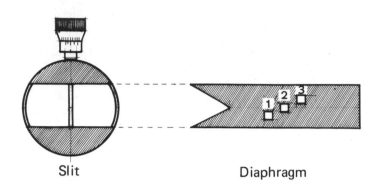

Slit   Diaphragm

Figure 1. Spectrographic slit and Hartmann diaphragm.

To facilitate identification of unknown lines of the sample, such as silicate minerals, it is advisable to use certain reference lines or line groups that are particularly characteristic and easily recognizable. These belong either to the major elements or to carbon. They are listed in Table I and can be obtained from the mixture described in Table II. The persistent lines of each element are easily identified from the complex mixture to which the trace elements are successively added first individually and then simultaneously.

The detection limit of the elements defined by the concentration in the analyzed medium giving a signal at least two times stronger than the background depends on several factors: excitation conditions, quantity of sample in the source, nature of analyzed medium (matrix effect), electrode size, optical characteristics of the spectrometer, exposure time and nature of the photographic emulsion. Table III shows the mean detection limits obtained in an anodic arc for some elements. The values listed are for two types of silicate (or silicoalumina) and calcium carbonate matrices. The influence of the chemical nature of the matrix on the detection limit is evident.

1.2 Semiquantitative Analysis

1.2.1 Principle and Definition

Semiquantitative analysis consists of evaluating the concentration of an element by comparing the intensity of a characteristic line (density of the photographic plate or direct reading) to that of the same line of a series of standard samples of similar composition. An internal standard is not used in semiquantitative analysis. However, it is advantageous to buffer the analyzed medium with a spectral buffer (alkali or alkaline-earth chloride, sulfate or carbonate) in order to stabilize the spectral emissions

Table I. Characteristic Lines and Line Groups in Qualitative Analysis

Elements	Number of Lines	Wavelengths (nm)
Boron	2	249.7-249.8
Silicon	6	250.7-251.4-251.6
		251.9-252.4-252.85
Aluminum	2	256.8-257.1
Aluminum	2	265.25-266.0
Magnesium	5	277.7-277.8-278.0
		278.1-278.3
Magnesium	2	279.55-280.2
Magnesium	1	285.2
Silicon	1	288.1
Iron	6	301.6-301.8-301.9
		302.05-302.1-302.1
Aluminum	2	308.2-309.3
Iron	3	310.0-310.3-310.1
Calcium	2	315.9-317.9
Copper	2	324.75-327.4
Sodium	2	330.2-330.3
Iron	2	330.6-330.6
Iron	2	344.1-344.1
CN	1 band head	359.0
CN	1 band head	388.3
Calcium	2	393.4-396.85
Aluminum	2	394.4-396.15
Manganese	3	403.1-403.3-403.45
Potassium	2	404.4-404.7
CN	1 band head	421.6
Calcium	1	422.7
Calcium	5	442.5-443.5-443.6
		445.5-445.6
Calcium	3	457.9-458.1-458.6
Titanium	4	498.2-499.1-499.95
		500.7
Calcium	3	526.4-526.6-527.0
Sodium	2	589.0-589.6
Lithium	1	670.8
Potassium	2	766.5-770.0
Sodium	2	818.3-819.5

Table II. Base Mixture

Sodium borate: $Na_2B_4O_7, 7 H_2O$	1.0 g
Silica: SiO_2	1.0 g
Alumina: Al_2O_3	0.5 g
Magnesia: MgO	0.5 g
Ferric oxide: Fe_2O_3	0.5 g
Calcium carbonate: $CaCO_3$	0.5 g
Copper oxide: $CuO O_2$	0.5 g
Manganese oxide: Mn	0.5 g
Titanium dioxide TiO_2	0.5 g
Lithium chloride: LiCl	0.5 g
Potassium chloride: KCl	0.5 g

Table III. Detection Limits of Some Elements in an Anodic Arc (10 amperes)

Elements	Wavelength (nm)	Detection (ppm)	
		Silicate	Calcium Carbonate
Ag	328.01	1	2
As	234.98	1000	–
B			
Ba	493.41	5	5
Be	313.04	5	5
Bi	306.77	100	10
Cd	326.11	300	200
Co	345.35	2	5
Cr	425.43	1	2
Cs	807.90	300	–
Cu	324.75	1	1
Ga	294.36	1	3
La	333.75	30	25
Li	670.78	1	1
Mn	279.80	3	3
Mo	317.03	1	5
Ni	341.46	2	5
Pb	283.33	10	10
Rb	780.02	10	30
Sn	284.00	5	10
Sr	460.73	10	10
Ti	398.98	30	10
Tl	276.79	50	10
V	318.54	10	10
Y	332.79	30	20
Zn	334.50	300	1000
Zr	339.20	10	10

and obtain a uniform thermal and electrical conductivity of the substance exposed to spectral excitation.

The lines are measured on a photographic plate or by direct reading with reference to the nearby spectral background and compared to standard spectra obtained from artificial mixtures covering a concentration range of the trace elements extending from 0 to 10,000 ppm.

The standards for a mineral matrix with the chemical composition of silico-alumina are prepared as described below.

(a) *Preparation of the base medium.* An oxide of spectroscopic purity forms the following chemical composition: 63% SiO_2, 20% Al_2O_3, 5% Fe_2O_3, 2% CaO, 2% MgO, 3.5% Na_2CO_3, 3.5% K_2SO_4, 1% TiO_2. The constituents are ground, mixed and ignited at 1250°.

(b) *Preparation of a mixture of trace elements.* The mixture, the composition of which is listed in Table IV, contains 0.100 g of each element.

Table IV. Mixture of Trace Elements (to obtain 0.100 g of each element)[a]

Compound	Weight	Compound	Weight	Compound	Weight	Compound	Weight
LiF	0.3738	Co_3O_4	0.1362	SnO_2	0.1270	As_2O_3	0.1321
RbCl	0.1414	NiO	0.1272	PbO	0.1077	Sb_2O_3	0.1197
CsCl	0.1266	CuO	0.1352	Tl_2O_3	0.1116	SeO_2	0.1463
BaO	0.1117	MoO_3	0.1500	ThO_3	0.1138	TeO_2	0.1251
SrO	0.1182	$K_2Cr_2O_7$	0.2824	ZnO	0.1215	BeO	0.2773
CeO_2	0.1228	Ag_2O	0.1074	CdO	0.1142	HgO	0.1080
H_2WO_4	0.1359	V_2O_5	0.1784	Bl_2O_3	0.1114	GeO_2	0.1441
Mn_2O_3	0.1436	La_2O_3	0.1173	Ga_2O_3	0.1344	Y_2O_3	0.1271
TiO_2	0.1666	ZrO_2	0.1351				

Total weight: 4.9181 g

[a]Compounds and weight in g.

(c) *Preparation of a series of standards.* The following series is prepared so as to contain 1%, 0.316%, 0.100%, 0.0316%, 0.0100%, 0.0032%, 0.0010%, 0.0003%, 0.0001%, (*i.e.* 10,000 to 1 ppm). The 1% standard is prepared by mixing 0.4918 g of the mixture of trace elements with 0.5082 g of the mixture of base elements (total = 1 g). The 0.316% standard is obtained by mixing 0.316 g of the 1% standard with 0.684 g of base. Each subsequent standard is prepared from the preceding one by diluting 0.316 g with 0.684 g of base.

As mentioned, it is advisable to prepare the base mixture from products of very high purity that do not contain the trace elements in the concentrations to be determined. The mixture of trace elements is generally prepared from oxides, sometimes chlorides or other compounds, but the products employed should be of a perfectly stable, well-known and nonhygroscopic chemical composition. Finally, the grinding and mixing procedures require high cleanliness to prevent contamination.

The following mixture can be used as a basis for plant analysis (plant ashes): 30% K_2O, 5% Na_2O, 10% CaO, 5% MgO, 20% SiO_2, 10% P_2O_5, 10% SO_3, 10% CO_2, 0.2% Fe_2O_3, 0.3% Al_2O_3. For this purpose, the following products are used: H_2KPO_4, K_2SO_4, K_2CO_3, MgO, SiO_2, Na_2CO_3, $CaCO_3$, FeO_3, Al_2O_3.

1.2.2 Semiquantitative Analysis Methods

Several methods are distinguished depending on the procedure of evaluating the spectral line intensity.

1.2.2.1 Method of Line Comparison. This is a *visual* or *photometric* study of the unknown spectrum in comparison with a series of standard spectra obtained from artificial samples of analogous composition as the analyzed products. The group of spectra is photographed on the same plate so that the density of the sample lines can be visually compared to that of the standard lines. If the excitation conditions are suitably reproducible, the estimate is made with an error of less than 30% in a range of 3-10,000 ppm.

A better precision (error smaller than 20%) can be obtained by densitometry of the lines. However, the curve of the photographic density of the line as a function of the logarithm of the concentration is linear only for a limited concentration range, beyond which the curve slope results in a more or less significant error. This can sometimes be remedied by photometry of another less sensitive line.

1.2.2.2 Comparison Method of Characteristics. This method resides in the variation of the characteristic of a line (see Chapter 2, Section 5.2) as a function of the concentration of the element responsible for the line in the emission source (Figure 2). Standardization consists of the curve representative of the variation (OC) of the characteristic as a function of concentration or its logarithm. Recording of the spectrum requires the use of a step sector (Chapter 2, Section 4.2.4) and a suitable projection system of the source on the spectrometer slit.

If the photographed spectrum has a spectral background at the site of the measured lines, this must be taken into account by determining the characteristics of the spectral background (Figure 3). The intensity of a line is measured by the value OC on the graph.

1.2.2.3 Sample Preparation. The most common method consists of placing the powdered sample into the well of an electrode serving as the anode or cathode in the dc arc. However, the object is to obtain the best reproducibility conditions.

The arc with a cathodic deposit (Chapter 2, Section 4.1.2) is often preferred because its stability is higher than that of an anodic arc, although the latter is more sensitive. However, the use of a spectral buffer is necessary. Generally, an alkai sulfate or carbonate (Na_2SO_4, K_2SO_4, $LiCO_3$) or sodium or lithium borate is used. The proportions between

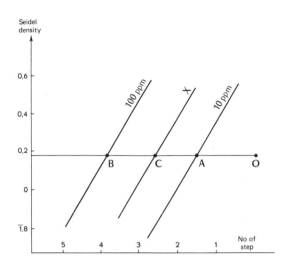

Figure 2. Semiquantitative analysis by comparing characteristic curves; spectral background absent.

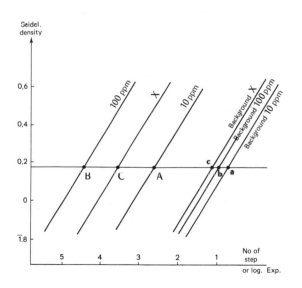

Figure 3. Semiquantitative analysis by comparing curves; spectral background present.

sample, spectral buffer and graphite generally are 1:1:2 to 1:2:4. The electrode containing the analyzed sample can also be prepared by pelletizing with graphite, in which case the proportions are 1:1:10. A special graphite should be used. A small amount of cellulose powder can also be added to improve the physical strength of the pellet.

1.2.3 Range of Determinations by Semiquantitative Analysis

The principal usable lines with the concentration limits for the most sensitive lines of each element are listed in Table V. Two values are given for each line: the first corresponds to the effective concentration to obtain a density equal to about two times the spectral background on the plate, representing the detection limit. The second value is the concentration producing an overexposure unfavorable for analytical accuracy (too strong bend of the curves; line density/concentration). (Mitchell, 1945; Pinta, 1973).

1.3   Quantitative Analysis

1.3.1 Definition and Properties

The principle of quantitative analysis was described in Chapter 2, Sections 3 and 5.3. The development of a practical and special method consists of finding the excitation conditions offering a maximum of stability, precision, accuracy and sensitivity. It has already been mentioned that there is often an antagonism between precision or reproducibility and sensitivity. Stability is the most important factor of precision. It is improved with the qualities of the generator and also the preparation of the sample. Thus, the stability of an electrical arc is improved by operating in a controlled argon or carbon dioxide atmosphere (Figure 4) that reduces the intensity of the cyanogen bands in the 350-400 nm region. The electrode geometry (size and shape) has an important influence, and the quality of the samples placed into the electrode and electrical operation of the arc (ignition, discharge current) are also essential factors. The same applies to the exposure, which can be determined either by the total sample consumption (volatilization of the total substance in the arc) and controlled by the color and appearance of the discharge or by a constant time shorter than the time for total consumption of the sample.

Table V.   Analysis Lines and Concentrations Range in ppm

Elements	Lines (nm)	Concentrations	Elements	Lines (nm)	Concentrations
Ag	328.0	1-30	Mo	313.2	3-300
	338.2	1-30		317.0	3-300
Al	308.2	10-30		281.6	300-
	309.2	10-30		287.1	1,000-
	257.5	100-3,000	Na	589.0	3-100
	236.8	300-10,000		819.4	300-3,000
B	249.7	30-1,000		818.3	1,000-
	249.7	100-3,000	Nb	309.4	100-3,000
Ba	455.4	10-100		319.5	100-3,000
	307.1	300-10,000	Ni	341.5	3-100
	233.5	1,000-		349.3	10-1,000
Be	234.8	3-300		341.4	300-
	313.1	10-100		320.0	1,000-
	249.5	100-1,000	Pb	283.3	10-1,000
	249.4	100-3,000		287.3	300-
Bi	306.7	10-100		282.3	300-
	289.8	300-10,000		240.2	1,000-
	293.8	300-10,000	Rb	780.0	30-300
	431.8	30-1,000		794.7	100-300
Ca	317.9	100-10,000	Sb	259.8	100-
	315.9	100-10,000		302.9	1,000-
	299.7	1,000-	Si	288.1	10-100
	326.2	100-3,000		258.2	30-1,000
Cd	346.6	300-3,000		250.7	30-1,000
	345.3	3-100		243.5	300-
Co	340.9	30-300		298.7	1,000-
	336.7	300-3,000	Sn	317.5	10-300
	428.9	1-30		284.0	10-1,000
Cr	304.0	100-3,000		242.1	1,000-
	278.0	100-3,000	Sr	346.4	100-
	299.4	1,000-		335.1	300-
	327.4	10-100	Ti	337.2	10-300
Cu	296.1	1,000-3,000		334.9	30-1,000
	282.4	1,000-3,000		316.8	30-3,000
	302.0	10-100		284.2	1,000-
Fe	259.9	100-3,000	Tl	322.9	300-
	262.8	1,000-		276.8	1,000-
	303.9	10-1,000		292.1	3,000-
Ge	326.9	100-3,000	U	424.1	30-30,000
	766.5	3-300		434.2	1,000-
K	769.9	3-300	V	318.4	10-100
	404.7	1,000-10,000		318.3	10-1,000
	279.5	3-100		291.5	100-10,000
Mg	280.2	3-30	W	429.4	300-
	277.9	30-1,000	Zn	334.5	300-
	278.1	100-3,000		328.2	1,000-
	279.0	1,000-	Zr	339.2	10-1,000
Mn	279.8	1-30		349.6	100-3,000
	256.7	30-1,000		271.2	1,000-
	260.5	100-1,000		306.1	3,000-
	293.3	300-			

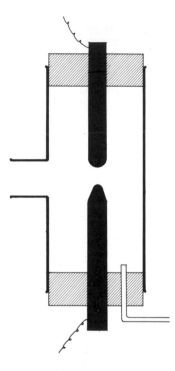

Figure 4. Arc in controlled atmosphere.

1.3.2 Spectral Buffer and Separating Methods

A suitable use of a spectral buffer and internal standard improves the precision as well as the accuracy of a spectral analysis. It is difficult to establish exact rules for their use. The most common spectral buffers for the analysis of silicates, rocks, ores, minerals, soils and plant ashes are lithium carbonate, lithium sulfate, potassium sulfate, sodium chloride, sodium sulfate and strontium carbonate always mixed with graphite. The sample-buffer-graphite ratios are 1:1:12-1:2:10. Sometimes volatile elements must be distinguished from sparingly volatile elements and different analytical conditions must be employed. Strontium carbonate can be used for volatile elements such as Cu, Mn, Fe and Mg, while yttrium oxide serves for less volatile elements such as Al, Si and Ti.

The detection limits are often considered inadequate when analysis is performed with the crude sample. Depending on the elements, these limits range between 1 and 10 ppm. To improve the sensitivity, a chemical concentration procedure is used. After dissolving the sample by alkaline fusion or acid attack, the trace elements are precipitated in the form of chelates (oxinates) in the presence of a suitable entraining agent, such as aluminum or indium (Mitchell, 1945; Pinta, 1973). The separated and ignited precipitate contains the trace elements (Cu, Mn, Ni, Co, Zn) in a base of alumina (or indium oxide), which will play the role of a spectral buffer in the subsequent spectrographic analysis.

The separation can also be made by extracting the chelates (oxinates, dithiocarbamates, dithizonates) into an organic solvent, such as chloroform. The extract is evaporated in the presence of a known quantity (25 mg) of a spectral buffer (lithium carbonate or sulfate). These extraction techniques are well known. Recently, Kantor *et al.* (1972) reviewed the extraction methods for trace elements for their determination by arc, spark or flame spectroscopy. Most frequently, the organic extract is evaporated on graphite powder, which is then introduced in the electrode.

1.3.3 Internal Standard

The choice of internal standard is important for the quality of an analysis (Winefordner, 1971; Addink, 1971; Slavin, 1971; Grove, 1971; Mika and Torok, 1973). This is a chemical element present in or added to the analyzed sample in a concentration analogous to that of the element to be determined and in a chemical form with similar properties in the arc. In other words, the volatility, excitation potential and wavelength of the analyzed line must be similar to those of the element or elements to be determined. Generally, however, spectral analysis involves a group of elements of different properties. The number of internal standards should be minimized as far as possible.

Molybdenum, palladium and germanium are often used as internal standards. Table VI lists the combinations used for spectroscopic analysis. Indium is often used for volatile elements such as Pb, Ga, Cu, Ag and Zn; palladium is used for elements of low volatility, such as Be, Ba, Sr, V, Cr, Ni, Co, Mo, Sc, Y, La and Zr. Another classification uses thallium, germanium, palladium and beryllium as a function of volatility (Table VII).

The internal standards are added to the spectral buffer sodium sulfate, resulting in the following composition: 0.26% Tl as sulfate, 0.50% Be as oxide BeO, 0.50% Pd as chloride $PdCl_2$, 1% Ge as oxide GeO_2. Then 50 mg of sample are mixed with 12.5 mg of this mixture. Lutetium can also be used for the nonvolatile elements, such as Ba, Be, Co, Cr, Cu, Mn, Mo, Ni, Sr, Ti, V and Zr, and cadmium and antimony can be used for the volatiles Ag, As, Ga, Ge, Pb, Sn and Zn.

In some cases, the spectral buffer also serves as an internal standard, such as in the analysis of refractories that can be made with germanium oxide as the buffer and internal standard (30 mg sample, 100 mg germanium

Table VI. Lines Pairs Used for the Determination of Trace Elements with Molybdenum, Germanium and Palladium as Internal Standards (wavelength in nm)

Element		Standard		Element		Standard	
Mg	297.55	Mo	281.61	Ga	294.36	Ge	303.91
Mn	279.48	Mo	281.61	Bi	306.77	Ge	303.91
Al	309.27	Mo	311.21	Sn	317.50	Ge	303.91
Cu	324.75	Mo	317.03	V	318.40	Ge	303.91
Ti	319.99	Mo	317.03	Mo	317.03	Ge	326.95
Sr	407.77	Mo	407.00	Cd	326.11	Ge	326.95
Pb	283.31	Ge	265.16	Zn	334.50	Ge	326.95
Cr	284.32	Ge	265.16	Co	345,35	Ge	326.95
				Ti	319.99	Pd	324.27
				Ni	305.08	Pd	324.27

Table VII. Lines Pairs Used with Thallium, Beryllium, Palladium and Germanium as Internal Standards (wavelengths in nm)

Element		Standard		Element		Standard	
As	245.65	Tl	276.79	Al	396.15	Be	332.13
Bi	306.77	Tl	351.92	Ti	365.35	Be	332.13
Cd	326.11	Tl	351.92	Zr	339.20	Be	332.12
Li	328.26	Tl	351.92	B	249.77	Pd	324.27
Pb	283.31	Tl	276.79	Ba	455.40	Pd	346.08
Rb	420.19	Tl	351.92	Co	345.35	Pd	340.46
Zn	334.50	Tl	351.92	Cr	298.65	Pd	342.27
Ag	320.87	Ge	303.91	Cr	425.43	Pd	346.08
B	249.77	Ge	259.25	Cu	327.40	Pd	324.27
Cu	327.40	Ge	303.91	Fe	302.06	Pd	324.27
Ga	294.36	Ge	275.46	Mn	279.48	Pd	324.27
Li	328.26	Ge	303.91	Mo	317.03	Pd	324.27
Mo	317.03	Ge	303.91	Ni	341.47	Pd	340.46
Sb	259.81	Ge	259.25	V	318.40	Pd	324.27
Sn	284.00	Ge	275.46				
Sr	460.73	Ge	303.91				

oxide and 300 mg graphite). The 282.9 nm germanium line serves for internal standardization of the elements Al, B, Ba, Ca, Co, Cr, Cu, Fe, Li, Mg, Mn, Mo, Na, Ni, Pb, Sr, Sn, Ti, V and Zn. In direct-reading spectrometry, when the internal standard is the same for several elements, only a single line generally needs to be measured. In practice, the sample to be analyzed is prepared by mixing a suitable amount of sample (10-50 mg) with a mixture (20-200 mg) containing the internal standard, spectral buffer and graphite. Preparation of this mixture is very important because it must be perfectly homogeneous and free of all contaminations.

2.   PRACTICAL APPLICATIONS

2.1   General Remarks

The applications of the method consisted of the analysis of natural media—rocks, ores, soils, plants; biological media—fluids and tissues; chemical and industrial products; and crude oil products and derivatives. As mentioned at the beginning of this book, the analysis of metals and alloys has not been attempted.

The analytical sample is in either a solid or liquid state. When the sample is organic, the most common method requires destruction of the organic material, ashing or acid attack. It is then necessary to assure that no loss of the elements occurs. Some exceptions to this rule are certain analyses of oils, crude oil products and biological media.

Depending on the nature of the sample (solid or liquid) or more accurately, the analyzed medium, the following practical methods are distinguished:

(a) Solid samples. Arc spectroscopy (or spectrometry) of the product deposited in the electrode or of a pellet is performed to determine trace elements (a few ppm). A dc anodic arc is most commonly used. Spark spectroscopy (or spectrometry) determines major elements and traces generally in amounts of more than 100 ppm.

(b) Liquid samples (or solutions). Spark spectroscopy (or spectrometry) uses the porous cup solution spark or rotative electrode solution spark. Plasma spectrometry uses the arc plasma jet or inductive coupled plasma. Flame emission spectrometry is generally performed with an air-acetylene or nitrous oxide-acetylene flame as the source.

The most common and most sensitive technique for the determination of trace elements is spectroscopy with a dc anodic arc. Ten to fifty mg of sample in powder form (granulometry of 50-100 μ) are mixed with 10-100 mg spectral buffer and 20-200 mg graphite powder in an agate mortar. The total sample or an aliquot is placed into an electrode (graphite rod of 6-7 mm) with a hole at the top. This gas is 2-4 mm in diameter and 1-4 mm in depth. The counterelectrode is a graphite rod, 6-7 mm diameter, the end of which is either formed into a 60° cone or is plane (see Chapter 2, Figure 14). The electrode gap ("entrode") is 1-4 mm, the excitation current is 6-25 A depending on the refractory nature of the sample and the temperature at the sample site is 2000-4000°C. At the moment of arc ignition with an auxiliary high-frequency spark, the temperature rises rapidly from 20 to 3000°C so that sample particles may be ejected and escape from excitation. This may occur when the sample contains products decomposing into gases or when air pockets have been trapped in the electrode during introduction of the sample. To avoid a sharp temperature rise to 3000°C, the arc current can be increased progressively or in steps. However, the volatilization of elements or their compounds may occur selectively in an electric arc. Therefore the compounds of Se, Te, Pb and Sn are volatilized and then excited within the first seconds after ignition inversely to elements passing through the stage of refractory oxides (Al, Ti, Zr, Ta, Nb) which are excited only 10-20 sec after ignition of the arc.

The curves of Figure 5 show the intensity variation of a line emitted by elements a, b, c in the arc during volatilization of the element. This must be kept in mind in photographic spectroscopy as well as in spectrometry with signal integration. If ignition occurs at time 0,

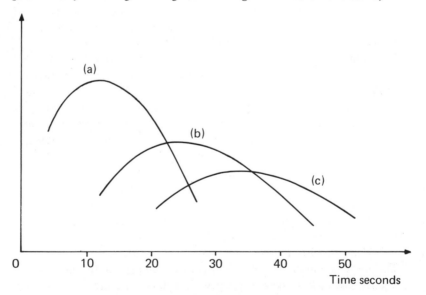

Figure 5. Volatilization of elements in direct current arc:
(a) volatile element, (b) slightly volatile element, (c) refractory element.

measurement of the element a requires an exposure of 0-30 sec, while element b requires an exposure of 30-60 sec after ignition.   For the simultaneous determination of a, b and c exposure will start at time 0 and terminate at 60 sec.

The selective volatilization phenomena are utilized to improve the sensitivity of a determination of volatile elements in a refractory matrix, as in the case of an impurity determination in alumina.   The sample is mixed with a volatile spectral buffer (for example, Ga_2O_5) in a ratio of 10:1 and the mixture is placed in the bottom of the hole (4 mm diameter, 8-10 mm depth) of an electrode of 6 mm o.d.   During ignition of the arc at 10 A, the volatile elements—Ag, As, Cd, Pb, Se, Sn and Te—distill while being entrained by gallium oxide and are excited in the first 20 sec.   Gallium also serves as the internal standard.

Another technique for introducing the sample in the electrode is pelletizing with graphite of matrix-forming quality: 100 mg of sample containing the spectral buffer and the internal standard are mixed with 1 g of graphite powder and pelletized in a 20-mm diameter mold in a hydraulic press under a load of 10-12 t.   The addition of a small amount of cellulose powder improves the endurance and strength of the pellet. This pellet is placed flat on a suitable support.   Discharge (dc arc, pulsed arc or spark) occurs between the circular horizontal surface of the pellet, which serves as the anode, and a graphite counterelectrode.   However, it can also be used as a "rotrode," where the discharge is produced between the section of the rotating pellet and a counterelectrode (Figure 6).

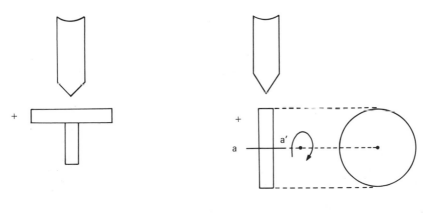

Fixed electrode   Rotating electrode

Figure 6.   Use of pellet electrode.

Table VIII. Comparative Detection Limits of Trace Elements by Anodic Arc Spectroscopy, Spark Spectroscopy, Plasma Spectrometry and Nitrous Oxide-Acetylene Flame Spectrometry[a]

Elements	Radiation Wavelength (nm)	Arc Spectrometry on Powder (ppm)	Spark Spectrometry Solution (μg/ml)	Plasma Spectrometry Solution (μg/ml)	Flame Spectrometry $N_2O\text{-}C_2H_2$ Solution (μg/ml)
Aluminum	309.3	30	0.8	0.002	0.005
Argent	328.1	1	0.6	0.004	0.008
Arsenic	228.8	1000	40	0.04	50
Barium	553.5	30	0.4	0.0001	0.002
Beryllium	313.0	5	0.01	0.0005	0.1
Bismuth	298.9	30	8	0.05	2
Boron	249.6	1	0.1	0.005	30
Cadmium	326.1	500	4	0.002	0.8
Calcium	422.7	4	0.4	0.00007	0.0001
Cesium	852.1	100	15	-	-
Chromium	425.4	10	2	0.001	0.004
Cobalt	345.4	10	2	0.003	0.03
Copper	324.7	3	0.05	0.001	0.01
Tin	284.0	30	15	0.3	0.3
Iron	372.0	30	0.5	0.005	0.03
Gallium	294.4	10	4	0.014	0.01
Lanthanum	324.5	30	5	0.003	2
Lithium	670.7	3	0.1	-	-
Magnesium	280.2	3	0.005	0.0007	0.005
Manganese	279.3	10	0.08	0.0007	0.005
Molybdenum	317.0	10	2	0.005	0.1
Nickel	341.5	10	4	0.006	0.02
Gold	267.6	50	200	0.04	4
Phosphorus	253.6	25	-	0.02	-
Platinum	265.9	100	7	0.08	2
Lead	283.3	50	10	0.008	0.1
Potassium	766.5	4	200	-	-
Rubidium	780.0	20	-	-	-
Silicon	288.1	30	-	0.01	5
Sodium	589.0	1	60	0.0002	0.0001
Strontium	460.7	10	0.2	0.00002	0.0002
Thallium	276.8	50	40	0.2	0.02
Titanium	365.3	30	0.4	0.003	0.2
Uranium		100	-	0.03	10
Vanadium	318.4	10	4	0.006	0.01
Zinc	334.5	100	2	0.002	0.01
Zirconium	339.2	30	2	0.005	-

[a]According to Pinta (1973); Fassel *et al.* (1974); Winefordner *et al.* (1970); and Omenetto *et al.* (1973).

Analysis with excitation from liquid samples or solutions is made with either a spark ("rotrode" or "porode"), a plasma (sample introduced into the plasma in atomized form with an inert gas, usually argon, by a pneumatic or ultrasonic method), and flame excitation.

The sensitivity of the methods is very different. To assist a possible choice, Table VIII lists the detection limits of some common elements. With regard to the precision of the analytical methods, those with solutions are more precise, particularly when used with direct reading. Furthermore, they allow more representative samples to be used. The precision of the different methods can be summarized as in Table IX.

Table IX. Precision of Different Methods.

Methods	Error in Precision (%)
Arc spectroscopy	
Semiqualitative method	30
Quantative method	5-10
Spark spectroscopy	
and spectrometry	2-5
Plasma spectrometry	2-5
Flame spectrometry	2-5

2.2 Analysis of Natural Minerals: Rocks, Ores, Soils

Analysis is made either with the ground-solid sample or a solution after perchloric-hydrofluoric acid attack. The general method for solid-state analysis is the following: the sample is ground to 100 μ, ignited at 450°, and an aliquot fraction of 50 mg is mixed with 50 mg potassium sulfate and 100 mg graphite containing 200 ppm of palladium. The mixture, or a portion of it, is placed into the hole of a graphite electrode. The following excitation conditions are used:

Electrode gap:	2 mm
Arc:	anodic dc, 12 A
Prearcing:	10 sec
Exposure time:	60 sec
Photographic plate:	Kodak spectrum analysis

The total discharge is projected on the spectrgraph slit. The analysis lines are listed in Table X.

Table X. Analytical Lines

Element	Wavelength	Internal Standard	Wavelength
B	249.7	Pd	324.3
Ba	455.4	Pd	346.1
Co	345.3	Pd	340.5
Cr	425.4	Pd	346.1
Cu	327.4	Pd	324.3
Li	328.3	Pd	324.3
Mn	279.3	Pd	324.3
Mo	317.0	Pd	324.3
Ni	341.5	Pd	340.5
Pb	283.3	Pd	324.3
Sn	284.0	Pd	324.3
Sr	460.7	Pd	346.1
Ti	399.0	Pd	340.5
V	318.5	Pd	324.3
Zr	339.2	Pd	340.5

The standards are prepared as described in Chapter 2, to obtain a range of 0-1000 ppm of the investigated trace elements. The composition of the base depends on the type of rock or soil investigated. For example, Table XI shows the composition of some classical base media. In common rocks and soils, the following trace elements are generally detectable and directly determinable since their concentration is higher than 10 ppm: B, Cu, Mn, Ni, Cr, Pb, V, Zr, Ga, Li, Sr, Ba and Ti. Some of them, such as Mo, Co, Bi, Cd, Be, Tl, As and Se, are present in small amounts, and chemical concentration and precipitation in the form of organic complexes or solvent extraction of the complexes is necessary to enrich the analyzed medium (Mitchell, 1945; Wark, 1954).

A literature study reveals the principal applications of spectroscopic analysis for rocks, soils, and ores. Not all studies can be cited and we have listed only a few of the most important that appeared between 1971 and 1975. The applications are listed in Table XII.

Today, emission spectroscopy is used primarily for semiquantitative multielemental analysis, a typical example being geochemical prospecting.

Two new directions of emission spectroscopy should be noted. One is developing in the USSR and uses a nitrous oxide-acetylene flame as the emission source. Prudnikov *et al.* (1973) have determined 20 elements in very small samples (1-10 mg) with detection limits of 0.01-0.001 ppm in the prepared solution. The second direction is oriented toward plasma for excitation of the elements from a sample solution

Table XI. Base Composition of Rocks and Soils (%)

	Basalts	Granites	Soils
SiO_2	51	70	63
Al_2O_3	20	15	20
Fe_2O_3	12	5	5
CaO	0.5	2.5	2
MgO	5	2	2
Na_2CO_3	2	2	3.5
K_2SO_4	2	2	3.4
TiO_2	2	1.4	1
MnO	0.5	0.1	0.1

(Sermin, 1973). This source would also permit the determination of phosphorus (213.6 nm) and sulfur (182.0 nm) with detection limits of 1.7 and 0.08 $\mu g/ml$. Kirkbright *et al.* (1972) and Yudelevich *et al.* (1972) believe that the sensitivity is 10-100 times higher with plasma sources than with a dc arc. It should be noted that these authors do not go through the stage of a solution but inject the sample (powder or rock) directly into the plasma. Jones *et al.* (1974) analyzed soil extracts with an inductive-coupled plasma, and determined 15 elements.

Finally, we should mention the application of the laser microprobe for spot analysis (Moenke-Blankenburg, 1973; Petrakiev, 1973).

2.3 Water Analysis

Water is analyzed directly without pretreatment with flame excitation sources (nitrous oxide-acetylene) or a plasma (inductive-coupled plasma). However, the concentrations of trace elements, particularly in natural water, require a separation with concentration of the extract. The most common method remains extraction in the form of a complex with ammonium pyrolidine dithiocarbamate (APDC) in methylisobutylketone (MIBK).

The best sensitivity, however, is obtained by spectroscopic analysis of a dry water extract. In this procedure, 500 ml-1 liter of water with a sulfuric acid addition of 0.5-2 ml and an alkali salt (50-100 mg of potassium sulfate) to serve as the spectrochemical buffer are evaporated to dryness. The amount of buffer depends on the content of dissolved inorganic salts. The dry extract is treated with a suitable quantity of graphite (1:1-1:4), possibly containing an internal standard, such as palladium. The product is analyzed in an anodic dc arc of 8-10 A.

Table XII. Spectrochemical Analysis of Rocks, Ores and Soils

Elements	Samples	Concentrations (ppm)	Excitation Source	Sample Treatment	Analyte Form	References
Au, Pt	Rocks	0.05-5	dc arc 8 A	Solubilization in Br/HB_2, Br/HCl; add Te and Rh; dry off solution in electrode.	Solid	Croft, 1971
K, Na, Li	Silicates	20	flame N_2O-C_2H_2	Solubilization with $HF/H_2SO_4/HNO_3$, fume to dryness, dissolve in HCl + Cs (ionization buffer).	Liquid	Hildon et al., 1971
Pt, Pd, Rh	Sulfide, ores	0.1-400	dc arc 20 A	After separation from matrix.	Solid	Khitrov et al., 1971
Rare earths	Rocks	0.1-300	dc arc 13.5 A	Fuse with Na_2-O_2; dissolve and precipitate R.E. with Y_2O_3.	Solid	Tripkovic et al., 1971a
Trace elements	Calcite, fluorites dolomites	100	dc arc 5 A		Solid	Nova-Spackova et al., 1971, Alvarez-Herrero et al., 1971a
Ni, Cr, Co, Cu, V	Rocks	0.2	dc arc 10-17 A	Mix with $BaCO_3$.	Solid	Laktionova et al., 1971
Nb, Ta, Zr, Hf, Be	Minerals and ores	0.2-30	ac arc 18 A	Attack with HF/H_2SO_4; add EDTA; take up elements on ion exchange resin; ash resin; mix with graphite.	Solid	Moroshkina et al., 1970
Trace elements	Soils	ppm	dc arc		Solid	Ecrement, 1971
Trace elements	Silicates	ppm	arc plasma	Powder sample (70 μm) is blown in arc plasma.	Solid	Guselnikov et al., 1971
Trace elements	Lunar dusts		spark	Fuse with borates.	Solid	Govindaraju, 1972
Noble metals	Ores	0.1-5	dc arc 20 A	10 mg powder.	Solid	Broadhead et al., 1972
Cs, Rb	Rocks	20-500	flame	HNO_3-H_2SO_4-HF and residue solubilization with HCl.	Liquid	Allen, 1972
As	Cu, Zn, ores	300-2000	ac arc 16 A	Powder sample mixed with graphite and dextrin; add Ni as internal standard.	Solid	Finkin et al., 1971
Au	Rocks		ac arc 5 A	Concentrate with cupellation.	Solid	Anan'ev et al., 1972

Analyte	Matrix	Concentration	Technique	Sample preparation	Form	Reference
B, Be, F, P, S, Zn	Rocks	ppm	plasma		Solid	Savinova et al., 1972
B, V	Rocks	1-200	dc arc	Mix with graphite and Ge_2O_3.	Solid	Nova, 1973
Ba	Rocks	20-2500	N_2O-C_2H_2 flame	HF-$HClO_4$; fuse residue with LiF + H_3BO_3 and dissolve in HCl.	Liquid	Rubeska et al., 1973
La, Y, Yb	Rocks	0.5-5000	dc arc	Mix (1:1) with graphite + 5% NaCl carrier + 0.1% La_2O_3.	Solid	Laktionova et al., 1972 Antic et al., 1973
Li, Co, Rb	Rocks	—	flame		Liquid	Shapkina et al., 1972
Al, Mn, Ca, Fe, Mg, Ti	Rocks	—	spark rotrode	Fuse with Li/Br borate solubilization with HCl	Liquid	Besnus et al., 1973
Mo	Rocks	0.1-10	arc		Solid	Delavault, 1973
Trace elements	Rocks	ppm	dc arc 1 A	Mix sample + graphite.	Solid	Avni et al., 1972
Trace elements	Beryl, wolframite	—	spark + laser		Solid	Moenke et al., 1973
Trace elements	Rocks	—	spark + laser		Solid	Petrakiev et al., 1973
Trace elements	Rocks	1-100	plasma		Solid	Sermin, 1973
Trace elements	Rocks	—	ac arc	Mix sample buffer (SiO_2:graphite:SrO, 4:5:1).	Solid	Golubeva, 1972
Trace elements	Rocks	—	flame N_2O/C_2H_2	HF-H_2SO_4 in Teflon® bomb.	Liquid	Prudnikov et al., 1973
Trace elements	Rocks	ppm	plasma		Solid	Yudelevich et al., 1972
Trace elements	Rocks	ppm	arc	Fuse with borate solubilization and separation of T.E. in cation exchange resin.	Solid	Govindaraju, 1973
S	Soils	2	plasma	Solubilization.	Liquid	Kirkbright et al., 1972
Trace elements	Soils	ppm	arc		Solid	Naqueda et al., 1973

Table XII, Continued

Elements	Samples	Concentrations (ppm)	Excitation Source	Sample Treatment	Analyte Form	References
As, Sb	Rocks	1-100	dc arc	Mix with $NH_4Cl:S:C$ (3:1:6).	Solid	Gerasimov, 1974
Au	Ores, rocks		ac arc	Extract in H_2O/C_6H_6/diantipyrinyl/heptane with 2.5 N HCl.	Solid	Petrov et al., 1973
Co, Ni	Pyrites		spark + laser		Solid	Smirnov et al., 1974
Noble metals	Ores		arc	Preconcentration.	Solid	Snodgrass, 1972
Se	Ores	10-100	glow discharge		Solid	Czakow, 1973
Nb, Ta	Rocks		dc arc 20 A	HF, mix residue with graphite.	Solid	Korsun et al., 1972
Trace elements	Sulfides		arc	Mix with graphite.	Solid	Liese, 1974
Trace elements	Ores		arc	Fuse with borate; form pellets.	Solid	Wittmann, 1972
Trace elements	Rocks		flame N_2O/C_2H_2	Solubilization.	Liquid	Busch et al., 1974
Trace elements	Sulfides		dc arc 30 A	Powder + graphite + quartz + sulfide base.	Solid	Mays et al., 1974
Trace elements	Soils		plasma	Soils extracts.	Liquid	Jones et al., 1974

*Teflon® is a registered trademark of E. I. duPont de Nemours and Company, Inc., Wilmington, Delaware.

The standards are prepared by the routine method, considering the nature of the water in order to determine the base of the medium and the spectral buffer that may be added.

If the analyzed water contains a mineral extract of 100 mg/l, the extract (after addition of the buffer) will be about 200 mg, *i.e.,* a concentration factor of 1000:0.2 = 5000. If the detection limit is 5 ppm for the sample in the arc, it will be 5:5000 = 0.001 ppm or 1 μg/l of water. Such limits are difficult to realize by direct analysis of the water either by emission (flame or plasma) or by atomic absorption.

The general practical conditions are: 10 ml of a potassium sulfate solution of 10 mg/ml are added to a water sample of 500-1000 ml. The solution is evaporated to dryness at 100°C, the final evaporation being performed in a platinum or Teflon dish. The residue is weighed and an aliquot (50 mg) is mixed with graphite powder (50 mg) containing 250 ppm palladium as internal standard. The mixture is placed into the hole (4 mm diameter, 3 mm depth) of a graphite electrode of 6 mm diameter. The spectroscopy conditions are an electrode gap of 3 mm, anodic arc of 9 A, prearcing for 10 sec and exposure for 90 sec.

The standards are prepared to cover a range of 0, 1, 3.2, 10, 32, 100, 316, 1000 ppm of the different trace elements in a suitable base. For river water, the following composition is used: 1 g SiO_2, 0.4 g MgO, 2 g Na_2CO_3 and 1 g $CaCO_3$. The reliability may be subject to an error of up to 25-30% in semiquantitative analysis and up to 10% in quantitative analysis.

Variants of this method have been proposed. Farhan (1972) adds sulfur or graphite to obtain a better arc stability. Pepin *et al.* (1973a, 1973b) analyzed mineral water, and LeRoy and Lincoln (1974) analyzed industrial effluents, determining 36 elements. Khitrov *et al.* (1972b) determined the trace elements directly in water using a rotrode and an ac arc.

Flame or plasma emission spectrometry has been used either after extraction of the trace lements or for direct analysis. Winge *et al.* (1973) used direct-reading spectrometry for the simultaneous determination of 20 elements among water pollutants using a sample of less than 5 ml. In fact, flame or plasma emission may require a concentration at the ppb or lower level. The most common method is extraction of APDC complexes into methylisobutylketone (MIBK): a 500-ml sample is adjusted to pH 3, treated with 10 ml of 3% APDC and 25 ml of MIBK, and agitated. After settling, the organic phase is separated and analyzed. In this case, the concentration factor is 50, making it easily possible to attain 0.2 ppb for the elements Ag, Co, Cr, Cu, Fe, Mn, Mo, Ni and Pb.

Trace elements can also be separated by precipitation of pH 10-12 in the presence of an entraining agent, such as Fe(III) and magnesium phosphate. The precipitate contains the heavy metals and is mixed with graphite and analyzed spectroscopically in an arc.

Water with a low degree of mineralization—10-50 mg of mineral salts per liter—may require a sample of 2-10 liters. In practice, it is not possible to evaporate such a volume to dryness without contamination. A simple method consists of binding the cations of 2-10 liters of water on an ion exchange column (Dowex or Amberlite) and eluting them subsequently with 50 or 100 ml hydrochloric acid. The eluate is evaporated to dryness and analyzed by emission spectroscopy.

Some applications are compiled in Table XIII. Classical emission spectrometry with excitation in a dc arc seems to have been abandoned since the development of atomic absorption methods. However, it remains a method of choice for simultaneous semiquantitative multi-elemental analysis.

2.4   Analysis of Plant and Biological Media

Plants are mineralized either by ignition at 450-500°C or by acid treatment (H_2SO_4-HNO_3). The dry extract is analyzed directly in an arc or redissolved to be analyzed with excitation in a spark, plasma or flame. The elements most often investigated in plant media are Fe, Cu, Mn, Zn, Mo, Co, Ni, Al, Sr, Ba, and more rarely Ag, V, Pb, Sn, Cr and Ti. Spectral analysis of the ash is not always sufficiently sensitive (Mo, Co, V, Ag, Pb, Sn, Cr). An enrichment is important. The ash (10-20 g of plant material) is brought into solution and the trace elements are separated by precipitation from oxine in the presence of aluminum as an entraining agent or by extraction with APDC in MIBK (see Section 2.3).

The analysis of biological media such as blood, urine, organs, tissue and fluids involves elements present naturally—Fe, Cu, Zn, Mn, Co—and those that may be present due to medical treatment or intoxication—Pb, Hg, Cr, Au, Li, As, Cd and Sn. Spectroscopic analysis is not suited for determination of all elements. Today, the methods of atomic absorption spectroscopy are often preferred to the emission techniques, although the latter retain their interest for semiquantitative multi-elemental analysis.

The classical conditions for emission analysis in the arc are as follows: for the majority of elements except the volatiles a sample of 1-2 g of the plant or 10-20 g of the biological tissue is ignited at 450°C. The ash or aliquot (50 mg) is mixed with 50 mg potassium sulfate as the spectral buffer and 100 mg graphite powder containing 250 ppm palladium. The mixture is placed into an electrode serving as the anode in

Table XIII. Spectrochemical Analysis of Waters

Elements	Samples	Concentrations	Excitation Source	Sample Treatment	Analyte Form	References
Trace elements	Water	0.05 µg/l	Arc	Evaporate to dryness.	Solid	Khitrov et al., 1972a
As	Water	5-50 ng/l	Plasma	Convert to arsine.	Liquid	Lichte et al., 1972
Trace elements	Water		Flame N_2O/C_2H_2	Solvent extraction with dithizone or APDC in ketones.	Liquid	Christian, 1971
Cd	River	2 ng/ml	Plasma	Extraction with Na-DDC.	Liquid	De Boer et al., 1973
Cu, Fe, Pb, Zn	Waters	ppb	dc arc 13 A	Evaporate to dryness; mix with C + S.	Solid	Farhan, 1974
Li	Water		Flame		Liquid	Presley et al., 1972
Mn	Water	5-500 ng/l	dc arc	Evaporate to dryness.	Solid	Pepin et al., 1973b
Trace elements	Natural water	ppm	ac arc + rotrode	Direct.	Liquid	Khitrov et al., 1972b
Trace elements	Water	ppm	Plasma		Liquid	Winge et al., 1973
Cs	Seawater	0.3 µg/l	Flame O_2/H_2	Extract with Na tetraphenylboron in hexane-cyclohexane.	Liquid	Folsom et al., 1974
Co, Cu, Ni	Water	trace	dc arc 7 A	Precipitation with Fe(III) and Mg phosphate at pH 10-12; mix (1:1) with graphite.	Solid	Lebedinskaya et al., 1973
Trace elements	Industrial effluents	trace	Stallwood jet arc	Evaporate to dryness; mix with graphite + GeO_2.	Solid	Le Roy et al., 1974
Trace elements	Water	µg/ml	dc arc	Evaporate to dryness; mix with NaCl, CsCl or AgF/NaCl + graphite.	Solid	Sahini et al., 1973

an arc. The excitation conditions are similar to those described for water analysis.

Generally, direct reading with spectrophotometry or spectrometry allows a determination of the elements Al, B, Ba, Cu, Fe, Li, Mn, Sr and Ti. The elements Ag, Bi, Cd, Co, Cr, Mo, Ni, Pb, Sn and V require separation, which is done after dissolving the ash by precipitation or extraction of complexes such as oxinates, carbamates and dithiocarbamates. Some applications are listed in Table XIV. Studies to be consulted on direct analysis: Dothie, 1971; Borovik-Romanova *et al.*, 1973; Mosier, 1972; and Webb *et al.*, 1973.

Harding-Barlow (1973) recently reviewed the application of a laser microprobe for the analysis of biological tissue. Treytl *et al.* (1972) have also used emission spectroscopy with a laser microprobe for the simultaneous determination of eight metals, obtaining excellent detection limits from 10^{-13} g for lead to 2.10^{-15} g for magnesium. A general review of the applications of the dc arc in an argon atmosphere for the analysis of trace elements in biological media was published by Gordon *et al.* (1973).

Among recent studies of multi-elemental emission spectrometry, we should note those of Neidermeier *et al.* (1974) on the simultaneous determination of 14 elements in a 2-ml serum sample, and of Alvarez-Herrero *et al.* (1971b) on the determination of 12 trace elements in human hair at concentrations of less than 0.01 μg/ml. In order to improve the precision, Hambidge *et al.* (1971) used a dc arc in an argon atmosphere to determine seven trace elements in 0.2 ml serum and urine. The plasma torch is also used for the excitation of Al, Cu, Fe, Mg, P and Pb in a few μl of blood with detection limits of 10^{-9} to 10^{-10} g (Greenfield *et al.*, 1972). Kniseley *et al.* (1973) also analyzed serum and blood with an inductive-coupled plasma and studied the conditions permitting an improvement of precision. The sample is diluted with water (1:10) and atomized in the coldest zone of the plasma. The detection limits realized with samples of 25-100 μl ranged between 5 and 50 μg/l.

Multi-elemental plant analysis was also used by Isaac *et al.* (1972) with a spark as the excitation source for the determination of Al, Cu, Fe, Mn, Na and Zn, and by Jones *et al.* (1974) for the determination of 15 trace elements in routine soil and plant analysis.

2.5 Analysis of Atmospheric Aerosols

Atmospheric aerosols contain 10-20 μg of solid particles per m^3 of air. With high air pollution, the concentration may reach 1000 μg/m^3. Analytical control of the environment has led to the development of new methods dealing particularly with the determination of Hg, Pb, Zn, Cu, Cr, Cd and As. However, in addition to these elements that are often

Table XIV. Spectrochemical Analysis of Vegetal and Biological Materials

Elements	Samples	Concentrations	Excitation Source	Sample Treatment	Analyte Form	References
Trace elements	Ash plant	ppm	Arc	Mix with graphite + Li_2CO_3.	Pellet	Dothie, 1971
Ag, Mg	Biological	ng level	Laser	None.	Solid	Marich et al., 1970
Li, K, Na	Neuron		Flame air–C_2H_2	Dry on Pt wire; put in flame.	Solid	Giacobini et al., 1970
Li	Biological	trace	Flame air–C_2H_2	Digest in HNO_3; add butanol + Na, K.	Liquid	Gorvat et al., 1971
Pb	Biological	0-1 mg/l	Hollow cathode	Digest in acid; add Li, K.	Solid	Prakash et al., 1971
Sn	Foetus	0-625 ppm	Stallwood jet	Oxidize with HNO_3; ash; mix with graphite + Li_2CO.	Solid	Theuer et al., 1971
Trace elements	Serum	0.01 mg/l	dc arc 10 A	Centrifuge, wet dissolve in HCl NH_4/Cl; wet and mix with graphite.	Solid	Niedermeier et al., 1971
Cr, Cu, Fe, Mg, Mn, Mo, Zn	Serum, urine, hair	trace	dc arc in argon	Ash (in low temperature asher); dissolve in HCl.	Liquid	Hambidge, 1971
Fe, Mn, Zn	Clover	trace	Flame	Wet-ash with HNO_3/$HClO_4$/H_2SO_4.	Liquid	Lopez et al., 1973
Zn	Plants	0.01-1 ppm	ac arc 7-12 A	Mix plant with (2/l) K_2SO_4/graphite.	Solid	Borovik-Eomanova et al., 1973
Trace elements	Plant ash	1 ppm	dc arc 13 A	Ash and mix with (1/9) $CaCO_3$/graphite.	Solid	Mosier, 1972
Trace elements	Aquatic plant	0.1-100 ppm	dc arc 6 A	Dry at 80°C grind; mix with graphite.	Solid	Cowgill, 1973
B, Cu, Fe, Mg, Mn, P, Zn	Plants	ppm	Spark, rotrode	Ash at 450°C; digest with HCl + Co + Li.	Liquid	Chaplin, 1974
Trace elements	Plants		Plasma		Liquid	Jones et al., 1974
Trace elements	Plants		Spark rotrode	Ash and dissolve in acid.	Liquid	Jones, 1974
Trace elements	Biological	ppm	dc arc	Ash.	Solid	Webb et al., 1973, Niedermeier et al., 1974
Pb	Block	0.05-1 ppm	dc arc	Extract with trichloracetic acid.	Solid	Steiner et al., 1972
Cu	Biological	ppm	dc arc	Ash; mix with sulfur.	Solid	Farhan et al., 1974

considered air pollutants, the more classical elements such as Si, Al, Fe, Ca, Mg, Ba in continental atmospheric aerosols and Na, K, Mg, B and Sr in oceanic air aerosols have also been investigated. Generally, the sample is obtained by filtering a known volume of air through a cellulose-membrane filter. The filter deposit is recovered either after igniting the filter or by solubilizing in an acid or an acid mixture (HNO_3 + $HClO_4$, HNO_3 + HCl). Solubilization is facilitated by ultrasonic agitation.

When the ignition route is used, a high risk of losses of volatile elements exists so precautions must be taken. If permitted by the nature of the filter, sample recovery is possible after dissolving the filter in a solvent.

The sample can also be collected on a glass-fiber filter (Zdrojewski et al., 1972). In this case, the filter can be solubilized in hydrofluoric acid, which is then eliminated by entraining the silica by volatilization and taking up the residue in nitric acid. However, this method is unsuited for the determination of Ba, Sr, Rb, Zn, Ni, Fe, Ca and As because of their possible presence in the glass fiber. Hasegawa et al. (1971a) also collected samples on glass-fiber filters, which were then mixed with sodium fluoride and graphite. The mixture is then introduced into an electrode for spectroscopy. In the electrode, 50 μl of a solution containing 30 μg/ml indium and palladium in 2 N nitric acid are added to serve as standards.

More elaborate sampling devices, such as cascade impactors, allow the collection of atmospheric particles with a simultaneous granulometric classification. A description of some of the sample treatment techniques follows: Imai et al. (1973) filter the particle sample on a Millipore membrane filter, which is then impregnated with ethyl alcohol and slowly mineralized for 2 h at low temperature in an igniter. The decolorized ash is mixed with a graphite and lithium carbonate buffer and analyzed by spectroscopy in a dc arc.

Sugimae (1974b) collects the air sample on a Gelman DM 800 filter (pore size 0.8 μm) for seven days at a flow rate of 30 l/min. The filter is solubilized in analytical-grade acetone, the solvent is evaporated and the residue weighed. An aliquot is mixed with double the weight of a spectroscopic buffer consisting of one part sodium fluoride and one part graphite containing 100 ppm indium and 20,000 ppm tantalum oxide. The standards consist of a base of 30% SiO_2, 10% Fe_2O_3, 3% CaO, 1% MgO, 1.5% Na_2O, 0.5% K_2O and 48% graphite with additions of the trace elements to cover the appropriate range (Na, K and Ca are added as carbonates). Table XV shows the analytical lines and the concentration range. The excitation source is a 15 A dc arc in an argon-oxygen atmosphere (10% O_2, 90% Ar) with photographic exposure for

Table XV. Analytical Line Pairs and Working Ranges

Element (nm)		Internal Standard (nm)		Range (ppm)
Be	313.0	Ta	301.3	1-100
Bi	306.8	In	303.9	2-200
Cd	326.1	In	303.9	2-200
Cr	302.2	Ta	301.3	10-1000
Cu	327.6	Ta	301.3	20-2000
Mn	304.4	Ta	301.3	50-5000
Ni	305.1	Ta	301.3	10-1000
Pb	283.3	In	303.9	50-5000
Sn	326.2	In	303.9	10-1000
Ti	295.6	Ta	301.3	20-2000
V	318.4	Ta	301.3	10-1000
Zn	328.2	In	303.9	50-5000

60 sec. The authors report a relative standard deviation of 10-13% depending on the elements. Sugimae (1974a) proposed another method in which samples are collected on a silver-fiber membrane. The total unit is then solubilized in nitric acid containing indium and cobalt as an internal standard, the solution is evaporated, mixed with graphite and finally analyzed by spectroscopy.

Another interesting method was published by Lander et al.(1971) for the analysis of polluted air. The authors collect the sample on a Schleicher and Schuell filter, type 589-1 H of 25 mm diameter. The volume of filtered air is determined so that the filter will contain more than 0.1 μg of each of the analyzed elements (Al, Cd, Ca, Cr, Cu, Fe, Pb, Mg, Ni, Si, Sn, Zn). The filter with the collected sample is folded in half and rolled into a cylinder so that it can be directly introduced into a graphite cylinder (4 mm i.d.) serving as the electrode in a high-voltage spark source (Figure 7). During discharge and spectroscopic exposure, the filter is slowly moved at a speed of 5 cm/min by means of a motor. Combustion of the filter in the spark is improved with an oxygen stream. The electrode gap is 2 mm, the exposure time 30 sec and the oxygen flow rate 30 l/hr. The determination range is between 0.3 and 10 μg for these elements. The coefficient of variation of the results ranges from 9 to 60% with a mean of ± 25% as a function of the concentration. The line profile is recorded spectrometrically during exposure by means of a computer.

The technique developed by Seeley et al. (1974) uses a porous graphite electrode (porous cup spectrographic electrode Ultra Carbon No. 202) shown in Figure 8 to filter the air particles. The analyzed air

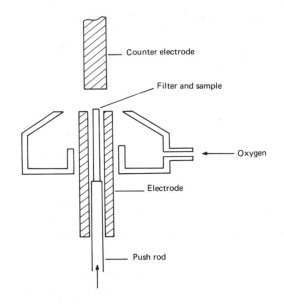

Figure 7. Direct analysis of filter with atmospheric particles by spark spectrography.

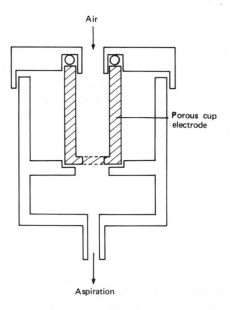

Figure 8. Filter cup electrode for atmospheric particle analysis by arc spectrography.

is aspirated through the electrode with a vacuum pump, while the flow rate of the filtered air (0.5-2 l/min) is measured with a rotameter. Before the sample is collected, 0.25 μg indium in solution is added to the electrode as an internal standard for spectroscopic analysis and dried in vacuum. The standards are prepared from solutions of suitable composition, which are evaporated to specific volumes (100-200 μl) in the electrodes. With the use of a 28 A dc arc in an argon atmosphere and an exposure of 20 sec, the following elements and their detection limits in ng are determined: Al(1), Be(0.1), Cr(0.5), Co(1), Hg(0.5), Mg(0.5), Mn(0.5), Mo(1), Ni(0.5), Pb(3), Ti(1), V(2), W(5) and Zn(3). Generally, the electrode impurities (Fe, Si, Ca, Mg, Al) are far lower than the quantities collected during sampling.

Munoz-Ribadeneira *et al.* (1974) have reviewed the collection and treatment methods for air samples to be analyzed by atomic absorption spectrometry. Generally, the same techniques can be used for emission spectroscopy or spectrometry. A few applications of multielemental analysis by emission spectrometry are reported in Table XVI, which also indicates the sample collection and treatment methods.

2.6 Analysis of Chemical and Industrial Products

Applications of emission spectroscopy in the analysis of industrial products appear to be decreasing every year in favor of atomic absorption methods. However, when multielemental analysis is involved, such as simultaneous determination of several elements, emission spectroscopy retains its interest. This is particularly true for purity controls: control of starting materials, production control and control of finished products and intermediates.

Quality control of chemicals, particularly products of "analytical grade," makes wide use of emission spectroscopy. Some applications, reported in Table XVII, involve the industrial silicates with an analysis of numerous elements, especially the rare earth oxides in which a search is made for traces of classical metals, as well as aluminum and titanium oxides. In many cases, chemical preconcentration of the trace elements is used either with separation of the matrix, such as by volatilizing silicon, or by precipitation of organic complexes. This precipitation can be carried out in the presence of graphite powder and an entraining agent.

Another field that makes wide use of emission spectroscopy is the analysis of crude oil products, in both new and used lubricating oils. Some applications are shown in Table XVIII.

The spark with a rotrode is widely used for multielemental analysis of Al, B, Cu, Cr, Fe, Mg, Mo, Ni, Pb, V and Si. Oils must be diluted

Table XVI. Spectrochemical Analysis of Air and Atmospheric Particles

Elements	Samples	Concentrations	Excitation Source	Sample Treatment	Analyte Form	References
Mn, Pb	Airborne particles	0.6-1.2 $\mu g/m^3$	dc arc 10 A	Filter on tape air sampler; put a disc of filter electrode cup.	Solid	Hasegawa et al., 1971a
Al, Cd, Hg, Ni, Pb	Air	0.1-4 $\mu g/m^3$	Spark	Filter; ash; fuse with $Li_2B_4O_7$, mix with graphite and pellet.	Solid	Morrow et al., 1971
Trace elements	Airborne particles	0.05-19 $\mu g/m^3$	dc arc	Filter on glass fiber; add NaF, mix with graphite + In and Pd (internal standards).	Solid	Hasegawa et al., 1971b
Trace elements	Airborne dirt	0.1-100 μg	Spark	Filter on paper.	Solid	Lander, 1971
Trace elements	Chimney effluents		Flame		Liquid	Rollier, 1972
Trace elements	Airborne particles	ng-$\mu g/m^3$	Arc	Collect on millipore filter; ash at low temperature; mix with In_2O_3 + Pd_2O_3 and graphite + Li_2CO_3.	Solid	Imai et al., 1973
Trace elements	Airborne particles		Arc	Collect in 5-stage Anderson sampler; extract with HNO_3; add LiCl buffer + In and Y as internal standards.	Solid	Lee et al., 1972a
Trace elements	Airborne particles	Trace levels	dc arc 10 A	Filter and mix with NaCl + graphite (+ Sr and Pd as internal standards).	Solid	Yamane et al., 1973
Trace elements	Air particles	Trace levels	dc arc	Filter through graphite electrode.	Solid	Seeley et al., 1974
Trace elements	Airborne particles	ng/cm^3	dc arc 10 A	Collect on Ag membrane; dissolve in HNO_3 + In, Co; add HCl + graphite powder; dry.	Solid	Sugimae, 1974a
Trace elements	Airborne dust	1-5000 $\mu g/g$	dc arc	Filter; dissolve membrane in acetone; decant; mix with NaF + graphite.	Solid	Sugimae, 1974b
Trace elements	Airborne dust	traces	dc arc 7 A	Filter; ash mix residue with carbon powder.	Solid	Baetsle, 1974

Table XVII Spectrochemical Analysis of Industrial Products

Elements	Samples	Concentrations	Excitation Source	Sample Treatment	Analyte Form	References
Ag, Au, Be, Bi, Cr, Cu, In, Mo, Pb, Sn, Tl, W	Silicates	ppm	dc arc 10 A in Ar/O$_2$	Mix with S + NaF.	Solid	Tripkovic et al., 1971a, 1971b
Ag, Al, Bi, Ca, Co, Cu, Fe, Mg, Mn, Ni, Pb, Sn, Ti	Cerium oxide	1-10 ppm	dc arc 15 A	Dissolve in HNO$_3$/H$_2$O$_2$; dry and dissolve in HNO$_3$/NaB$_2$O$_3$; extract in tri-butyl phosphate; evaporate with graphite powder.	Solid	Osumi et al., 1971a
Ag, Al, Au, Bi, Co, Cr, Cu	Silica	ppm	Hollow cathode lamp	Dissolve in HF/HNO$_3$; evaporate in graphite cathode.	Solid	Kusovlev et al., 1971
Al, As, B, Ca, Cd, Co, Cu, Fe, Mg, Mn, Pb, Sn	Yttria	ppm	dc arc 25 A	Ignite at 800°C; add Ga$_2$O$_3$ + AgF (2:1); load 50 mg in carrier distillation electrode.	Solid	Osumi et al., 1971b
Al, Ca, Co, Fe, Si	Magnesia		ac arc	Mix with graphite (1:9).	Solid	Florian et al., 1971
Trace elements	Alumina		dc arc 7/14 A	Mix with carbon + NaCl (1:1) + PbO$_2$ 1%.	Solid	Svangiradze et al., 1971
Trace elements	Silica	ppm	Hollow cathode lamp	Dissolve in HF/HNO$_3$; evaporate in carbon cathode.	Solid	Kusovlev et al., 1971
Al, B, Cd, Cr, Cu, Fe, In, Mg, Mn, Pb, Si, P	Th compounds	trace levels	dc arc 12 A	Ignite at 900°C; mix with AgCl or NaF carriers.	Solid	Chan et al., 1972
Trace elements	Yttria	ppm	dc arc 10 A	Mix (3:1) with carrier NaF/AgCl/graphite (1:4:5).	Solid	Naik et al., 1973
Al, Bi, Ca, Fe, Mg, Pb, Sb, Sn	Boron nitride	2-1000 ppm	dc arc 10 A	Mix (1:1) with graphite + 2% NaF (carrier) + 1% La$_2$O$_3$ (internal standard).	Solid	Vengsarkar et al., 1972

Table XVII. Continued

Elements	Samples	Concentrations	Excitation Source	Sample Treatment	Analyte Form	References
Trace elements	Titania	ppm	Arc	Mix with 10% AgCl carrier.	Solid	Zakhariya et al., 1974
Trace elements	Organic reagents	ng/g	dc arc 10 A	Dissolve in H_2O; adjust to pH 7-7.5 with NH_4OH; add carbon powder and 1% thioacetamide; filter; dry and mix with NaCl; put in electrode.	Solid	Bozhevolnov et al., 1974
Trace elements	Alkali halides	ng/g	dc arc	Dissolve; add cupferron, DDC and tartaric acid at pH 8-9; extract in $CHCl_3$; evaporate on graphite powder.	Solid	Pavlenko et al., 1974
Trace elements	Semiconductor material	traces	Laser and spark		Solid	Mohaupt et al., 1974
Trace elements	Silicon	ppb	dc arc	Dissolve in HF/HNO_3; dry; dissolve in H_2O and evaporate on electrode.	Solid	Zilbershtein et al., 1973
Trace elements	Zinc compounds	0.005-5 ppm	dc arc	Dissolve and precipitate impurities with dimethyl glyoxime, Na DDC and 1-nitroso-2-naphtol in presence of graphite powder; filter and mix with NaCl; put in electrode.	Solid	Bykova et al., 1973

Table XVIII. Spectrochemical Analysis of Petroleum Products

Elements	Samples	Concentrations	Excitation Source	Sample Treatment	Analyte Form	References
Ag, Al, Fe,	Engine oils	0-15 μg/ml	Flame H_2/A_2	Dilute in CCl_4.	Liquid	Miller et al., 1971
Ag, Al, Cu, Fe, Mg	Lub. oils		ac arc		Liquid	Lelong, 1971
Al, Cr, Cu, Fe, Mg, Pb, Sn	Lub. oils		Spark on rotrode	None.	Liquid	Leistner et al., 1971
Trace elements	Oils	trace	Arc		Liquid	Barr et al., 1972
Cu, Mn, Ni, Pb, Ti, V	Used oils	>0.01 μg/ml	ac arc 14 A	Treat with HNO_3 + hexane at 50°C; mix charred residue with graphite; heat at 500 C; put in electrode.	Solid	Kuznecova et al., 1973
Trace elements	Petroleum and bitumen	μg/g	ac arc 8 A	Add H_2SO_4; heat in stages at 600°C; treat ash with HCl; evaporate to low bulk; put 0.03 ml in NaCl/polystyrene-treated electrode.	Solid	Klemm, 1972
As, Cu, Pb	Petroleum	trace $>$ ng/g	ac arc 12 A	Evaporate in electrode cavity; add buffer (graphite + 2% NaCl) to counter electrode.	Solid	Shmulyakovskii et al., 1973
Al, Cr, Cu, Fe, Mg	Used oils		dc arc plasma jet	Dilute oil with xylene (1:1).	Liquid	MacElfresh et al., 1974
Al, B, Cu, Cr, Fe, Pb, Si	Lub. oils	ppm	Spark on rotrode	Float 0.5-ml sample on H_2O in cup of rotrode.	Liquid	Woods, 1973
Ca, Cr, Fe, Mo, Ni, V	Petroleum products	ppm	Arc	Mix with carbon powder containing 1.1% NaCl, 0.8% $SrCl_2$, 0.8% Ga_2O_3, 0.7% AgCl.	Solid	Biktimirova, 1973
Trace elements	Crude oils	ppm	dc arc	Mix oil with graphite + sulfur (4:1); put 10 mg to electrode hole	Solid	Farhan et al., 1973

with a suitable solvent, such as xylene. The most sensitive methods include the use of evaporation at a suitable temperature, in which the
residue is mixed with a spectral buffer to be analyzed in a dc arc. A
risk of losses of the volatile elements does exist.

Kuznecova *et al.* (1973) studied a method for the separation of
metals from oils. The sample is treated with nitric acid and hexane, and
asphalt compounds containing the metals are then separated, carbonized
and analyzed directly or after ashing. The Ni and V losses are reportedly
negligible.

To reduce ignition losses, Eskamani *et al.* (1973) convert the metals
(Fe, Ni, V, Pb) into sulfonates by adding magnesium sulfonate to the
sample. Sulfonation is performed at low temperature, after which the
sample is ignited at $650°C$ and analyzed. This method is also applied to
crude-oil products, particularly gasoline.

The choice of the spectral buffer is important in arc spectroscopy.
Biktimirova *et al.* (1973) showed that a buffer of $NaCl:SrCl_3:Ga_2O_3:AgCl$
(see Table XVIII) improves the sensitivity of the spectral lines of V, Ca
and Cr ten times and those of Ni and Mo five times.

Excitation with a plasma source can also be considered. MacElfresh
et al. (1974) analyzed old oils with excitation in an arc plasma (plasma
jet dc arc). The sample is diluted with xylene (1:1), and Al, Cu, Fe, Mn
and Mg are determined by spectroscopy with respective detection limits
of 0.01, 0.11, 0.08, 0.40 and 0.003 ppm.

However, one of the major problems in emission spectroscopy of
crude oils and derivatives is preparation of the standards, particularly
when the sample is excited in the liquid-organic phase. Organic solutions
must have the physical properties of the samples and a perfectly known
chemical composition with regard to the investigated elements. Oil-soluble
organometallic compounds of well-defined composition are used. Patents
have been obtained, particularly by MacKinney *et al.* (1974), and the
properties of the products proposed by these authors have been studied
by the U.S. National Bureau of Standards. Golightly *et al.* (1973) also
studied the preparation and application conditions of synthetic standards
consisting of oil-soluble organometallic compounds. Metal caprates have
been used by Hearn *et al.* (1971) and MacElfresh *et al.* (1974) to
analyze oils and other organic media by emission and atomic-absorption
spectrometry. The caprates are prepared by adding a solution of a selected
metal salt to an aqueous ammonium caprate solution. The precipitated
caprate is filtered, washed and vacuum-dried on phosphorus pentoxide.
The most common compounds are listed in Table XIX.

These compounds are soluble in oils and organic solvents. The
National Bureau of Standards has proposed individual stock solutions for

Table XIX. Organometallic Compounds

Al	Aluminum 2-ethylhexanoate
Ba	Barium cyclohexanebutyrate
B	Methyl borate
Cd	Cadmium cyclohexanebutyrate
Ca	Calcium 2-ethylhexanoate
Co	Cobalt cyclohexanebutyrate
Pb	Lead cyclohexanebutyrate
Li	Lithium cyclohexanebutyrate
Mg	Magnesium cyclohexanebutyrate
Mn	Manganous cyclohexanebutyrate
Hg	Mercuric cyclohexanebutyrate
Ni	Nickel cyclohexanebutyrate
K	Potassium erucate
Ag	Silver 2-ethylhexanoate
Na	Sodium cyclohexanebutyrate
Sr	Strontium cyclohexanebutyrate
Sn	Dibutyltin bis (2-ethylhexanoate)
Zn	Zinc cyclohexanebutyrate

oil analysis under the designation CONOSTAN. Each contains 5000 ppm of each of the following elements in the form of sulfonates: Ag, Al, B, Mg, Mn, Mo, Ba, Be, Bi, Na, Ni, P, Ca, Cd, Co, Pb, Sb, Si, Cr, Cu, Fe, Sn, Ti, V, K, La, Li, Y and Zn, as well as multielement solutions (12 and 20) covering a range of 10-900 ppm. These standards are used particularly in emission spectrometry with a spark on a rotrode, plasma source and nitrous oxide flame.

REFERENCES

Addink, N. W. H. *DC Arc Analysis* (New York: Macmillan, 1971).

Allen, W. J. F. *Anal. Chim. Acta* 59:111 (1972).

Alvarez-Herrero, C., J. A. de la Pena-Blasco, J. J. Alonso-Pascual and F. Burriel-Marti. *16th CSI*, Heidelberg (1971a).

Alvarez-Herrero, C. and F. Burriel-Marti. *16th CSI*, Heidelberg (1971b).

Anan'ev, V. S., V. A. Sidorov and L. N. Zaichenko. *Trudy tsent. nauchno-issled. gornorazved. Inst. tsvet. redk. blagorod. Metall.* (1972), p. 171.

Antic, E., P. Caro and G. Schiffmacher. *17th CSI*, Florence, Italy (September 1973).

Avni, R., A. Harel and I. B. Brenner. *Appl. Spectrosc.* 26:641 (1972).

Baetsle, L. H. and N. Kemp. *Serving Science and Industry* (New York: Philips, 1974).

Barr, D. R. and H. J. Larson. *Appl. Spectrosc.* 26:51 (1972).

Besnus, Y. and R. Rouault. *Analusis* 2:111 (1973).
Biktimirova, T. G. and L. G. Mashireva. *Zav. Lab.* 39:1086 (1973).
Borovik-Romanova, T. F. and E. A. Belova. *Zh. Anal. Khim.* 28:1828 (1973).
Bowen, B. C. and H. M. Foote. *Petro. Rev.* 28:680 (1974).
Bozhevolnov, E. A., I. N. Bykova, T. G. Manova and V. S. Silakova. *Zav. Lab.* 40:170 (1974).
Broadhead, K. G., B. C. Piper and H. H. Heady. *Appl. Spectros.* 26:461 (1972).
Burridge, J. C. and R. O. Scott. *17th CSI*, Florence, Italy (September 1973).
Busch, K. W., N. G. Howel and G. H. Morrison. *Anal. Chem.* 46:575 (1974).
Bykova, I. N., T. G. Manova, V. G. Silakova and G. P. Boznyakova. *Zh. Anal. Khim.* 28:1481 (1973).
Chan, K. Y., Y. Lu and T. P. Tam. *Chemistry, Taipe* 17 (1972).
Chaplin, M. H. and A. R. Dixon. *Appl. Spectros.* 28:5 (1974).
Christian, G. D., C. E. Matkovich and L. W. Schertz. *U.S. Nat. Tech. Inform. Serv. PB Rept.* No. 207867 (1971); *Chem. Abst.* 77:79405g (1972).
Cowgill, U. M. *Appl. Spectros.* 27:5 (1973).
Croft, P. E. A. *16th CSI*, Heidelberg (1971).
Czakow, J. *Chem. Anal.* 18:891 (1973).
De Boer, F. J. and P. W. J. M. Boumans. *17th CSI,* Florence, Italy (September 1973).
Delavault, R. E. and D. B. Marshall. *Can. J. Spectros.* 18:10 (1973).
Dothie, H. J. *16th CSI*, Heidelberg (1971).
Ecrement, F. *Met. Phys. Anal.* 7(2):128 (1971).
Eskamani, A., M. S. Vigler, H. A. Strecker and N. R. Anthony. *4th International Conference on Atomic Spectroscopy,* Toronto, Ontario, Canada (October-November, 1973).
Farhan, F. M. and M. Makhani. *Analusis* 1(1):46 (1972).
Farhan, F. M. *Analusis* 2:49 (1973).
Farhan, F. M. and H. Pazandeh. *32nd Congress of GAMS,* Paris (December 1974).
Fassel, V. A. and R. N. Kniseley. *Anal. Chem.* 46:1110A (1974).
Finkin, K. Z., Y. M. Dubosarskaya and L. S. Kolosova. *Zav. Lab.* 4:431 (1971).
Florian, K., V. Jurcikova and M. Martherny. *Chemicke Zvesti* 25:421 (1971).
Folsom, T. R., N. Hansen, G. J. Parks and W. E. Weitz. *Appl. Spectros.* 28:345 (1974).
Gerasimov, K. S., E. S. Kostyukova and Y. D. Raikhbaum. *Zav. Lab.* 40:392 (1974).
Giacobini, E., S. Novmark and M. Stepita-Klacko. *Acta Physiol. Scand.* 80:528 (1970).
Golightly, D. W. and J. L. Weber. National Bureau of Standards (U.S.) Technical Note No. 751 (January 1973).
Golubeva, E. D. *Zh. Prikl. Spektrosk.* 17:567 (1972).
Gordon, W. A., K. M. Hambidge and M. L. Franklin. *Prog. Anal. Chem.* 6:23 (1973).

Gorvat, F. Yu., Yu. S. Ryabukhkin and V. N. Letov. *J. Anal. Chem. USSR* 26:1975 (1971).
Govindaraju, K. *Analusis* 1(1):40 (1972).
Govindaraju, K. 4th International Conference on Atomic Spectroscopy, Toronto, Ontario, Canada (October-November 1973).
Greenfield, S. and P. B. Smith. *Anal. Chim. Acta* 59:341 (1972).
Grove, E. L. *Analytical Emission Spectroscopy* (New York: Dekker, 1971).
Guselnikov, A. A. and A. K. Rusanov. *Zh. Prikl. Spektrosk.* 15:11 (1971).
Hambidge, K. M. *Anal. Chem.* 43:103 (1971).
Harding-Barlow, I. and R. C. Rosan. In *Microprobe Analysis,* C. A. Andersen, Ed. (New York: Wiley-Interscience, 1973), Chapter 14.
Hasegawa, T. and A. Sugimae. *Japan. Anal.* 20:840 (1971a).
Hasegawa, T. and A. Sugimae. *Japan. Anal.* 20:1406 (1971b).
Hearn, W. E., R. A. Mostyn and B. Bedford. *Anal. Chem.* 43:1821 (1971).
Hildon, M. A. and W. J. F. Allen. *Analyst* 96:480 (1971).
Imai, S., K. Ito, A. Hamaguchi, Y. Kusaka and M. Warashina. Bunseki Kagaku *(Japan Analyst)* 22:551 (1973).
Isaac, R. A. and J. B. Jones. Conference, Pittsburg (1972).
Janssens, M. and R. Dams. *Anal. Chim. Acta* 65:41 (1973).
Jones, J. B. 88th Annual Meeting of the Association of Official Analytical Chemists, Washington, D.C. (October 1974).
Jones, J. B., F. Brech and R. L. Crawford. 88th Annual Meeting of the Association of Official Analytical Chemists, Washington, D.C. (October 1974).
Kantor, T., L. Polos and L. Bezur. *Magy. Kem. Lap.* 27:313 (1972).
Khitrov, V. G. and G. E. Belousov. *Zh. Prikl. Spektrosk.* 14:5 (1971).
Khitrov, V. G. and G. E. Belousov. *J. Anal. Chem., USSR* 27:1223 (1972a).
Khitrov, V. G. and G. E. Belousov. *Zh. Anal. Khim.* 27:1357 (1972b).
Kirkbright, G. F., A. F. Ward and T. S. West. *Anal. Chim. Acta* 62:241 (1972).
Klemm, W. *Z. Angew. Geol.* 18:202 (1972).
Kniseley, R. N., V. A. Fassel and C. Butler. *Clin. Chem.* 19:807 (1973).
Korsun, V. I., E. A. Pometun and D. N. Pachadzhanov. *Zh. Prikl. Spektrosk.* 17:872 (1972).
Kusovlev, I. A., V. L. Sabatovskaya, O. A. Sverdlina, L. S. Horkina, M. S. Vinogradova and N. M. Kuzmin. *Zav. Lab.* 9:1071 (1971).
Kuznecova, A. P., Z. I. Otmahova and G. A. Kataev. *Zav. Lab.* 39:957 (1973).
Laktionova, N. V., L. V. Ageeva and L. V. Simonova. *Zh. Anal. Khim.* 26:554 (1971).
Laktionova, N. V., L. V. Ageeva and A. V. Karyakin. *Zh. Anal. Khim.* 27:2358 (1972).
Lander, D. W., R. L. Steiner, D. H. Anderson and R. L. Dehm. *Appl. Spectros.* 25(2):270 (1971).
Lebedinskaya, M. P. and V. T. Chuiko. *Zh. Anal. Khim.* 28:2413 (1973).
Lee, R. E. and S. S. Goranson. *Environ. Sci. Technol.* 6:1019 (1972a).
Lee, R. E., S. S. Goranson, R. E. Enrione and G. B. Morgan. *Environ. Sci. Technol.* 6:1205 (1972b).
Liese, H. C. *Appl. Spectrosc.* 28:135 (1974).

Leistner, C. J. and L. A. Dugas. *16th CSI*, Heidelberg (1971).
Lelong, H. *16th CSI*, Heidelberg (1971).
Le Roy, V. M. and A. J. Lincoln. *Anal. Chem.* 46:369 (1974).
Lichte, F. E. and R. K. Skogerboe. *Anal. Chem.* 44:1480 (1972).
Lopez, P. L. and E. R. Graham. *Soil Sci.* 115:380 (1973).
MacElfresh, P. M. and M. L. Parsons. *Anal. Chem.* 46:1021 (1974).
MacKinney, C. N. and W. K. Pollard. British Patent 1,345,682 (1974).
Marich, K. W., P. W. Carr, W. J. Treytl and D. Glick. *Anal. Chem.* 42:
    1775 (1970).
Mays, R. E., G. K. Czamanske and C. Heropoulos. *Appl. Spectros.* 28:462
    (1974).
Miller, R. L., L. M. Fraser and J. D. Winefordner. *Appl. Spectros.* 25(4):
    477 (1971).
Mitchell, R. L. *Spectrographic Analysis of Soils, Plants and Related Ma-
    terials* (London: Commonwealth Bureau of Soils Science, 1945).
Mitteldorf, A. T. and D. O. Landon. *Appl. Spectrosc.* 10:12 (1956).
Mika, J. and T. Torok. *Analytical Emission Spectroscopy*, Vol. 1
    Fundamentals (London: Butterworth, 1973).
Moenke-Blankenburg, L., H. Moenke, K. Wiegand, W. Quillfeldt, J. Mohr,
    W. Grassme and W. Schron. *17th CSI*, Florence, Italy (September 1973).
Mohaupt, G. and G. Patzmann. *Jena Rev.* 252 (1974).
Moroshkina, T. M. and L. V. Vanaeva. *Zh. Anal. Khim.* 25:2374 (1970).
Morrow, N. L. and R. S. Brief. *Environ. Sci. Technol.* 5:786 (1971).
Mosier, E. L. *Appl. Spectros.* 26:636 (1972).
Munoz-Ribadeneira, F. J., M. L. Nazario, and A. Vega. 7th Materials
    Research Symposium, USA (October 1974).
Naik, R. C. and P. D. Kornik. *Z. Anal. Chem.* 265:349 (1973).
Naqueda, C. *17th CSI*, Florence, Italy (September 1973).
Niedermeier, W., J. H. Griggs and R. S. Johnson. *Appl. Spectrosc.* 25(1):
    53 (1971).
Niedermeier, W., J. H. Griggs and R. J. Webb. *Appl. Spectros.* 28:1 (1974).
Nova-Spackova, A. *16th CSI*, Heidelberg (1971).
Nova, A. *17th CSI*, Florence, Italy (September 1973).
Omenetto, N., L. M. Fraser and J. D. Winefordner. *Appl. Spectros. Rev.*
    7(2):147 (1973).
Osumi, Y., A. Kato and Y. Miyake. *Z. Anal. Chem.* 255:103 (1971a).
Osumi, Y., A. Kato and Y. Miyake. *Japan Analyst* 20:1393 (1971b).
Pavlenko, L. I., O. M. Petrukhin, Y. A. Zolotov, A. V. Karyakin, G. N.
    Gavrilina and I. E. Tumanova. *Zh. Anal. Khim.* 29:933 (1974).
Pepin, D., A. Gardes, J. Petit, J. A. Berger and G. Gaillard. *Analusis* 2:337
    (1973a).
Pepin, D. and A. Gardes. *Analusis* 2:549 (1973b).
Petrakiev, A. and L. Georgieva. *17th CSI*, Florence, Italy (September 1973).
Petrov, B. I., Y. A. Makhnev and V. P. Zhivopistsev. *Zh. Anal. Khim.*
    28:911 (1973).
Pinta, M. *Detection and Determination of Trace Elements*, 5th ed.
    (Ann Arbor, Michigan: Ann Arbor Science Publishers, 1973).
Prakash, N. J. and W. W. A. Harrison. *Anal. Chim. Acta* 53:421 (1971).
Presley, B. J., Y. Kolodny, A. Nissenbaum and I. R. Kaplan. *Geochim.
    Cosmochim. Acta* 36:1073 (1972).

Prudnikov, E. D. and Y. S. Shapkina. *Zh. Anal. Khim.* 28:1257 (1973).
Rollier, M. A. *Chim. Ind.*, Milan 54:895 (1972).
Rubeska, I. and M. Miksovsky. *Chemicke Listy* 67:213 (1973).
Sahini, V. E., M. Craiu and E. Ivana. *Rev. Roum. Chim.* 19:165 (1973).
Savinova, E. N., A. V. Karyakin and T. P. Andrccva. *Zh. Anal. Khim.* 27:777 (1972).
Seeley, J. L. and R. K. Skogerboe. *Anal. Chem.* 46:415 (1974).
Sermin, M. *Analusis* 2:186 (1973).
Shapkina, Y. S. and E. D. Prudnikov. *Univ. Geol. Geogr.* 18:116 (1972).
Shmulyakovskii, Y. E., A. A. Baibazarov, F. P. Khapaeva, T. K. G. Biktimirova, and L. M. Zamilova. *Khimya Tekhnol. Topl. Masel* 18:55 (1973).
Slavin, M. *Emission Spectrochemical Analysis* (New York: Wiley-Interscience, 1971).
Smirnov, V. I., N. I. Eremin, V. E. Kel'ch, V. M. Okrugin and N. E. Sergeeva. *Jena Rev.* 240 (1974).
Snodgrass, R. A. *J. S. Afr. Chem. Inst.* 25:268 (1972).
Steiner, R. L. and D. H. Anderson. *Appl. Spectros.* 26:41 (1972).
Sugimae, A. *Appl. Spectros.* 28:458 (1974a).
Sugimae, A. *Anal. Chem.* 46:1123 (1974b).
Svangiradze, R. R., I. L. Visokova, T. A. Mozgovaya and O. A. Petrova. *Zav. Lab.* 4:430 (1971).
Theuer, R. C., A. W. Mahoney and H. P. Sarett. *J. Nutr.* 101:525 (1971).
Timperley, M. H. New Zealand Institute of Chemistry Annual Conference (August 1974).
Treytl, W. J., J. B. Orenberg, K. W. Marich, A. J. Saffir and D. Glick. *Anal. Chem.* 44:1903 (1972).
Tripkovic, M., L. Amirshahi and M. Dmitrovic. *16th CSI*, Heidelberg (1971a).
Tripkovic, M. and N. Cosovic. *16th CSI*, Heidelberg (1971b).
Vengsarkar, B. R., I. J. Machado and S. K. Malhotra. Rep. Bhadha Atom. Res. Centre, Barc-632 (1972).
Wark, W. J. *Anal. Chem.* 26:203 (1954).
Webb, J., W. Niedermeier, J. H. Griggs and T. N. James. *Appl. Spectros.* 27:342 (1973).
Winefordner, J. D., V. Svoboda and L. J. Cline. C. R. C. Critical Reviews, in *Anal. Chem.* (August 1970).
Wincfordner, J. D. *Advances in Analytical Chemistry and Instrumentation Series*, Vol. 9 (New York: Wiley-Interscience, 1971).
Winge, R. K., V. A. Fassel, R. N. Kniseley and W. L. Sutherland. Symposium of Water Quality Parameters, Ontario, Canada (November 1973).
Wittmann, A. *Cent. Doc. Sider. Circ. Inf. Tech.* 29:1063 (1972).
Woodriff, R. and J. F. Lech. *Anal. Chem.* 44:1323 (1972).
Woods, H. P. *Appl. Spectros.* 27:490 (1973).
Yamane, Y., M. Miyazaki and H. Nakazawa. *Bunseki Kagaku (Japan Analyst)* 22:1135 (1973).
Yudelevich, I. G., V. L. Kustas, G. V. Lazebnaya, A. F. Fedyashina and G. V. Poleva. *16th CSI*, Heidelberg (1971).
Yudelevich, I. G., A. S. Cherevko and N. G. Skobelkina. *J. Anal. Chem., USSR* 27:1982 (1972).

Zakhariya, N. F. and O. P. Trulina. *Zh. Anal. Khim.* 29:1170.

Zdrojewski, A., N. Quickert, L. Dubois and J. L. Monkman. *Int. J. Environ. Anal. Chem.* 2:63 (1972).

Zilbershtein, K. I., S. S. Legeza and M. P. Semov. *Zh. Anal. Khim.* 28:1323 (1973).

CHAPTER 4

FLAME ATOMIC ABSORPTION SPECTROMETRY

1. GENERAL REMARKS

1.1 Introduction

Under certain conditions every chemical substance can absorb the radiation it emits itself. This principle, which summarizes the Kirchhoff law formulated in 1859, interested chemists only 100 years later when Walsh (1955) in Australia and Alkemade (1969) in Europe established the principle for the determination of a chemical element by means of its atomic absorption spectrum.

The "reversal of spectral lines" (self absorption) was a well-known phenomenon to the spectrographic analyst: the intensity I of the spectral lines in an excitation source increases with the atomic concentration C, but the curve I = f(C) (Figure 1) ceases to be linear. When C increases, I passes

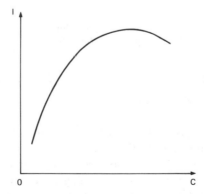

Figure 1. Reversal of spectral lines. Variation of intensity I
as a function of concentration C.

133

through a maximum and then decreases. This is the line reversal also observed in the electric arc, in flames and in the spark. It is particularly marked with atoms such as magnesium, zinc, sodium and lithium.

Since the first papers by Walsh in 1955, atomic absorption has undergone considerable development. More than 1000 papers discussing it are published every year, and the method has appeared in all branches of industry. We will briefly review the basic principles of this technique, the instruments, and finally some of the applications.

1.2 The Wood Experiment—Resonance Lines

When an excited atom changes from one energy level E_h to a lower level E_b, it emits a spectral radiation of frequency ν, and the energy variation E_h-E_b is determined in the form of a quantum of radiative energy $h\nu$ (Figure 2):

$$E_h\text{-}E_b = h\nu \tag{1}$$

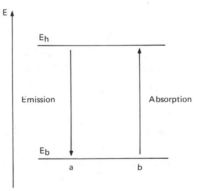

Figure 2. Emission-absorption. (a) Emission of a line of frequency $\nu = (E_h\text{-}E_b)/h$. (b) Absorption of the same line.

Inversely, when an atom in the neutral (or ground) state E_b is exposed to radiation of frequency ν, it can absorb a quantum $h\nu$ and pass into state $E_h = E_b + h\nu$.

When an atomic gas is subjected to any excitation, its atoms are in their normal lowest energy state $E_b = 0$. The only lines that can be absorbed under these conditions are those corresponding to an electron transition during the emission process, leading to the lowest energy level or ground state. The Wood experiment illustrates the phenomenon as shown schematically in Figure 3.

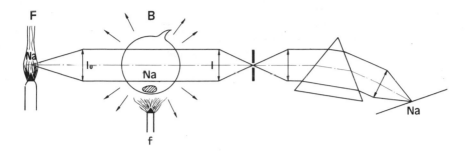

Figure 3. The emission-absorption phenomenon. Wood experiment. (Reproduced from Pinta, 1971, by permission of Masson et Cie, Ed.)

A flask contains a lump of sodium at low pressure, which is vaporized by heating on flame f. If the flask is then exposed to the flux emitted by flame F containing sodium [which emits the doublet D (589.0/6 nm)], the sodium in the flask then emits a yellow light in all directions. This is called the fluorescence emission. The sodium atoms in the flask, which are initially neutral, absorb a quantum $h\nu$ of exciting energy and change from the ground state E_b to the energy level E_h, resulting in resonance of the sodium atoms. The exciting line is called the resonance line. The second observation that can be made in this experiment is that the exciting line is absorbed by neutral sodium atoms in the flask. If the line is studied with a spectrometer before vaporizing the sodium in the flask, the intensity of the line observed in the spectrometer is I_0. When the flask is heated, the intensity I_0 of the line decreases and becomes I. This is the reversal phenomenon $(I < I_0)$ or atomic absorption.

1.3 Atomic Absorption and Chemical Analysis

We will consider an atomic gas consisting of a population of atoms in thermal equilibrium capable of emitting a radiation of frequency ν, *i.e.*, of absorbing a quantum of energy $h\nu$, and we will assume that a parallel energy beam containing a radiation of frequency ν and intensity I_0 passes through the atomic gas (Figure 4). The absorption of the incident radiation by the neutral atoms is

$$A = \log \frac{I_0}{I} \tag{2}$$

This expression is called the absorbance.

Absorption occurs in the direction of the incidence radiation. The resulting emission does not compensate this absorption because it is omnidirectional. Finally, the absorption can be measured by classical

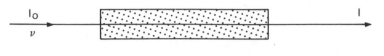

$$N = N_O + N_1$$

Figure 4. Absorption of a radiation by an atomic population.

spectrophotometric procedures. It permits the quantitative determination of the elements introduced into the atomizing source.

A relation exists between I_0 and I that depends on the absorption:

$$I = I_O \, e^{-k\nu N_O l} \tag{3}$$

or:

$$\log \frac{I_O}{I} = k_\nu N_O l$$

The absorption coefficient is k_ν at frequency ν of an atomic population with a concentration of N_O atoms per cm^3 in path.

1.4 Thermodynamic Equilibria; the Boltzmann Law

In the flame the total number of atoms N is distributed into N_O atoms in the ground state and N_1 excited atoms: $N = N_O + N_1$. It can be assumed that thermodynamic equilibrium exists in the flame because the atoms and molecules all have the same mean velocity. The Boltzmann distribution law applies and can be written as follows:

$$\frac{N_1}{N_O} = \frac{g_1}{g_O} \, e^{-\frac{E_1}{KT}} \tag{4}$$

In this formula, g_1 and g_O are the statistical weights of the excited and ground atomic states ($g = 2J + 1$, where J is the internal quantum number), E_1 is the excitation energy, K the Boltzmann constant and T is the absolute temperature. We should specify that the statistical weight is the number of possible coincident states of the same energy of an atom in excitation state i characterized by the three quantum numbers n, L_i, J_i.

A few values of the ratio N_1/N_O at 2000, 2500 and 3000°K are listed in Table I (Mavrodineanu *et al.,* 1965). A table in the appendix also contains the excitation energies (or potentials) of the principal resonance lines of the most common elements.

Table I. Values of the Ratio of Excited Atoms/Total Atoms (N_1/N_0) at Thermal Equilibrium as a Function of the Flame Temperature

Chemical Species	nm	g_1/g_0	Excitation Energy (in eV)	N_1/N_0		
				T = 2000°K	T = 2500°K	T = 3000°K
Na	589.0	2	2.10	0.99×10^{-5}	1.14×10^{-4}	5.83×10^{-4}
Ba	553.6	3	2.24	6.83×10^{-6}	3.19×10^{-5}	5.19×10^{-4}
Sr	460.7	3	2.69	4.99×10^{-7}	11.32×10^{-6}	9.07×10^{-5}
V	437.9	–	3.13	6.87×10^{-9}	2.50×10^{-7}	2.73×10^{-6}
Ca	422.7	3	2.93	1.22×10^{-7}	3.67×10^{-6}	3.55×10^{-5}
Fe	372.0	–	3.33	2.29×10^{-9}	1.04×10^{-7}	1.31×10^{-6}
Co	352.7	–	3.51	6.03×10^{-10}	3.41×10^{-8}	5.09×10^{-7}
Ag	338.3	1	3.66	5.85×10^{-10}	4.11×10^{-8}	6.99×10^{-7}
	328.1	2	3.78	6.03×10^{-10}	4.84×10^{-8}	8.99×10^{-7}
Cu	324.7	2	3.82	4.82×10^{-10}	4.04×10^{-8}	6.55×10^{-7}
Mg	285.2	3	4.35	3.35×10^{-11}	5.20×10^{-9}	1.50×10^{-7}
Pb	283.3	3	4.375	2.83×10^{-11}	4.55×10^{-9}	1.34×10^{-7}
Au	267.6	1	4.63	2.12×10^{-12}	4.60×10^{-10}	1.65×10^{-8}
Zn	213.9	3	5.795	7.45×10^{-15}	6.22×10^{-12}	5.50×10^{-10}

Thus, it will be noted that the number of excited atoms is very small compared to the total number of atoms. Practially all atoms are in their electronic ground state, $N_0 \cong N$, particularly in low-temperature flames (the air-butane flame gives 2200°K and the oxygen-hydrogen flame 2900°K).

1.5 Ionization in the Flame

1.5.1 Ionizable Metals

Some metals, particularly the alkali and alkaline-earth metals, have relatively low ionization potentials (the necessary energy to remove one electron from the atom) and consequently form ionized atoms in addition to neutral atoms in the flame. Generally, the production of ionized atoms compared to neutral atoms increases with the flame temperature.

Equilibrium exists between the neutral atoms M_0 and the corresponding ions M^+:

$$M_0 = M^+ + e^-$$

This is the ionization equilibrium or the SAHA equilibrium governed by the law of mass action and defined by:

$$k_s = \frac{[M^+]\,[e^-]}{[M_o]}$$

with k_s = ionization constant. We will return later to the ionization equilibria in the flame and their consequences in atomic absorption. Note that the ionization phenomena reduce the number of atoms in the ground state further, the effect increasing with higher flame temperature. This results in a reduction of atomic absorption. Table II shows the influence of the flame temperature on the ionization of alkali and alkaline-earth metals (Amos and Willis, 1966). This phenomenon is reduced by introducing not only the analyzed element but another ionizable element in large amounts, *i.e.*, one capable of furnishing free electrons captured by the ionized atoms of the measured element. These will then return to the ground state.

Table II. Percentage of Ionized Atoms

	Ionization Potential (eV)	Air-Propane Flame (2200°K)	Oxygen-Hydrogen Flame (2900°K)	Nitrous Oxide-Acetylene Flame (3200°K)
Lithium	5.37	0.01	1	68
Sodium	5.12	0.3	5	82
Potassium	4.32	2.5	31	98
Rubidium	4.16	13.5	44	99
Cesium	3.87	28.3	69	100
Calcium	6.11	–	1	43
Strontium	5.69	–	2.7	71
Barium	5.21	–	8.6	92

1.6 Practical Determination of Atomic Absorption

We have seen that if I_0 and I are the incident and transmitted intensities of an absorbable line of frequency ν through a population of atoms of thickness 1 in a concentration $c = N_0$ absorbing frequency ν, we have the relation of Equation 3:

$$I = I_0 e^{-k_\nu\, cl}$$

where k_ν is the absorption coefficient of the atomic medium at frequency ν. (This expression is analogous to that describing the absorption of molecules in solution.) The value of k_ν depends neither on the intensity I_0 nor on the thickness 1, but is a function only of the nature of the medium and the frequency ν.

If the variation of k_ν is recorded as a function of ν (or of λ), a curve like that in Figure 5 will be produced. The curve, which is also similar to the molecular absorption curves and characterizes the absorption line, has a sharp maximum at the resonance radiation frequency ν_0 of the absorbing atoms. The value $\Delta\nu = AB$, measured at midheight of the ordinate to the maximum, characterizes the absorption line width and is commonly called the half-width of the absorption line. The shape of this curve can be compared to that representative of the profile of an emission line of intensity I defined by the variation of I_ν as a function of ν with a maximum at frequency ν_0.

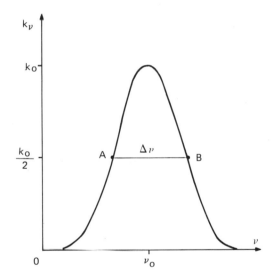

Figure 5. Variation of the atomic absorption coefficient k_ν as a function of frequency ν.

Since an atomic absorption line, like the emission lines, is a monochromatic phenomenon, it should be expressed on the graph by an infinitely flattened curve, *i.e.,* one with a zero width. At least this is what one might expect with the use of a spectrograph having a theoretical monochromatic resolution. An improvement of the dispersion and spectral resolution of the spectrophotometer apparently reduces the value $\Delta\nu$ but in no case are infinitely small values or zero obtained. In fact, the spectral lines (emission) have a natural width, d, that is not zero, resulting from the fact that every atom can have an energy different from the most probable mean energy at any moment.*

This formula leads to line widths approximately 10^{-5} nm (0.0001 Å), an entirely theoretical value because in practice several causes of line broadening interfere: the Lorentz effect (pressure resulting from molecular collisions), the Doppler-Fizeau effect (relative velocity of atoms), Stark effect (influence of the electrical field), Zeeman effect (influence of the magnetic field), and absorption by neutral atoms. In practice, the line width is approximately 0.002-0.006 nm. The same applies to the absorption lines, which attain widths of 0.001-0.01 nm.

The preponderant causes of line broadening are the Doppler and Lorentz effects, while the other causes are much less important. Table III lists some absorption line widths with Doppler broadening $\Delta\lambda_D$ and Lorentz broadening $\Delta\lambda_L$ at different temperatures (according to Mavrodineanu *et al.*, 1965).

Table III. Absorption Line Widths

Element	Atomic Mass	λ nm	T = 2000°K		T = 3000°K	
			$\Delta\lambda_D$	$\Delta\lambda_L$	$\Delta\lambda_D$	$\Delta\lambda_L$
Na	23	589.6	0.0039	0.0032	0.0048	0.0027
Ca	40	422.7	0.0021	0.0021	0.0026	0.0012
Sr	88	460.7	0.0016	0.0026	0.0019	0.0021
Ba	137	553.6	0.0015	0.0032	0.0018	0.0026
Cu	63	324.7	0.0013	0.0009	0.0016	0.0007
Ag	108	328.1	0.0010	0.0015	0.0012	0.0013
Fe	56	372.0	0.0016	0.0013	0.0019	0.0012
Co	59	352.7	0.0013	0.0016	0.0016	0.0013

The consequence of absorption line broadening is that if the atoms are in thermal equilibrium at the flame temperature T, it can be demonstrated that the absorption coefficient k_ν is a function of ν (or λ):

$$k_\nu = k_o \, e^{-\left[\frac{\lambda-\lambda_0}{\Delta\lambda_D} \, 2\sqrt{Ln\,2}\right]^2}$$

where Ln is the symbol of the Neperian logarithm. Furthermore, $\Delta\lambda_D$ is related to T and the atomic mass M of the element by the formula:

---

*$d = \lambda^2/Tct$, where c = speed of light, t = lifetime of the atom in its mean energy state.

$$\Delta\lambda_D = 1.67 \, \frac{\lambda}{c} \, \sqrt{\frac{2\,RT}{M}} = 7.16 \times 10^{-7} \, \lambda_0 \, \sqrt{\frac{T}{M}}$$

where c is the speed of light and R the gas constant. From these two expressions we obtain the result that $\Delta\lambda_D$ is proportional to $\sqrt{T}$. Therefore, k_ν, the value measured in atomic absorption, is proportional to $1/\sqrt{T}$. Thus, temperature variations are relatively unimportant in absorption, much less so than in direct emission in which the emitted intensity is proportional to the number of excited atoms N^*, *i.e.*, to $e^{-E/kT}$ (Equation 4).

1.7 Measurement of the Absorption Coefficient

1.7.1 Integrated Absorption Coefficient

The total area between the curve (Figure 5) and the abscissa de-fining the atomic absorption, resulting from the population of investigated neutral atoms, is the integrated absorption coefficient: the integral $\int k_\nu \, d\nu$ which is a measure of the number of absorbing atoms $\int k_\nu \, d\nu = K \cdot N_0$. K is a numerical coefficient defined by the absorption line under consideration.

Unfortunately, $\int k_\nu$ can be measured only with spectrometers of very high resolution, which cannot be used for routine analysis.

1.7.2 Central Absorption Coefficient

Rather than determining the integral $\int k_\nu \cdot d\nu$, the absorption co-efficient k_0 at the curve maximum is measured. The relation is demonstrated:

$$k_0 = b \, \frac{2}{\Delta\nu} \cdot k\nu \, d\nu$$

where b is the numerical factor depending on the physical conditions, particularly the Doppler and Lorentz widths of the absorption lines and the line oscillator strength. Therefore

$$k_0 = b \, \frac{2}{\Delta\nu} \cdot K \, N$$

or

$$k_0 = K' \cdot N$$

1.7.3 Practical Measurement

Consequently, k_0 is measured by transmitting monochromatic radiation of frequency λ with the smallest possible width through the atomic gas. For this purpose, a radiation of frequency ν with a width smaller than 0.002 nm should be used.

Starting with a continuous polychromatic source, this spectral resolution is obtained only with spectrographs with a resolving power of 100-200,000. These are exceptional instruments; the dispersion of a conventional monochromator is 0.01-0.02 nm at 300 nm (resolving power $\simeq$ 20,000). The record of the transmitted flux I_0/I is shown schematically in Figure 6 for different dispersions.

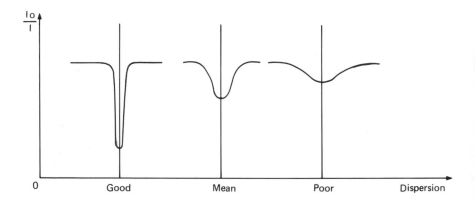

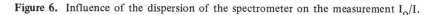

Figure 6. Influence of the dispersion of the spectrometer on the measurement I_0/I.

In practice, one starts with a radiation source producing a resonance line of the element to be analyzed. The profile of the incident line is shown in Figures 7 and 8, curve 1 and that of the absorbed line is given by curve 2. The minimum of this curve is more pronounced if the line width $\Delta\nu$ is the smaller (Figure 7). Thus, an absorption measurement at

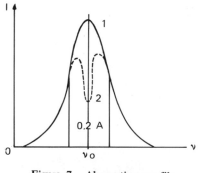

Figure 7. Absorption profile of a narrow line.

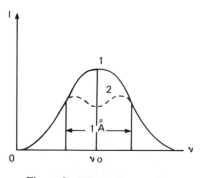

Figure 8. Absorption profile of a broad line.

frequency ν_o amounts to determining the above-defined central absorption coefficient k_o.

1.7.4 Choice of Incident Radiation

While the Kirchhoff law states that an absorption at the same wavelength should correspond to every emitted radiation, the range of application is practically limited to a small number of lines. Thus, a radiation produced by an electron transition $E_{i \to j}$ is capable of absorbing only the atoms emitting the line corresponding to the same transition.

An atom can exist in different well-defined energy states, E_i, the lowest level of which is called the "ground state," which has zero energy, $E_o = 0$. The resonance lines result from the transition into the ground state. According to Equation 1, the frequency is:

$$\nu = \frac{E_i - E_o}{h}$$

In a given atomizing source, we generally have a distribution of N atoms present as N_o atoms in the ground state, $N_1, N_2, \ldots, N_i, N_j \ldots$ atoms at excitation levels $E_o, E_1, E_2, \ldots, E_i, E_j \ldots$ The ratio of atoms in the ground state to atoms in state E_j is N_j/N_o and is given by Equation 4.

When atomic absorption is used for chemical analysis, it is important to work with the largest possible number of atoms. For example, in conventional flames (air-acetylene), it was seen that the atoms in the ground state were predominant. Consequently, the measurement must be made on the basis of the resonance lines (ground state). This rule is not always applicable. In fact, in some cases (Table II), the percentage of ionized atoms may be large for the alkaline-earth metals, particularly in the nitrous-oxide-acetylene flame. Thus, for barium, which has 42% ionized atoms, the atomic absorption determination can be made either on the basis of the resonance line λ 553.5 nm absorbed by the atoms in the ground state or on the basis of the ion line λ 455.4 nm that can be absorbed by the ionized atoms.

However, under conventional operating conditions, the absorption measurements concern atoms in the ground state. In other words, the resonance lines are used as incident radiations. A given atom can generally produce several resonance lines or several doublets (Table IV).* Each of these lines corresponds to an electron transition leading to the zero-energy ground state. Evidently, the radiations with the lowest excitation potentials will be absorbed the most. Thus, for a given element, a lower detection limit is obtained from resonance lines having the lowest excitation energy. The line Na 589 nm allows the detection of 0.005 μg Na/ml and the line Na 330 nm will reveal

*See the appendix for the table of resonance lines of atoms.

Table IV. Resonance Lines and Excitation Potential

Wavelength	nm	Excitation Potential (eV)	Detection Limit (µg/ml)
Na	589.0/589.6	2.1	0.005
	330.3/330.2	3.7	0.5
Li	670.8	1.8	0.005
	323.3	3.8	2.00
	274.1	4.5	
Ca	422.7	2.9	0.05
	396.9	9.4	
	327.4	3.8	0.2
	324.7	3.8	0.1

0.5 µg Na/ml. The two resonance lines of lithium, 670.1 nm and 323.3 nm, allow detection limits of 0.005 and 2 µg Li/ml when an aqueous solution is vaporized in the flame.

Despite everything, then, the resonance lines are not the only useful ones. Other lines also show absorption phenomena. Under certain energy conditions, especially in the flames, a relatively large population of atoms can exist in a nonzero energy state, E_i, corresponding to a level near the ground state. The atomic absorption measured from such atoms often is considerable, as is the case for iron. See, for example Table V, which shows nonresonance lines giving rise to sensitivities higher than those obtained with resonance lines.

For another example: to determine silicon, the 251.6 nm line corresponding to an electron transition not leading to the ground state is used because the principal resonance line (212.4 nm) is quite weak in emission as

Table V. Iron Lines that can be Used in Atomic Absorption with the Corresponding Energy Level and Excitation Potential

λ nm	Energy Level in Kayser cm^{-1}	Excitation Potential (eV)	Detection Limit in Air-Acetylene Flame (µg/ml)
372.0	0-26875	3.3	1
271.9	0-36767	4.6	0.4
252.7	416-39970	4.9	0.6
252.3	0-39626	4.9	0.2
248.8	416-40594	4.6	0.2

well as absorption under our experimental conditions. The Si 251.6 nm line permits the detection of 0.5 μg Si/ml. However, as a general rule and when high sensitivity is required in atomic absorption, the resonance lines resulting from electron transitions to the ground state are most frequently used.

2. PRACTICAL INSTRUMENTS

2.1 Absorption Spectrophotometer

The most sensitive atomic absorption measurements are realized in practice from the resonance lines produced in hollow cathode lamps associated with monochromators of mean dispersion (10-20,000). The instrument is shown schematically in Figure 9. The analyzed substance is atomized in a

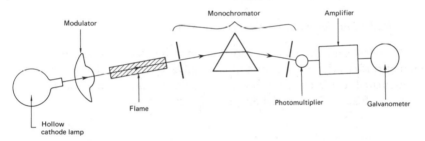

Figure 9. Diagram of an atomic absorption spectrometer.
(Reproduced from Pinta, 1971, by permission of Masson et Cie, Ed.)

flame of air-propane, air-acetylene, nitrous oxide-acetylene, as necessary. The radiation transmitted through the atomic gas is I_ν and the central absorption coefficient is measured in the spectrophotometer:

$$k_o = \log \frac{I_o}{I}$$

Under certain conditions, proportionality exists between k_o and the atomic concentration:

$$\log \frac{I_o}{I} = K.c.$$

an expression analogous to the Beer law. More generally, we write:

$$c = f(k_o)$$

The emission phenomena counteract the absorption and can contribute to a bending of the curve $c = f(k_o)$:

(a)   The flame contains excited and thus emitting atoms (direct emission).

(b)   The neutral atoms reemit in all directions when they absorb their resonance lines (fluorescence emission).

The first phenomenon is remedied by modulating the incident source as well as the measuring detector to the same frequency. The flux emitted by the atoms excited in the flame is continuous and thus practically does not interfere with the measurement. As for the second phenomenon, we have seen it is very weak in the direction of the incident beam (Figure 9).

2.2   Components of Instrumentation

2.2.1 Radiation Generator or Emission Source

Hollow cathode lamps are most common because they emit resonance lines of sufficiently fine structure of the analyzed element. The emitted flux is modulated to a frequency identical to that of the amplifier of the measuring detector. Figure 10 shows the diagram of such lamps.

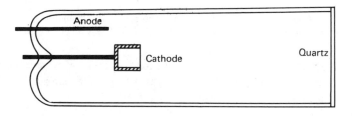

Figure 10. Diagram of a hollow cathode lamp.

The interior of a bulb contains two electrodes, an anode and a cathode in the form of a hollow cylinder consisting of an alloy containing the metal from which the resonance line is to be produced. An inert gas (argon, helium, neon) fills the bulb at low pressure (1 mm Hg). When a suitable potential difference is applied between the electrodes (a few hundred volts), the gas is ionized and current flows between the electrodes. The current is then adjusted to a few milliamperes and the voltage to a few dozen volts. Under these conditions, the constituent elements of the cathode are excited. The partial pressure of the cathodic gas is low as is the temperature, and the resulting emission lines have very small widths, which is desirable in atomic absorption spectrometry.

The intensity and width of the emitted lines increase with the current intensity of the lamp. The absorbance measured on the basis of the atomization of a given quantity of atoms depends on the lamp current as shown by the graphs of Figure 11. Curve A, obtained with a cadmium solution

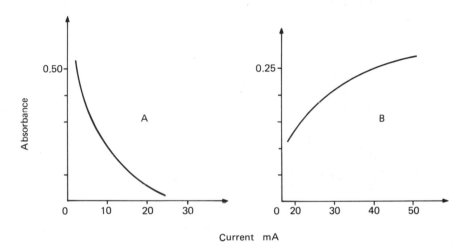

Figure 11. Influence of the supply current of hollow cathode lamps on the absorbance. A: Cd 228.8 nm–20 µg/ml. B: Pb 283.3 nm–100 µg/ml.

(20 µg/ml), expresses the theoretical conditions most often found, although other cases (B) have also been observed.

The life of a hollow cathode lamp depends largely on the nature of the cathodic alloy. Lamps with volatile elements (alkali metals, arsenic, antimony, bismuth, mercury, tin and lead) have a relatively short life, a few dozen to several hundred hours. This is remedied with discharge lamps; that is, the classical discharge lamp for alkali metals, mercury and "electrodeless" discharge lamps for the other elements. This involves a quartz tube containing a volatile compound of the element at low pressure from which the resonance line is to be excited. The tube is placed into the resonance cavity of a high-frequency field (2450 MHz).

Under these conditions, emission is obtained with a high line intensity of the metal vapor (Dagnall *et al.,* 1967; Aldous *et al.,* 1969; Browner *et al.,* 1969). This type of "electrodeless" lamp is being used increasingly for such elements as As, Sb, Bi, Pb, Se, Te and Cd. They are also recommended for atomic fluorescence spectrometry.

2.2.2 *The Atomization Source*

Flame sources and thermal sources are distinguished. The flame sources constitute the most widely used atomization mode at this time. The solution is introduced in the form of a mist by a pneumatic method, with vaporization by the combustion support or by ultrasonic atomization. The most common flames are the following:

$$\text{Air} - C_2H_2: \quad T = 2600°K$$
$$N_2O - C_2H_2: \quad T = 3200°K$$
$$\text{Air} - H_2: \quad T = 2300°K$$

Generally, the flame has an elongated flattened form (10 x 0.5 cm) to offer a suitable optical path length; the length of the atomizing can be increased with the device of Figure 12.

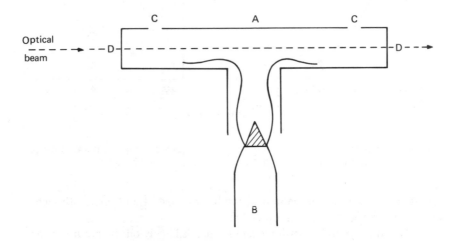

Figure 12. Burner attached to an adapter.
A: T adapter (2.5-3 cm diameter). B: Conventional burner.
C: Gas evacuation. D: Quartz windows.

More recently, thermal atomization has been devised by heating the sample by the Joule effect in an inert (argon) or reducing atmosphere (hydrogen), either in a graphite tube or on a filament of carbon, tantalum, tungsten or another refractory conductor. These devices are described in Chapter 5. They permit detection limits of 10^{-11}-10^{-9} g (absolute value) to be obtained for elements such as Pb, Cu, Cd, Mo, Mn, Zn, Fe, Co and Ni from sample solutions of 0.01-0.2 ml. Interest is centered on microanalysis and trace analysis (0.1-1 ppm).

The choice of the type of flame as well as its application conditions—vaporization of the analysis solution and combustion of the flame—is very important in atomic absorption. The role of the temperature on the atomization and ionization equilibria has already been discussed. Table VI shows the operating conditions and properties of the flames used in atomic absorption.

If we consider a solution of element M in concentration c (mol/l) in the analysis solution, c is measured on the basis of the atoms M_O produced in the flame for which the atomic absorption is determined. A relation exists between the concentration N_O of atoms M_O (in atoms/ml) in the flame and the concentration c that was established by Winefordner and Vickers (1964):

$$N_O = 3.10^{21} \cdot \frac{n_{298}}{n_T} \cdot \frac{F \, \epsilon \, \beta}{Q \, T} \; c$$

For an element M, we have:

N_O = number of atoms M per ml in the flame

c = concentration of element M in the solution in mol/l

n_{298} and n_T = number of moles of gas in the flame at room temperature (298°K) and the flame temperature T

F = solution aspiration rate (ml/min)

ϵ = vaporization efficiency

β = atomization efficiency

Q = gas flow rate

T = flame temperature

In this formula, three factors are particularly important. Factor Q is always high. In fact, it is necessary to operate with a sufficiently high gas flow rate so that the gas velocity at the burner outlet will be higher than the combustion rate of the fuel-combustion support mixture in order to prevent an explosion by a backflash. Consequently, the residence time of the atom in the effective part of the flame is very short, approximately 10^{-4} sec.

The vaporization efficiency is always far lower than unity for pneumatic systems and may be closer to it with ultrasonic systems. Therefore, only a fraction of the quantity of elements present in the solution is actually transformed into the atomic state.

The increase of the number of atoms N_O with an increase of the solution flow rate is fixed within rather narrow limits by the burner design. For example, an increase of F is often obtained by increasing the gas flow rate and is accompanied by a decrease of ϵ and of flame stability. Generally, the flow rates vary from 1 to 10 ml/min.

All of these factors contribute to limiting the atomic density in the flame and thus notably reduce the sensitivity and detection limits. Finally, it should be pointed out that the zone of maximum atomic density in a flame depends on the ratio of fuel gas-combustion support

Table VI. Combustion Characteristics of Common Flames

Fuel	Combustion Support	Reaction	Energy (k cal)	Temperature ($^\circ$K)
City gas	Air	$CG + 0.98\ O_2 + 3.9\ N_2 \rightarrow$ $CO_2 + H_2O + 3.9\ N_2$	108.8	1980
Butane	Air	$C_4H_{10} + 6.5\ O_2 + 26\ N_2 \rightarrow$ $4\ CO_2 + 5\ H_2O + 16\ N_2$	687.9	2170
Propane	Air	$C_3H_2 + 5\ O_2 + 20\ N_2 \rightarrow$ $CO_2 + 4\ H_2O + 20\ N_2$	530.6	2200
Acetylene	Air	$C_2H_2 + O_2 + 4\ N_2 \rightarrow 2\ CO + H_2 + 4\ N_2$	106.5	2600
		$C_2H_2 + 5/2\ O_2 + 10\ N_2 =$ $2\ CO_2 + H_2O + 10\ N_2$	300.0	
Acetylene	O_2	$C_2H_2 + O_2 \rightarrow 2CO + H_2O$	106.5	3300
		$C_2H_2 + 2.5\ O_2 \rightarrow 2\ CO_2 + H_2O$	300.0	
Acetylene	N_2O	$C_2H_2 + 2\ N_2O \rightarrow 2\ CO + H_2 + 2\ N_2$	106.5	3200
		$C_2H_2 + 5\ N_2O \rightarrow 2\ CO_2 + H_2O + 5\ N_2$	401.5	
Acetylene	50% O_2 50% N_2	$C_2H_2 + N_2 + O_2 \rightarrow 2\ CO + H_2 + N_2$	106.5	3090
H_2	Air	$H_2 + 1/2\ O_2 + 2\ N_2 \rightarrow H_2O + 2\ N_2$	58.0	2300
H_2	O_2	$H_2 + 1/2\ O_2 \rightarrow H_2O$	58.0	2900

(see Section 4.2.2). The position of the optical axis of the incident beam as well as its width must be taken into account.

2.2.3 Dispersive System and Measuring Instruments

The principle of the classical spectrophotometer is shown in Figure 9. The single-beam instruments are based on this design. The intensity difference between the incident radiation I_0 and the transmitted radiation I, e.g., (I_0 - I) or the logarithm of this difference log (I_0 - I) or an expression such as log I_0/I, is measured on the detector.

A better measuring stability is obtained with the double-beam system (Figure 13), in which the beam I_0 emitted by the hollow cathode lamp is split into two. We thus measure an expression of the form log (I_0 - I)/I_0. If I_0 varies as a function of source fluctuations, for example, the measurement is practically not affected.

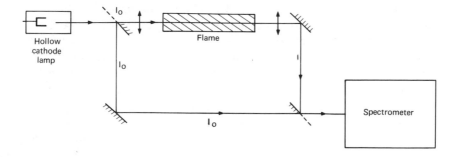

Figure 13. Diagram of the double-beam spectrometer.

Finally, in some double-channel systems (Figure 14), two elements, X and Y, can be measured simultaneously. One (Y) is used as the internal standard, being present in known and predetermined quantity in the sample. In fact, the measurement consists of the absorbance ratio A_x/A_y, which is a function of concentration C_X and is practically not influenced by vaporization fluctuations (variations of the gas flow rate).

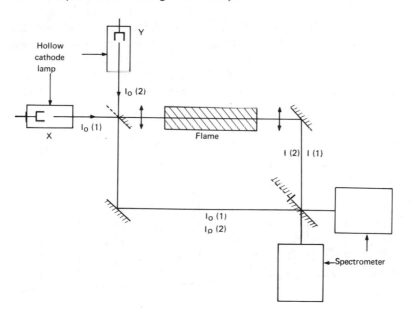

Figure 14. Diagram of the double-beam, double-channel spectrometer (Instrumentation Laboratory).

With regard to the spectrometer, the major two parameters are the slit width, which defines the band pass, and the photoelectric detector. An increase of the slit width improves the detection limit. Generally, the slit width has a value of 50-20 μ; too narrow a slit produces too weak a flux to be suitably measurable, while too large a slit interferes with resolution (Table VII).

Table VII. Influence of the Slit Width on the Limit of Detection of Nickel

Slit Width (mm)	Band Pass (nm)	Detection Limit (μg/ml)
0.03	0.32	5
0.05	0.32	5
0.075	0.39	6
0.10	0.47	6
0.20	0.77	6
0.30	1.18	7
0.40	1.61	7
0.60	2.47	10
1.0	No resolution	50
1.5	No resolution	100

The operating conditions are defined by the manufacturer. In any case, the spectrometer is calibrated by a sequence of absorbance readings, for example, with a solution of 1 and 5 μg Cu/ml, the recording of which is shown in Figure 15. This test reveals the drift of the incident flux I_o and of the absorbance reading after a suitable time: 30 min or 60 min or more if necessary.

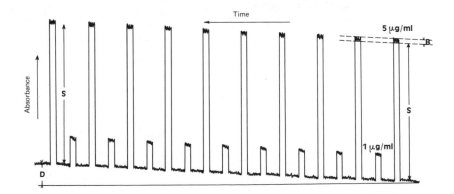

Figure 15. Recording of the absorbance of a series of copper solutions of 5 and 1 μg/ml concentration for 60 min. S: Absorbance signal of the 5 μg/ml solution. B: Background D: Drift of the instrument after 1 hr of operation. (Reproduced from Pinta, 1971, by permission of Masson et Cie, Ed.)

3. PHYSICAL AND CHEMICAL EQUILIBRIA IN THE FLAME

3.1 Introduction

The combustion of the flame gases results from chemical reactions. Some of these are shown in Table VI but were stoichiometric reactions. The combustion of a gas such as acetylene with air can lead not only to the formation of CO_2 and H_2O but also to a series of radicals and molecules, the principal ones being OH, C_2H, C_4H_2, OH*, CO* and C_2*. Combustion of acetylene with nitrous oxide yields:

$$C_2H_2 + 5 N_2O = 2 CO_2 + H_2O + 5 N_2$$

and

$$C_2H_2 + 2 N_2O = 2 CO + H_2 + 2 N_2$$

It may also lead to the radicals NH and CN which impart a highly marked reducing character to the flame in the red zone immediately above the blue cone.

When a solution is introduced into a flame in the form of a spray, a series of physical and chemical reactions occurs, leading to atomization. In particular, these reactions involve fusion, volatilization, dissociation or decomposition, resulting in the formation of free atoms and recombinations, especially with the combustion products. They yield oxides and hydroxides, reducing reactions in the presence of reducing compounds and radicals that can favor the formation of free atoms, and finally reactions leading to atomic ionization or deionization.

Under certain flame conditions, thermodynamic equilibria, governed by the law of mass action, are the result each time. A knowledge of these equilibria allows an understanding and correction of the interactions that interfere with the absorbance measurement of an element in a complex medium (matrix effect). The general scheme of these reactions is summarized in Figure 16.

3.2 Study of Various Reactions

3.2.1 Transformation into the Gas Phase

We assume a salt of formula MA in solution. A first step leads to the volatilization of compound MA with the following reactions:

(1) $MA = M^+ + A^-$: initial solution, liquid-gas aerosol

(2) $MA_{(solution)} = MA_{(solid)}$: desolvatation in the flame, formation of a solid-gas aerosol

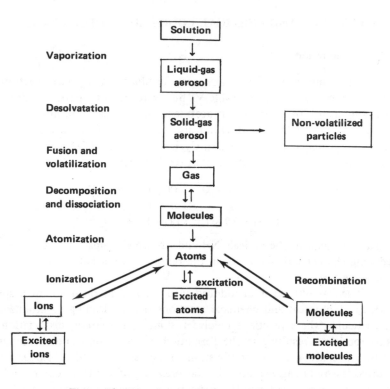

Figure 16. Reactions leading to atomization.

(3) $MA_{(solid)} = MA_{(liquid)}$: fusion (liquid-gas aerosol)
(4) $MA_{(liquid)} = MA_{(vapor)}$: volatilization.

We can also have $MA_{(solid)} = MA_{(vapor)}$ (sublimation).

3.2.2 Decomposition and Dissociation

The following sequence of reactions will occur from a gas phase in a simple medium:

(5) $MA_{(vapor)} = M^{\circ} + A^{\circ}$ (dissociation)
(6) $MA_{(vapor)} = M^{*} + A^{*}$ (excitation)
(7) $MA_{(vapor)} = M^{+} + A^{-}$ (ionization)
(8) $MA_{(vapor)} = MA^{*}$ (molecular excitation).

All of these reactions are based on the assumption of a direct decomposition of the initial salt into atoms. In fact, these are simple theoretical reactions. We will see later that this is not always the case and that the formation of intermediate compounds may be observed.

3.2.3 Reactions Between the Studied Element and the Flame Constituents

These involve all possible reactions with the compounds and free radicals that may form by combustion. The major ones are listed below:

(9) $M + O = MO$ (oxide formation)

(10) $M + OH = MO + H$ (oxide formation)

(11) $M + H_2O = MO + H_2$ (oxide formation)

(12) $M + OH = MOH$ (hydroxide formation)

(13) $MO + CO_2 = MCO_3$ (carbonate)

(14) $MO + C = M + CO$ (reduction).

All of these reactions are actually equilibria resulting from the nature of the flame and the combustion conditions (gas flow rate ratio). Still other reactions can be imagined with radicals such as CH, C_2H, CN and NH (see Section 3.1).

3.2.4 Reactions Between the Investigated Element and Any Other Element of the Complex Medium

The general case of analysis is to determine an element from a complex matrix, leading to atomization of a large quantity of impurity elements. For example, let M' be one of these elements. The following reactions are observed:

(15) $M + M' + O \rightleftharpoons MOM'$ (double oxide)

(16) $MA + B \rightleftharpoons MB + A$ (anionic exchange)

(17) $MB \rightleftharpoons M + B$.

Finally, the group of the preceding reactions can be summarized in the following schematic form:

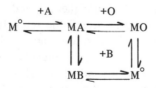

These reactions will now be discussed on the basis of practical cases.

3.3 Volatilization Phenomena

3.3.1 Alkali Metals

We will use sodium chloride as an example. Atomization passes through the following stages: desolvatation, fusion and volatilization. It is the vapor of NaCl that will be dissociated in the flame. We should recall that the melting and boiling points of sodium are 1100 and 1750°K, *i.e.*, temperatures lower than those of conventional flames. Dissociation occurs generally with alkali halides, and is involved in the gas phase of the initial salt. The reaction Na + OH = NaOH can then occur.

3.3.2 Alkaline-Earth Metals

We will now consider the case of calcium chloride $CaCl_2$. The following reactions are observed in the condensed and then in the vapor phase:

(1) $CaCl_2, H_2O_{(liquid)} \rightleftharpoons CaCl_2 \ _{(vapor)}$

(2) $CaCl_2, H_2O \rightleftharpoons CaO + HCl$

Since the boiling point of $CaCl_2$ is lower than that of CaO, reaction (1) primarily favors atom production:

$$CaCl_2 \rightleftharpoons Ca^\circ + Cl \quad \text{and} \quad CaO \rightleftharpoons Ca^\circ + O$$

Thus, atomization can take place by two different routes. The most favorable path is the volatilization of $CaCl_2$, followed by decomposition, while oxide formation leads to a nonvolatile product under the flame conditions.

With calcium nitrate, we will have: $Ca(NO_3)_2 \rightleftharpoons CaO + NO_3$ and then $CaO \rightleftharpoons Ca + O$. However, the decomposition of CaO occurs before volatilization (CaO has an m.p. of 2850°K). Consequently, the atomization efficiency of calcium in a given flame will be better from the chloride than from the nitrate. In a nitrate medium, atomization depends on the decomposition temperature of the nitrate as well as on the dissociation of the oxide. Unfortunately, precise data to explain and predict the phenomena are often lacking. When the melting and volatilization points are known and these are higher than the flame temperature, a relatively slow volatilization rate results. Volatilization can not be reached, and the residence time of the neutral atom may then be too short.

Another example cited by Alkemade (1969a) involves magnesium sulfate, which decomposes in the condensed phase at 1160°K and yields the oxide MgO, which can sublime at 3040°K, a temperature slightly lower than the melting point (3070°). Consequently, sublimation will be practically complete in hot flames and will be very slow in cool flames.

We will now consider the atomization of calcium salts in the presence of an aluminum salt. In a hydrogen chloride medium, we have:

$$CaCl_2, H_2O \rightleftharpoons Ca^\circ + Cl_2$$

$$CaCl_2, H_2O \rightleftharpoons CaO + HCl$$

$$CaO \rightleftharpoons Ca^\circ + O$$

and $$AlCl_3, 6H_2O \rightleftharpoons Al_2O_3 + HCl$$

finally $$CaO + Al_2O_3 \rightleftharpoons CaAl_2O_4 .$$

In a nitrate medium we have:

$$Ca(NO_3)_2, 3H_2O \rightleftharpoons CaO + NO_2 + 3 H_2O$$

$$Al(NO_3)_3, 9 H_2O_{(vapor)} \rightleftharpoons Al_2O_3 + H_2O$$

$$CaO \rightleftharpoons Ca^\circ + O$$

and $$CaO + Al_2O_3 \rightleftharpoons CaAl_2O_4 .$$

In a nitrate medium, the formation of calcium atoms results from the decomposition of CaO. However, under these conditions, the number of atoms is smaller than with calcium chloride. Furthermore, the presence of an aluminum salt interferes with calcium atomization because of the formation of a thermally stable double oxide of calcium and aluminum, which thus blocks a part of the calcium atoms. The interference with atomization of calcium due to the presence of aluminum is greater in a nitrate than in a hydrogen chloride medium. In other words, for an equal initial calcium concentration, the number of neutral calcium atoms formed in the flame in the presence of large amounts of aluminum is greater in a hydrogen chloride than in a nitrate medium.

The interference curves for calcium due to aluminum are shown in Figure 17. In a sulfate medium, the calcium and aluminum compounds are still more stable, and even greater interference results.

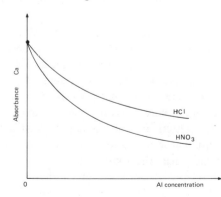

Figure 17. Interference of the calcium absorbance due to aluminum in a hydrochloric acid and nitric acid medium.

3.3.3 Practical Consequences

Volatilization followed by atomization depends on reactions that can occur between the given element (or its compounds) with the matrix constituents. These reactions can occur in the vapor phase (see above) as well as in the condensed phase (at the moment of evaporation). Some examples of condensed-phase reactions are shown in Table VIII (Alkemade, 1969b).

Table VIII. Examples of Condensed-Phase Reactions
(Capable of Interfering with Volatilization)

Types of Reaction	Examples	Melting Point of Formed Product ($^{\circ}K$)
$MA \rightarrow MA_{vapor}$	$NaCl \rightarrow NaCl_{vapor}$	1075
$MA + H_2O \rightarrow M_xO_y + A$	$MgCl_2, H_2O \rightarrow MgO$	3070
	$MgSO_4 \rightarrow MgO$	3070
	$CaC_2O_4 \rightarrow CaO$	2850
	$Al(NO_3)_3 \rightarrow Al_2O_3$	2320
$MA + HA' \rightarrow M_xA_y'$	$Al(NO_3)_3 + HF \rightarrow AlF_3$	1310
	$CaCl_2 + H_3PO_4 \rightarrow Ca_3(PO_4)_2$	1940
$MA + organic\ compound \rightarrow MOrg$	$MgCl_2 + EDTA \rightarrow Mg\text{-}EDTA$	dec.
	$MgCl_2 + oxine \rightarrow Mg\text{-}oxine$	dec.
$MA + M'A' \rightarrow M_xM'_yO_z + H_2O$	$CaCl_2 + AlCl_3 \rightarrow CaAl_2O_4 + H_2O$	1820

MA: initial investigated salt; M & M': metals; A & A': anions; H: hydrogen.

A knowledge of the formed compounds is important. Unfortunately, their physical properties are often not well known, particularly their boiling points. In the absence of known boiling points, a knowledge of the melting point can determine the choice of reactions. The formation of low-melting and thus sparingly volatile products reduces the atomization efficiency. For example, it may be of interest to convert aluminum nitrate, which forms the refractory oxide Al_2O_3, into a fluoride, because aluminum fluoride is much more volatile than the oxide, by adding hydrofluoric acid. The action of organic compounds—EDTA or oxine— is of interest only with magnesium and aluminum because they form more volatile compounds than the oxide, which would form from de- composition of the inorganic magnesium and aluminum salts. Further- more, these organic compounds are easily dissociated in flames, yielding atoms. Finally, we should note that the majority of metals in hydrochloric acid or nitrate solution are volatilized with the formation of oxides, so the oxide quantity is often preponderant.

3 4   Dissociation Equilibrium

3.4.1 Principal Reactions

The three reactions occurring in a flame when a metal salt MA is introduced are:

(1) $MA \rightleftharpoons M° + A°$

(2) $MO \rightleftharpoons M° + O$

(3) $M + H_2O \rightleftharpoons MOH + H$

These equilibria depend on the temperature and, in the case of reactions 2 and 3, on the nature of the gases and gas flow rate of the flame. Generally, diatomic molecules CaO and triatomic molecules (LiOH) are observed in the flame. Complex molecules, such as nitrates or sulfates, are generally not stable.

The elements Na, Cu, Tl, Ag and Zn are almost completely atomized. Except in reducing flames with a high C_2H_2 concentration, alkaline earths form monoxides (CaO). In an air-acetylene flame, the alkaline–earth oxides have the following degrees of dissociation: 0.8% Ba, 1.44% Mg, 8.55% Ca and 19.6% Sr (Huldt *et al.*, 1950). The metals such as La, U and Ti yield very stable oxides in an air-C_2H_2 flame. The oxides are reduced in a hot flame:

$$O_2 - C_2H_2, \quad -N_2O - C_2H_2$$

and in the reducing flame.

Hydroxide formation is observed from alkali and alkaline-earth metals in the hydrogen-air flame in the following increasing order:

$$Na < K < Rb < Cs < Li$$

and

$$Sr < Ca < Mg < Ba.$$

In an air-acetylene flame, dissociation is about 20% for lithium and 100% for sodium. In an hydrogen-air flame we can also note hydroxide formation with In, Ga, Cu and Be.

3.4.2 Law of Mass Action and Dissociation Constant

The equilibria are governed by the law of mass action:

$$MA = M + A$$

where A is the anion or oxygen. The dissociation constant K is given by:

$$(1) \quad K = \frac{[M] \ [A]}{[MA]}$$

where $[M]$ is the number of free atoms per cm^3.

K depends on the flame temperature and its nature (gas and flow rates), and is independent of the quantity of salt introduced into the flame. K can also be expressed as a function of the partial pressures of the constituents (in atm):

$$K = \frac{\rho_M \times \rho_A}{\rho_{MA}}$$

We should recall that K (also known as the equilibrium constant) is calculated by the Van't Hoff equation:

$$\log K = \frac{\Delta H}{RT} + constant$$

where R is the ideal gas constant = 1.987 cal/mol and ΔH is the heat of vaporization or the reaction enthalpy (in cal/mol).

In the case of diatomic molecules, ΔH is equal to the dissociation energy E_d and thus K is essentially a function of E_d (see Table XII listing the dissociation energy of some compounds). Using alkaline-earth halide solutions, Huldt and Lagerqvist (1950) calculated the percentage of free atoms observed in an air-acetylene flame (Table IX). From this it can be seen that only a small quantity of metal is present in the form of free atoms, while the balance exists essentially in the form of the oxide. For manganese, there is as much metal as oxide and for copper, ten times more oxide than metal.

Table IX. Dissociation of Alkaline-Earth Halides in an Air-Acetylene Flame

Elements	Salts	Concentrations (mol/l)	Wavelengths (nm)	Percentage of Free Atoms
Mg	$MgCl_2$	10^{-1}	285.3	1.44
Ca	$CaCl_2$	10^{-3}	422.6	8.55
Sr	$SrCl_2$	10^{-4}	460.7	19.6
Ba	$BaCl_2$	10^{-3}	455.4	0.84

The dissociation constants of the alkali metal halides in flames at 2000 and $2500°K$ are listed in Table X (Rubeska, 1969). The number of free atoms is directly proportional to the total number of molecules (MA)

Table X. Dissociation Constants of Alkali Halides

	Temperature = 2000°K Air-City Gas Flame	Temperature = 2500°K Air-Acetylene Flame
NaF	4.0×10^{-9}	–
NaCl	6.3×10^{-6}	2.2×10^{-4}
NaBr	1.1×10^{-4}	1.4×10^{-3}
KCl	2.5×10^{-6}	7.9×10^{-5}
KBr	–	5.4×10^{-4}
RbF	4.0×10^{-9}	4.7×10^{-5}
RbCl	6.3×10^{-6}	3.8×10^{-4}
RbBr	4.0×10^{-5}	5.8×10^{-3}
CsCl	–	2.6×10^{-5}
CsBr	–	2.3×10^{-4}

introduced into the atomizing source. The degree of dissociation is defined by:

$$(2) \quad \alpha = \frac{[M]}{[M] + [MA]}$$

and if dissociation is complete, $\alpha = 1$.

By combining (1) and (2), we obtain:

$$\alpha = \frac{1}{1 + [A]/K}$$

This expression shows that an excess of *anions* [A] decreases the degree of dissociation and thus reduces the number of free atoms.

If A is oxygen (as is frequently the case for elements forming stable monoxides such as Fe, Ca, Al and Ba), we have:

$$M + O \rightleftharpoons MO.$$

Alkemade (1969a) reported the values of the degrees of dissociation calculated according to the data of Parsons et al. (1966) (see Table XI). We find from this that the oxygen-hydrogen ratio has little influence on iron but a much greater effect on aluminum and barium.

In summary, by starting with a salt, the compounds forming in the flame either in the gas or in the solid particle state are chlorides or oxides. These are then decomposed into atoms, provided the flame energy is higher than the dissociation energy of the compounds formed in the flame (Table VI).

Table XI. Degree of Dissociation of Some Oxides in Reducing and Stoichiometric Oxygen-Hydrogen Flames

	FeO	CaO	AlO	BaO
Dissociation energy (eV)	4.3	4.0	5.5	5.0
Reducing O_2/H_2 flame, $2640^\circ K$ 1.3×10^{15} O_2/cm^3	0.95	0.60	0.08	0.03
Stoichiometric O_2/H_2 flame, $2640^\circ K$ 1.3×10^{16} O_2/cm^3	0.7	0.2	0.008	0.003

Table XII (based on Gaydon, 1968) summarizes the values of the dissociation energies (in kcal and eV) of the most common metal halides and oxides.

3.4.3 Double Oxides

We have seen that molecular combinations can form in the condensed phase:

$$xM_1 + yM_2 + zO = M_{1x}M_{2y}O_z$$

or also:

$$xM_1 + yM_2 = M_{1x}M_{2y}$$

The degree of dissociation increases in the same way with temperature. It is also a function of the dissociation constant of the complex and thus of its nature. Numerous examples are known. Thus, in an air-butane flame we have:

$$CaCl_2 + H_3PO_4 \rightleftharpoons Ca_2P_2O_7 + HCl$$

$$CaCl_2 + SiO_3Na_2 \rightleftharpoons CaSiO_3$$

The same applies to H_2SO_4 and HF. The effect is weaker in an air-C_2H_2 flame. The same reactions are observed with magnesium.

We should also note the effect of hydrofluoric acid on aluminum and titanium, but since the Al or Ti fluorides are more volatile than the oxides, absorption will be enhanced. Other examples can be cited for the air-acetylene flame:

$$CaX + MoY + O \rightarrow CaMo_2O_7$$

$$CaX + AlY + O \rightarrow CaAl_2O_4$$

Table XII.   Dissociation Energies of Metal Compounds

Molecule	kcal	eV	Molecule	kcal	eV
AgCl	74	3.2	MgF	105.5	4.5
AgF	83.8	3.6	MgO	94	4.1
AgO	46	2	MnCl	85.3	3.7
AlCl	117	5.1	MnF	120	5.2
AlF	158	6.85	MnO	96	4.1
AlO	106	4.6	MoO	116	5.0
AsO	113	4.9	NaCl	98	4.25
AuCl	81	3.5	NaF	114	4.9
BaCl	115	5.0	NiCl	88	3.8
BaF	115	5.0	NiF	88	3.8
BaO	133	5.75	NiO	97	4.2
BiCl	72.3	3.13	PbCl	71	3.1
BiF	61	2.65	PbF	69	3
BiO	71	3.1	PbO	89.3	3.87
CaCl	105		RbCl	102	4.4
CaF	125	5.4	RbF	120	5.2
CaO	100	4.3	SbCl	85	3.7
CdCl	48.8	2.1	SbF	104	4.5
CdO	88	3.8	SbO	92	4
CrCl	86.5	3.75	SeO	100	4.3
CrF	92	4	SiCl	104	4.5
CrO	101	4.3	SiF	115	5.0
CsCl	105	4.55	SiO	187	8.1
CsF	123	5.33	SnCl	74	3.2
CuCl	83	3.6	SnF	90	3.9
CuF	81	3.5	SnO	124.5	5.40
CuO	95	4.1	SrF	126	5.45
FeCl	83	3.6	SrO	97	4.2
FeO	99	4.3	TaO	195	8.4
KCl	100.5	4.36	ThO	196	8.5
KF	117	5.07	TiO	166	7.2
LiCl	113	4.9	TlCl	87.5	3.8
LiF	137	5.95	TlF	104	4.5
LiO	81	3.5	VO	147.5	6.4
LiOH	105		ZnCl	49	2.1
MgCl	81	3.5	ZnO	65	2.8

These compounds are thermally stable in the flame. The same applies to magnesium and strontium.

In a nitrous oxide-acetylene flame, we have:

$$Al X + V Y + O \rightleftharpoons AlV_2O_4 \text{ or } AlV_2O_6$$

but here the resulting absorbance for Al is higher than that resulting from $AlX + H_2O \rightleftharpoons Al_2O_3 + HX$. It may be validly assumed that the compound AlV_2O_4 is more volatile than Al_2O_3. Actually, there is no reason why the double oxides are systematically more stable than simple oxides of the same elements.

The following reaction occurs between calcium and titanium:

$$Ca\,X + Ti\,Y + O \rightleftharpoons CaTiO_3 \text{ (perovskite)}$$

$CaTiO_3$ is more stable than CaO. Analogous reactions occur with Mg, Sr and Ba.

Between calcium and iron, we also have:

$$Ca\,X + Fe\,Y + H_2O \rightleftharpoons Ca_2Fe_2O_5 \text{ (ferrite)}$$

between chromium and iron:

$$Cr\,X + Fe\,Y + H_2O \rightleftharpoons FeCr_2O_4$$

between iron and titanium:

$$Fe\,X + Ti\,Y + H_2O \rightleftharpoons FeTiO_3.$$

All of these compounds, which are stable in the flames, are possible causes of interactions. Thus, for example, titanium and iron reduce the absorbance of calcium. Iron, cobalt and nickel reduce the absorbance of chromium, and titanium that of iron. The degree of these interferences depends on the flame composition, particularly on the gas ratio. The values of the depression of chromium absorbance by iron, cobalt and nickel are quite characteristic (Table XIII) (Rubeska, 1969).

Table XIII. Relative Depression of Chromium Absorbance by Fe, Co and Ni

Interfering Substance (1 mg/ml)	Normal Flame[a]	Reducing Flame[b]
Fe	8	64
Co	23	53
Ni	35	46

[a]Three slot burner, air-acetylene ratio = 6.5.
[b]Three slot burner, air-acetylene ratio = 5.3.

The case of lanthanum is characteristic:

$$La + Al + O \rightleftharpoons Al\,La\,O_3$$

Here again, it must be noted that $AlLaO_3$ is more volatile than Al_2O_3. The presence of lanthanum contributes to enhancing the absorption of aluminum.

In conclusion, it must be considered that the presence of an impurity element is capable of modifying not only the volatilization of the compound of the investigated element but also its dissociation. This rule can be used beneficially to improve the absorbance as well as to correct certain interferences. Thus, the addition of a lanthanum salt to the analysis solution provides for a correction by eliminating the interferences of aluminum with the absorbance of alkaline-earth elements. The lanthanum salt (generally the chloride) is called a spectrochemical buffer. It should be present in a quantity 10-20 times greater than that of the interfering substance.

3.5   Ionization Equilibrium

3.5.1 Ionization in the Flames

The flame energy is generally sufficient to displace and eliminate one of the outer electrons from neutral atoms. Ionization then occurs. In fact, ionization is produced when the flame energy is higher than the ionization potentials of the elements. Some ionization potentials are listed in Table II with the percentages of ionized atoms (N^+) compared to the number of neutral atoms ($N^°$) in flames of different temperatures. The ionization energies are entirely compatible with the energy available in the flames (see Table VI). We find that in a nitrous oxide-acetylene flame the percentage of ionized atoms can become preponderant for K, Rb and Cs.

Ionization therefore depends on the flame temperature as well as on the chemical composition of the matrix, which may also contain ionizable elements capable of releasing electrons as well as ionizable molecules. Thus we are dealing with physicochemical equilibria that add to the dissociation equilibria. It is important, however, to stress that ionization reactions occur only in the vapor phase, while dissociation reactions can also occur in the solid or liquid phase.

3.5.2 Physical Laws

The ionization equilibria are governed by the Saha law:

$$M^° = M^+ + e^-$$

This reaction follows the dissociation and oxidation reactions:

$$MX = M^° + X^°$$

and

$$M^° + O = MO$$

where $M^°$ = neutral atom and M^+ = ionized atom.

The law of mass action applies:

$$(1) \quad K_i = \frac{[e^-] \cdot [M^+]}{[M^°]}$$

where K_i is the ionization constant of the respective atom.

We also define the degree of ionization α_i:

$$(2) \quad \alpha_i = \frac{[M^+]}{[M^+] / [M^\circ]}$$

In flames, an electroneutrality equilibrium exists:

$$(3) \quad [M^+] = [e^-]$$

By combining these three expressions, we obtain:

$$(4) \quad \frac{\alpha_i^2}{1-\alpha_i} = \frac{K_i}{[M^\circ] + [M^+]}$$

or if

$$[M] = [M^\circ] + [M^+] :$$

$$\frac{\alpha_i^2}{1-\alpha_i} = \frac{K_i}{[M]}$$

where $[M]$ is the total free atom concentration in the flame. It can be assumed that under given conditions $[M]$ is proportional to the concentration of the element in the analysis solution. We should note that K_i increases with temperatures and α_i thus increases in the same direction.

However, the degree of ionization of a given element is modified by the presence of another ionizable metal, and we have the reactions:

$$M = M^+ + e^-$$

and

$$M' = M'^+ + e^-$$

The electroneutrality equilibrium is written as follows:

$$[M^+] + [M'^+] = [e^-]$$

If M' is an element of the matrix and present in large amounts, the release of a large number of electrons $[e^-]$ shifts the ionization equilibrium $M^\circ \rightleftharpoons M^+ + e^-$ in a direction favorable to the formation of neutral atoms. Element M, then, is subject to an ionization interference due to the presence of metal M'. The ionization interactions enhance the absorbance of the measured element.

3.5.3 Practical Consequences

The elements most sensitive to ionization interferences are the alkali and alkaline-earth metals (Tables II and XV), which do not have very high ionization potentials. The ionization interferences are particularly notable in a hot flame (nitrous oxide-acetylene).

Elements such as the rare earths, aluminum and titanium can yield a high percentage of ionized atoms in a nitrous oxide-acetylene flame. Some values have been reported by Manning (1966) (Table XIV).

Table XIV. Percentages of Ionized Atoms of Elements Requiring a
Nitrous Oxide-Acetylene Flame

Elements	Percentage of Ionized Atoms
Rare earths	35-80
Thorium	50
Uranium	45
Yttrium	25
Aluminum	15
Titanium	15
Hafnium	10
Vanadium	10
Zirconium	10

Table XV also lists some ionization values in a nitrous oxide-acetylene flame as a function of the ionization potential as reported by Amos and Willis (1966). Consequently, the content of ionizable elements in the matrix must be at least approximately known. The interference curve of the absorbance of an ionizable element by another ionizable element generally has the shape of those shown in Figure 18 (Riandey, 1971). The curve tends toward a plateau: beyond a certain concentration the interference is constant. A correction for ionization interferences is made either by adding a suitable quantity of the interfering substance to be in the plateau of the curve or by adding an ionization buffer in such an amount that the degree of ionization of the analyzed element will be at the level of the plateau of the curve.

Table XV. Percentage of Ionized Atoms of Some Elements as a Function of
Their Ionization Potential

Metal	Ionization Energy (eV)	Concentration in Solution (μg/ml)	Percentage of Ionization
Be	9.32	2	0
Mg	7.64	2	6
Yb	6.2	15	20
Ca	6.11	5	43
Sr	5.69	5.3	71
Ba	5.21	30	92

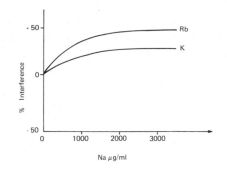

Na μg/ml

Figure 18. Interference in rubidium and potassium absorbance due to sodium.

3.5.4 Ionization of Molecules in Flames

The metal M may be in the presence of an ionized molecule Y^+ in the flame:

$$M^\circ + Y^+ \rightleftharpoons Y + M^+$$

and electron transfer from the metal to the molecule occurs, resulting in a reduction of the absorbance of metal M. The molecules Y consist primarily of positive ions resulting from flame combustion. $C_3H_3^+$ and H_3O^+ have been detected in relatively large amounts in the reaction zone of hydrocarbon flames.

The alkali metal elements are particularly sensitive to these electron transfers, and ionized hydroxides have been reported in flames. Thus, with barium, the following reaction has been observed:

$$Ba + OH^* \rightarrow BaOH^+ + e^-$$

which leads to the formation of an ionized hydroxide. With lead:

$$Pb + OH^* \rightarrow PbOH^+ + e^-$$

where OH* is the excited hydroxyl radical.

The presence of ionized molecules, of which there may be a considerable number in the flame reaction zone, can modify the ionization equilibria of the investigated elements. Again, the alkali metals are most sensitive to these effects:

$$SrOH^+ + Na^\circ \rightleftharpoons SrOH + Na^+$$

4.   PRACTICAL METHODS AND ANALYTICAL PROCEDURE

Elemental analysis by atomic absorption is a comparative analysis with the use of suitable standards. In the following, we describe the general principles that allow a determination of the optimum analytical conditions.

4.1 Preparation of Solution

The sample solution is prepared so as to eliminate any interfering elements by bringing the analyzed element to a concentration compatible with the determination range and finally producing a constant acid concentration (for example, 1% HCl).

4.2 Instrumental Conditions

The optimum conditions for an analysis depend on a certain number of factors (Table XVI) (Pinta, 1971):

    Radiation source and analysis lines
    Nature and composition of the flame
    Quality of the spectrometer and detector.

The operating conditions of the hollow cathode lamp, particularly the current, must be defined in advance. The line is generally a resonance line (Table XVI). The absorbance is a function of the lamp current and the optimum conditions vary with the nature of the cathode (Figure 11).

Table XVI. Analytical Data

Element	Line (nm)	Flames		Detection Limit (μg/ml)	Analytical Range (μg/ml)	Sensitivity 1% Absorption (μg/ml)
Ag	328.1	air-C_2H_2 air-C_3H_8	lean	0.01	0.1-10	0.2
Al	309.3	N_2O-C_2H_2	rich	0.5	5-500	1.0
As	193.7	air-H_2	lean	3	20-200	2.8
Au	242.8	air-C_2H_2 air-C_3H_8	rich	0.1	0.5-5	0.3
Ba	553.6	air-C_2H_2	rich	1	10-200	
	455.4	N_2O-C_2H_2	rich	1	–	–
Be	234.9	N_2O-C_2H_2	lean	0.02	0.2-10	0.03
Bi	223.1	air-C_2H_2	lean or sto.	1	10-100	0.7
Ca	422.7	air-C_2H_2 N_2O-C_2H_2	rich	0.01	1-10	0.08
Cd	228.8	air-C_2H_2 air-C_2H_2	lean	0.1	0.5-5	0.03
Co	240.7	air-C_2H_2	lean	0.02	1-20	0.2
Cr	357.9	air-C_2H_2 air-H_2	rich sto.	0.10	2-20	0.22
Cs	852.1	air-C_2H_2 air-C_3H_8	lean	0.1	8-80	0.16
Cu	324.7	air-C_2H_2	lean	0.01	0.5-20	0.1
Fe	248.3	air-C_2H_2	lean	0.02	0.2-20	0.004
Ge	265.1	N_2O-C_2H_2	red.	0.5	–	2

Table XVI, Continued

Element	Line (nm)	Flames		Detection Limit (μg/ml)	Analytical Range (μg/ml)	Sensitivity 1% Absorption (μg/ml)
Hg	253.7	air-C_2H_2	lean	10	20-200	–
K	766.5	air-C_2H_2 air-C_3H_8	lean	0.01	1-10	0.01
La	392.8	N_2O-C_2H_2	lean	2.0	–	–
Li	670.8	air-C_2H_2 air-C_3H_8	lean	0.005	1-10	0.02
Mg	285.2	air-C_2H_2	sto.	0.003	0.1-5	0.008
Mn	279.5/8	air-C_2H_2	lean	0.01	0.1-10	0.06
Mo	313.3	air-C_2H_2 N_2O-C_2H_2	rich	0.2	1-20	0.8
Na	589.0/6	air-C_2H_2 air-C_3H_8	lean	0.001	0.3-30	0.004
Ni	232.0	air-C_2H_2	lean	0.01	0.1-20	0.16
Pb	283.3	air-C_2H_2 air-H_2	lean or sto.	0.5	5-50	0.6
Pd	244.7	air-C_2H_2	sto.	0.3	1-10	0.3
Pt	214.8	air-C_2H_2 air-C_3H_8	sto.	2	10-100	–
Rb	780.0	air-C_2H_2 air-C_3H_8	lean	0.02	2-20	0.04
Rh	343.5	air-C_2H_2 air-C_3H_8	sto.	0.3	4-40	0.3
Ru	349.8	air-C_2H_2 N_2O-C_2H_2	sto.	0.25		0.3
Sb	217.6	air-C_3H_8	red.	0.2		0.6
Se	196.1	air-H_2	lean	0.5		0.5
Si	251.6	N_2O-C_2H_2	rich	1.0	20-200	2
Sn	224.6	air-H_2	rich	0.1	1-25	0.16
Sr	460.7	air-C_2H_2	rich	0.02	2-20	0.06
Te	214.3	air-C_2H_2 air-H_2	rich			0.4
Ti	364.3	N_2O-C_2H_2	rich	0.1	20-500	2.2
Tl	276.8	Air-C_2H_2	lean	1		0.1
V	318.4	N_2O-C_2H_2	rich	1.0	5-100	0.4
W	400.1	N_2O-C_2H_2	rich	10		20
Zn	213.9	air-C_2H_2	lean	0.01	0.1-10	0.025
Zr	360.1	N_2O-C_2H_2	rich	1.0		10

In the atomization source, *i.e.*, the flame, the atomic distribution depends on the ratio of fuel-combustion support (C_2H_2-air). Table XVI shows the classical flame conditions.

Figure 19 shows the distribution of strontium atoms in an oxidizing and a reducing flame. Consequently, it is important to select the optimum flame position relative to the optical axis and to collimate the diameter of the incident beam correctly. These conditions increase the analytical sensitivity.

With regard to the spectrometer, the practical application conditions have been discussed earlier (Section 2.2.3).

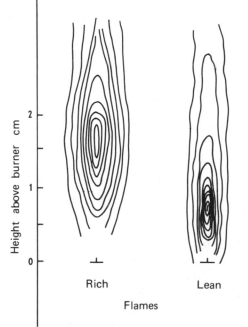

Height above burner cm

Rich

Lean

Flames

Figure 19. Distribution of strontium atoms in the flame.

4.3 Search for Interferences. Calibration. Sensitivity

A search for interferences is important. It is made with artificial solutions so that curves such as those in Figures 20 and 21 can be constructed. The influence of each impure element on the element to be analyzed is studied in succession. The correction for interferences is made as

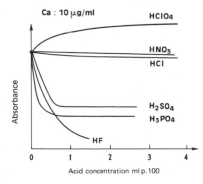

Ca : 10 μg/ml $HClO_4$

HNO_3

HCl

Absorbance

H_2SO_4

H_3PO_4

HF

Acid concentration ml p. 100

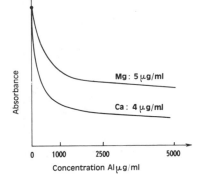

Absorbance

Mg : 5 μg/ml

Ca : 4 μg/ml

Concentration Al μg/ml

Figure 20. Influence of acids on the absorbance of calcium.

Figure 21. Influence of aluminum on the relative absorbance of calcium and magnesium.

described above, but most of the analytical errors are eliminated by suitable calibration.

The following different calibration methods exist:

Calibration in simple solutions: example: 0-2 μg/ml of Mg and 1% HCl.

Complex solutions: 0-2 μg/ml of Mg in the presence of the matrix elements (Al, Fe, Ca), 1% HCl.

Simple solutions + buffer: example: 0-2 μg/ml Mg + 1000 μg/ml La + 1% HCl.

Complex solutions + matrix elements + buffer.

The concentration C_x is determined from the absorbance-concentration curve. Interpolation of the concentration C_x is also possible by calculation on the basis of two similar standards C_1 and C_2:

$$C_x = C_1 + (C_2 - C_1) \frac{A_x - A_1}{A_2 - A_1}$$

where C = concentration and A = absorbance.

Another calibration method consists of metered additions that correct the matrix effects (Figure 22). The absorbances of x, x + 1 and x + 2 μg/ml of the analyzed element—A_x, A_{x+1} and A_{x+2}—are measured and the straight line passing through the corresponding points intersects the abscissa O. OX is the desired concentration. The method of metered additives is based on the assumption that the absorbance-concentration curve is linear.

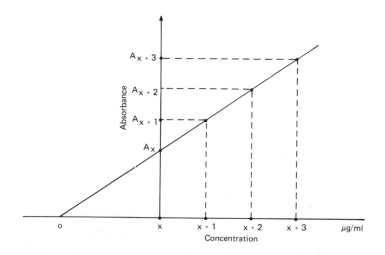

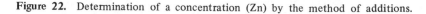

Figure 22. Determination of a concentration (Zn) by the method of additions.

The sensitivity of an analysis by atomic absorption spectrometry is defined by:

$$S = \frac{\Delta A}{\Delta C}$$

where ΔA is the absorbance variation for a concentration variation ΔC. It is expressed in $\mu g/ml$ of an element, giving an absorption equal to 1%. The sensitivity depends on the instrumental conditions as well as on the analyzed medium. In fact, we have seen the influence of the matrix and the resulting interferences with absorbance.

The sensitivity limit (or detection limit) is defined by the lowest measurable concentration, *i.e.,* producing an absorbance equal to two times the variation of the background of the analysis solution. The analysis range is the interval of concentration in which the analytical precision has an acceptable value. The lowest concentration generally is 5-10 times the sensitivity limit and the highest concentration is determined by the curvature of the calibration curve. Beyond this concentration, the precision decreases markedly. Table XVI lists the mean values of the limits of detection and the analysis ranges.

4.4 Controlling the Validity of a Method

Finally, the validity of a method is controlled by the following tests.

(a) Sensitivity control: Determination of $S = \Delta A/\Delta C$ (variation of absorbance compared to the concentration variation). The sensitivity can vary with the concentration, and consequently defines the analysis range.

(b) Precision control: The standard error of a series of measurements ($n \geqslant 12$) is determined:

$$\sigma = \sqrt{\frac{\Sigma(x_0 - x)^2}{n - 1}}$$

as well as the coefficient of analytical variation: C.V. $= \sigma \times 100/x$.

(c) Control of accuracy: The error of accuracy is defined by the difference between the found and the actual values. It generally results from poorly corrected matrix effects. In practice, accuracy is controlled by analyzing standard samples of known composition similar to the analysis samples. For rocks, ores, industrial materials such as cement, glasses and ceramics, and metals and alloys there are standard samples that are sold together with their analytical data. The method of metered additions also permits a control of the accuracy of the method in some cases.

REFERENCES

Aldous, K. M., R. M. Dagnall and T. S. West. *Analyst* 94:347 (1969).

Alkemade, C. Th. J. In: *Flame Emission and Atomic Absorption Spectrometry*, J. A. Dean and T. C. Rains, Eds. (New York: M. Dekker, 1969a), p. 101.

Alkemade, C. Th. J. In: *Analytical Flame Spectroscopy*, R. Mavrodineanu Ed. (London and New York: MacMillan, 1969b).

Amos, M. D. and J. B. Willis. *Spectrochim. Acta* 22:1325 (1966).

Browner, R. F., R. M. Dagnall and T. S. West. *Anal. Chim. Acta* 45:163 (1969).

Dagnall, R. M., K. C. Thompson and T. C. West. *Atomic Absorp. Newslett.* 6:117 (1967).

Gaydon, A. G. *Dissociation Energies and Spectra of Diatomic Molecules* (London: Chapman, 1968).

Huldt, L. and A. Lagerqvist. *Arkiv Fysik* 2:333 (1950).

Manning, D. C. *Atomic Absorp. Newslett.* 5:127 (1966).

Mavrodineanu, R. and H. Boiteux. *Flame Spectroscopy* (New York: J. Wiley, 1965).

Parsons, M. L., N. J. McCarthy and J. D. Winefordner. *Appl. Spectrosc.* 20:223 (1966).

Pinta, M. *Spectrométrie d'absorption atomique* (Paris: Masson, 1971).

Riandey, C. In: *Spectrométrie d'absorption atomique*, M. Pinta, Ed. (Paris: Masson, 1971).

Rubeska, Y. In: *Flame Emission and Atomic Absorption Spectrometry*, J. A. Dean and T. C. Rains, Eds. (New York: M. Dekker, 1969).

Rubeska, Y. *Meth. Phys. Anal.* Numéro spécial 3ème CISAFA, Paris (September 1971), p. 61.

Walsh, A. *Spectrochim. Acta* 7:108 (1955)

CHAPTER 5

NONFLAME ATOMIC ABSORPTION SPECTROMETRY

1.   GENERAL REMARKS

Knowledge of the principles of conventional atomic absorption as well as the physicochemical and chemical mechanisms leading to atomization and the fields of application of the method are assumed in this chapter. In fact, nonflame atomic absorption spectrometry is a special method of atomic absorption complementing the classical flame technique but its increasing use, particularly in the field of microanalysis of trace elements, requires its special treatment. The following pages describe and expand on the new perspectives offered by the method compared to flame atomic absorption spectrometry.

Some of the principles described in the previous chapter can be reviewed briefly. The flame is the classical source of atomization from solutions. The most common flames are:

$$\text{Air-}C_2H_2, \text{ temperature } 2300^{\circ}C$$
$$N_2O\text{-}C_2H_2, \text{ temperature } 2950^{\circ}C.$$

Absorption is generally measurable for concentrations greater than 0.01 μg/ml (0.001-1 μg/ml depending on the elements). Numerous interactions limit the dissociation equilibrium

$$MA \rightleftharpoons M^{\circ} + A^{\circ}$$

In a complex medium, the detection limit is often greater than 0.1 (matrix effect).

The mechanism of atomization in the flame can be summarized as follows: In the region of the flame front:

Evaporation
Fusion, volatilization
Dissociation: $MA \rightleftharpoons M^{\circ} + A^{\circ} + M^* + M^+$
Oxidation: $M^{\circ} + O \rightleftharpoons MO$

The products rapidly pass into the flame, so the lifetime of the atoms $M°$ is very brief, *i.e.,* a few thousandths of a second. The atoms recombine in the cold zones, and can give rise to simple and double oxides and hydroxides, MO, MOH, MOM', as well as to stable compounds $[Ca_3(PO_4)_2]$. Consequently, the atomization yield in flames is always low.

Other methods have been investigated for decomposing the analytical sample. In particular is electrical heating in a furnace and an inert atmosphere, which was first studied by L'vov (1970). The first furnace used to produce atomic vapors was that designed by King (1908) to vaporize carbon as well as other refractory compounds. In 1961, L'vov (1961) adapted this method for the production of atomic vapors for atomic absorption spectrometry.

Numerous types of furnaces for producing atoms by rapid volatilization of a small quantity of sample heated to high temperature quickly appeared. One of these is the Massmann carbon furnace (1968) used by Manning *et al.* (1970). The carbon filament is named as a function of its dimensions (*i.e.,* carbon rod, graphite rod, mini-Massmann) and has been studied by West *et al.* (1969) and Matousek (1971). A tantalum strip placed into an inert atmosphere was studied by Takeuchi *et al.* (1972), Hwang *et al.* (1972), and Riandey *et al.* (1974). More recently, Renshaw (1973) investigated a tantalum-clad carbon furnace.

Some atomization sources are based on other forms of volatilization: a hollow cathode and luminescent discharge; the laser, investigated by Mossoti (1967); and electron bombardment (Rousselet *et al.*, 1968).

2. APPARATUS AND METHOD

The apparatus consists of a classical atomic absorption spectrophotometer but with a nonflame atomization source replacing the atomizer-burner system (Figure 1). It involves a furnace heated by the Joule effect (a few

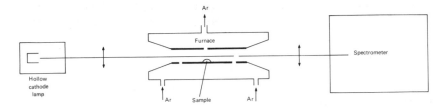

Figure 1. Nonflame atomic absorption: hollow cathode lamp; furnace; spectrometer.

dozen to a few hundred amperes). This heating element, in which the liquid or solid sample is deposited, may consist of various refractory metals, such as tantalum, platinum or tungsten; of graphite or vitreous carbon; or of metal-clad or metal-coated graphite. Graphite is being used increasingly. The proposed systems differ primarily in their geometry. Some of the units are shown in Figure 2.

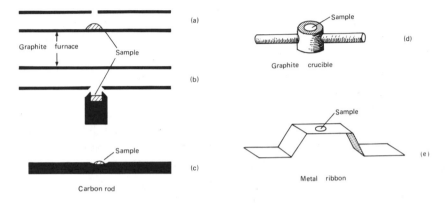

Figure 2. Nonflame atomization design.

After introducing the solid sample (a few mg) or liquid sample (a few dozen μl) into the furnace, heating that leads to atomization is carried out in an inert atmosphere in several time- and temperature-programmed steps. The first step is *drying,* during which the sample is desolvatized. This step is performed at the boiling point of the solvent and allows its evaporation. Subsequently, analysis is made with the absolute quantity of sample independent of its concentration.

A *decomposition* step follows. It serves to decompose the salt of the investigated element, driving off the anions; and to destroy the organic materials, with simpler extraction of the element from the solid film forming a deposit at the bottom of the furnace. This step can also serve as a chemical pretreatment. In the case of a solid sample, acid attack can be performed *in situ.* Thus, this phase of analysis simplifies the matrix to the maximum. A maximum temperature exists for this step, beyond which there is a loss of atoms by volatilization.

High-temperature *atomization* follows, during which the chemical compound which binds the metal (salt, oxide, double compound) is vaporized and then dissociated into neutral atoms capable of absorbing the resonance radiation emitted by the corresponding hollow cathode.

A typical analysis follows this procedure:

(a) *charging* of the sample into the furnace either in solid form (a few mg) or as a liquid (10-100 μl);

(b) *drying* of the sample at 100°C, if the solution is aqueous, for 1.5 sec/µl of solution;

(c) *decomposition* at an intermediate temperature between 100°C and 1800°C for 0.5-5 min depending on the composition of the medium and the nature of the analyzed element;

(d) high-temperature *atomization* between 1800°C and 2600°C for 5-10 sec. This energy is sufficient for the element, in metallic state, in the form of an oxide salt, or in a chemical combination with another element of the matrix, to be released as neutral atoms that absorb the cathodic beam;

(e) high-temperature *cleaning* to prevent memory effects, followed by cooling of the heating element.

The entire heat-treatment program is conducted in an inert gas atmosphere, generally by argon sweeping. However, the gas stream can be suspended at the time of the atomizing step. An example of the absorbance recording is shown in Figure 3 as a function of time. The signals at 100 and 800° result

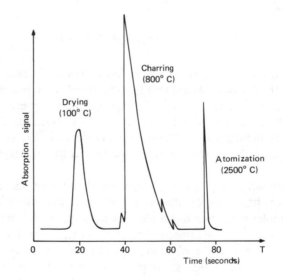

Figure 3. Scanning of absorbance. (1) desolvatation; (2) decomposition; (3) atomization.

from nonspecific absorbance, and only the peak at 2500°C is specific for the analyzed element. Compared to a flame, the residence time of atoms in the light beam is 100-1000 times longer than in the heating chamber of the furnace— 10^{-1} sec in the furnace, 10^{-3} to 10^{-4} sec in the flame (L'vov, 1970). Refractory compounds can then be decomposed more easily.

Because the entire sample is introduced into the furnace the absorption reading varies with time. The specific signal that can be compared to a flash is extremely brief, only a few tenths of a second depending on the volatility and quantity of the element. It is recorded graphically as a peak. The accuracy of the readings is lower than that obtained with a flame, with a relative standard error of 5-10% being common. A part of these errors (1-3%) results from measurements of the small volumes used. Accuracy can be improved by using an automatic sample injection device (Slavin, 1975) and by measuring the peak area (Baudin et al., 1972).

It may happen that the signal read at the atomization temperature is perturbed by stray absorptions from solid or liquid particles that were incompletely decomposed in the decomposition step and that will superimpose on the characteristic atomic absorption of the analyzed element. Most often, these are fumes persisting at the time of atomization and they are difficult to eliminate during programming. In this case, it is necessary to follow a reading by a correction for nonspecific absorption. An auxiliary source that produces a continuous background is used, allowing measurement of the nonspecific absorption in the vicinity of the atomic line. The sources utilized are hydrogen lamps (190-280 nm) or deuterium lamps (190-325 nm). Nonflame atomic absorption apparatus is generally equipped with these correctors of nonspecific absorption.

Careful planning of the heat-treatment program should allow the selective atomization of the analyzed element from a suitable matrix, with the best sensitivity and under conditions as free as possible from interference. In fact, at the atomization temperature of the sample, its volatilization should be selective and not perturbed by matrix constituents.

3. INTERFERENCES

The principal interferences encountered are:

(a) *Stray emission from the furnace*: These may saturate the photomultiplier. This effect is reduced by a suitable collimation of the optical beam.

(b) *Spectral interferences*: Superposition of molecular bands or atomic absorption lines (the latter are rare). Direct atomic emission is practically eliminated by modulating the incident optical beam.

(c) *Chemical interferences:* The components of the sample may have a volatility close to that of the analyzed element, which may lead to nonspecific absorptions. This applies, for example, to the determination of lead in sea water. The latter contain 30 mg/ml of sodium chloride and 3×10^{-5} to 9×10^{-3} μg Pb/ml (Riley, 1965). Since lead is a volatile element, the maximum decomposition temperature is insufficient for complete destruction of the sodium chloride. A chemical separation is necessary.

(d) *Physical interferences*: During atomization of a complex sample, fumes may be generated due to compounds that are not eliminated during the decomposition step. A stray absorption is superimposed on the signal and interferes with the atomic absorption reading. This effect results from the fact that the atomic vapors formed in the furnace chamber produce a cloud of particles by condensation that diffracts the incident beam.

(e) *Practical consequences*: The major elements of the solution can modify the absorbance, but analysis is still possible. It is necessary to study the matrix effect before analysis to select a suitable standard containing the principal interfering elements, or to use an interference corrector as in the case of lanthanum in flame atomic absorption.

(f) *Correction for nonspecific absorptions*: A background correction device is indispensable for eliminating the above-mentioned stray absorption.

Generally, the background corrector consists of a deuterium arc emitting 190-325 nm on a continuous background. The deuterium and cathodic radiations are focused on the center of the furnace with equivalent energies. During the reading, the nonspecific absorptions attenuate the two beams in the same way, while the analyzed element absorbs only the cathodic beam because of its narrow line width. With a suitable electronic system, the signal due to the element is distinguished from nonspecific absorption.

In practice, the correction is valid for optical densities that may range up to 1. The background correction is not always effective. This is true, for example, for the analysis of elements whose analytical line is not located in the wavelength region of the background corrector (190-325 nm) and whose high volatility does not allow a sufficient decomposition temperature to be used.

4.   ATOMIZATION SOURCE

4.1   Carrier Gases

An inert or reducing atmosphere is indispensable for preventing rapid oxidation of the heating element. Various carrier gases (argon, nitrogen, helium, hydrogen) have been investigated by Hwang *et al.* (1972), Donega *et al.* (1970), and Manning *et al.* (1970). The gas flow rate is not critical, but it should not be less than 2 l/min in the case of tantalum cells and 1 l/min in the case of graphite tubes so the furnace will have a sufficient life.

In contrast, the nature of the carrier gas may be important. Since hydrogen is a reducing agent, it should facilitate the reduction of oxides, but experiments have shown that the measured signals are weaker than with an inert gas (Donega *et al.,* 1970; Hwang *et al.,* 1972; Takeuchi *et al.,* 1972). L'vov (1970) showed that the diffusion coefficient of the atoms in the carrier gas depends not only on temperature but also on the size and weight of the molecules of the furnace atmosphere. In the final analysis, argon is used most commonly and in its absence, nitrogen.

The gas flow can be interrupted during atomization. Since the atomic vapor is no longer entrained by the argon sweep, the atoms remain in the light beam for a longer time. An increase in sensitivity is then observed, especially with graphite tubes where the detection limit is doubled and sometimes multiplied by a factor of 5.

4.2 Heating Element

Any material with a melting point above 2600°C can be used as the heating element. The most common elements are either tungsten or tantalum boats (Instrumentation Laboratory, Jarrel Ash) or graphite tubes (Perkin-Elmer, Beckman, Techtron). In contrast to tantalum, graphite serves as a reducing catalyst and facilitates the reduction of oxides that may have formed or prevents their formation.

On the other hand, it can also form carbides stable at high temperatures, which does not occur with tantalum. This is true for elements such as vanadium, iron, molybdenum and calcium. The presence of these carbides requires thorough high-temperature cleaning between analyses to prevent a memory effect from falsifying the readings. Difficulties appear during the analysis of certain elements with the tantalum boat because this metal often contains impurities that may interfere with analysis. This is the case for chromium, nickel, cobalt and iron, for example. A purification of tantalum is possible in the case of chromium by a series of high-temperature heat treatments to volatilize it completely.

Graphite seems to be sufficiently free of impurities. However, some difficulties do appear, particularly in the analysis of nickel, which presents a very special problem. In order to suppress stray absorptions that are often observed with new furnaces, the furnace must be brought to maximum temperature at least ten times before any analysis. The stray absorptions result from carbon particles volatilized at the moment of the atomization step. A molecular band, C_2, which was also observed by Dieperro *et al.* (1971), absorbs the cathodic nickel beam at 232 nm.

Aging of the heating element by oxidation causes a sensitivity loss and especially poor reproducibility. Therefore the furnace must be changed periodically. Since tantalum is less refractory and more oxidizable than graphite, it has a shorter life: 50 instead of 100 atomizations.

L'vov's studies (1970) show that the dimensions of a graphite tube have a great influence on the detection limits. The sensitivity increases inversely to the square of the tube diameter. However, the volume that the sample vapor can occupy decreases in proportion to the square of the diameter. An elongation of the tube, increasing the absorption space, leads to higher sensitivities. Renshaw (1973) has shown that the barium absorbance sensitivity is multiplied by a factor of 20 if the interior of the graphite tube is coated with tantalum. The formation of highly stable barium carbide and the diffusion of gas atoms into the graphite pores is prevented. A similar result is obtained with an internal pyrolytic carbon coating or with the use of an element made entirely of pyrolytic carbon (L'vov *et al.*, 1967).

5. DETECTION LIMITS. PRECISION

Because of its principle itself, nonflame atomic absorption allows the analysis of elements in concentrations much lower than those determined in classical atomic absorption. Analysis is made for the absolute quantity of the element present in the furnace, independent of its concentration in the solution, since the solvent is evaporated during the drying step. Generally, the measurable quantities are less than a nanogram. Dilution by the combustion gases of the flame is then prevented. Analysis is rapid when the solid sample is placed directly into the furnace, and contaminations produced during dissolution are eliminated. The possibilities of this method therefore are significant, but it must not be forgotten that problems of contamination and homogeneity of the sample still arise in trace analysis.

Since the analysis is made with a few microliters or milligrams, the results can be determined with a relative standard error of 5-10%.

6. ATOMIZATION MECHANISM IN A SIMPLE MEDIUM— DETERMINATION OF OPTIMUM ANALYTICAL CONDITIONS

6.1 Principle of the Method

First, the determination of the optimum analytical conditions in a simple medium, *i.e.,* a medium containing only the analyzed element evidently associated with an anion, will be investigated. Subsequently, the matrix effect will be considered.

In nonflame atomic absorption, the reactions and chemical equilibria occurring are analogous to those observed during flame atomization, but the degree of some of these reactions may be different. The following main reactions take place during atomization:

$$MA_{gas} \rightleftharpoons M° + A° \qquad \text{(dissociation)}$$
$$M° + O \rightleftharpoons MO \qquad \text{(oxidation)}$$
$$MO + C \rightleftharpoons M + CO \qquad \text{(reduction)}$$
$$M + C \rightleftharpoons MC \qquad \text{(carburization)}$$

These reactions depend on the nature of the compound of element M (chloride, nitrate, sulfate) and the atomization conditions (time, temperature). Atomization can also result from volatilization, followed by decomposition of the salt, oxide or carbide. In graphite furnaces, carburization phenomena are particularly marked with the elements Mo, V, Cr, Fe, Ca and Sr: the higher the stability of the carbide, the more difficult and incomplete will be the volatilization and thus the decomposition.

In addition to the above reactions, we must note the ionization-deionization equilibria:

$$M° \rightleftharpoons M^+ + e^-$$

observed in nonflame atomization, particularly with alkali and alkaline-earth metals. Consequently, it is necessary to determine the decomposition and heating conditions that will lead to the best atomization yield.

A systematic study of the heating and time program during decomposition and atomization is necessary to determine the optimum sensitivity, precision and accuracy conditions (Welz, 1973; Caillot, 1974; Riandey *et al.*, 1975). Drying is practically identical in all analyses: 100°C for 20 sec for 10 μl.

In our study the influence of the decomposition temperature on the absorbance will be studied first, while the atomization temperature is kept constant at an arbitrary value selected from known literature data. Subsequently, the inverse variation will be investigated, *i.e.*, variable atomization temperature with a constant decomposition temperature. The selected decomposition time is 60 sec and the atomization time is 10 sec. This choice is determined experimentally so that all elements can be analyzed in a simple as well as a complex medium.

The variation of these temperatures is shown graphically in the form of two separate curves (Figure 4). Curve D shows the variation of the absorbance as a function of the atomization temperature. The salient points of these two curves represent the changes of state of the investigated metal. A comparison of these temperatures with the physical constants of compounds that might be present in the furnace makes it possible to describe the most probable atomization mechanism (Caillot 1974, Pinta *et al.*, 1975).

6.2   Decomposition

The maximum decomposition temperature (M) marks the start of volatilization of the molecule MX or MO in which the analyzed element is

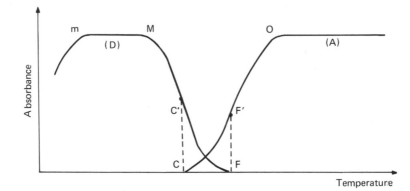

Figure 4. Typical program curves. (D)–decomposition temperature is variable, and atomization temperature is constant; (A)–atomization temperature is variable and decomposition temperature is constant.

bound. Beyond this value, a fraction of the molecules is volatilized during the decomposition step and will be lost at the time of atomization. The temperature of point F marks the end of the decomposition curve. Beyond this temperature, all molecules are volatilized during the decomposition step. Point m for the start of decomposition noted here is observed primarily in a complex medium.

The salts deposited at the bottom of the furnace form a solid film approximately 1μ thick (L'vov, 1970). At the moment of atomization, the molecules must be extracted from this film and decomposed. One of the most important objectives of the decomposition cycle is the degradation of this film in order to facilitate the ejection of molecules from the investigated element.

Below this value (m), heating is insufficient to destroy this aggregate of substances. The rate of temperature increase in the atomization step is too high to counteract this lack of initial degradation.

The choice of the decomposition temperature leads to various phenomena: if n is the number of molecules, MX, of the metallic salt or oxide in the furnace after drying and A the maximum absorbance, all volatile molecules are degraded into free atoms when the atomization temperature is optimum. Letting α be the fraction of molecules volatilized during decomposition and β the fraction of molecules not volatilized after decomposition, the interpretation of temperature θ is as follows:*

(a) If $\theta_1 < m$: incomplete volatilization

$$n \, MX(s) \to n \, MX(l) \to (n\text{-}\beta) \, MX(g) + \beta \, MX(l) \, (n\text{-}\beta) \, MX(g) \to (n\text{-}\beta) \, X + (n\text{-}\beta) \, M \to (n\text{-}\beta) \, A.$$

---

*s = solid, l = liquid, g = gas, A = absorbance.

(b) If $m < \theta_1 < M$: complete volatilization and dissociation:

$n \, MX(s) \rightarrow n \, MX(l) \rightarrow n \, M\dot{X}(g) \rightarrow n \, X + n \, M \rightarrow A$

(c) If $M < \theta_1 < F$: loss by volatilization during the decomposition step:

$n \, MX(s) \rightarrow n \, MX(l) \rightarrow \alpha \, MX(g) + (n-\alpha) \, MX(l) \quad (n-\alpha) \, M(l) \rightarrow (n-\alpha) \, MX(g) \, \rightarrow$
$(n-\alpha) \, X + (n-\alpha) \, M \, (n-\alpha) \, M \rightarrow (n-\alpha) \, A$

(d) If $\theta_1 > F$: complete loss by volatilization during the decomposition step

$n \, MX(s) \rightarrow n \, MX(l) \rightarrow n \, M\dot{X}(g) \rightarrow A = O.$

6.3   Atomization

The decomposition temperature is chosen to be slightly lower than M. The optimum atomization temperature (point O) corresponds to complete decomposition of molecules MX into free atoms. Beyond this value, every element is in the form of free atoms; below it, the temperature is insufficient to decompose the aggregate of gas molecules. The temperature of point C marks the start of the atomization curve, *i.e.*, the start of dissociation of the gas molecules. The temperatures of points M and C should coincide as should those of points F and O if the gas molecules were to decompose immediately after volatilization.

The choice of the atomization temperature θ_2 relative to temperatures O, C and F leads to the following results: let γ and γ' be the number of undissociated molecules MX for $\theta_2 < O$ and for $\theta_2 < F$, respectively; δ and δ' the number of nonvolatilized molecules MX for $\theta_2 < F$ and for $\theta_2 < C$, respectively. A is the absorbance corresponding to the atoms M.

(a)   If $\theta_2 < C$: incomplete volatilization and no dissociation:

$n \, MX(s) \rightarrow n \, MX(l) \rightarrow (n-\delta')MX(l) + \delta' \, M\dot{X}(g)$
$\delta' \, M\dot{X}(g) \rightarrow A = O$

(b)   If $C < \theta_2 < F$: incomplete volatilization and dissociation:

$n \, MX(s) \rightarrow n \, MX(l) \rightarrow (n-\delta) \, MX(l) + \delta \, MX(g)$
$\delta MX(g) \quad \gamma' \, M\dot{X}(g) + (\delta - \delta')X + (\delta - \gamma')M \rightarrow (\delta - \gamma)A$

(c)   If $F < \theta_2 < O$: incomplete dissociation;

$n \, MX(s) \rightarrow nMX(l) \rightarrow nMX(g)$
$n \, MX(g) \rightarrow \gamma \, M\dot{X}(g) + (n-\gamma)X + (n-\gamma)M \rightarrow (n-\gamma)A$

If $\theta_2 > O$:   complete volatilization and dissociation :

$n \, MX(s) \rightarrow n \, MX(l) \rightarrow n \, MX(g) \rightarrow n \, X + n \, M \rightarrow A.$

The number of atoms produced at a given instant is simultaneously influenced by the following three parameters:

> sensitivity (d in g for 1% absorption)
> heat of vaporization (ΔHv in cal/mol)
> stability constant (K_d in atm).

Only qualitative laws can be determined. The positions of points F' and C' on the decomposition and atomization curves, respectively, correspond to the temperatures of points F and C. The slope of the curve MC' is a function of ΔHv; the slope of the curve $C'F$ is a function of ΔHv and d; the slope of curve CF' is a function of d and K_d; and the slope of curve $F'O$ is a function of K_d. The volatilization rate is related with the slope of curve MC'. The stronger the slope, the more readily volatilization takes place. The stability constant is given by the slope of curve $F'O$. The higher the slope, the lower the stability of the molecule.

Therefore the best analysis conditions for a drying temperature of 100°C are:

$$\text{decomposition } \theta_1 \leqslant M, \text{ atomization } \theta_2 \geqslant O$$

The time of each of these three steps is a function of the volume, the medium and the analyzed element. The decomposition time must be studied carefully. The rate of temperature increase of the tube is the maximum rate compatible with the equipment (approximately 150-200°C/sec).

6.4   Reaction Mechanisms Producing Free Atoms.
Examples of Applications

The maximum decomposition temperature, M, and the optimum atomization temperature, O, are a function of the melting and boiling temperatures of compound MX. Consequently, compounds having low melting and boiling points will be decomposed into atoms more easily. On the other hand, for refractory compounds with high melting and boiling points, a high atomization temperature will have to be used, but the influence of the medium will be less marked because the maximum permissible decomposition temperature will be higher. Starting from a salt in solution, different processes can lead to the formation of atomic vapors. The following cases are possible:

   (a) via the salt, which is volatilized and then dissociated
   (b) via the oxide which is volatilized and then dissociated
   (c) via the metal which is volatilized
   (d) via the metal and carbide: the metal is reduced into carbide that is volatilized and then decomposed
   (e) via the salt and the oxide: the salt is converted into oxide that is volatilized and then dissociated.

6.4.1 Atomization Processes Involving the Salt or Metal: Example of Cadmium Chloride

Solution used: 100 μg/ml of cadmium chloride. The temperature curves shown in Figure 5, plotted point by point, were obtained by successive

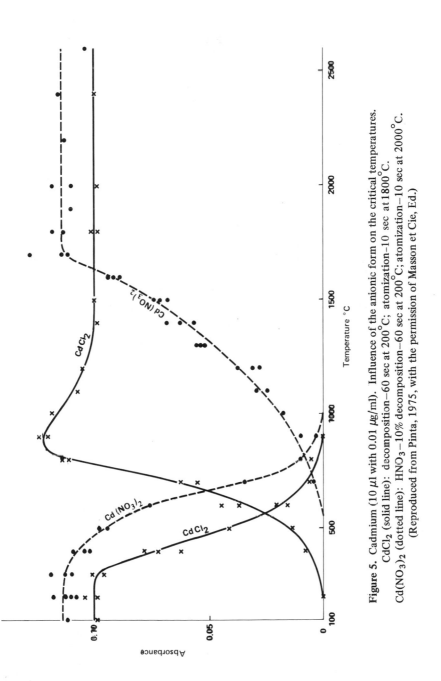

Figure 5. Cadmium (10 μl with 0.01 μg/ml). Influence of the anionic form on the critical temperatures. CdCl₂ (solid line):   decomposition–60 sec at 200°C;   atomization–10 sec at 1800°C. Cd(NO₃)₂ (dotted line): HNO₃–10% decomposition–60 sec at 200°C; atomization–10 sec at 2000°C. (Reproduced from Pinta, 1975, with the permission of Masson et Cie, Ed.)

treatments of 10 μl fractions of a solution containing 0.01 μ/ml cadmium chloride (Pinta *et al.*, 1975). A correspondence exists between the following temperatures: M, C and the melting temperature of cadmium (300°C), F, O and the boiling temperature of cadmium (800°C). The temperature of M, C (300°C) corresponds to the melting of metallic cadmium at the start of cadmium chloride volatilization and its almost instantaneous decomposition into free cadmium atoms (see Figures 4 and 5).

The steep slopes of curves MF and CO show that cadmium is highly volatile and cadmium chloride has low stability. The temperatures F and O (800°C) correspond to the boiling of cadmium. At that temperature the vapor pressure of cadmium is 760 mm Hg, the absorbance is optimum, and all molecules of $CdCl_2$ are decomposed into atoms.

Therefore, the atomization of cadmium takes place via the salt, which is molten and then volatilized and decomposed into free cadmium atoms. It may also take place via the metal: the salt is reduced into metallic cadmium that is molten and then volatilized into free atoms. It is difficult to determine which of these two routes is involved.

The best analysis conditions for a cadmium chloride solution are drying at 100°C, decomposition at 150°C and atomization at 1800°C. The atomization process can be schematically summarized as follows:

- 100°C desolvatation of the salt: $Cd^{2+} + 2\,Cl^- \rightleftharpoons Cd\,Cl_2$ (solid)
- 300°C start of volatilization and decomposition of the salt:

$$CdCl_2(s) \rightleftharpoons CdCl_2(g) \rightleftharpoons \alpha\,Cd(g)$$

or reduction of the salt, melting of the metal and start of cadmium volatilization:

$$CdCl_2(s) \overset{C}{\rightarrow} Cd(s) \rightarrow Cd(l) \rightarrow \beta Cd(g)$$

- 500°C complete melting of the salt but incomplete decomposition:

$$CdCl_2(s) \rightarrow CdCl_2(l) \rightarrow CdCl_2(g) \rightarrow \delta Cd(g) \rightarrow \delta A$$

- 800°C complete volatilization and decomposition:

$$CdCl_2(l) \rightarrow CdCl_2(g) \rightarrow Cd(g) \rightarrow A$$
$$\text{or} \qquad Cd(l) \rightarrow Cd(g) \rightarrow A.$$

- Beyond 1500°C: sublimation of the oxide that is entrained by the gas stream without decomposition.

6.4.2 Atomization Process Involving the Oxide: Example of Cadmium in Nitric Acid Medium

The curve plotted point by point is obtained by successive analyses of 10-μl fractions of a 0.01 μg/ml cadmium chloride solution in 10% HNO_3 (Figure 5). Correspondence exists between the temperatures of the following points: M, C and the start of volatilization of cadmium oxide (600°C-700°C), and O and sublimation of cadmium oxide (1600°C). The temperatures of the characteristic points of these curves (Figures 4 and 5) are higher than those observed in aqueous medium. Therefore, the atomization processes are different in a neutral and in an oxidizing medium. Volatilization of cadmium oxide begins at 600°C (point M) and decomposition at 700°C (point C).

The strong slope of curve MF indicates rapid volatilization of CdO, while the weak slope of curve CO shows slow decomposition of CdO. The temperature O (1800°C) corresponds to complete decomposition of CdO, requiring a temperature of 300°C above its sublimation point. Therefore, the atomization of cadmium in an oxidizing medium takes place from cadmium oxide that is volatilized without passing through the liquid phase and is then decomposed into free atoms.

The best analysis conditions for a cadmium solution in an oxidizing medium are: drying at 100°C, decomposition at 400°C and atomization at 2000°C. The atomization process can be schematically summarized as follows:

- $100°C$ desolvatation and shift of the equilibrium:

$$CdCl_2 + 2\ HNO_3 \rightarrow 2\ HCl + Cd(NO_3)_2$$

  and oxide formation:

$$Cd(NO_3)_2 \rightarrow CdO(s) + 2\ NO_2\uparrow + O\uparrow$$

- $600°$ start of volatilization of the oxide CdO:

$$CdO(s) \rightarrow CdO(g)$$

- $700°C$ start of decomposition of the oxide CdO(g):

$$CdO(g) \rightarrow \beta Cd(g) \rightarrow \beta A$$

- $900°C$ complete melting of the oxide CdO but incomplete decomposition:

$$CdO(s) \rightarrow CdO(l) \rightarrow \gamma Cd(l) \rightarrow \gamma Cd(g) \rightarrow \gamma A$$

- $1600°C$ sublimation of the oxide CdO but complete decomposition is slowed down by the excess of oxidant.

- $1800°C$ complete decomposition of the oxide CdO:

$$CdO(s) \rightarrow CdO(g) \rightarrow Cd(g) \rightarrow A.$$

6.4.3 Lead

A similar study shows that the atomization of lead takes place via the oxide that is molten, volatilized and then decomposed into lead atoms (Figure 6) (Pinta et al., 1975). The best analysis conditions for

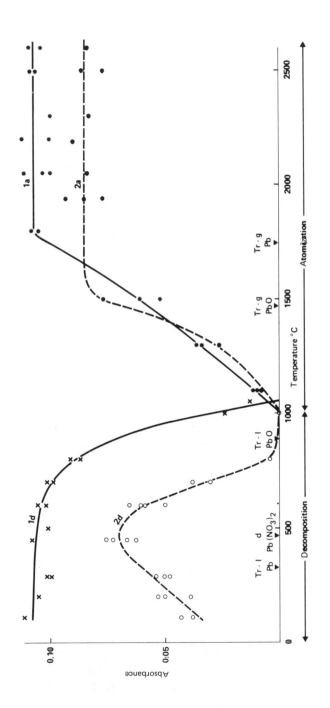

Figure 6. Lead [(NO₃)₂ Pb] (10 μl). Effect of the decomposition temperature with constant atomization [temperature] and inversely. Curves 1d and 2d: drying—20 sec at 100°C; decomposition—1 min at variable T°; atomization—6 sec at 1950°C. Curves 1a and 2a: drying—20 sec at 100°; decomposition—1 min at 450°C; atomization—6 sec at variable T°. 1—(d and a): Pb 0.2 μg/ml; 2—(d and a): Pb 0.2 μg/ml + Ca 1000 μg/ml. Correction for nonspecific absorptions necessary. Tr: solid-liquid (l) and liquid-gas (g) transition temperatures. (Reproduced from Pinta, 1975, with the permission of Masson et Cie, Ed.)

a lead nitrate solution are: drying at $100°C$, decomposition at $500°C$ and atomization at $2000°C$ (Caillot, 1974).

The atomizing process can be summarized as follows:

- $100°C$ desolvatation and oxide formation:

$$Pb^{2+} + 2\ NO_3^- \rightarrow Pb(NO_3)_2(s) \overset{470°C}{\rightarrow} PbO(s)$$

- $600°C$ start of volatilization and decomposition of the oxide PbO:

$$PbO(s) \rightarrow \alpha PbO(g) \rightarrow \alpha Pb(g) \rightarrow \alpha A$$

- $900°C$ complete melting of the oxide PbO:

$$PbO(s) \rightarrow PbO(l) \rightarrow \beta PbO(g) \rightarrow \beta Pb(g) \rightarrow \beta A$$

- $1800°C$ complete volatilization and decomposition:

$$PbO(l) \rightarrow PbO(g) \rightarrow Pb(g) \rightarrow A.$$

6.4.4 Copper

Starting with 10 μl of a 10-μg-Cu/ml copper nitrate solution, the copper atomization process successively involves the two oxides that are volatilized and then decomposed into copper atoms. Cupric oxide, CuO, is reduced into cuprous oxide, Cu_2O, which in turn is decomposed into copper atoms. The best analysis conditions of a copper nitrate solution are drying at $100°C$, decomposition at $900°C$ and atomization at $2400°C$.

The atomizing process can be summarized as follows:

- $100°C$ desolvatation and oxide formation:

$$Cu^{2+} + 2\ NO_3^- \rightarrow Cu(NO_3)_2 \rightarrow Cu_2O + CuO$$

- $1000°C$ start of volatilization of the copper oxides:

$$CuO(s) + Cu_2O(s) \rightarrow CuO(g) + Cu_2O(g)$$

- $1500°C$ complete melting of the mixture of the two oxides:

$$CuO(s) + Cu_2O(s) \rightarrow CuO(l) + Cu_2O(l) \rightarrow \alpha CuO(g) + \alpha Cu_2O(g)$$

$$\alpha CuO(g) + \alpha Cu_2O(g) \rightarrow \beta Cu(g) \rightarrow \beta A$$

- $2200°C$ complete volatilization and decomposition of the oxides:

$$CuO(l) + Cu_2O(l) \rightarrow CuO(g) + Cu_2O(g) \rightarrow Cu(g) \rightarrow A.$$

This complete volatilization before the boiling point of copper is reached can be explained by the volatilization and decomposition of the copper oxides, which are less refractory than metallic copper.

6.4.5 Nickel

The atomization of 10 μl of a 0.1-μg/ml nickel chloride solution involves volatilization of the chloride, which is converted into oxide NiO and then dissociated into free atoms at about 2400°C. The best analysis conditions of a nickel chloride solution are drying at 100°C, decomposition at 1000°C and atomization at 2500°C. Cobalt chloride shows exactly the same curves as nickel, and the atomization process is similar to that of nickel.

6.4.6 Molybdenum

The temperature curve plotted point-by-point was obtained by successive analyses of 10-μl fractions of a solution containing 0.2 μg/ml molybdenum solubilized in aqua regia (Figure 7) (Pinta *et al.*, 1975). The characteristic temperatures of each of the compounds capable of existing in the furnace are entered in Figure 7.

In contrast to the elements cited earlier, molybdenum is highly refractory, and the furnace does not allow its very high melting temperature of 2610°C to be reached. The temperature of M and C (1800°C) are close to the decomposition temperature of the dioxide MoO_2 and correspond to the start of its volatilization. The dioxide vapor MoO_2 consists of higher oxides MoO_2, MoO_3, $(MoO_3)_2$ (Margrave, 1967) having little stability. Their successive reduction leads to formation of the monoxide MoO, the physical constants of which are not known but which should have a melting point of about 2400°C according to the curves obtained. Thus, atomization of a molybdic acid solution takes place via the dioxide MoO_2, which volatilizes in the form of complex oxides. These are decomposed into monoxide and then into free atoms. The best analysis conditions of a molybdenum solution are drying at 100°C, decomposition at 1700°C and atomization at 2600°C.

Note: Molybdenum forms a stable carbide, making complete atomization difficult. It is imperative that the furnace be cleaned after each or several high-temperature heating cycles in order to prevent memory effects.

6.5 Conclusions and Atomization Processes

According to the results obtained, classification of the atomization processes can be attempted. Three different atomization processes, which may include subcategories, may be involved in the formation of free atoms.

> (a) via the *metal salt*, which is volatilized and then dissociated. This applies to the volatile salts of cadmium ($CdCl_2$) and potassium

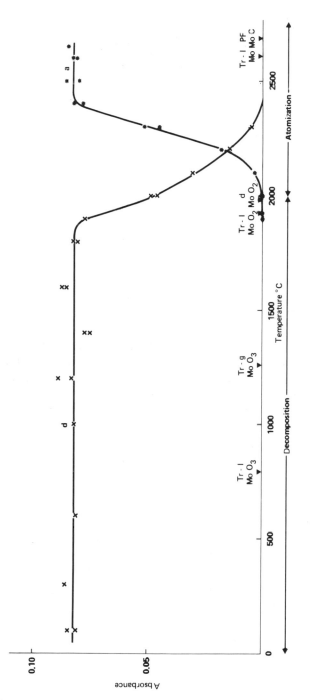

Figure 7. Molybdenum (H_2MoO_4) (20 μl with 0.1 $\mu g/ml$). Absorbance as a function of the decomposition temperature with constant atomization (temperature) and inversely. Curve d: drying–30 sec at 100°C; decomposition–30 sec at variable T°; atomization–10 sec at 2600°C; cleaning. Curve a: drying–30 sec at 100°C; decomposition–30 sec at 1800°C; atomization–10 sec at variable T°; cleaning. Tr: Solid-liquid (l) and liquid-gas (g) transition temperatures. (Reproduced from Pinta, 1975, with the permission of Masson et Cie, Ed.)

(KCl). This group also includes more complex processes involving the salt that is converted into an oxide, that is in turn volatilized and then dissociated, as in the case of nickel ($NiCl_2$-NiO) and cobalt ($CoCl_2$-CoO).

(b) via the *metallic oxide,* which is volatilized and then dissociated. This is true for cadmium in an oxidizing medium (CdO), lead (PbO), vanadium (VO) (this group also includes processes involving several metallic oxides), copper (Cu_2O-CuO), molybdenum [$(MoO_3)_2$, MoO_3, MoO_2, MoO] and aluminum (Al_2O_3, AlO).

(c) via the *metal* and the *carbide*: the metal is reduced into carbide, which is volatilized and then decomposed. This is the case for strontium (Sr-SrC_2) and perhaps also chromium (Cr-Cr_3C_2).

A fourth process must also be considered, particularly in the case of iron, in which there is a succession of all possible compounds: the salt, oxides, metal and carbide.

A fraction of the chlorides $FeCl_3$ and $FeCl_2$ is volatilized during decomposition because the flat segment of this curve declines uniformly. Then the chlorides are converted into oxides Fe_2O_3 and FeO, which are reduced into metallic iron. The latter is reduced into the carbide Fe_3C. The formation of iron carbide makes its complete atomization difficult, and cleaning of the furnace is necessary. A few test results are summarized in Table I (Caillot, 1974). The first column lists the salt in solution, the second the metallic compound volatilized and decomposed during atomization. Columns three and four indicate the temperatures used for a maximum atomization yield. The fifth and sixth list the absolute detection limits obtained in a gas flow and after stopping the gas flow. The seventh column indicates the oscillator forces (Price, 1972) and the last column shows the wavelength utilized for each of the elements.

The melting and boiling points of the elements are not of great significance. However, the melting point of the compound in which the element is bound is important for the determination of the maximum decomposition temperature. There seems to be no correlation between the oscillator forces noted here and the observed detection limits.

Studies of temperature variations are of principal importance to determine adequate heating programs on one hand and the chemical combination giving rise to atomization on the other. We will see in the following that the atomization process depends not only on the initial salt but also on the medium of the sample.

Table I. Recapitulation of Atomization Processes and Heating Conditions

Elements (salt)	Process	Temperature (°C)		Absolute Detection Limit (g)		Oscillator Force	λnm
		Decomposition	Atomization	With Gas Sweep	With Gas Flow Suspended		
$CdCl_2$	$CdCl_2$ or Cd	200	1800	3.3×10^{-12}	1.5×10^{-12}	1.2	228.8
$Cd(NO_3)_2$	CdO	400	2000	3.3×10^{-12}	1.5×10^{-12}	1.2	228.8
KCl	KCl	600	2400	6×10^{-13}	2.5×10^{-13}	0.69	766.5
$NiCl_2$	$NiCl_2$-NiO	1000	2500	6×10^{-11}	5×10^{-11}	0.095	232.0
$CoCl_2$	$CoCl_2$-CoO	1100	2500	5.5×10^{-11}	3.5×10^{-11}	0.22	240.7
$Pb(NO_3)_2$	PbO	500	1800	7×10^{-11}	2.5×10^{-11}	0.21	283.3
NH_4VO_3/HCl	VO	1600	2600	6.5×10^{-10}	7.5×10^{-10}	0.66	318.4
H_2MoO_4	Mo_xO_y	1700	2600	8×10^{-11}	7.5×10^{-11}	0.2	313.2
$AlCl_3$	Al_xO_y	1500	2400	6.5×10^{-11}	9×10^{-12}	0.23	309.3
$Cu(NO_3)_2$	Cu_xO_y	900	2400	5.5×10^{-11}	3×10^{-11}	0.74	324.7
$SrCl_2$	$Sr\text{-}SrC_2$	1100	2500	1.5×10^{-11}	1.5×10^{-11}	1.54	460.7
$K_2Cr_2O_7$	$Cr\text{-}Cr_3\ C_2$	1300	2400	2.5×10^{-11}	1.5×10^{-11}	0.34	357.9
$FeCl_3$	Fe_xO_y Fe-Fe_3C	1200	2300	4.5×10^{-11}	2×10^{-11}	0.34	248.3

7. MECHANISM OF ATOMIZATION
IN A COMPLEX MEDIUM

7.1   Matrix Effect on Atomization

The composition of a sample is always complex. Before an analysis is performed, the influence of the elements present in the sample as well as the elements introduced by the chemical treatment of the sample must be studied. Two perturbing effects must be distinguished:

> (a) Absorbance can be influenced by an excess amount of acid, modifying the chemical equilibria. It is also influenced by the nature of the salt in solution.
> (b) The sequence of reactions leading to atomization can be modified by the qualitative and quantitative nature of the matrix, *i.e.,* of the analyzed medium.

The following discussion will consider the influence of acids and their concentrations and some typical matrices on the basis of a few practical examples. Compared to the simple medium, the atomization mechanisms may also be modified as demonstrated by the decomposition and atomization curves. The determination of new characteristic temperatures should make it possible to reduce interactions and thus to improve the precision and accuracy of the analysis.

7.2   Influence of Acids

Two identical absorbances are not necessarily obtained in the same sequence of reactions. A comparison of the curves showing the absorbance variation as a function of the heating temperatures in an aqueous and a concentrated acid medium shows the dependence of mechanism on the nature of the salt in solution. The results obtained have two different aspects, depending on whether the acid influenced the mechanism or not.

The characteristic temperatures of the curves for the absorbance variation as a function of temperature as well as the absorbance are not modified in the following cases: presence of 1% HCl or 10% HNO_3 in a copper solution; presence of 10% HCl or 10% HNO_3 in a vanadium solution; presence of 5% HCl or 10% HNO_3 in a cobalt solution. Regardless of the nature of the salt, the atomization mechanism is the same. These elements are volatilized slowly and their optimum atomization temperature is high: 2300-2500°C. In all cases, decomposition of the oxide gives rise to the formation of free atoms.

However, there are cases when the mechanism is modified, as for example, with cadmium and lead. For these the absorbances are unchanged, but the characteristic temperatures of the curves of the absorbance variation as a function of temperature are different. This also applies to chromium, for which the absorbance is enhanced in the presence of an acid, while the characteristic temperatures of the absorbance curves are analogous.

The influence of hydrochloric and nitric acids on cadmium is shown in Figure 5. As we have seen, cadmium atomization in an oxidizing medium takes place via the oxide, while in hydrochloric acid medium it occurs via the chloride. Furthermore, we should note that an excess of the corresponding acid does not modify the decomposition and atomization temperatures. Consequently, the acid influences the atomization mechanism much more by its qualitative nature than by its concentration.

The effects of hydrochloric and nitric acids on lead are much less pronounced than in the case of cadmium. A comparison of the characteristic temperatures of the curves obtained for different lead solutions shows that atomization of a lead chloride solution involves volatilization of lead chloride, which is decomposed into free lead atoms absorbing the cathode-ray beam. In contrast, in the case of a nitric acid solution (HNO_3 in excess), the maximum decomposition temperature corresponds to melting of the lead oxide, and the atomization temperature is higher than the boiling points of lead oxide and metallic lead (see Figure 6). Decomposition of the oxide is slow and difficult, as shown by the weak slope of the atomization curve. Therefore, atomization of a lead nitrate solution involves the lead oxide that is volatilized and then decomposed into free atoms.

The presence of 1% sulfuric acid in a chromium solution produces a 10% enhancement. The observation of such an enhancement is very rare and especially difficult to understand because the matrix interactions are either negligible or depressive. Clark (1973) noted a similar effect in the case of zinc in the presence of sulfuric acid. The characteristic temperatures of the absorbance variation curves as a function of temperature are analogous, so the only hypothesis that can be advanced is that decomposition of chromium carbide must be more complete.

The results obtained show that for the majority of investigated elements, the anions do not have an effect on the read signals because adaptation of the heating temperatures to the medium allows an elimination of anionic interference effects for the majority of elements (only chromium in sulfuric acid medium is an exception).

7.3   Influence of the Matrix Elements

7.3.1 Procedure

The effects of cations Fe, Al, Ca, Si, K, Na and Cs present in high concentration in matrices such as soils, rocks, plants and water take several forms. The same is true for the anions, depending on whether the absorbance or atomization mechanism is disturbed.

No enhancement or depression greater than 5% has been observed in the following cases:

Matrices: $AlCl_3$, $FeCl_3$, $CaCl_2$, KCl of 1 $\mu g/ml$-100 $\mu g/ml$ on vanadium (NH_4VO_3/HCl 2 $\mu g/ml$)

Matrices: $FeCl_3$, $CaCl_2$, $AlCl_3$ of 1 $\mu g/ml$-1000 $\mu g/ml$ on copper [$Cu(NO_3)_2$ 0.08 $\mu g/ml$]

Matrices: $FeCl_3$, $CaCl_2$, KCl of 1 $\mu g/ml$-1000 $\mu g/ml$ on chromium ($K_2Cr_2O_7$ 0.05 $\mu g/ml$) (aluminum is an exception, since it produces an enhancement; see Section 7.2.5)

Matrices: $AlCl_3$, $FeCl_3$, $CaCl_2$ of 1 $\mu g/ml$-300 $\mu g/ml$ on nickel ($NiCl_2$ 0.1 $\mu g/ml$)

Matrix: KCl of 1-1000 $\mu g/ml$ on cobalt ($CoCl_2$ 0.1 $\mu g/ml$).

Thus, regardless of the matrix, the atomization of vanadium, chromium, nickel and cobalt is not modified. Therefore, no interference exists either in the condensed or the vapor phase, since the matrix atoms and those of the investigated element are present simultaneously at the moment of atomization, as we have noted earlier.

Perturbations were observed with the following elements: cadmium, lead, chromium, nickel and cobalt in complex media. Results of investigations of a few special cases will be presented.

7.3.2 Cadmium

The absorbance of cadmium solutions containing an aluminum, iron or calcium matrix and analyzed under normal programming conditions (decomposition at 150°C, atomization at 1800°C) shows significant depressions (starting with 100 $\mu g/ml$ aluminum, 1 $\mu g/ml$ iron). Figure 8 shows the curves for the absorbance variation of cadmium as a function of temperature in the presence of different matrices (Pinta *et al.*, 1975, Caillot, 1974): $CaCl_2$ with 1000 μg Ca/ml; $AlCl_3$ with 1000 $\mu g/ml$ Al/ml; and $FeCl_3$ with 100 μg Fe/ml.

These curves should be compared to those of Figure 5. A judicious choice of the decomposition and atomization temperatures makes the depression and enhancement effects negligible. The optimum analysis conditions for cadmium in the presence of calcium can be derived from the curves of Figure 8: decomposition at 150° and atomization at 1800°C. There is no interference regardless of the quantity of calcium present in solution. Volatilization of cadmium chloride (given by the slope of the decomposition curve) is unchanged, but decomposition into free atoms is extended by 400°C compared to the pure medium.

According to the curves of Figure 8, the analysis conditions for cadmium in the presence of aluminum are: decomposition at 600°C and atomization at 1900°C. The presence of aluminum increases the stability of cadmium chloride as soon as the aluminum concentration attains 1 $\mu g/ml$. The cadmium

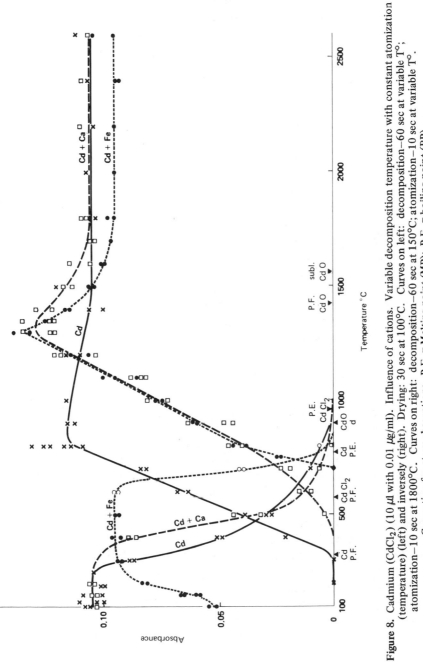

Figure 8. Cadmium ($CdCl_2$) (10 μl with 0.01 $\mu g/ml$). Influence of cations. Variable decomposition temperature with constant atomization (temperature) (left) and inversely (right). Drying: 30 sec at 100°C. Curves on left: decomposition—60 sec at variable $T°$; atomization—10 sec at 1800°C. Curves on right: decomposition—60 sec at 150°C; atomization—10 sec at variable $T°$. Correction for stray absorptions. P.F. = Melting point (MP); P.E. = boiling point (BP). Reproduced from Pinta, 1975, with the permission of Masson et Cie, Ed.)

chloride is converted into the oxide in the condensed phase. Since the oxidation reactions are endothermic, the decomposition curve has a rising slope with temperature up to the beginning of volatilization of cadmium oxide. Therefore, the presence of aluminum modifies the atomization process of cadmium chloride, which is converted into oxide. The oxide is volatilized and then decomposed into free atoms. Samples containing up to 100 μg/ml of aluminum should be analyzed under normal conditions for cadmium: decomposition at 150°C and atomization at 1800°C.

However, for a concentration higher than 100 μg/ml of aluminum, *i.e.*, for a ratio Al/Cd = 10,000, analysis conditions must be modified to decomposition at 600°C and atomization at 1900°C. Aluminum interferes only in the condensed phase, since the reaction involved is a conversion of chloride into oxide during decomposition.

The optimum analysis conditions for cadmium in the presence of iron are decomposition at 500°C and atomization at 1800°C. The presence of iron, like that of aluminum, modifies the nature of the cadmium compound so that the chloride is converted into oxide. Oxidation is more rapid than in the presence of aluminum, since it is complete at 300°C. The slope of the decomposition curve near the vertical shows a very rapid volatilization of cadmium oxide. Decomposition into free atoms is facilitated by the presence of iron at the moment of atomization. The presence of iron modifies the atomization mechanism of cadmium chloride, which is converted into oxide. It is the oxide CdO that is volatilized and then decomposed into free atoms.

7.3.3 Lead

When analyzed under normal programming conditions for lead (decomposition at 500°C and atomization at 200°C), lead solutions containing an aluminum matrix, iron or calcium show depressions starting at 1 μg/ml of iron or calcium and at 100 μg/ml of aluminum. Consequently, the curves showing the absorbance variation as a function of temperature must be studied for a given concentration of the interfering element.

Figure 6 shows the curve of lead in simple solution (0.2 μg Pb/ml) and in the presence of calcium chloride with 1000 μg Ca/ml. Examination of these curves shows that with a judicious choice of the temperatures for the two heating steps, the depressions are about 20%, while they become more pronounced under the programming conditions for lead from pure solutions.

The optimum analysis conditions for lead in the presence of calcium are decomposition at 450°C and atomization at 1800°C. The presence of calcium increases the volatility of lead, since it starts at a temperature 200° lower than the usual level, but the atomization mechanism is different. Lead nitrate is converted into chloride in the condensed phase, followed by volatilization and decomposition of the chloride into free atoms. Up to a calcium concentration of 10 μg/ml, the interferences are negligible, but beyond 100 μg/ml the readings show a 20% depression.

The optimum analysis conditions for lead in the presence of aluminum are determined in the same way: decomposition at 800°C and atomization at 2000°C. Starting at a concentration of 1 μg/ml, the presence of aluminum increases the stability of lead nitrate, but as in the case of cadmium the nitrate is converted into lead oxide. This oxidation is complete only at 600°C as in cadmium. The presence of aluminum modifies the atomization mechanism of lead nitrate, which is converted into the oxide.

The optimum analysis conditions for lead in the presence of iron are decomposition at 600°C and atomization at 1900°C. As in the case of aluminum, the presence of iron increases the stability of lead nitrate, but, like cadmium, the nitrate is converted into lead oxide. This oxidation is more rapid than in the presence of aluminum. As with cadmium, the slope of the decomposition curve is near the vertical, showing a rapid volatilization of lead oxide. Decomposition into free atoms is facilitated by the presence of iron at the moment of atomization.

In conclusion, the depressive effect that can reach and exceed 30% can be reduced by a judicious choice of the heating temperatures. The interactions of calcium and aluminum with lead have a depressive effect limited to 20%. For an iron matrix, the depression varies from 0 to 20%, depending on the quantity of iron present.

7.3.4 Nickel

A study of nickel shows that the influence of aluminum is negligible; the atomization process is the same as in a simple medium. However, with silicon, volatilization of the nickel compound takes place in two stages, at 1100°C and 1600°C. The temperature of 1600°C corresponds to melting of silicon oxides, SiO and SiO_2. Volatilization of nickel chloride is increased by the presence of silicon, and its conversion into oxide is more complete. The standards should contain silicon and be analyzed under the following temperature conditions: decomposition at 1000°C and atomization at 2600°C.

7.3.5 Chromium

The presence of sulfuric acid or aluminum causes an enhancement of absorbance without modifying the curves of the absorbance variation as a function of temperature (Figure 9). This enhancement is probably due to a more complete volatilization of chromium carbide. Therefore, the standards should contain aluminum and be analyzed as in an aqueous medium, with decomposition at 1200°C and atomization at 2400°C.

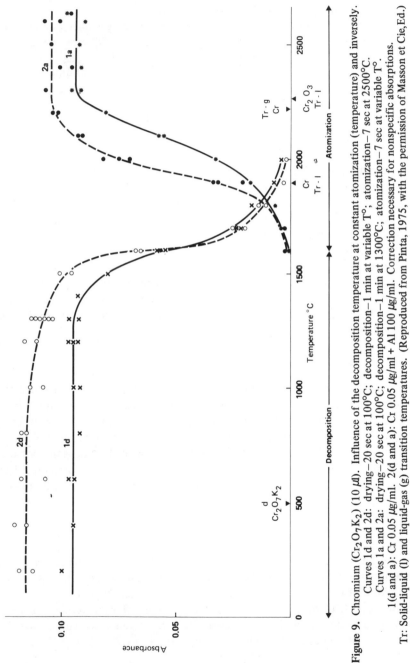

Figure 9. Chromium ($Cr_2O_7K_2$) (10 μl). Influence of the decomposition temperature at constant atomization (temperature) and inversely. Curves 1d and 2d: drying—20 sec at 100°C; decomposition—1 min at variable T°; atomization—7 sec at 2500°C. Curves 1a and 2a: drying—20 sec at 100°C; decomposition—1 min at 1300°C; atomization—7 sec at variable T°. 1(d and a): Cr 0.05 $\mu g/ml$. 2(d and a): Cr 0.05 $\mu g/ml$ + Al 100 $\mu g/ml$. Correction necessary for nonspecific absorptions. Tr: Solid-liquid (l) and liquid-gas (g) transition temperatures. (Reproduced from Pinta, 1975, with the permission of Masson et Cie, Ed.)

7.3.6 Aluminum

Aluminum is an element different from those studied before. Its low ionization potential, E_i = 5.93 eV, suggests an ionization effect. The presence of alkali elements, which are readily ionizable and the ionization of which was reported by Langmyhr et al. (1973), permits a demonstration of the ionization of aluminum in a graphite tube. In the presence of alkalies, the released electrons cause a decline of the ionization of aluminum, enhancing the measured signal. However, this enhancement can be neglected in the analyses.

Some authors report other cases of ionization: Baudin (1972) noted that technetium, which has an ionization potential higher than that of aluminum (E_i = 7.28 eV), is enhanced by the presence of sodium, and Langmyhr (1973) reports the ionization of cesium and rubidium.

8.   CONCLUSIONS

The above considerations, based on some specific cases, show that a knowledge of the chemical composition of the matrix allows a determination of optimum operating conditions, i.e., a reduction of interference effects and improvement of the analysis precision. It is the volatile elements that are perturbed the most, particularly from the standpoint of the atomization mechanism. The perturbations occur primarily in the condensed phase, since the effects of acids or matrices are negligible at a suitable decomposition temperature although the atoms of the matrix and the analyzed element are present simultaneously at the moment of atomization. We attempted to show this, but not everyone is in agreement, especially Clark et al. (1973) who concluded that the perturbations occur in the gas phase. Their results differ from those we obtained, but the working conditions were also different and may explain this discrepancy. The decomposition curves are often more perturbed than the atomization curves and, depending on the medium, the differences observed are the result of different compounds giving rise to atomization.

The elements can be divided into three groups: volatile, sparingly volatile and refractory (Caillot, 1974). In general, the volatile elements have a maximum permissible decomposition temperature below 900°C because the melting points of their salts or oxides are low (below 900°C). Decomposition of the chemical compound in which the element is bound into free atoms does not require an atomization temperature higher than 2000°C. Among the investigated elements, cadmium, lead and potassium form part of this group, and Cs, Rb, Li, Na and Zn can also be added. The risks of interference are greater. A number of analyses remains impossible to perform because the low permissible decomposition temperature does not permit sufficient destruction of the matrix. For example, in sea water, the natural concentrations of cadmium and lead can not be analyzed (Segar, 1972).

The atomization mechanisms differ as a function of the matrix. An example of this is cadmium, which has a very different atomization mechanism in hydrochloric acid than in nitric acid medium. Depressions are observed in the case of lead, which is more difficult to volatilize in the presence of matrices based on aluminum, iron and calcium. Heating temperatures to analyze these elements are very critical, and optimum analysis conditions must be determined for each investigated medium.

The elements of sparing volatility are those that require the choice of a maximum decomposition temperature, ranging between $900°C$ and $1500°C$, because of the volatility of their salts or oxides. The decomposition of their compounds (salt, oxides or carbide) requires a temperature higher than $2000°C$ in the atomization cycle.

Among the investigated elements, Al, Sr, Cu, Ni, Co, Fe and Cr can be placed into this category as can Ge, Ag, Au and Mn. For these elements, the risks of interference are small. The majority of analyses can be performed even in a complex medium; for example, chromium in urine (Schaller, 1973), copper and iron (Segard, 1972), and chromium in sea water. Generally, the atomization mechanisms are not perturbed. However when the readings are perturbed it results most often from enhancement: chromium is enhanced in the presence of sulfuric acid and aluminum, and cobalt in the presence of iron. The high maximum decomposition temperatures of compounds of these elements facilitate most of the analyses. The heating temperatures are less critical than for volatile elements.

The refractory elements are those whose compounds have a maximum permissible decomposition temperature higher than $1500°C$. The atomization process involves volatilization and decomposition of their highly refractory oxides. This high decomposition temperature permits a simplification of numerous matrices by means of this preheating to high temperature. The atomization temperature must be higher than $2500°C$ to allow decomposition of the highly refractory oxides or carbides into free atoms. Among the investigated elements, vanadium and molybdenum form part of this group, as do platinum and silicon. These elements cause no interference or modification of the atomization process. However, Schramel (1973) observed perturbations during the analysis of vanadium in the presence of a complex matrix. All analyses are feasible provided the sensitivity permits. For example, vanadium can only be analyzed in polluted sea water (Segar, 1972) since the detection limit for vanadium is too low. The high maximum decomposition temperatures of these elements allow most analyses to be performed because a large part of the matrix is decomposed during the decomposition stage.

Regardless of the analyzed element and matrix, it is necessary to determine the optimum heating conditions for each element as a function of the medium of the sample.   For a series of similar samples, the optimum conditions can be determined in one of them.   If the temperature conditions found are very different from those of the standard solution, this allows the conclusion that a significant matrix effect is present.   There seems to be no matrix effect for the refractory elements.   Therefore, it is possible to use simple standard solutions (vanadium, molybdenum) that contain the same acid in the same concentration.

In contrast, the problem is more critical for volatile elements.   The above studies show that in the case of cadmium, perturbations are small when the medium allows an increase of the maximum decomposition temperature.   For this purpose, cadmium standards should contain aluminum or iron in a concentration of 100 μg/ml or should be in 1% nitric acid medium.   Therefore, for the analysis of cadmium, the matrix effects seem to be minimized by placing the standards and samples into nitric acid solution.   The oxidation of cadmium allows the choice of a sufficiently high decomposition temperature to destroy the matrix more easily.

On the other hand, an analysis of lead in the presence of calcium is more difficult because the maximum decomposition temperature is very critical.   The standards must contain calcium at a concentration similar to that of the samples.   Iron and aluminum, like calcium, have a depressive effect on the lead analysis.   However, when the sample contains several interfering elements simultaneously, the overall effect is not equal to the summation of specific interactions.   Generally, the most perturbing element dominates in its effect if present in sufficient concentration.   Consequently, lead standards should contain 100-1000 μg/ml of calcium.

Elements of sparing volatility can be determined most often in comparison to simple standards.   However, since chromium is enhanced by the presence of sulfuric acid or aluminum, it is useful to add sulfuric acid to the standards and to the samples in order to increase the sensitivity for chromium.   Analysis of cobalt in the presence of iron requires a different heating program.   It is advantageous to add iron to the standards when the samples contain more than 100 ppm iron.   When the total composition of the samples is not known or the results obtained seem different from those expected, calibration is possible by the method of additions (Riandey et al., 1975).   The dilution and additions must be selected so that the absorbances fall into the straight part of the calibration curve.   This is the only method giving the exact desired value because the matrix effects will be compensated by the medium being the same in the samples and the standards.   The method of these metered additions is also used to control the validity of a classical calibration curve, even if it is not straight.

REFERENCES

Baudin, G. and M. Chaput. *Bull. Soc. Fr. Ceram.* 96:39 (1972).
Caillot, C. "Study of Atomization Processes by Electrothermal Methods and Their Perturbations," Thesis, Faculty of Science, Paris (1974).
Clark, D., R. M. Dagnall and T. S. West. *Anal. Chim. Acta* 63:11 (1973).
Dipierro, S. and G. Tessari. *Talanta* 18:707 (1971).
Donega, H. M. and T. E. Burgess. *Anal. Chem.* 42-13:1521 (1970).
Hwang, J. Y., C. J. Mokeler and P. A. Ulluci. *Anal. Chem.* 44-12:2018 (1972).
King, R. B. *Astrophys. J.* 27:353 (1908).
Langmyhr, F. J. and Y. Thomassen. *Z. Anal. Chem.* 264:122 (1973).
L'vov, B. V. *Spectrochim. Acta* 17:761 (1961).
L'vov, B. V. and G. G. Lebedev. *Zhur. Prike. Spectrosk.* 7:264 (1967).
L'vov, B. V. *Atomic Absorption Spectrochemical Analysis* (London: A. Hilger, 1970).
Manning, D. C. and F. J. Fernandez. *Atom. Absorp. Newsletter* 9-3:65 (1970).
Margrave, J. L. *The Characterization of High Temperature Vapors* (London: J. Wiley and Son, 1967).
Massmann, H. *Methodes Physiques d'Analyses* 4-2:193 (1968).
Matousek, J. P. *Amer. Lab.* (June 1971).
Mossoti, V. G., K. Lagua and W. D. Hagenah. *Spectrochim. Acta* 23B:197 (1967).
Pinta, M. and C. Riandey. *Analusis* 3(2):86 (1975).
Price, W. J. *Analytical Atomic Absorption Spectrometry* (London: Heynden and Son, 1972).
Renshaw, G. D. *Atom. Absorp. Newsletter* 12:158 (1973).
Riandey, C. and M. Pinta. *Analusis* 3:179 (1973).
Riandey, C., P. Linhares and M. Pinta. *Analusis* 3:303 (1975).
Riley, J. P. and G. Shirrow. *Chemical Oceanography* (London) 2:343 (1965).
Rousselet, F., M. L. Girard and C. Amiel. *C. R. Acad. Sci.,* Paris, ser. C 266:1682 (1968).
Schaller, K. H., H. G. Essing, H. Valentin and G. Schacke. *Atom. Absorp. Newsletter* 12:147 (1973).
Schramel, P. *Anal. Chim. Acta* 67:69 (1973).
Segar, D. A. and J. G. Gonzalez. *Anal. Chim. Acta* 58:7 (1972).
Slavin, M. 4e ICAS, Melbourne (August 1975).
Takeuchi, T., M. Yanagisawa and M. Suzuki. *Talanta* 19:465 (1972).
Welz, B. Communication XVIIth C.S.I., Florence (September 1973).
West, T. S. and X. K. Williams. *Anal. Chim. Acta* 45:27 (1969).

CHAPTER 6

ATOMIC ABSORPTION SPECTROMETRY:
ANALYTICAL APPLICATIONS

1. INTRODUCTION

The object of this chapter is to summarize the principal methods of application in atomic absorption, with more particular attention to studies published since 1970. We will not review the practical applications of atomic absorption; in the past dozen years or so, the number of publications has become considerable and has involved increasingly specialized areas. While the development of instrumentation has allowed an improvement of sensitivity and accuracy as well, the most interesting and new developments have occurred primarily in the preparation of the analytical sample. The combination of chemical separation methods with atomic absorption spectrometry offers very broad perspectives. This applies particularly to the use of separation methods for hydrides (AsH_3, SbH_3, SeH_2 and TeH_2) associated with atomic absorption that today allows the analysis of elemental As, Sb, Se and Te with a far higher sensitivity than do the classical methods.

Another field in which the development of atomic absorption has been remarkable in recent years is the nonflame method, which offers two possibilities: microanalysis, which is of particular interest in biology, and trace analysis, in which the sensitivity gain obtained in comparison to the flame methods is often 10-100, as we have seen in Chapter 5.

In this chapter the conventional methods are presented as a function of the analyzed medium: natural mineral media, water, plant and biological, atmospheric particles, industrial products and petroleum products. In each section we will show the major outlines of the methods for sample preparation as well as the characteristics of the resulting techniques—especially the sensitivity and accuracy—without going into the details. The nonflame methods are also reviewed.

2.   ANALYSIS OF NATURAL MINERAL MEDIA

2.1   Preparation Methods

Dissolution requires destruction of the mineral, followed by acid solubilization. Two types of techniques are used today:

(a) Melting of the sample with a suitable flux and solubilization of the fusion obtained in an acid.

(b) Digestion of the sample with the use of a percholoric and hydrofluoric acid mixture and taking up the residue in dilute acid.

The first method permits complete dissolution of the sample, while the second consists of eliminating silica (volatilization in the form of SiF_6H_2).

2.1.1 Solubilization with Melting of the Specimen

The specimen—rock, mineral, ore, soil—is mixed with a flux (an alkaline hydroxide or carbonate, alkaline or alkali earth borate) brought to melting (1000-1200°C) in a graphite crucible, and solubilized in a 5 or 10% acid. The method proposed by Jeanroy (1973-74) includes the use of a flux consisting of strontium metaborate or a mixture of strontium carbonate-boric acid in a suitable proportion of 100 g $SrCO_3$ and 83.8 g H_3BO_3 to form the metaborate $Sr(BO_2)_2$. The sample is 100 mg for silica-rich specimens and 50 mg for those rich in iron oxide ($Fe_2O_3 > 10\%$). It is mixed with 1.0 g of the flux mixture and placed into a graphite crucible and brought to 1100-1200°C in an argon atmosphere for about 10 min in an induction furnace. The advantage of the latter is that it permits observation of the bead and control of its final homogeneity. It is then poured in the form of a melt into 100 ml 5% nitric acid. After 30 min of agitation, solubilization is complete, and the solution is brought to 200 ml.

The disadvantages inherent in the method are the following.

(a) Certain minerals such as chromite and ilmenite may be difficult to melt and consequently leave an unmodified residue.

(b) When the melting product is brought to too high a temperature, a reduction of the compound into metal may occur, with a subsequent formation of carbide with the graphite. A fraction of the specimen may thus escape melting.

(c) The high temperature may lead to a loss of volatile compounds.

(d) Purity of the reagents is essential. In particular, it is advisable to use a high-purity strontium salt if calcium is to be determined.

(e) Finally, dilution of the sample (0.1 g in 200 ml) limits the detection threshold. While it is 0.1 μg/ml for the analysis solution, it will be 200 ppm for the specimen.

The method has the following advantages:

(a) The possibility of determining all elements of the specimen from a simple dilution using the same solution.

(b) The strontium in the flux mixture serves as a suppressant in interactions in certain analyses. For example, it allows a correction of the interactions of Si, Al and Fe with calcium and magnesium with no need to add lanthanum.

Calibrations are made from solutions containing the same strontium borate base as the sample solutions, an air-acetylene flame is suitable for elemental Fe, Mg, Na, K, Mn, Cu, Ni and Co. The elements Si, Al and Ti are determined with a nitrous oxide-acetylene flame as well as calcium because the presence of strontium alone is not sufficient for a complete correction of the interactions resulting from the presence of Si and Al in particular. The combination of a strontium-based buffer and the $N_2O-C_2H_2$ flame provides a suitable correction for these interactions because calcium is present at a level of 0.01-1% in iron and aluminum silicates. The sensitivity values of the method, expressed in percent of oxide of the sample giving a 1% absorption are 0.80% SiO_2, 0.26% Al_2O_3, 0.017% Fe_2O_3, 0.006% MnO, 0.0013% MgO, 0.006% CaO, 0.0014% Na_2O, 0.0024% K_2O, 0.42% TiO_2, 0.0019% CuO, 0.018% NiO and 0.016% Co_3O_4. Other elements, such as Cr, Mo, Pb and Zn, can be added to this list if their concentrations in the specimen are greater than 200 ppm.

2.1.2 Solubilization by Acid Digestion

The most effective methods make use of hydrofluoric acid combined with perchloric acid or nitrohydrochloric acid with volatilization of the silica. In some special cases, other acid mixtures can be used. This applies to soils that are generally solubilized in a mixture of H_2SO_4-HNO_3-HCl, and to carbonates for which simple hydrochloric acid attack is often sufficient. Some trace elements in soils are assimilated by plants. These elements are easily extractable by suitable reagents, i.e. solution of ammonium acetate, of sodium or ammonium oxalates, acetic acid, citric or hydrochloric acid or EDTA. A few examples are listed in Table I.

The hydrofluoric-perchloric acid digestion (or hydrofluoric-nitric-hydrochloric acid) merits a description because of its widespread application: Bernas (1969), Langmyhr et al. (1968) and Lerner et al. (1971). Two methods will be described: the first comprises solubilization of the

Table I. Determination of Trace Elements after Silica Separation (2 g of sample in 100 ml)

Element	λ nm	Flame	Interactions	Correction	Detection Limit	Analysis Range, ppm of Sample
Co	241	air-C_2H_2	lean	Fe	La 5000 μg/ml	0.2 μg/ml	10-500
Cr	358	air-C_2H_2	rich	Al - Fe	La 5000 μg/ml	0.2 μg/ml	5-500
Cu	325	air-C_2H_2	lean	–	–	0.1 μg/ml	2.5-250
Li	671	air-C_2H_2	rich	Na,K,Al,Fe	Cs,Na,La 1000 μg/ml	0.2 μg/ml	10-500
Mn	279.5	air-C_2H_2	–	Al - Fe	La 5000 μg/ml	0.2 μg/ml	20-1000
Ni	232	air-C_2H_2	lean	Fe	La 5000 μg/ml	0.2 μg/ml	10-500
Pb	283	air-C_2H_2	rich	–	–	0.1 μg/ml	10-500
Sr	461	N_2O-C_2H_2	rich	Al - Fe	La 5000 μg/ml	0.2 μg/ml	100-2000
V	318	N_2O-C_2H_2	rich	Al	Al 1000 μg/ml	0.5 μg/ml	100-2000
Zn	214	air-C_2H_2	lean	Al,Fe,Ca,Na	Al,Fe,Ca,Na	0.05 μg/ml	10-500

silica and the second its elimination. Digestion can be performed in a
boat in air, or better, in a Teflon* bomb.

The method involving dissolution of the silica is (Agemian *et al.*,
1975): The specimen (100 mg) is attacked with 4.0 ml concentrated
nitric acid, 1.0 ml 60% perchloric acid and 6.0 ml 48% hydrofluoric acid
in a Teflon bomb. The bomb is closed, sealed and heated to 140°C for
3 hr 30 min. After cooling, the mixture is placed into a polyethylene beaker
with 4.8 g boric acid and 30 ml water to solubilize the fluorides, and is
then brought to the mark at 100 ml. The standards must contain the
acids in the above concentrations. The elements Be, Cd, Cr, Co, Cu, Li,
Mg, Pb and Zn are determined directly on the solution in air-acetylene
flame. For the chromium analysis, ammonium chloride is added to the
solutions (to 2% w/v) in order to correct for the interference of iron.
For Mo and V, 500 mg Al/l must be added in the form of $AlCl_3$. Mn,
Fe and Al require a dilution of 10 from the initial solution. Ba, Ca and
Sr are determined in a nitrous oxide-acetylene flame and in the presence
of an ionization buffer (3000 mg Na/l). Finally, Na and K require a
medium buffered with 1000 mg Cs/l. The detection limits are relatively
high if one considers the concentration of the elements in the sample in
a dilution of 1000, which results from dissolution (100 mg in 100 ml).

In some cases, perchloric acid can be replaced by nitrohydrochloric
acid (HCl-HNO_3 3:1), if there is a risk of explosion if the sample contains
organic materials; in this case, the attack is milder. The method with elimination of silica allows us to start with a larger sample (1-2g), particularly when
a silicate is involved. The sample is placed into a Teflon boat with 10 ml of
65% perchloric acid and 10 mil of 40% hydrofluoric acid, and is heated on a
water bath up to complete volatilization of the fluorosilicic acid. Several additions of hydrofluoric and perchloric acid are needed. After elimination of silica,
the residue is heated to 130°C to eliminate the excess acids and finally taken
up in hydrochloric acid, solubilized and adjusted to 100 ml in 5% HCl medium.

To facilitate decomposition of the mineral, the digestion may be started
in a hermetically sealed Teflon bomb and heated to 100°C for 1 hr. After
cooling, the residue is transferred into a Teflon boat and the digestion is continued as above. Under these conditions, the sample solution is easily prepared
from a 2-g sample adjusted to a final volume of 100 ml (or even 50 ml, if
necessary). The sensitivity is thus greatly improved compared to methods
involving complete solubilization. Numerous trace elements can be determined
in this solution. However, the complexity of the medium is the cause for interaction with certain elements. Table I shows the most favorable spectrometric
conditions with the recommended flame type, the nature of the interactions
and their correction as well as the detection limit and analysis range.

---

*Registered trademark of E.I. duPont de Nemours & Company, Inc., Wilmington, Delaware.

The standards are solutions containing the analyzed elements to cover a range of 0.1-10 μg/ml, including the principal elements of the matrix as well as the interaction suppressors and acid (5% HCl). Most often a lanthanum concentration of 0.5% (5000 μg/ml) is suitable.

2.1.3 Enrichment Methods

The sensitivity limits offered by the above method are not always sufficient. Often it is necessary to separate the trace elements to be analyzed from the base medium. The most common methods are based on extraction of metal complexes formed with dithiocarbamates in a suitable solvent. If a multielemental analysis is made, an attempt will be made to extract all of the trace elements, except for the major matrix elements. The most common chemical reagents are sodium dithiocarbamate, ammonium pyrrolidine dithiocarbamate and hydroxy-8-quinoline. The sample solution is buffered to a suitable pH and agitated with a water-immiscible solvent, the most common of which are methylisobutylketone, ethylacetate, diisobutylketone and n-butylether. The solvent must be compatible with the flame combustion. Recently, Busev et al. (1972) proposed hexamethyleneammonium hexamethylenedithiocarbamate for the extraction of elemental Cu(II), Sn(II), Pb(II), As(III), Sb(III), Se(IV), Te(IV), Fe(III), Co(II), Ni(II) and Mo(VI). This reagent was also used by Alimarin et al. (1972a, 1972b, 1973).

However, for the analysis of natural silicates, these reagents have the disadvantage that they extract not only the trace elements but also a large amount of iron, which limits the enrichment efficiency.

On the other hand, specific extraction methods for a given element have also been proposed. Silver is separated from ores by extraction of the iodide in isoamyl alcohol (Fishkova et al., 1971) or with a solution of trialkylbenzene-ammonium chloride in benzene (Galanova et al., 1974). Gold is separated from ore solutions by precipitation with selenic acid in the presence of hydroxylamine (Bazhov et al., 1972), or by extraction with an 0.1 M dibutylsulfide solution in toluene (Galanova, 1972). The elements forming volatile hydrides, such as As, Sb and Te, are separated from the ore solutions by reduction with stannous chloride and zinc powder (Terashima, 1974). Molybdenum is selectively extractable with ammonium thiocyanate in the presence of ascorbic acid and stannous chloride (Kim et al., 1974). The beryllium from rocks, clay and ores can be extracted from their solution in the form of acetylacetonate in chloroform (Terashima, 1973b). Antimony can be extracted after oxidation of Sb(III) into Sb(V) with bromine water in methylisobutylketone (Ng, 1970).

2.2   Practical Applications of Flame Atomic Absorption

Although a number of methods have been proposed in recent years, we do not intend to report on all of them, since there are several hundred. Some have been listed in Table II, element by element (column 1) for the classical mineral media (column 2) of rocks, soils and ores. The wavelength for analysis is listed in column 3, and the analyzable concentration range is shown in column 4, although precise data are sometimes lacking and only the lower limit is indicated. The indication of "trace" applies to concentrations of a few dozen to a few hundred ppm. Column 5 lists the recommended flame, and column 6 briefly describes the principal phases of chemical treatment: dissolution and possible extraction of the analyzed trace elements. Some methods refer to one or two specific elements—Ag, Au, As, Hg, Se—and generally involve a special extraction technique. Others seek to determine a group of elements: Co, Cu, Mn, Ni, Pb and Zn. In that case, the analysis is made directly with the solution or after complete extraction (APDC & MIBK).

In selecting an analytical method, reference can also be made to the techniques described or reported in the following paragraphs dealing with the analysis of water, plant and biological media and industrial products.

2.3   Practical Applications of Nonflame Atomic Absorption

Chapter 4 presented the conditions for the application of nonflame atomic absorption spectrometry. This is primarily the domain of microanalysis (5-50 μl of solution, 0. 5 -5 mg of solid sample) and also of trace analysis at levels of ppb for solutions and a tenth ppm or so for solids. At the present time, most nonflame methods require dissolution of the samples. However, Langmyhr et al. (1972, 1973, 1974a, 1974b, 1974c) developed methods allowing a direct analysis of specimens of rocks, finely ground silicates, particularly in application to alkali elements (Li, Rb, Cs) and volatile elements (Ag, Cd, Pb, Zn). The sensitivity depends on the volatility of the elements. Thus, cadmium is determined directly between 0.05 and 0.3 ppm, while zinc, which is less volatile, is determined between 32 and 166 ppm. A few applications are described in Table III. When the sample is dissolved by acid digestion, generally 0.5-2 g in 50-100 ml, direct analysis of the solution allows detection limits of 0.1 to a few ppm, depending on the elements. The object is to determine the optimum conditions for atomization (see Chapter 5). The nature of the associated ion during decomposition is especially important. Thus, the best conditions require a nitric acid medium for lead and gold, and a hydrochloric acid medium for cobalt and vanadium (Riandey et al., 1975). This must be taken into account during dissolution of the sample by the methods described in Section 2.1.2.

Table II. Applications of Flame Atomic Absorption: Rocks, Ores and Soils Analysis

Element	Sample	λnm	Concentration	Flame	Sample Preparation	Reference
Ag	Ores	328.1	0.5-10 ppm	air-propane	Extract either as iodide or DDC into isoamyl alcohol.	Fishkova et al., 1971
Cr	Rocks		4-1000 ppm	air-C_2H_2	Decomposition with HF and $HClO_4$; add NH_4Cl.	Rubeska et al., 1972
Se	Rocks, soils	196.0	1-5000 ppm	N_2-H_2	Dissolve in HNO_3-HCl_4; coprecipitate Se with As; redissolve in HNO_3.	Severne et al., 1972a
Se, Te	Soils	Se 196.0 Te 214.2	5-5000 ppm	N_2-H_2	Dissolve in HNO_3-$HClO_4$; coprecipitate Se with As; redissolve in HNO_3.	Severne et al., 1972b
Cd, Pb	Soils	Cd 228.8 Pb 283.3	0.1-1.5 ppm 10-200 ppm		Extract with $0.05 M$ HCl + $0.025 M$ H_2SO_4 or NH_4 acetate.	Kahn et al., 1972a
Co, Cu, Ni	Soils	Co 240.7 Cu 324.7 Ni 232.0	trace, ppm	air-C_2H_2	Total extract or available extract, and extraction of trace elements with hexa-methylene ammonium dithiocarbamate.	Busev et al., 1972
Mo	Rocks, soils	313.3	3-20 ppm	N_2O-C_2H_2	Dissolve in HF-HCl or HF-HCl-H_2SO_4; evaporate and dissolve in dilute HCl.	Van Loon, 1972
As, Se	Sulfide ores	As 193.7 Se 196.1	>0.17 ppm	A_2-H_2-air	Add excess $SrCl_2$ as interference suppressant.	Nakahara et al., 1973a
Au	Ores	242.8	5 ppm	air-butane	Ignite; treat with HCl-HNO_3; precipitate Au with selenic acid and NH_2OH; dissolve in HCl-HNO_3.	Bazhov et al., 1972
Ba	Silicates	553.5	>20 ppm	N_2O-C_2H_2	Decompose with HF-$HClO_4$; evaporate and dissolve in HCl-KCl, or fuse with LiF-H_3BO_3; dissolve in HCl.	Rubeska, 1973
Cr	Phosphates, rocks	357.9	60-750 ppm	N_2O-C_2H_2 or air-C_2H_2	Decompose with HNO_3-$HClO_4$; oxidize with $KMnO_4$; adjust pH 1-2; pass through ion exchange column; add H_2SO_4; evaporate to have Cr 0.5-4 μg/ml.	Johnson et al., 1972
Co,Ni,Cr, Fe,Mn	Ores, silicates	—	>75 ppm	N_2O-C_2H_2	Dissolve with HF-HNO_3-$HClO_4$, or fuse with Na_2O_2-KNO_3; add ionization buffer.	Pearton et al., 1972

Element	Material	Wavelength (nm)	Range	Flame	Procedure	Reference
Cs,Li,Rb, Pb,Zn	Silicates, rocks	—	1-100 ppm	air-C_2H_2	Mix with NaCl or Na_2CO_3 buffer; transfer to iron screw, treated with 95% acetone - 5% H_2O; insert screw in flame.	Govindaraju et al., 1972 Govindaraju, 1973
Sb	Galena	217.6	2.5-12 ppm	air-C_2H_2	Extract with HNO_3, H_2SO_4; add 3 M HCl oxidize Sb(III) to Sb(IV) with Br water extract with MIBK.	Ng, 1970
Trace elements	Standard rocks	—	ppm	air-C_2H_2	Dissolve in acid by Teflon bomb technique.	Fernandez, 1973
Trace elements	Phosphate rocks	—	ppm	air-C_2H_2 N_2O-C_2H_2	Digest with HF-HCl-HNO_3; add H_3BO_3; dilute.	Hendel, 1973
Trace elements	Clays	—	ppm	air-C_2H_2	Fuse with $LiBO_2$; dissolve in HNO_3; add 1% La.	Brown et al., 1973
As	Soils	197.2	0-25 ppm	air-C_2H_2	Digest with HCl; add KI and $SnCl_2$ solution; adjust to 33°C; add Zn metal.	Melton et al., 1973
Cd	Soils		0.5-90 ppm		Extract Cd soluble with HNO_3.	John et al., 1972
Cr,Pb, Ni,Zn	Soils		traces		Extract with 10% HNO_3 at 100°C.	Baumslag et al., 1972
Cu,Zn	Soils		0.5-130 ppm	air-C_2H_2	For total element, digest in $HClO_4$; for extractable: treat with NH_4 acetate-HCl-citric acid-NH_4 oxalate; filter; ignite and digest with $HClO_4$.	Macias, 1973
Trace elements	Soils	—	ppm		Extract with 1.25 N NH_4 acetate + 0.03 N NH_4F.	McIntosh, 1973
Au	Ores	242.8	ppm	air-propane	Dissolve in HCl-HNO_3 (3:1); extract Au into 0.1 M dibutyl sulfide in toluene; mix (1:1) with C_2H_5 OH; atomize.	Galanova et al., 1972
Ag	Ores	328.1	>0.005 μg/ml	air-propane	Dissolve in acid extract with trialkyl-benzene-NH_4Cl into C_6H_6; dilute (1:1) with C_2H_5OH.	Galanova et al., 1974
As,Sb	Silicates	As 193.7 Sb 217.6	>0.04 ppm	A_2-H_2	Dissolve in acid; add MnO_4K; fume and retreat with H_2SO_4; evaporate; dilute with HCl; add KI + $FeCl_3$ + $SnCl_2$, and Zn powder; pass AsH_3 in flame.	Terasima, 1974

Table II, Continued.

Element	Sample	λnm	Concentration	Flame	Sample Preparation	Reference
Au,Pt	Ores, slag	—	0.03-10 ppm		Fuse with Na_2CO_3-$Na_2B_4O_7$ (2:1) containing NiO + S. Crush NiS button; digest in HCl; filter, treat residue with HCl-H_2O_2.	Robert et al., 1972
Be,V,Ba,Rb	Rocks		Be: 0.04-50 ppm / V: 5-500 ppm	N_2O-C_2H_2	Dissolve in acid; evaporate; dissolve residue in HCl; extract with acetylacetonate in $CHCl_3$.	Terashima, 1973a
Cu,Zn,Fe	Rocks		trace		Treat with HF-$HClO_4$; dissolve in HCl; filter on resin; elute Cu,Fe chloro-complexes with 0.2 N HCl and Zn with 0.005 N HCl.	Koester, 1973
Mo,V	Geological materials		ppm		Filter acid solution on Dowex 1-8; elute with CH_3OH + 6 M HCl (9:1); evaporate to analysis volume.	Korkisch et al., 1973
As	Soils	193.7	0.2-1.2 ppm	A_2-H_2+air	Digest in acid; dissolve residue in HCl; add KI + H_2SO_4; add Zn powder.	Nakamura et al., 1973b
Fe,Mn,Cu, Zn,Ni,Cr Co	Soils		traces	Air-C_2H_2	Extract with NH_4 acetate-KCl; add 1000 μg Sr/ml; add in excess Ca to reduce Si interference on Fe.	Aoba et al., 1973
Cd	Soils		traces	Air-C_2H_2	Digest in acid and extract with APDC into MIBK.	Yamasaki, 1973
Co,Cu,Pb, Zn,Mn	Soils, sediments		Co: 1-20 ppm / Cu: 10-100 / Mn: 500-5000 / Pb: 5-120 / Zn: 10-200	Air-C_2H_2	Extract with 0.1 M NH_2OH HCl in HNO_3; determine Mn,Fe,Zn in diluted solution; extract Co,Ni,Cu,Pb with APDC into MIBK.	Chao et al., 1973
Mo	Soils, rocks	313.3	1-500 ppm	N_2O-C_2H_2	Digest in HF-$HClO_4$; evaporate; dissolve in HCl; add NH_4CNS + ascorbic acid + $SnCl_2$; extract with MIBK.	Kim et al., 1974
20 trace elements	Lake sediments		ppm		Digest in acid in Teflon bomb; dissolve residue with boric acid and water.	Agemian et al., 1975
Cu,Co,Ni, Mn,Zn,Pb	Geological materials		trace	Air-C_2H_2	Direct introduction of suspension of ground powders (>44 μm) in water (5%).	Willis, 1975

Table III. Applications of Nonflame Atomic Absorption: Rocks, Ores and Soils Analysis

Element	Sample	Concentration	λnm	Atomization	Sample Preparation	Reference
Ag, Au	Geological material	Au $0-1.2\times10^{-9}$ g, Ag $0-1\times10^{-10}$ g		Carbon rod	Solubilization and extraction of Au as chloro complex into MIBK, Ag, with APDC into MIBK.	Bratzel et al., 1972a, 1972b
Be	Rocks	0.1-22 ppm	234.8	Graphite tube	Decomposition with HF-HClO$_4$ in Teflon bomb; evaporate; dissolve in H$_2$SO$_4$.	Sighinolfi, 1972
Hg	Rocks	0.005-0.2 ppm	253.7	Cold vapor	Heat sample in H$_2$ in SiO$_2$ tube at 950°C; collect Hg in cold trap.	Aston et al., 1972
Hg	Rocks, soils	1-1000 ng	253.7	Cold vapor	Heat at 500°C in SiO$_2$ tube; in air stream, collect Hg on Au wire.	Kato, 1971
Trace elements	Rocks, ores			Graphite tube	Direct atomization of solid.	Langmyhr, 1973
Hg	Soils, sediments	$>50\times10^{-9}$ g	253.7	Cold vapor	Extract with 2:1 H$_2$SO$_4$-HNO$_3$ at 0°C; oxidize at 50°C with KMnO$_4$-K persulfate; dilute in 6 N HCl.	Iskandar et al., 1972
Ag,Pb,Zn,Cu	Rocks	0-50 ppm		Tantalum ribbon	Solubilization in HF-HClO$_4$; evaporate and dissolve in HNO$_3$; heat solution on Ta ribbon at: Cu,Pb 200-450-2000°C, Cr 200-1000-2750°C, Zn 200-1050°C, Ag 200-2000°C.	Riandey et al., 1973
Ag,Cu,Zn	Rocks	$>2.10^{-11}$ g		Laser pulse	Mix with graphite + KBr (1:5:2); press to make disc (15x1 mm), atomize with laser.	Vulfson et al., 1973
Cs	Rocks	>1 ppm		Graphite tube + Ta loop	Grind and atomize solid sample at 2000°C.	Langmyhr et al., 1973
Hg	Minerals	>0.1 ng/g	253.7	Cold vapor	Combust in O$_2$; filter gases through Ag/Mn oxides, Co/Mn oxides, Na$_2$CO$_3$; collect Hg in Au filter; release Hg by heating.	Lidums, 1973
Hg	Geological samples	>1 ng/g	253.7	Cold vapor	Digest in acid; reduce with acid ascorbic and SnCl$_2$; extract Hg with air; trap in Au wire; release Hg by heating to 850°C.	Head, 1973

Table III, Continued

Element	Sample	λnm	Concentration	Atomization	Sample Preparation	Reference
Hg	Geological samples	253.7	>1 ng/g	Cold vapor	Digest with oxidant; reduce with $SnCl_2$; collect Hg on Ag mesh; heat in air stream.	Huffman, 1972
Trace elements	Rocks		0.1-100 ppm	Graphite tube	General method without prior dissolution.	Welz, 1972a
Ag,Bi,Cd	Sulfide ores	Ag 328.1 Bi 223.1 Cd 228.8	4-24 ppm 2-35 ppm 1-300 ppm	Graphite tube	(1) Direct volatilization in furnace. (2) Digest in acid and evaporate.	Langmyhr et al., 1974a
Ag,Cd,Pb, Zn	Silicate rocks	Ag 328.1 Cd 228.8 Pb 283.3 Zn 307.6	0.03-0.3ppm 0.05-0.3ppm 4-41 ppm 32-166 ppm	Graphite tube	Grind and volatilize directly in furnace.	Langmyhr et al., 1974b
Hg	Ores	253.7	0.05-25 ppm	Cold vapor	Digest in acid in Teflon bomb; cool; add NH_2OH-HCl; dilute and reduce with $SnCl_2$.	Brandvold et al., 1974
Te	Rocks	214.3	ng/g	Graphite furnace	Digest in HF and aqua regia; dissolve in $6\,M$ HCl; extract with MIBK; atomize $50\,\mu l$ at 600 and 2000°C.	Beaty, 1974
Trace elements	Geological material	–	ppm	Laser evaporation	Atomization of sample with laser.	Krichever et al., 1973
Cd	Soils	228.8	0.02-0.2 ppm	Graphite tube	Digest in HCl; adjust pH 4 with NaOH; extract with APDC into MIBK.	Dudas, 1974a
Co,Ni	Soil extracts	–	0-50 ng/ml	Graphite tube	Extract with APDC into MIBK.	Dudas, 1974b
Hg	Soils	253.7	ng/g	Cold vapor	Heat to 700°C in Ar; absorb gases in H_2SO_4/$KMnO_4$; reduce with $SnCl_2$.	Gerasimova et al., 1974
Au,Co,Pb, V	Silicates	–	Au >0.06 ppm Co 0.5-10 Pb 0.2-10 V 10-100	Graphite tube	Au: media HNO_3; decompose 60 sec 1000°C; atomize 15 sec 2300°C; Co: media HCl 5%, decompose 60 sec 800°C, atomize 10 sec 2500°C; Pb: media HNO_3, decompose 600°C, atomize 2000°C, V: media HCl, decompose 60 sec 1600°C, atomize 15 sec 2500°C.	Riandey et al., 1975

Among the nonflame methods, one of the most widespread consists of a mercury analysis after separation in the form of an atomic mercury vapor at room temperature. [Mercury extracted from the medium by heating to 900-1000°C in a hydrogen atmosphere (Aston et al., 1972; Kato, 1971) heats only to 500° in an air stream.] Mercury can be directly entrained into the absorption cell or also bound in the form of an amalgam on a gold support, which is then heated to release the mercury. However, in the classical methods, mercury is liberated by chemical reduction. The sample is solubilized in an oxidizing acid medium, and then addition of reducing agents ($SnCl_2$ and ascorbic acid) leads to the formation of free mercury which is entrained by an air stream and bonded to a gold or silver wire. The latter is finally heated to 800° and the mercury is collected in the absorption cell. The detection limit is 1 ppb. The elements As, Sb and Te, the compounds of which can be reduced into hydrides (AsH_3, TeH_2), can also be determined by decomposing the hydride in the absorption furnace at suitable temperatures, 800-1200°C.

The main difficulties of the nonflame atomic absorption methods are:

(a) The representativeness and precision of the sample, the latter always amounting to a few microliters.

(b) The risk of contamination, since the quantities of elements involved in an analysis are 10^{-10} to 10^{-12} g. Contamination by the reagents, apparatus and laboratory atmosphere is often extensive. Work must be carried out in a room in which dust has been removed from the atmosphere by filtering the air on an 0.5 μ filter.

3. WATER ANALYSIS

3.1 Range of Application

Atomic absorption remains a method of choice for water analysis. In the following paragraphs, we will consider the different types of water that are analyzed most often: natural water, spring and river water, sea water, wastewater and industrial effluents. The major elements in the water are Na, K, Ca and Mg, for which the analysis is classical; these will not be discussed here. The object of analysis is trace elements, particularly those considered to be environmental pollutants resulting from human, industrial and agricultural activity, i.e., cadmium, chromium, arsenic, lead, mercury and selenium. Atomic absorption in its various forms has made an important contribution to the knowledge of these elements in our environment and thus in water.

A direct analysis of water of low hardness in most cases allows the elements (Fe, Cu, Ni, Co, Mo, Zn, Pb, Sr, Li) to be determined at a level of 50-100 μg/l. These detection limits are not always sufficient, and enrichment methods are employed.

3.2 Preparation and Analysis of Water Samples

While the concentration of the analyzed trace elements is compatible with the sensitivity of atomic absorption, a few precautions must nevertheless be taken so that the atomic absorption reading will be valid. The water must be clean, which often requires filtration on a Millipore-type filter. River water in particular can contain a colloidal suspension of clay and silica, the chemical constituents of which (Si, Al, Fe) can cause interference with the elements to be determined. Furthermore, if the anions present are unknown, the water should be acidified to an excess of 1% with, for example, hydrochloric acid. This destroys the carbonates and makes the anionic medium more uniform. Finally, care must be taken in storing water of low hardness. Water evolution may always be anticipated, which becomes manifest either by an enrichment of certain elements derived from a contamination or by a depletion, subsequent to precipitation and bonding to the vessel walls. In general, the acid medium prevents this second cause of error.

The sensitivity of atomic absorption is not always sufficient to reach concentrations of ppb that are frequent in natural water. Separation or enrichment methods must be used for the trace elements. By extraction of organic complexes from a suitable solvent, it is possible to concentrate the trace elements in the solution by a factor of 10-50. The analytical sensitivity is improved to the same degree or even more, since absorbance is always increased for an identical concentration when the reading is taken in an organic medium rather than an aqueous medium. However, with the advance of nonflame methods, the improvement of sensitivity compared to classical flame atomic absorption might allow us to avoid extraction of the trace elements. In the following, we present the principle of the method used most commonly today to separate the trace elements by extraction of complexes with ammonium pyrrolidine dithiocarbamate (APDC) and methylisobutylketone (MIBK). The extract can be analyzed directly in an air-acetylene or nitrous oxide-acetylene flame and can also be analyzed in the furnace by a nonflame procedure. In the latter case, a considerable sensitivity gain is realized (100-1000 depending on the specific case) when compared to a direct water analysis by atomic absorption.

The practical applications cited above and presented in Tables IV and V include numerous examples of utilization of these methods.

The extraction method for trace elements with APDC in MIBK is performed as follows: a 500-ml volume is adjusted to pH 3 with 10% of pH 3 buffer and may be treated with a few drops of hydrochloric acid containing 10 ml of 3% APDC solution, and by mechanical agitation using a Vibromixer for 15 min with 25 ml of MIBK. After 3 min of settling, 15 ml of the organic phase is collected. This solution is ready for the atomic absorption reading. We should note that in the extract, the concentration factor is $500/15 \cong 33$. The analyses should be made within 30 min after extraction if possible. This is especially true for manganese, since these solutions are not stable. The calibration sample is prepared from synthetic aqueous samples extracted as above. The ranges are: 0-10 $\mu g/l$ of Co, 0-4 $\mu g/l$ of Cu, 0-10 $\mu g/l$ of Fe, 0-20 $\mu g/l$ of Mn, 0-50 $\mu g/l$ of Mo and 0-2 $\mu g/l$ of Zn. Detection limits are 1 $\mu g/l$ Co, 0.4 Cu, 1 Fe, 2.5 Mo and 0.2 Zn. Analytical conditions are also applicable to the elements Ag, Bi, Cd, Cr, Ni, Pb, Se, Te and V. This method has been studied by many authors: Malissa et al. (1955), Morrison et al. (1957), Ecrement (1971) and Koirtyohann et al. (1973). The APDC/MIBK extract can then be analyzed by atomization in an electrical furnace and a graphite tube (see Table V).

3.3 Practical Applications of Flame Atomic Absorption in Water Analysis

Table IV summarizes the principle of a few important applications in water analysis. In most cases, a chemical extraction is necessary to reach the ppb level ($\mu g/l$), particularly for elements usually present in water—Cd, As, Pb and Cr. This separation is indispensable for sea water where the matrix is preponderant. A survey of the literature shows the importance ascribed to elements such as Cd, Hg, As, Se and Pb, which have been the subject of a large number of publications. Mercury certainly is at the top of the list. All methods are based on the extraction of mercury in the form of the atomic vapor at normal pressure and temperature. The methods are discussed in Section 3.4. Arsenic, selenium and other elements capable of forming volatile hydrides by reduction are presently determined with high sensitivity in a hydrogen-air flame where the hydride is entrained by an argon stream.

The classical methods involve a first reduction of arsenic to bring it to valence III with the use of stannous chloride in an acid medium (HCl + K I). In a second step, the arsenous ion is reduced to a valence of -3 (AsH_3) by nascent hydrogen obtained by adding zinc pellets or powder to the sample. An argon blanket entrains the arsenic into the flame. In the presence of organic materials, the arsenic compounds that might be present in organic form must be mineralized. For this purpose, the sample can be treated with a mixture of HNO_3-H_2SO_4-$HClO_4$, evaporated and taken up in hydrochloric acid. Caldwell et al. (1973) has described the working conditions of the method to determine traces of As and Se at a ppb level. However, reduction of the element into volatile hydrides with the use of sodium borohydride ($NaBH_4$) in a single step is becoming increasingly popular. The reaction is quite severe and requires some precautions: the borohydride must be correctly introduced either in the form of solution or of pellets. Automated methods

Table IV. Applications of Flame Atomic Absorption: Water Analysis

Element	Sample	λnm	Concentration	Flame	Sample Preparation	Reference
Al	Water	309.3	10-300 µg/l	N_2O-C_2H_2	Extract to MIBK.	Fishman et al., 1972
Cd	Sea water	228.3	1-100 µg/l	air-C_2H_2	Precipitate with $SrCO_3$; extract with Na DDC/MIBK.	Owa et al., 1972
Co,Cr,Cu, Fe,Cd,Be	Wastewater	—	0-5 mg/l	air-C_2H_2	Direct; effect of any miscible solvent.	Hicks et al., 1972
Fe,Co,Cu, Mn,Ni,Pb, Zn	Sea water	—	>0.6 µg/l	air-C_2H_2	Buffer to pH 6: extract chelate into butyl acetate.	Tsalev et al., 1972b
Cu,Fe,Ni, Pb	Sea water	—	0.2-7.2 µg/l	air-C_2H_2	Organic Cu:Millipore filter; digest with $HClO_4$-HNO_3; extract with $CHCl_3$ Inorganic Cu: extract with 5% DDC/MIBK.	Duce et al., 1972
Al	Water	309.3	>9 µg/l	N_2O-C_2H_2	Extract Al complex to C_6H_6 at pH 5.	Hsu et al., 1972
As,Bi,Se	Water; effluents	—	>0.1 µg/l	Ar-H_2	Acidify with HCl; add Na BH_4; pass hydride to flame with argon.	Schmidt et al., 1973
As	Water	193.7	>0.7 µg/l	Ar-H_2	Acidify with HCl; add KI + $SnCl_2$ + Zn pellet; pass hydride to flame with argon.	Yamamoto et al., 1973a
Cd	Water	228.8	>0.02 µg/l	air-C_2H_2	To 250 ml add buffer (NH_4Cl, NH_4OH), 0.1 g Zn acetate, 10 ml 1% solution of 2-mercaptol-benzo-thiazole in butylacetate.	Doolan et al., 1973
Cd,Pb	Water	—	>4 µg/l	—	Add $(NH_4)_2SO_4$ extract with DDTC/MIBK.	Shiraishi et al., 1972
Cd,Cu,Pb, Zn	Water	—		air-C_2H_2	Adjust pH 2; extract with APDC/$CHCl_3$; evaporate with HCl/$HClO_4$; dissolve in H_2O; extract with APDC/$CHCl_3$.	Okusu et al., 1973
Co,Cu,Cr, Mn,Zn,Fe	Sea water	—	>1 µg/l	air-C_2H_2	Adjust pH 5; extract with Na DDC/butan-1-ol + 4-methylpentan-2-one.	Jones et al., 1973
Co,Cu,Fe, Ni,Zn	Water	—	5-70 µg/l	air-C_2H_2	Adjust pH 4-5; extract with PAN/MIBK.	Mizuno, 1973

Element(s)	Sample	Wavelength	Detection limit	Flame	Procedure	Reference
Cr,Cu,Zn, Fe	Water	Cr 357.8 Cu 324.8 Zn 213.6	1-40 μg/l	air-C$_2$H$_2$	Extract with APDC/MIBK. For Cr, oxidize to Cr(VI) with Br water.	Goulden et al., 1973
Cu,Cr,Ni, Pb, Zn	Water	–	trace	air-C$_2$H$_2$	Direct; or extract with APDC/MIBK.	Willey et al., 1972
Trace elements	Sea water	–	μg/l	air-C$_2$H$_2$	Add buffer to pH 6; extract with HMA-HMDTC into butylacetate.	Tsalev et al., 1972b
Trace elements	Sea water	–	trace	air-C$_2$H$_2$	Concentrate by ion-exchange on Chelex-100; extract with APDC/MIBK.	Le Meur et al., 1973
Al,Fe	Natural water	–	30-800 μg/l	N$_2$O-C$_2$H$_2$	Extract Al-oxine into MIBK.	Beck et al., 1974
B	Sea water	249.7	>0.1 μg/l	N$_2$O-C$_2$H$_2$	Concentrate and extract with 2-ethyl-1, 3-hexanediolinto MIBK.	Spielholtz et al., 1974
Ba	Water	553.5	>10 μg/l	air-C$_2$H$_2$	Pass on Dowex F500 W-8K; separate Ca, Mg, Sr by washing with DCTA as pH 7.2; eluate Ba with 4 N HNO$_3$.	Sixta et al., 1973
Cr,Cu,Pb	Sea water	–	μg/l	air-C$_2$H$_2$	Adjust pH 3.5; add APDC; heat to 80°C; cool and extract into MIBK.	Kubo et al., 1973
Pb	River water	283.3	>6 ng/l	air-C$_2$H$_2$	Acidify with 0.2 N HCl; add 0.3 M KI; shake with Amberlite La-1/xylene.	Goto, 1974
Se	Water	196.1	>2 μg/l	N$_2$-H$_2$	Add CaCl$_2$ and 0.1 M HCl; K MnO$_4$; boil add NaOH; dry; redissolve in HCl; add NH$_4$Cl, SnCl$_2$; pass N$_2$ stream to carry Se H$_2$ in flame.	Lansford et al., 1974

Table V. Applications of Nonflame Atomic Absorption: Water Analysis

Element	Sample	λnm	Concentration	Atomization	Sample Preparation	Reference
Ag,Al,Cd,Co, Cr,Fe,Mn, Ni,Pb,Sb,V	Natural Waters	–	μg/l	Graphite tube	Automatic introduction of sample.	Pickford et al., 1972
As	Wastewater	193.7		Heated absorption cell	Convert into arsine.	Hwang et al., 1972
Cd	Water	228.8	0.1-1 μg/l	Carbon rod	Direct.	Dolinsek et al., 1972a
Cu,Fe,Mn, V	Sea water	–	Cu 1-2 μg/l Fe 10-100 μg/l Mn 0.5-3 μg/l V 0.2-7 μg/l	Graphite rod	Direct.	Segar et al., 1972
Se	Water	196.0	26 pg	Carbon rod	Digest with HNO_3-$HClO_4$.	Baird et al., 1972
Cd,Pb,Ag	Natural water	–	traces	Tantalum ribbon	–	Ratonetti et al., 1973
Cd,Cu,Mn, Ni,Pb,Mo	Natural water, brines	–	μg/l	Graphite tube	–	Edmunds et al., 1973
Cd,Cu	Sea water		0.03-1.6 μg/l	Graphite rod	Adjust pH 8 with NH_4OH; extract with APDC/MIBK.	Shigematsu et al., 1973a
Hg	Water	253.7	0.08-2.4 μg/l	Cold vapor	Inorganic Hg: add H_2SO_4; reduce with $SnCl_2$. Organic Hg: add $CdSO_4$; reduce with $SnCl_2$ to remove inorganic Hg; add NaOH; repeat.	Kamada et al., 1973a
Hg	Sea water	253.7		Cold vapor	Acidify with H_2SO_4; extract with dithizone.	Gardner et al., 1973
Hg	Sea water	253.7	>0.1 μg/l	Cold vapor	Oxidize with $KMnO_4$; reduce with $SnCl_2$.	Gilbert et al., 1973
Al,Mn,Cr, Cu	Natural water	–	μg/l	Graphite tube		Welz, 1972a

Element	Sample	Wavelength	Concentration	Technique	Method	Reference
Cd,Cu,Cr, Mn,Pb,Fe	Water	—	trace	Graphite tube	Direct with standard addition method.	Barnard et al., 1973
Trace elements	Sea water	—	trace	Graphite tube	Adjust pH 2-5; extract with APDC/MIBK.	Paus, 1973
Be	Mineral water	234.8	>0.05 µg/l	Graphite tube	Direct.	Janouškova et al., 1974
Cd	Sea water		>3x10^{-13} g	Graphite tube	Atomize at 750°C.	Lundgren, 1974
Cd	Wastewater	228.8	>6x10^{-13} g	Graphite tube	Dry at 170°C; decompose at 330°C; atomize at 1400°C.	Yasuda et al., 1974
Cd	Sea water	228.8	0.5-4.5 µg/l	W Wire	Deposit Cd on W wire by electrolysis.	Lund et al., 1974
Cd,Pb,Zn	Sea water	228.8	0-11 ng/l	Graphite tube	Deposit on hanging Hg drop by electrolysis.	Jensen et al., 1974
Cu	Sea water	324.7	0.5-20 µg/l	Graphite tube	Dry; ash at 1000°C before atomization.	Ediger et al., 1974
Cu	Sea water	324.7	0.2-6 µg/l	Graphite tube	Filter on Millipore; treat with persulfate; pass through chitosan column. Eluate with 1 M H_2SO_4 or 1,10-pheananthroline.	Muzzarelli et al., 1974a
Ag	Sea water	253.7	10-100 µg/l	Cold vapor	Extract with dithizone.	Chester et al., 1973
Hg	Sea water	253.7	10-20 ng/l	Cold vapor	Add $SnCl_2$; pass argon and collect Hg on Au foil; heat.	Olafsson, 1972
Hg	Sea water	253.7	3 ng/l	Cold vapor	After reduction by standard method, trap Hg in liquid N_2 in U-tube; heat.	Fitzgerald, 1974
Hg	Natural water	253.7	1-100 µg/l	Cold vapor	Add $K_2S_2O_8$ + H_3PO_4; heat in sealed ampoule at 120°C.	Alberts et al., 1974
Pb	Sea water	283.3	1.7-8 µg/l	Graphite tube	Extract with DDC/MIBK.	Shigematsu et al., 1973b
As,Sb	Industrial effluents; river water	As 193.7 Sb 217.5	0-200 µg/l 0-100 µg/l	Graphite tube	Use 20 µl; dry at 270°C; ash at 330°C; atomize at 2300°C.	Yasuda et al., 1974b

Table V, Continued

Element	Sample	λnm	Concentration	Atomization	Sample Preparation	Reference
V	Sea water	318.4	0.2-12 μg/l	Graphite tube	Adjust pH 4, separate V on chitosan column.	Muzzarelli et al., 1974b.
Al,Cr,Cu, Fe,Hg, Mn,Pb	Natural and wastewater		ppb	Graphite tube		Welz et al., 1973
Mo	Sea water		0-10 μg/l	Graphite tube	Collect Mo by extraction on p-amino benzyl cellulose or chitosan column at pH 2-5.	Muzzarelli et al., 1973
Hg	Water	253.7	>10 ng/l	Cold vapor	Add $KMnO_4$ + HNO_3 + H_2SO_4; reduce $KMnO_4$ excess with NH_2OH, HCl, reduce Hg^{2+} with $SnCl_2$.	Montiel, 1972

have been proposed for the analysis of As, Sb and Se in water: Kwok-Tai (1973) developed a method for the analysis of arsenic and antimony at a level of 5-300 μg/l; Pierce *et al.* (1976) studied an automatic method in which the final reading is taken in the furnace (see Section 3.4).

3.4 Application of Nonflame Atomic Absorption to Water Analysis

The number of publications on nonflame atomic absorption increases every year. A few characteristic applications are listed in Table V. A direct water analysis (without extraction) often leads to detection limits of 0.01-0.1 μg/l for water that is not highly mineralized and 0.2-5 μg/l for sea water in which the mineral load is 35 g/l. However, the natural content of numerous trace elements such as Cd, Cr, Co, Ni, Pb and V in sea water is often less than 1 μg/l, and chemical extraction becomes necessary. The classical methods make use of the system APDC/MIBK (Table V). A few special cases merit attention. Cadmium is separated from the sea water matrix by electrolysis on a tungsten wire (Lund *et al.*, 1974) and atomization by electrical heating of the wire. Jensen *et al.* (1974) also use electrolysis with a dropping mercury electrode; atomization is performed in a graphite furnace.

Many publications deal with the analysis of mercury, which is separated from the analyzed medium by reducing into gaseous Hg$^{\circ}$ at normal pressure and temperature. The atoms are entrained into a quartz-window cell where atomic absorption is read. In water, however, mercury is either in the form of the mineral or combined in an organic molecule (for example, ethyl- or methylmercury). The mineral mercury is easily separated by reduction with stannous chloride in sulfuric acid medium and then entrained by an air stream into the absorption cell. However, the organic forms very often dominate. Water is treated at elevated temperature with permanganate up to complete decomposition of the organics, and the reducing agent, SnCl$_2$, is then added. One of the principal difficulties of analyzing organic mercury is the stability of certain compounds that are difficult to oxidize. Among the possible oxidants are potassium permanganate, potassium persulfate and hydrogen peroxide, while the reducing agents include stannous chloride, hydroxylamine hydrochloride and stannous sulfate. To improve the sensitivity, some authors use extraction with dithizone, particularly for analysis in sea water, followed by binding of the mercury either by condensation in liquid nitrogen (Fitzgerald, 1974) or by forming an amalgam on a gold support.

Among the nonflame methods of special interest, there is also the analysis of arsenic and selenium separated from the medium in the form of hydrides (see Section 3.3). Pierce *et al.* (1976) developed an automated method for the analysis of surface water; the authors use a Technicon apparatus for introducing the reagents and an atomic absorption spectrometer (IL 151) equipped with a furnace for decomposition of the hydride into the atoms. The conditions of the method make it possible to reach concentrations of a few ppb (μg/l).

In spite of the considerable development of electrothermal atomization methods, numerous problems are still unsolved. The nature of the furnace construction material (graphite, carbon, tungsten, tantalum, molybdenum) is still under discussion and the advantages and disadvantages of each are not yet fully known. McIntyre *et al.* (1974) compared tantalum and molybdenum as atomization support for the analysis of Co, Ni and Cu and recommended molybdenum, since it is more inert with respect to these metals, giving a better sensitivity and weaker interelement effects. Rattonetti (1974) used tantalum for the analysis of elemental Cd, Pb, Ag and In in rain water; he observed a certain number of interferences.

Another important problem is the geometry of the atomizing device: tube, boat, ribbon, filament, cup. Unfortunately there seems to have been no interest in a systematic study. While graphite remains most common despite everything, there is still hesitation concerning the dimensions to be used: diameter, length, wall thickness. Finally, injection of the samples still remains a delicate matter and a source of error in the precision. Electrothermal atomization often leads to nonspecific absorptions that increase the reading value. Many correction devices exist (deuterium, hydrogen lamp), but their application is limited to the spectral range and the optical density value that can be corrected. The correction of specific absorptions was discussed by Edmunds *et al.* (1973) and by Welz and Wiedeking (1973) in connection with water analysis.

A final application to be noted concerns the analysis of suspended sediments, particularly in sea water. The methods are similar to those used for the analysis of atmospheric particles (see Section 5).

4. ANALYSIS OF PLANTS AND BIOLOGICAL MATERIALS

4.1 Field of Application

A certain number of elements existing in the plant and biological medium play an essential role in the state of traces. This is true for elements Fe, Zn, B and Mn in a level of a few dozen ppm and of Cu, Co and Mo in a level of a few tenths to a few ppm. We can also add Se, an element necessary for animal nutrition at a tenth of a ppm, which becomes toxic in the ppm level. Furthermore, numerous toxic elements can be assimilated by animal and plant tissues—Cr, Pb, Hg, As, Sb, Te and Tl. Environmental pollution is a major problem; atomic absorption has already allowed the control problem of these chemical elements to be solved in many cases. In other cases, including certain nonflame atomic absorption techniques, the analysis of biological media requires microanalytical and highly sensitive methods. Finally, we should note the control of elements used in therapy at a trace level, such as Cr, Bi, As, Au and Ag.

This chapter treats plant and biological media simultaneously. Chemically, these are media with an organic matrix (90-99%), except for products such as serum or urine. The atomic absorption analysis techniques as well as the concentration levels are similar.

4.2 Preparation, Dissolution, Analysis

The dissolution of plant media takes place either by ashing at 450-500°, followed by acid redissolution (0.1 N HCl) or by acid digestion (HNO_3-H_2SO_4 or $HClO_4$-HNO_3-H_2SO_4), followed by evaporation of the excess acid and taking up in hydrochloric acid again.

The ashing method is performed as follows (Pinta, 1973). The sample (1-2 g) is ashed at 450°C for 2 hr, cooled, solubilized in concentrated hydrochloric acid and filtered. The insoluble material (SiO_2) is treated with hydrofluoric acid up to complete elimination of the silica, evaporated, taken up in hydrochloric acid and combined with the above filtrate to be brought to 100 ml (0.1 N HCl). The major elements (Na, K, Ca, Mg), as well as the oligo-elements (Fe, Cu, Mn, Zn) are determined classically with calibration in a simple medium with no need to add interaction suppressors. The lanthanum salts, added to correct the PO_4/Ca and PO_4/Mg interactions are generally useless for Fe, Cu, Mn and Zn.

The methods involving ashing have the risk of a loss of volatile elements (As, Sb, Se, Te, Cd and Hg). It is advisable to make use of acid digestion. Among the numerous methods, the following can be considered: the sample (0.5-2 g) is treated with 5 ml water and 5 ml nitric acid in a Kjeldahl flask for 16 hr at normal temperature. 10 ml perchloric acid and 5 ml sulfuric acid are added and the mixture is gently heated to complete decomposition of the organics. The mixture is then evaporated to dryness, taken up in hydrochloric acid (0.1 N) and brought to 100 ml. To avoid extensive acid consumption (contamination source), a reflux condenser should be attached on the Kjeldahl flask. The method is widely used for elements such as Cd, Hg, Se and As.

Acid digestion can also be performed in a Teflon bomb under pressure: 0.5 g of sample is treated with HNO_3, H_2SO_4 and HF. Perchloric acid involves risks of explosion in some cases (products with a high fat content). Tables VI and VII show numerous applications.

Biological tissue can also be solubilized with ashing (500-700°C), but with risks of losses (Mg, Zn, Cu). Dissolution by the wet method is often preferred: 0.1-0.2 g dried and homogenized (or lyophilized) organ material is digested in 1 ml concentrated HNO_3, 1 ml concentrated HCl and 6 ml water at 100°C with agitation. The residue is centrifuged and brought to 10 or 25 ml. For elements firmly bound to the organic material (Cu, Fe), the attack may not be strong enough. Nitroperchloric acid digestion is then preferred. The sample (0.1-0.5 g) is digested with 1.5 ml HNO_3, 10 ml water and 1.5 ml $HClO_4$ at room temperature for a few hours and is then brought to boiling with agitation. The sample should not be evaporated to dryness, particularly with products having a high fat content.

In addition to the classical methods, more specific techniques have been proposed for the dissolution of biological tissue: mineralization by acid digestion in a bomb and low-temperature calcination under oxygen and at low pressure (Hauser et al., 1972a, 1972b). Paus (1972) determined Cd, Cu, Fe, Hg, Pb and Zn in biological materials.

Jackson et al. (1972) proposed a method for the solubilization of tissue with quaternary ammonium hydroxide (Soluene): 100-300 mg of tissue are solubilized in 0.5-1.0 ml of Soluene and the solution is diluted 3-4 times with toluene. Murthy et al. (1973) determined Zn, Cu, Cd and Pb in organs after simple solubilization in alcoholic tetraethyl-ammonium hydroxide solution. The authors obtained complete recovery of the elements. Calibration is preferably conducted by the addition method because of the viscosity and surface tension of the solutions. For blood analysis, deproteinization with trichloroacetic acid (TCA) is often recommended.

Separation and enrichment methods are frequently used for the determination of contents less than 1 ppm. This applies particularly to elements such as Cd, Pb, Co, Mo, Se, Hg and Sb. APDC/MIBK extraction (see Sections 2.1.3 and 3.2) is widely used for the analysis of plant and biological media, particularly to detect elemental Cd, Co, Cu, Pb and Zn (Tables VI and VII), as well as for gold in blood and urine. Some elements require more specific methods: extraction of bismuth from the acid sample solution at pH 3 with piperidine dithiocarbamate in MIBK and extraction of molybdenum in the oxinate form in MIBK. Dithiocarbamate is also used for the extraction of Cd, Pb and Tl (Table VII). Mercury is the most commonly investigated element. There have been many publications on mercury analysis in recent years (see below, Section 4.4, Applications of Nonflame Methods).

Table VI. Applications of Flame Atomic Absorption: Vegetal and Biological Materials

Element	Sample	λnm	Concentration	Flame	Sample Preparation	Reference
Cd,Pb	Plants	Cd 228.8	0.1-0.5 ppm	air-C_2H_2	Extract overnight with 1:1 HNO_3.	Kahn et al., 1972b
Co	Plants		0-0.5 ppm	air-C_2H_2	Ash; dissolve in 0.1 N HCl; extract with APDC/MIBK.	Gelman, 1972
Pb	Plants			air-C_2H_2	Ash at 450°C.	Smith, 1972
Pb	Plant and animal	217.0			Ash; dissolve; coprecipitate with $SrSO_4$; convert to carbonate; dissolve in HNO_3.	Hoover, 1972
Ba	Bones, tissues	553.5		N_2O-C_2H_2	Ash at 500-550°C; dissolve in HNO_3; add K as ionization buffer.	Tanaka et al., 1972
Bi	Tissues, urine	223.1	10-100 ppm	air-C_2H_2	Digest with HNO_3-$HClO_4$-H_2SO_4-H_2O_2; extract into piperidine DTC/MIBK at pH 3.	Hall et al., 1972a
Cd,Pb	Blood	Cd 228.8 Pb 283.3	2.5-3 µg/l 20-600 µg/l	air-C_2H_2 + Ta boat	Ash 0.5 ml in low temp. O_2 asher for 16 hr.	Hauser et al., 1972a, 1972b
Cu,Fe,Mn, Zn	Tissues	Cu 324.7 Fe 248.3 Mn 279.5 Zn 213.9	0.7-5 ppm 1.5-150 ppm 1-30 ppm	air-C_2H_2	Dissolve in soluene; dilute 1:3 in toluene.	Jackson et al., 1972
Li	Blood	670.8	50 mg/l	air-C_2H_2	Dilute with H_2O; add Na + K.	Arroyo et al., 1971
Cd,Pb,Zn	Teeth			air-C_2H_2	Dissolve in HNO_3 extract with APDC/MIBK.	Kaneko, 1971
Pb	Blood, urine	283.3		air-C_2H_2	Deproteinize with TCA; extract with APDC/MIBK.	Yamauchi, 1971
Fe,Cu,Zn Mn	Plants	–	ppm levels	air-C_2H_2	Ash at 450°; dissolve in HCl.	Pinta, 1973
Cd	Human hair	228.8	0.2-8 ppm	air-C_2H_2	Digest with HNO_3; evaporate; dissolve in HNO_3.	Sorenson et al., 1973
Mo	Plants	–	>0.025 ppm	N_2O-C_2H_2	Ash; dissolve; extract Mo with oxine/MIBK.	Stupar et al., 1974
Rb	Plants	780.0	ppm	air-propane	Digest in HNO_3; dilute in N HNO_3 + 500 µg/ml Ca + 1000 µg/ml La.	Hafez et al., 1973

Table VI, Continued

Element	Sample	λnm	Concentration	Flame	Sample Preparation	Reference
Cd	Foodstuffs		0-0.3 ppm	air-C_2H_2	Digest with HNO_3 + V_2O_5; dilute; add H_2SO_4; evaporate; add H_2O_2 until clear citric acid; adjust pH 7 with NaOH; transfer to Chelex-100 ion-exchange column; elute Na with $(NH_4)_2SO_4$ and Cd with $2\,N\,H_2SO_4$.	Baetz et al., 1974
Cd,Pb	Fish	Cd 228.8	0.005-0.1 ppm	air-C_2H_2	Extract with Na DDC/MIBK at pH 5.5-8.5.	Childs et al., 1974
Co,Cu	Animal foods	Co 240.7	0.2-1 ppm	air-C_2H_2	Ash; dissolve in 0.1 N HCl; extract with APDC/MIBK.	Popova et al., 1974
Pb	Beverages	217.0		air-C_2H_2	Acidify with HCl; add acid ascorbic; extract with DEADC into xylene.	Snodin, 1973
Pb	Animal foods	–	trace	air-C_2H_2	Extract from HCl solution with dithizone/$CHCl_3$.	Oelschlaeger et al., 1973
Cu,Zn	Edible oils, fats	–	0-0.5 ppm	air-C_2H_2	Preconcentrate by refluxing with HCl-EDTA.	Jacob et al., 1974
As	Biological material	193.7	0-50 μg/l	Ar-H_2	Dissolve; reduce with KI/$SnCl_2$/Zn/HCl to arsenic.	Kamada et al., 1973b
Cd,Pb	Hair	–	trace	air-C_2H_2	Wash; degrease; ash dissolve in HCl/HNO_3 and extract with $HClO_4$.	Eads et al., 1973
Cd,Pb,Zn	Teeth	–	trace	air-C_2H_2	Dissolve with HNO_3/H_2O_2; adjust to pH 3.0 dilute; add APDC + MIBK; centrifuge.	Yamamoto et al., 1973
Cd,Pb,Se	Fish	–	trace	air-C_2H_2	Remove fats with hexane, extract residue with H_2O.	Lunde, 1973
Trace elements	Serum		trace	N_2O-C_2H_2 + Mo cup	Dry in Mol; cup and atomize at 1900°C.	Mitchell et al., 1974b
As,Sb,Se,Te	Foods	As 197.2 Sb 217.6 Se 196.0 Te 214.3	$\geqslant$7 ppb $\geqslant$6 ppb $\geqslant$16 ppb $\geqslant$20 ppb	N_2-H_2	Digest with HNO_3-H_2SO_4-$HClO_4$; evaporate; dissolve in HCl; add $NaBH_4$ + NaOH to form hydride.	Fiorino et al., 1974, Fiorino et al., 1976.
Au	Blood, urine	243.8	0-6 ppm	air-C_2H_2	Extract with APDC/MIBK.	Harth et al., 1973

Table VII. Applications of Nonflame Atomic Absorption: Vegetal and Biological Material

Element	Sample	λnm	Concentration	Atomization	Sample Preparation	Reference
Hg	Plants			Cold vapor	Dry at 80°C; grind and wet ash.	Smith, 1972
Ba	Hair, hand swab	553.5		Graphite tube	Hair: direct. Swab: extract Ba from solution with TTA	Renshaw, 1972
Cd	Blood	228.8	3 μg/l	Graphite crucible	Dilute 0.1 ml (1+9) with H_2O; atomize 2 μl.	Dolinsek et al., 1972b
Co,Cr,Ni	Blood, serum	Co 240.7 Cr 357.9 Ni 232.0	0.2 μg/l	Ta strip	Dilute.	Ullucci et al., 1972
Cr	Plasma, urine	357.9	0.2-8 μg/l	Graphite tube	Digest 200 μl with $HClO_4/H_2O_2$; dissolve in 0.3 N HCl.	Davidson et al., 1972
Hg	Biological material	253.7	μg/l		Inorganic Hg: reduce with $SnCl_2$; organic Hg: treat with $SnCl_2 + CdCl_2$.	Magos, 1971, Magos, 1972
Hg	Muscle	253.7	0.0005-4ppm	Hot vapor	Decompose in O_2 flask; dissolve in HCl reduce with NH_2OH, HCl; collect Hg on Ag wire; heat.	Okuno et al., 1972
Pb	Urine	217.0	0-200 μg/l	Carbon rod	Dilute (1:1) with conc. HNO_3; add 1 μl to rod.	Kubasik et al., 1972b
Pb	Blood	217	0.2-1 mg/l	Carbon rod	Dilute (1:2) with triton x 100; add 1 μl to rod.	Kubasik et al., 1972a
Pb	Blood	217	0.05 mg/l	Carbon crucible	Dilute (1:9) with H_2O; use 2 μl for analysis.	Dolinsek et al., 1972b
Pb	Hair, hand swab	283.3		Graphite tube	Hair: direct. Swab: extract Pb from solution with APDC/MIBK.	Renshaw, 1972
Sb	Hair, hand swab	218		Graphite tube	Hair: direct. Swab: extract Sb from solution as $(SbCl_6)^-$ H^+	Renshaw, 1972
Al,Cr,Cu, Cd,Mn,Ni	Blood	–	1-100 ng	Graphite tube	Direct.	Hudnik, et al., 1972
Al,Cr,Cu, Fe,Mg,Mn	Biological samples	–		Ta strip		Takeuchi et al., 1972

Table VII, Continued

Element	Sample	λ nm	Concentration	Atomization	Sample Preparation	Reference
Hg	Cereals	—		Cold vapor	Digest with H_2SO_4-HNO_3 with V_2O_5 (catalyst).	Malaiyandi et al., 1972
Pd,Cd,Cu,Zn	Oysters	—	0.001 ppm	Graphite tube		Willis et al., 1974
Co	Animal foods	—	0.01-0.2 ppm	Graphite tube	Ash at 500°C; extract with 1-nitroso-2-naphtol.	Hageman et al., 1974
Cu	Oils	—	0.001 ppm	Graphite tube	Ash at 490°C in O_2/N_2 (2:1).	Kundu et al., 1974
Hg	Foods	—	0.005-0.4 ppm	Cold vapor	Digest with HNO_3-H_2SO_4; add MnO_4K; NH_2OH, HCl and $SnCl_2$.	Gomez et al., 1974
Pb	Milk	283.3	trace	Graphite tube	Dilute with H_2O (x 10).	Dujmovic et al., 1974
Al	Serum	—	0.02-0.1 ppm	Graphite tube		Fuchs et al., 1973
As,Hg	Urine	—	Cd 20 ng Hg 5 ng	Graphite tube	Reduce with $NaBH_4$. Trap AsH_3 in liquid N_2-cooled U-tube.	Toffaletti et al., 1974
Au,Co,Li	Blood plasma	Au 242.8 Li 670.8	0.6-10 ppm	Carbon rod	—	Maessen et al., 1974
Cd,Pb,Tl,Zn	Biological materials	—	ppm	Graphite tube	Digest with HNO_3 (urine), or ash (blood) add La(NO3)3 to overcome Fe interference.	Machata et al., 1973
Cd	Biological materials	228.8	0.001-0.003 ppm	Graphite tube	Digest with $HClO_4$ to destroy partly organic matter.	Schumacher et al., 1974
Cd,Pb	Dental samples	Cd 228.8 Pb 217.0	0.09-2.2 ppm	Graphite tube	Grind, ash and atomize.	Langmyhr et al., 1974
Co	Biological tissues	240.7	>0.04 ppm	Graphite tube	Ash at 900°C before atomization.	Lundgren et al., 1974
Hg	Fish	253.7	trace	Cold vapor	Digest with HNO_3-H_2SO_4; add $KMnO_4$; reduce.	Kivalo et al., 1974
Hg	Fish	253.7	0.1-1 μg/g	Cold vapor	Digest with HNO_3-H_2SO_4; add HNO_3-$HClO_4$; reduce and volatilize Hg.	Antonacopoulos, 1974

Element	Material	Wavelength	Range	Atomizer	Procedure	Reference
Cr	Blood serum plasma	–	2 μg/l	Graphite tube	Add HNO_3 + H_2O_2 in tube, followed by sample.	Anand et al., 1974
Cu	Blood, plasma	324.7	10 μg/l	Graphite tube	Dilute; heat on atomizer to 2500°C.	Montaser et al., 1974
Hg	Urine	253.7	0-40 μg/l	Cold vapor	Add H_2SO_4-K MnO_4; stand 16 hr; centrifuge and reduce with $SnCl_2$-$CdCl_2$.	Least et al., 1974
Hg	Air	253.7		Cold vapor	Digest with H_2SO_4-K MnO_4; reduce with $SnCl_2$.	Eads et al., 1973
Hg	Urine	253.7	5-150 μg/l	Cold vapor	Add HNO_3 + antifoaming agent + $SnCl_2$; pass vapor through $Mg(ClO_4)_2$ before analysis.	Patriarca et al., 1973
Hg	Orchid leaves	253.7	–	Cold vapor	Digest with HNO_3-$HClO_4$-H_2SO_4; reduce with H_2SO_4-$NaCl$-$SnCl_2$.	Rains et al., 1972
Mo	Plants	313.3	–	Graphite tube	Ash at 500°C 5 hr; dissolve in 5% HCl.	Henning et al., 1973
Au	Serum	242.8	1-10 ppm	Graphite rod	Treat rod with xylene; add 2 μl serum; dry; char; ash; atomize.	Aggett, 1973
Cd	Biological materials	228.8	trace	Ta ribbon	Urine: acidify. Blood: add saponin, formamide, APDC, MIBK; soak hair, nail in 0.3% triton X-100; wash; dry; digest in HNO_3/H_2O_2.	Ullucci et al., 1973
Cd,Pb,Tl	Urine	Cd 228.8 Pb 217.0 Tl 276.8	0-10 μg/l 0-400 μg/l 0-40 μg/l	Graphite rod	Adjust pH 7 with buffer; add Na DDC + $CaCl_2$ + MIBK; extract.	Kubasik et al., 1973
Hg	Biological materials	253.7	ppb	Cold vapor	Digest in acid and reduce.	Cumont, 1976
Cr	Urine	357.8	1-10 μg/l	Graphite rod	Ash and atomize directly in stream N_2.	Schaller et al., 1972
Cr	Urine	357.8	3-38 μg/l	Graphite tube	In tube: dry at 230°C; ash at 1300°C, atomize at 2700 C (20 μl).	Ross et al., 1973
Hg	Urine	253.7	0.2-30 μg/l	Cold vapor	Digest 16 hr with H_2SO_4-K MnO_4 at 0°C; dilute; centrifuge; reduce with $SnCl_2$.	Kubasik et al., 1972c
Hg	Biological materials	253.7	0.02-1 μg/l	Cold vapor	Digest with HNO_3-H_2SO_4; add K MnO_4 and NH_2OH-HCl; sweep with air-stream.	Hollerer et al., 1973

Table VII, Continued

Element	Sample	λ nm	Concentration	Atomization	Sample Preparation	Reference
Hg	Organic materials	253.7		Cold vapor	Treat with $O_2 + O_3$ in bubbler; dilute with H_2O_2; reduce with $NaBH_4$.	Lopez-Escobar et al., 1973
Pb	Blood	—	>0.05 mg/l	Carbon rod	Treat carbon rod with 0.5 μl xylene; add blood (0.5 μl) and 0.5 μl xylene.	Rosen et al., 1972
Trace elements	Human hair	—	ppm	Graphite rod	—	Dick et al., 1973
Trace elements	Plasma, serum	—	ppm		Separate protein fractions by electrophoresis on cellulose acetate membranes.	Delves, 1973
Trace elements	Blood, urine	—	ppm	Graphite tube	Extract with Na DDC/MIBK for Pb, Tl, Cd, Hg.	Berman, 1973
Co	Plants	—	—	Graphite tube	Ash; treat with 0.1 N HCl; extract with APDC/CH_2Cl_2.	Soerensen, 1974
Pb	Leaves	283.3	—	Graphite tube	Dry at 105°C; apply to furnace with H_2O to form slurry; decompose at 650 C; atomize at 2000°C.	Brady et al., 1974
Hg	Meats	—	0.001-0.06 ppm	Cold vapor	Digest with HNO_3-H_2SO_4-K_2O_5; reduce.	Kirkpatrick et al., 1973
Hg	Fish protein	253.7	0.05-30 ppm	Hot vapor	Digest in H_2SO_4; add K MnO_4, then H_2O_2 dropwise; reduce with NH_2OH-$SnSO_4$-NaCl; flush with air.	Archer et al., 1973
Cd,Cu,Mn, Pb,Zn	Plasma, human tissues	Cd 228.8 Cu 324.7 Mn 279.5 Pb 217.0 Zn 213.9	0.35 ng/l 1.0 ng/l 1.8 ng/l 6.6 ng/l	Graphite tube Flame air-C_2H_2	Solubilize with alcoholic solution of tetramethyl ammonium hydroxide at 70°C.	Gross et al., 1974
Pb	Plasma and whole blood		0-70 μg/l 180-1300 μg/l	Graphite rod	Direct: use 1 μl sample.	Rosen et al., 1974

Element	Sample	Conc.	Wavelength	Atomizer	Procedure	Reference
Se	Biological samples	60 μg/l		Graphite tube	Digest in HNO_3-$HClO_4$; separate Se by reduction and precipitation with ascorbic acid; redissolve.	Ihnat et al., 1974
Sn,Mn	Serum	trace	460.7	Graphite tube	Digest with HNO_3-$HClO_4$; evaporate and redissolve.	Beck et al., 1974
Te	Biological samples	>0.03 ppm	214.3	Sample boat in flame		Cheng et al., 1974
Pb, Tl	Blood, urine	ppm	Tl 276.7	Graphite tube	Direct method, with background correction.	Hauck, 1973
Trace elements	Biological tissues	trace		—	Digest with HNO_3; add CH_3CO_2H; adjust pH 3 with NH_4OH; extract with APDC/MIBK; evaporate organic phase; dissolve with HNO_3; dilute.	Segar et al., 1973
Trace elements	Serum	trace		Mo cup + flame N_2O-C_2H_2	—	Mitchell et al., 1974b
Trace elements	Biological materials	trace		Delves cup	Digest in cup with H_2O_2-HNO_3-H_2SO_4 or HNO_3-$HClO_4$; preheat at 90 and 300°C.	Mitchell et al., 1974a
Hg	Fish	ng/g	253.7	Cold vapor	Digest with HNO_3-H_2SO_4; add K MnO_4-persulfate K; reduce with $SnCl_2$. Or: digest with H_2SO_4; add K MnO_4-H_2O_2; reduce with $SnSO_4$.	Barber, 1972
Hg	Fish	ng/g	253.7	Cold vapor	Combust in O_2; pass gases through heated Ag-Mn oxides; collect Hg on cold Co-Mn oxides; heat in stream of N_2 to evolve Hg; collect on Au filter, reheat.	Lidums, 1973

4.3 Applications of Flame Atomic Absorption

Table VI lists some of the many characteristic applications published in recent years. The *Annual Reports on Analytical Atomic Spectroscopy,* edited annually by Hubbard (1971-1972) and Woodward (1973, 1974, 1975) and published by The Society for Analytical Chemistry, contains a complete literature review covering the field of atomic absorption spectrometry. There are also the reviews of Winefordner and Vickers, *Flame Spectroscopy,* which appear annually in April in *Analytical Chemistry.*

In Table VI the majority of biological media are mineralized by acid digestion, and the residue is then taken up in hydrochloric acid. Direct analysis of this solution allows detection with a limit of a few ppm (Fe, Cu, Mn, Zn, Ni, Co, Pb, Cd, Cr, Sr) for solids and a few mg/l for liquid media. These detection limits are inadequate for toxic elements (Pb, Cd, Cr). Interest in the determination of As, Sb, Se and Te in plant and biological media is increasing. The sample is dissolved by acid digestion up to complete destruction of the organics. The elements are then reduced to a lower valence [As (V) $\rightarrow$ As (III)] by stannous chloride and then, in the form of hydride, by nascent hydrogen (Zn + HCl). The hydride is entrained by an inert gas (argon or nitrogen) into a hydrogen flame (Ar-H_2 or N_2-H_2). A few dozen ppb ($\mu g/kg$) of As, Sb, Se and Te can thus be detected.

Another increasingly popular method consists of the direct reduction of the element into hydride (As^{+5} $\rightarrow$ As^{-3}) by means of sodium borohydride (see also Section 3.3 and 3.4). Fiorino *et al.* (1976) proposed a method for the determination of As, Sb, Se and Te consisting of entraining the hydrides in a separate N_2-H_2 flame. However, this method requires a prereduction by sodium iodide Sb [Sb (V) $\rightarrow$ Sb (III)]. The sample solution (acid digestion in HNO_3-H_2SO_4-$HClO_4$) containing about 1 g in hydrochloric acid medium (20 ml) is treated with 0.5 ml 10% sodium iodide. After 1 min, a sodium borohydride solution (4 g $NaBH_4$ + 10 g NaOH in 100 ml water) is introduced with a special metering device at a flow rate of 24-32 ml/min for 15 sec (*i.e.* 6-8 ml). The hydride formed is entrained into the flame. A water rinse at 30 ml/min (5 sec) eliminates traces of borohydride remaining in the hydride feed tubes. A complete blank test is indispensable. For the determination of Se and Te, it is important not to perform the sodium iodide prereduction that may already entrain a portion of the elements in the form of SeH_2 and TeH_2 and elemental form. The detection limits are listed in Table VI. The method might also be extended to elemental Bi, Ge and Sn. Finally, for As, Sb, Se and Te, the authors recommend the use of an EDL lamp as an emission source instead of the classical hollow-cathode lamps.

4.4 Applications of Nonflame Atomic Absorption

In the analysis of biological media, improvement obtained with non-flame methods has been apparent when compared to the classical atomic absorption methods. The two reasons for this progress are the increased sensitivity and the possibility of working with very small amounts of sample. However, the analytical precision is clearly inferior because of the complexity of the electrothermal decomposition and the atomization program, as well as the significant risks of contamination, particularly for elements Cu, Zn, Fe, Pb and Cd, since the analysis requires absolute quantities of 10^{-10} to 10^{-12} g. In many cases, direct analysis of the sample solution allows detection limits of 0.1-1 ppm to be reached, such as for Cu, Co, Mo, Zn and Mn.

In biological media, higher sensitivities are necessary. These often require preliminary separation, particularly in toxicity problems (Pb, Cd, Hg, Se, As and Tl). The methods are similar to those described earlier (Section 4.3), but the sample sizes are often smaller.

Some special applications should also be noted (Table VII). Thus, a mercury analysis in plants and biological media has been the subject of numerous publications. The classical methods are based on acid digestion (HNO_3-H_2SO_4 or HNO_3-$HClO_4$-H_2SO_4), followed by reduction with stannous chloride to liberate the mercury in the atomic state, which is then entrained into the absorption cell with an air stream. The sample must be completely oxidized. The oxidative power of acid digestion can be improved by adding a catalyst, such as vanadium oxide, or an oxidant, such as $KMnO_4$, H_2O_2.

In some techniques (Table VII), mercury is liberated by ashing in air (Smith, 1972), in an oxygen atmosphere (Okuno *et al.*, 1972), or in oxygen and ozone (Lopez-Escobar *et al.*, 1973). It is sometimes useful to distinguish inorganic and organic forms of mercury. Magos (1971) liberates inorganic mercury by simple reduction of the sample solution with stannous chloride. Organic mercury is then liberated after treatment with $CdCl_2$ + $SnCl_2$. While stannous chloride is the most commonly used reductant, some authors recommend hydroxylamine hydrochloride (Okuno *et al.*, 1972; Hollerer *et al.*, 1973) or sodium borohydride (Lopez-Escobar *et al.*, 1973). Finally, we should note that the liberated mercury can be directly conducted into the absorption cell by an air sweep or bound on a gold or silver support in the state of an amalgam to be volatilized in the absorption cell by low-temperature heating.

Among the recently proposed methods, there is the analysis of elemental As, Sb and Se liberated from the sample in hydride form (AsH_3, SeH_2) and decomposed into neutral atoms in an absorption furnace.

Fiorino *et al.* (1976) (see Section 4.3) proposed a decomposition method for the elements, followed by reduction of the As, Sb, Se and Fe compounds into hydrides, which are finally decomposed into atoms in a hydrogen flame. Furthermore, Pierce *et al.* (1976) recently developed a method for water analysis (see Section 3.4) in which the Se and As determination involves a decomposition of the hydrides in a graphite tube. Substitution of the flame by a furnace in such cases should lead to improved atomization conditions, particularly with regard to temperature. In fact, these authors demonstrate that a temperature of $850°C$ is perfectly suited for the decomposition of AsH_3 and SeH_2.

The electrothermal decomposition and atomization methods were applied directly on the crude sample by Renshaw (1972) for the detection of Ba, Pb and Sb in hair by placing the sample directly into the graphite furnace. Hudnik *et al.* (1972) analyzed blood directly for the determination of Al, Cr, Cu, Cd, Mn and Ni and Aggett (1973) applied these methods for the analysis of gold.

Urine can also be analyzed directly (see Table VII). Langmyhr *et al.* (1974c) used the same method to detect cadmium in dental specimens, and Lundgren *et al.* (1974) determined cobalt in biological tissue. Mitchell *et al.* (1974a, 1974b) published several methods for the direct analysis of serum and biological tissue using the Delves technique: decomposition and atomization of the sample in a cup heated in a flame. These methods are undeniably very attractive, since sources of contamination are reduced by eliminating a pretreatment of the sample. On the other hand, the question may arise as to whether the sample taken is actually representative. Only a series of repeated analyses will show the real validity of the sampling method. Another difficulty resides in the preparation of standards.

5. ANALYSIS OF THE ATMOSPHERE

5.1 Field of Application

The use of atomic absorption for air analysis has made considerable progress in recent years because of the problems raised by air and environmental pollution. The most common technique consists of filtering a given volume of air (a few cm^3 to several hundred m^3) and analyzing the residue on the filter. Very often, however, very small quantities of materials, in the microanalytical range (0.1-3 mg) are involved, and the nonflame methods take on greater importance. In some cases, a direct analysis of the gaseous medium (air) is possible, such as for the determination of Pb and Hg. Thilliez (1967) determined Pb in the atmosphere of factories by using the air to aid combustion. He conducted it into a flame over which a T-tube was placed, the horizontal segment of which, measuring 100 cm, served as an atomization chamber (diameter of 1 cm). The detection range is 1-100 μg Pb/m^3. For mercury, the same author (1968) conducted trapped and filtered air at a flow rate of 1-2 m^3/hr into the horizontal arm of a 100 cm T-tube (1 cm diameter), which again constituted the atomization chamber (detection range = 15-100 μg Hg/m^3). Atomic absorption also allows the determination of metal carbonyls (Fe, Ni) in illuminating gas (Maman *et al.* 1967, Schwab *et al.,* 1968).

Among the elements of greatest interest today are mercury, cadmium, lead and beryllium. In most cases, their analysis is made in dust and suspended particles in the air, which are separated by filtration by various techniques. Detection limits are in the $\mu g/m^3$ order when a flame is used as the atomizing source and 0.1 $\mu g/m^3$ with the nonflame method. Regardless of the method employed, the quantity of filtered air particles is always very small and the risks of contamination are considerable. The use of nonflame methods in particular requires a dust-free laboratory. The air is injected into the laboratory after passing through an 0.4-μ filter, and the pressure is continuously maintained at 10-15 mm of water above the atmospheric level.

5.2 Sample Preparation and Analytical Methods

We will not discuss the sampling methods. Highly sophisticated and perfectly efficient filters are offered on the market to collect particles corresponding to a controlled volume of air which is filtered at an equally controlled flow rate. The quantity of air to be filtered evidently depends on the particle content of the atmosphere. It may vary from a few m^3 for a highly polluted atmosphere to a few hundred m^3 for a high-altitude oceanic air. In fact, each sample or series of samplings must be accompanied by a control that will be subjected to the same analytical conditions. Filter packaging requires the greatest care to prevent contamination. In the most classical methods, use is made of Whatmann or Millipore filters or the like. The physical properties of the filter, particularly its porosity, are important. The efficiency of trapping atmospheric particles was studied especially by Lech *et al.* (1974) for a filter range of 0.015-0.15 μm.

Most methods employ dissolution. Melting with sodium hydroxide in a silver crucible allows a determination of Si, Al, Fe, Ca and Mg. A sample of 5 cm^2 is placed into a silver crucible with about ten sodium hydroxide pellets. The crucible is first heated slowly to 100°C and then brought to red heat in an electrical furnace for about 10 min. After cooling, the residue is solubilized in water containing nitric acid and brought to 25 ml (so as to have a 1% excess of acid).

However, trace analysis requires acid digestion with an elimination of silica. The procedure is as follows: a 10-cm^2 disk is placed into a Teflon bomb of suitable size with 2 ml 50% nitric acid and heated for 10 hr on a water bath at 60°C. The bomb is then subjected to ultrasonic agitation for 5 min. The solution is evaporated at 50°C in vacuum and the residue taken up in 2 ml 1% nitric acid. The classical trace elements (Cd, Cu, Ni, Mn, Pb and Zn) can be determined in this solution. Some elements are analyzed by nonflame methods. The following conditions are used:

Copper: (nonflame method in the Perkin-Elmer 300 SG spectrometer) Sample size 10 μl, λ 327.4 nm, drying for 20 sec at 100°C, decomposition for 45 sec at 900°C, atomization for 6 sec at 2450°C; detection limit 3.10^{-11} g Cu, sensitivity 0.002 μg Cu for 1% of absorption, precision 10%.

Nickel: (nonflame method in the Instrumentation Laboratory IL 455 spectrometer) Sample size 20 μl, λ 232.0 nm; background correction necessary; drying for 40 sec at 250°C, decomposition for 65 sec at 1000°C, atomization for 5 sec at 2800°C; detection limit 2.10^{-10} g Ni, sensitivity 0.01 μg Ni for 1% absorption, precision 10%.

Lead: (nonflame method, Perkin-Elmer 300 SG spectrometer) Sample size 10 μl, λ 283.3 nm; background correction necessary; drying for 20 sec at 100°C, decomposition for 40 sec at 450°C, atomization for 6 sec at 2000°C; detection limit 2.10^{-11} g Pb, sensitivity 0.002 μg Pb for 1% absorption, precision 5-10%.

Zinc: (flame method) The quantity of zinc in atmospheric dust is often too large to be determined by a nonflame method. After previous dissolution, flame analysis can be performed with very good reproducibility by atomizing an exactly metered volume of solution (500 μl). The atomization time is 8 sec. The detection limit is 10^{-8} g Zn, sensitivity 0.005 μg Zn for 1% absorption and the precision is 3%. In all cases calibration is made with simple solutions in 1% nitric acid.

The nonflame method is also recommended for Cd, Be and As, while flame analysis is used for Al, Mn, Fe, V, Mg, Ca, etc.

A few special methods should be noted. With regard to sampling, several authors filter the air sample through a porous graphite crucible (Woodriff *et al.*, 1972), which is then placed into the graphite furnace or through the graphite tube constituting the furnace (Siemer *et al.*, 1973a, 1973b), or using the "carbon rod" crucible (Siemer *et al.*, 1974a, 1974b). Nullens *et al* (1974) placed a filter fragment into the furnace to determine several trace elements. This method is attractive, but mediocre precision may be anticipated.

A mercury analysis requires special conditions. The best sensitivity is obtained by binding mercury on a suitable support (gold or silver screen) as a first step and then heating to liberate the mercury in the absorption cell (Table IX). The detection limit is 0.3 $ngHg/m^3$.

5.3 Practical Applications

A few of the most important applications are briefly reviewed in Tables VIII and IX. Table VIII lists flame atomic absorption methods and Table IX the nonflame techniques. The much more extensive development of nonflame methods compared to the conventional techniques is typical.

Practically all methods are based on an acid destruction of the filter that served to collect the atmospheric particles. When the filter is cellulose, nitric acid is used, with an addition of hydrochloric acid, sulfuric acid, hydrofluoric acid, hydrogen peroxide or perchloric acid, depending on the authors. In some methods (Table IX), a glass filter is used, which is then destroyed in hydrofluoric acid. Destruction of the filter by ashing is rarely used.

Special mention must be made of mercury analysis. We have seen the possibilities of direct analysis in working-place air in Section 5.1. The sensitivity is not sufficient for a systematic study of mercury in the atmosphere. It is necessary, therefore, to bind a sufficient quantity of mercury on a gold or silver support in order to liberate it in the absorption cell by heating. Among the methods cited in Table IX, that of Lech et al. (1973) merits attention. The authors filter the air in a porous graphite cup, with an inside coating of gold designed to amalgamate the mercury. The effective air volume is 50-1000 cm^3 and the filtration flow rate is a few cm^3/sec. The cup is then heated electrically to liberate the mercury in the absorption cell. The detection limit is less than 0.3 ng (0.6 $\mu g/m^3$).

Finally, we should note that the methods for atmospheric particle analysis are also applicable to the analysis of suspended sediments in water, particularly sea water. A suitable volume of water (0.1-10 liters) is filtered on a Millipore or Nuclepore filter. The filter is washed and treated by similar methods. The nature of the filter can determine the acid digestion conditions. Thus, some filters, such as the Nuclepore, are soluble in chloroform, making it possible to eliminate a support and to solubilize the suspended product with nitric or hydrochloric acid. Presley et al. (1972) studied the separation of trace elements in suspended sediments of fjord water and thus determined Co, Cu, Fe, Li, Mn, Ni and Sr. Segar (1973a) can also be consulted.

Table VIII. Applications of Flame Atomic Absorption: Atmosphere Analysis

Elements	Samples	λ nm	Concentrations	Flame	Sample Preparation	References
Cd,Pb	Air	Cd 228.8 Pb 283.3		Air-C_2H_2	Collect; treat with HF; heat; dissolve in HNO_3; H_2O; filter.	Zdrojewski et al., 1972
Pb	Air	217.0	0.5 vpm	Air-C_2H_2	Wet ash residue with HNO_3-H_2O_2.	Szivos et al., 1972
Al,Mn, Pb,V	Aerosols, chimney effluents	—		N_2O-C_2H_2	Comparison AA, EF, RA.	Rollier, 1972
Ca,Cu,Fe, Mg	Airborne dust	Ca 422.7 Cu 324.7 Fe 248.3 Mg 285.2	1-6 $\mu g/m^3$ 0.16-1.5 $\mu g/m^3$ 0.9-4.5 $\mu g/m^3$ 0.6-2.4 $\mu g/m^3$	Air-C_2H_2	Collect on cellulose filter pad; ignite in Pt; dry; ash at 550°C; fuse with Na_2CO_3; dissolve in HCl; add 1% La for Ca and Mg determination.	Hoschler et al., 1973
Cd,Cu,Fe, Ni,Pb,Zn	Airborne dust			Air-C_2H_2	Filter; treat with H_2SO_4; ash 30 min at 300°C and 60 min at 500°C; dissolve in HF-HNO_3; evaporate; redissolve.	Kometani et al.,1972
Cd,Cu,Mg, Mn,Pb,Zn	Air	—		Air-C_2H_2 N_2O-C_2H_2	Filter; treat with HCl-HNO_3; evaporate; wash; dilute to volume.	Severs et al., 1972
Pb	Air	—	0.1-4 $\mu g/m^3$ (organic Pb) 0.4-10 $\mu g/m^3$ particulate Pb)	Air-C_2H_2	Collect particulate Pb on filter; extract with acid; collect organic Pb in iodine monochloride solution; extract with APDC into MIBK.	Purdue et al., 1973
Pb	Airborne dust	217.0 283.3		Air-C_2H_2	Collect on glass filter; treat with HF; evaporate; dissolve in HNO_3.	Zdrojewski et al., 1972a
Pb,Zn	Air, waste gases, automobile exhausts	Pb 283.3 Zn 213.9		Air-C_2H_2	Dissolve in HCl-HNO_3; filter.	Hermann, 1973
Al,Mn,V, Pb	Airborne dust			N_2O-C_2H_2 air-C_2H_2	Filter (2 m^3/min), dissolve in HCl-HNO_3-$HClO_4$ (5:3:2).	Gallorini et al., 1973

Element	Sample			Flame	Treatment	Reference
Be	Air particles			$N_2O-C_2H_2$		Zdrojewski et al., 1972b
Cd	Air particles	228.8	2.5 ng/m^3	Air-C_2H_2	Filter through glass fiber; treat with HF; evaporate; dissolve in HNO_3; dilute; filter.	Quickert et al., 1973
Pb	Air	—	—	Air-C_2H_2	Digest filter with acid.	Can. Res. Dev., 1973
Trace elements	Airborne particulate	—	trace	Air-C_2H_2	Filter and treat by low-temperature r.f. oxidation; dissolve in acid; determine elements directly or after DDTC/MIBK extraction.	Tsuji et al., 1972
Trace elements	Airborne particulate	—	0-100 $\mu g/m^3$	Air-C_2H_2 Air-H_2	Filter and extract with HNO_3 (1:5); dilute.	Stupar et al., 1974
Trace elements	Airborne dust	—	—	Air-C_2H_2	Filter; ash at 450°C with H_2SO_4; repeat extract with 2 N HCl.	Munoz-Ribadeneira et al., 1974

Table IX. Applications of Nonflame Atomic Absorption: Atmosphere Analysis

Elements	Samples	λ nm	Concentrations	Atomization	Sample Preparation	References
Hg	Air	184.9	1-30 $\mu g/m^3$	Hot carbon rod	Pass sample over hot carbon.	Robinson et al., 1972a
Hg	Air	253.7	0.003-5 $\mu g/m^3$	Cold vapor	Collect Hg on fine mesh gold screen.	Foote, 1972
Pb	Air	283.3	>5x10^{-12} g	Graphite furnace	Filter through graphite crucible; place crucible in furnace.	Woodriff et al., 1972
As,Cd,Hg, Pb	Air	—	—	Graphite rod	Monitoring method; real time determination.	Robinson, 1972b
Cd,Pb	Air	—	0-100 $\mu g/m^3$	Graphite tube	Collect on 0.22-μ Millipore filter; add H$_3$PO$_4$ solution.	Matousek, 1973
Cd	Air	228.8	0.02-0.035 $\mu g/m^3$	Carbon bed	Atomize on carbon bed at 1400°C.	Robinson et al., 1973
Hg	Air	253.7	0-20 ng/m^3	Cold vapor	Collect on thin film Au; atomize by heating at 500°C in stream of N$_2$.	Scullman et al., 1972
Hg	Air	253.7	15 ng-10 $\mu g/m^3$	Cold vapor	Collect on Ag wool; atomize by heating at 400°C in carrier gas flow.	Long et al., 1973
Pb	Airborne dust	217.0 283.3	0.1 $\mu g/m^3$	Graphite tube	Collect on 0.22-μ Millipore filter; add H$_3$PO$_4$; dry.	Matousek et al., 1973
Pb	Airborne dust	—	10^{-10} g	Tantalum ribbon	Collect on cascade impactor.	Roques et al., 1973
Pb	Airborne dust	283.3	0-5 $\mu g/m^3$	Graphite tube	Collect on filter; extract with HNO$_3$.	Janssens et al., 1973
Pb	Airborne dust	—	1-10 $\mu g/m^3$	Graphite tube	Collect through 0.22-μ pore filter in graphite.	Amos et al., 1973
Trace elements	Air	—	trace	Tantalum ribbon	Collect on Millipore filter.	Smith et al., 1973
Trace elements	Air	—	—	Carbon rod		Parker et al., 1973

Trace elements	Air					
As	Air	—	—	Carbon rod	Collect through porous graphite tube.	Siemer et al., 1973a
	Airborne dust	193.7	—	Flameless method SiO$_2$ tube	Filter through glass fiber; digest in HNO$_3$-H$_2$SO$_4$; add NaBH$_4$; pass AsH$_3$ to heated SiO$_2$-absorption tube.	Vijan et al., 1974
Be	Air particulates	234.8	2-20 μg/m^3	Graphite cup	Filter through porous graphite cup; atomize in graphite cup.	Siemer et al., 1973b
Be	Air particulates		>0.1 ng/m^3	Graphite tube	—	Zdrojewski et al., 1972b
Cd	Air		>0.2 ng/m^3	Graphite tube	Filter and digest in mineral acid.	Can. Res. Dev., 1973
Cd	Air particulates	228.8	>0.2 ng/m^3	Graphite tube	Filter through glass fiber; treat with HF; evaporate; dissolve in HNO$_3$; put in graphite tube; dry at 95°C; ash at 330°C, atomize at 1900°C.	Quickert et al., 1973
Cd	Air	228.8	8 ng/m^3	Graphite rod	Filter through 0.22-μ Millipore; treat with H$_3$PO$_4$.	Brodie et al., 1974
Cd	Air	228.8	50-200 ng/m^3	Graphite tube	Filter; extract residue with 0.1 M HNO$_3$ by ultrasonic agitation.	Janssens et al., 1974
Hg	Air	253.7	1-20 ng/m^3	Graphite tube	Collect on Au-plated graphite cup; heat to release Hg.	Lech et al., 1973
Hg	Air	253.7	10 ng/m^3	Graphite tube	Collect on Au-plated graphite cup; heat to release Hg; heat to 850°C.	Siemer et al., 1974a
Hg	Air	253.7	50-1500 ng/m^3	Cold vapor	Draw sample through charcoal; heat in N$_2$ stream; pass over quartz wool and Ag wool; heat Ag to release Hg vapor to 10 cm cell.	Scaringelli et al., 1974

Table IX, Continued

Elements	Samples	λ nm	Concentrations	Atomization	Sample Preparation	References
Hg	Air	253.7	1-10 mg/m^3	Cold vapor	Pass through 0.1% L-cysteine solution to absorb nonmetallic Hg; then 5% H_2SO_4 + 0.5% K MnO_4 to absorb metallic Hg; separate inorganic and organic Hg from L-cysteine solution by $HCl\text{-}C_6H_6$ treatment and add H_2SO_4 + K MnO_4; reduce 3 solutions with $SnCl_2$.	Tomikichi, 1973
Trace elements	Airborne particulates		μg/m^3	Graphite tube	Filter through atomizer graphite tube.	Siemer et al., 1974b
Trace elements	Airborne particulates		trace	Graphite tube	Study of preparation of air particulate standards.	Lech et al., 1974
Trace elements	Atmospheric dust		trace	Graphite tube	Place filter directly in furnace.	Nullens et al., 1974

6. ANALYSIS OF INDUSTRIAL AND MISCELLANEOUS PRODUCTS

6.1 Field of Application

Atomic absorption methods are developing in all fields of industrial control. Trace analysis involves chemical products of every type, including *analytical reagents* and certain ultrapure *industrial materials,* such as those used for semiconductors. Other industrial products for which atomic absorption is used include *glasses* and *ceramics.* This method is also employed for *pharmaceutical products* (Rousselet *et al.,* 1973) either for purity controls (toxicity studies) or to verify the formula, such as Zn in insulin, Co in vitamin B_{12} preparations, and Cu and Mn in vitamin tablets (see Table X). Atomic absorption is also employed for the control of *agricultural products,* such as fats, animal feeds, various beverages and vegetable oils. Another important application is for the control of certain toxic or undesirable elements in industrial products, such as Co, Cr, Ni, Fe, Mn in *asbestos.*

Despite the diversification of applications, it is typical in this area to find that emission spectroscopy methods (arc or spark) still dominate the number of publications appearing annually, particularly in connection with purity controls. The example of *petroleum products* is characteristic. Despite the advances of nonflame atomic absorption methods, which are particularly suited for the analysis of such media, 35% of the publications dealt with emission spectrometry and 65% with atomic absorption.

With regard to sample preparation, no general method can be given. Reference will be made to the preceding sections dealing with applications, particularly with mineral, plant and biological media and water.

6.2 Practical Applications

Some examples of applications listed in Table X concern *industrial products,* such as chemicals, agricultural products, glasses and ceramics and pharmaceuticals. The methods with flame atomization are still used most often, while nonflame techniques are reserved primarily for high sensitivity requirements.

Table XI shows the applications in *petroleum products* analysis, for crude and refined oil, lubricating oils, used oils and fuel oil. The most commonly analyzed elements are Pb, V, Ag, Na, Ni and Fe. The important position held by nonflame methods in the analysis of petroleum products is evident.

Table X. Application of Atomic Absorption in Industrial Products and Miscellaneous

Elements	Samples	λ nm	Concentrations	Atomization	Sample Preparation	References
Co,Fe,Ni	Alkali and alkaline earth chlorides	Co 240.7 Fe 249.2	20 $\mu g/g$ 20 $\mu g/g$	Flame	—	Hohn et al., 1972
Cu,Fe,Mg, Mn,Zn	AgCl			Flame air C_2H_2	Dissolve in NH_4OH; extract as 8-hydroxyquinolate into MIBK.	Edwards et al., 1971
Al,Cu,Fe, Mn,Si	Forensic samples		0-30 $\mu g/g$	Graphite tube	Weigh directly into Ta boat and atomize.	Kerber et al., 1973
Co,Ni	Chemicals	240.7	>0.3 ppm	Flame air-C_2H_2	Prepare 2% aqueous solution.	Yudelevitch et al., 1972
Co	Vitamin B_{12} preparation	240.7		Flame air-C_2H_2	Dissolve in H_2O, C_2H_5OH or HCl.	Kidani et al., 1973
K,Mg,Na	CaI_2		ppm	Flame air-C_2H_2	Prepare 5% solution.	Fidel'man et al, 1971
Zn	Pharmaceutical products	213.8		Flame air-C_2H_2	Dilute liquid samples with HCl; extract creams, ointments, pastes with ether; dissolve residue in HCl.	Moody et al., 1972
Trace elements	Pharmaceutical products		ppm	Flame	Review.	Smith, 1973
Ba,Sr	Glass			Flame air-C_2H_2	Dissolve in HF/HCl; add lanthanum.	Sedykh et al., 1972
Trace elements	Glass, silica		0.01-100 ppm	Graphite tube	Dissolve and evaporate in graphite furnace.	Fuller, 1972a
Mo,Zn,Fe	Fertilizers		trace	Flame	Decompose with aqua-regia or $HClO_4$-HF.	Miwa et al., 1972
Trace elements	Pharmaceutical products		trace	Flame air-C_2H_2 N_2O-C_2H_2 air-H_2	Dissolve in H_2O or HNO_3.	Gomiscek et al., 1972
Ag	Glass		trace	Graphite tube	Place sample directly in graphite furnace.	Bath et al., 1973
Ag,Co,Fe,Li, Mg,Ni,Pb, Sb,Sr,Zn	Glass		>10 ppm	Flame air-C_2H_2	Dissolve in HF-HCl, $HClO_4$, HNO_3; dilute.	Sedykh, 1972

Element	Material	Wavelength	Range	Technique	Procedure	Reference
Cu	High purity silica		0-0.01 ppm	Graphite tube	Dissolve in HF; volatilize SiF_6H_2; place in furnace; add an excess HF before analysis.	Fuller, 1972b
Pb	Ceramic glazes	217.0	trace	Flame air-C_2H_2	Extract with cold 4% acetic acid for 24 hr; evaporate to low bulk; add HNO_3.	Fey et al., 1972
Trace elements	Glass		trace	Flame	Decomposition in bomb.	Bernas, 1973
Trace elements	Glass			Graphite tube		Fuller, 1972c
Cd	Fungicides	326.1	100-500 ppm	Flame air-C_2H_2	Digest in HNO_3; filter.	Jung et al., 1974
Cu,Mn	Vitamin tablets	324.7		Flame	Dissolve in 10% HCl; dilute.	Chae et al., 1973
Fe	Polymers	248.3	0.5-2 ppm	Flame	Ash; dissolve in HCl-HNO_3.	Ramirez-Muñoz, 1974
Hg	Chemical reagents	253.7	trace	Cold vapor	Prepare alkaline solution (NaOH) reduce with $SnCl_2$.	Vitkun et al., 1974
Pb	Forensic samples		trace	Flame	Method for estimation of firing distance in gunshot incidents.	Krishnan, 1974
Cu,Fe,Ca, K,Mg,Na	Rare earth-oxides		0.1-1 ppm	Graphite crucible		Guzeev et al., 1973
Cu,Zn	Glass powder		>0.2 ppm	Laser atomization	Mix with carbon powder + KBr (1:5:2).	Vulfson et al., 1974
Rb	Glass	780.0	trace	Flame	Dissolve in HF; evaporate; redissolve in H_2O.	Demarin et al., 1973
Rh,Ir,Pt	Glass	343.5	1-100 ppm	Graphite tube	Dissolve with $HF/HClO_4$; separate with DEDC/MIBK at pH 6.	Adriaenssens et al., 1974
Trace elements	Glass; alkali carbonate		>0.001 ppm	Graphite tube		Fuller et al., 1974
Co,Cr,Ni, Fe,Mn	Asbestos		2-14000 ppm	Flame air-C_2H_2	Dissolve in HF in Pt-ware; evaporate; dissolve in HNO_3; evaporate; redissolve in HCl.	Roy-Chowdhury et al., 1973
Hg	Vaccines	253.7	50 $\mu g/ml$	Flame air-C_2H_2	Extract with APDC into MIBK at pH 1 to determine Timerosol content.	Ribeiro, 1971
Cr,Cu,Fe	Oils		>10 ng/g	Graphite tube	Dilute with MIBK (1:1).	Prevot et al., 1973
Trace elements	Pharmaceutical products			Flame, graphite tube		Rousselet et al., 1973

Table XI. Application of Atomic Absorption: Petroleum Products

Elements	Samples	λ nm	Concentrations	Atomization	Sample Preparation	References
Ag,Cu	Oils	—	Ag 0.1-1.3 μg/ml Cu 2.5-8 μg/ml	Carbon rod	Direct.	Alder et al., 1972
Pb	Petrol	283.3-217.0	0.1-12 μg/ml	Carbon rod	Direct.	Bratzel et al., 1972a
Ag,Cu,Fe, Ni,Pb,V	Crude oils and jet engine oils	—	Trace	Carbon rod	Dilute with xylene.	Hall et al., 1972b
Trace elements	Oils	—	Trace	Carbon rod	Dilute with isooctane.	Wozniak et al., 1972
Ag,Cu,Fe, Ni,Pb	Crude and lubricating oils	—	Trace	Carbon rod + Air-H_2 sheath	Dilute with MIBK.	Hall et al., 1973
Al,Cd,Co, Mg,Na,Sn, Zn	Crude and lubricating oils	—	Trace	Graphite tube	Direct.	Chakrabarti et al., 1973a, 1973b
Fe,Ni,Pb, V	Petroleum products	—	ppm	Graphite rod	Mix with Mg sulfonate; ashing acid; heat gently in stages to 650°C; dissolve residue in HNO_3.	Eskamani et al., 1973b
Na,Ni,V	Fuel oils	—	ppm	Flame air-C_2H_2 and N_2O-C_2H_2	Dilute with isopropyl alcohol and MIBK; use standard additions method.	Ropars, 1973
Pb	Petroleum products	283.3	0.1-10 μg/ml	Flame air-C_2H_2	Extract Pb alkyl with iodine monochloride solution; boil; add KI; extract with MIBK.	Campbell et al., 1972
Pb	Gasoline	—	600-800 μg/g	Flame air-C_2H_2	Dilute with isooctane.	Quickert et al., 1972
Pb	Gasoline	—	>0.01 μg/ml	Flame air-C_2H_2	React with Br_2; extract with HNO_3.	Hodkova et al., 1973
Se	Petroleum products	—	Trace	Graphite rod	Enclose sample in gelatin capsule and combust in oxygen bomb.	Eskamani et al., 1973a
V	Fuel oil	—	320 μg/g	Flame N_2O-C_2H_2	Dilute with xylene.	Betancourt et al., 1973

Element	Sample	Wavelength (nm)	Concentration range	Method	Remarks	Reference
Trace elements	Petroleum products	—	0.001-5 µg/g	Flame air-C_2H_2 and N_2O-C_2H_2	Ash with Mg or K sulfonate; add few drops H_2SO_4; evaporate; dissolve in dilute mineral acid.	Vigler et al., 1973
Trace elements	Crude oils	—	ng/g	Flames and graphite rod	Dilute with xylene; comparison of flame and nonflame methods.	Araktingi et al., 1973
Ag, Cu, Mg, Ni, Pb, Sn	Engine oils	Ag 328.1, Cu 324.7, Mg 285.2, Ni 232.0, Pb 283.3, Sn 224.6	0.2-0.9 µg/ml, 2.5-9 µg/ml, 1.7-3.2 µg/ml, 0.6-2.5 µg/ml, 0.4-4.6 µg/ml, 0.2-2.5 µg/ml	Graphite rod	Dry to 250°C: ash at 345°C (Mg, Pb), 380°C (Cu, Ni, Sn), 440°C (Ag); atomize at 1580°C (Pb); 1620°C (Ag); 1700°C (Sn); 1720°C (Cu); 1990°C (Mg, Ni).	Chuang et al., 1974
Pb	Gasoline	283.3	0.007-0.1 g/gal	Flame air-C_2H_2	Add I_2 (0.5% solution in n-octanol); extract PbI_2 into HNO_3 (0.8% w/v).	Heistand et al., 1974
Pb	Gasoline	283.3		Flame air-C_2H_2	Extract with acid; use aqueous $Pb(NO_3)_2$ standards.	Bowen et al., 1974
Pb	Gasoline		>1 ng/g	Graphite tube	Dilute with MIBK; add halide to overcome matrix variations.	Kashiki et al., 1974
V	Crude oils	318.4	0-300 µg/g	Flame N_2O-C_2H_2 Graphite rod	Comparison of methods.	Chakrabarti et al., 1973
V	Petroleum feedstocks	—	0.1-1 µg/g	Flame N_2O-C_2H_2		Rice et al., 1973
Pb, Zn	Lubricating oils	—	trace	Flame air-C_2H_2	(a) Dilute with MIBK or, (b) ash and dissolve in HNO_3 or (c) digest oil with hot 50% HNO_3 and dilute (preferred method).	Harrison et al., 1974
Trace elements	Oil field brines	—	µg/ml	Flame air-C_2H_2, N_2O-C_2H_2, graphite tube	Dilute; adjust pH 2.4; extract with APDC/MIBK; use standard addition method and background correction.	Fletcher et al., 1974
Trace elements	Petroleum products	—	1 ng/g-5 µg/g	Flame air-C_2H_2 N_2O-C_2H_2	Ash with Mg sulfonate; add concentrated H_2SO_4, fume; ash at 650°C; dissolve residue in dilute HNO_3.	Vigler et al., 1974

REFERENCES

Adriaenssens, E. and P. Knoop. *Anal. Chim. Acta* 68:37 (1974).
Agemian, H. and A. S. Y. Chau. *Anal. Chim. Acta* 80:61 (1975).
Aggett, J. *Anal. Chim. Acta* 63:473 (1973).
Alberts, J. J., J. E. Schindler, R. W. Miller and P. W. Carr. *Anal. Chem.* 46:434 (1974).
Alder, J. F. and T. S. West. *Anal. Chim. Acta* 58:331 (1972).
Alimarin, I. P., N. I. Tarasevich and D. L. Tsalev. *Zh. Anal. Khim.* 27:647; *J. Anal. Chem.* USSR 27:569 (1972).
Amos, M. D., K. G. Brodie and J. P. Matousek. *Pittsburg Conf. Anal. Chim.* Cleveland, March 1973.
Anand, V. D. and M. C. Lancaster. *26th Nat. Meeting Amer. Assoc. Clin. Chem.* Las Vegas, August 1974.
Antonacopoulos, N. *Chem. Mikrobiol. Technol. Lebensm.* 3:8 (1974).
Aoba, K. and K. Sekiya. *Engei Shikenjo Hokoku,* Series A, 12:79 (1973).
Araktingi,Y.E. and C. L. Chakrabarti. *4th ICAS,* Toronto, October 1973.
Archer, M. C., S. R. Tannebaum, D. I. C. Wang and B. R. Stillings. *J. Agric. Fd. Chem.* 24:1595 (1973).
Arroyo, M., J. F. De Paz and M. C. Coca. *Revta Clin. Esp.* 123:433 (1971).
Aston, S. R. and J. P. Riley. *Anal. Chim. Acta* 59:349 (1972).
Baetz, R. A. and C. T. Kenner. *J. Assoc. Off. Anal. Chem.* 57:14 (1974).
Baird, R. B., S. Pourian and S. M. Gabrielian. *Anal. Chem.* 44:1887 (1972).
Barber, R. T., A. Vijayakumar and F. A. Gross. *Science* 178:636 (1972).
Barnard, W. M. and M. J. Fishman. *Atom. Abs. Newsl.* 12:118 (1973).
Bath, D. A. and R. Woodriff. *4th ICAS,* Toronto, October 1973.
Baumslag, N. and P. Keen. *Archs. Environ. Health* 25:23 (1972).
Bazhov, A. S. and E. A. Sokolova. *Zh. Anal. Khim.* 27:2442 (1972).
Beaty, R. D. *Atom. Bas. Newsl.* 13:38 (1974).
Beck, K. C., J. H. Reuter and E. M. Perdue. *Geochim. Cosmochim. Acta* 38:341 (1974).
Bek, F., J. Janouskova and B. Moldan. *Atom. Abs. Newsl.* 13:47 (1974).
Berman, E. *4th ICAS,* Toronto, October 1973.
Bernas, B. *Anal. Chem.* 40:1683 (1968).
Bernas, B. *Amer. Lab.* August:41 (1973).
Betancourt, O. J. and A. Y. Mc Lean. *J. Inst. Petrol.* 50:223 (1973).
Bowen, B. C. and H. M. Foote. *Petrol. Rev.* 28:680 (1974).
Brady, D. V., J. G. Montalvo, J. Jung and R. A. Curran. *Atom. Abs. Newsl.* 13:118 (1974).
Brandvold, L. A. and S. J. Marson. *Atom. Abs. Newsl.* 13:125 (1974).
Bratzel, M. P. and C. L. Chakrabarti. *Anal. Chim. Acta* 61:25 (1972a).
Bratzel, M. P., C. L. Chakrabarti, R. E. Sturgeon, M. W. Mc Intyre and H. Agemian. *Anal. Chem.* 44:372 (1972b).
Brodie, K. G. and J. P. Matousek. *Anal. Chim. Acta* 69:200 (1974).
Brown, G. and A. C. D. Newman. *J. Soil. Sci.* 24:339 (1973).
Busev, A. I., V. M. Byrko, L. A. Lerner and V. I. Migunova. *Zh. Anal. Khim.* 27:607 (1972).
Caldwell, J. S., R. J. Lishka and E. F. Mc Farren. *J. Amer. Water Works Assoc.* 65:731 (1973).

Campbell, K. and J. M. Palmer. *J. Inst. Petrol.* 58:193 (1972).
Can. *Res. Dev.*, 47, September/October, 1973.
Chae, Y. S., J. P. Vacik and W. H. Shelver. *J. Pharm. Sci.* 62:1838 (1973).
Chao, T. T. and R. F. Sanzolone. *J. Res. USGS.* 1:681 (1973).
Chakrabarti, C. L. and Y. E. Araktingi. *56th Canadian Chem. Conf.* Montreal, June 1973a.
Chakrabarti, C. L., Y. E. Araktingi and G. Hall. *56th Canadian Chem. Conf.* Montreal, June 1973b.
Chakrabarti, C. L. and G. Hall. *Spectrosc. Letter.* 6:385 (1973c).
Cheng, J. T. and W. F. Agnew. *Atom. Abs. Newsl.* 13:123 (1974).
Chester, R., D. Gardner, J. P. Riley and J. Stoner. *Mar. Pollut. Bull.* 4:28 (1973).
Childs, E. A. and J. M. Gaffke. *J. Assoc. Off. Anal. Chem.* 57:360 (1974).
Chuang, F. S. and J. D. Winefordner. *Appl. Spectrosc.* 28:215 (1974).
Cioni, R., F. Innocenti, and R. Mazzuoli. *Atom. Abs. Newsl.* 11:102 (1972).

Cumont, G. *Anal. Chem.* 53:634 (1971).

Davidson, I. W. F. and W. L. Secrest. *Anal. Chem.* 44:1808 (1972).
Delves, H. T. *4th ICAS,* Toronto, October 1973.
Demarin, V. T., A. D. Mirishli, N. K. Rudnevskii, A. I. Pikhtelev, V. P. Ryabschikova and A. N. Tumanova. *Tr. Khim. Tekhnol.* 3:89 (1973).
Dick, D. and R. K. Skogerboe. *Pittsburg Conf. Anal. Chem.,* Cleveland 1973.
Dolinsek, F. and J. Stupar. *2nd Inter. Symp. Anal. Chem.,* Ljubliana, June 1972a.
Dolinsek, F., J. Stupar and I. Glazer. *4th Yugoslav Conf. in Applied Spectroscopy, Zagreb,* October 1972b.
Doolan, K. J. and L. E. Smythe. *Talanta* 20:241 (1973).
Duce, R. A., J. G. Quinn, C. E. Olney, S. R. Piotrowicz, B. J. Ray and T. L. Wade. *Science* 176:161 (1972).
Dudas, M. J. *Atom. Abs. Newsl.* 13:109 (1974a).
Dudas, M. J. *Atom. Abs. Newsl.* 13:67 (1974b).
Dujmovic, M., P. Adrian and E. Knozinger. *Zh. Anal. Chem.* 272:177 (1974).
Eads, E. A. and C. E. Lambdin. *Environ. Res.* 6:247 (1973).
Ecrement, F. In: *Atomic Absorption Spectrometry,* Vol. II, M. Pinta, Ed. (Paris: Masson, 1971).
Ediger, R. D., G. E. Peterson and J. D. Kerber. *Atom. Abs. Newsl.* 13:61 (1974).
Edmunds, W. M., D. R. Giddings and M. Morgan-Jones. *Atom. Abs. Newsl.* 12:45 (1973).
Edwards, J. W., G. D. Lominac and R. P. Buck. *Anal. Chim. Acta* 57:257 (1971).
Eskamani, A., H. A. Strecker and M. S. Vigler. *4th ICAS,* Toronto, October 1973a.
Eskamani, A., M. S. Vigler, H. A. Strecker and N. R. Anthony. *4th ICAS,* Toronto, October 1973b.

Fernandez, F. J. *Atom. Abs. Newsl.* 12:93 (1973).

Fey, R. and G. Becker. *Z. Lebensm. Unters. Forsch.* 150:87 (1972).

Fidel'man, B. M. and E. S. Zolotovitskaya. *Metody Anal. Galo. Shchel. Metal. Vys. Chist.* 2:11 (1971).

Fiorino, J. A., J. W. Jones and S. G. Capar. *88th Annual Meeting Assoc. Off. Anal. Chem.* Washington, D.C., October 1974.

Fiorino, J. A., J. W. Jones and S. G. Capar. *Anal. Chem.* 48:121 (1976).

Fishkova, N. L. and T. M. Kazarina. *Zavod. Lab.* 37:1447 (1971).

Fishman, M. J. *Atom. Abs. Newsl.* 11:46 (1972).

Fitzgerald, W. F. *168th ACS National Meeting,* Atlantic City, September 1974.

Fletcher, G. E. and A. G. Collins. *U.S. Bur. Mines Rep. Invest.* RI-7861 (1974).

Foote, R. S. *Science* 177:513 (1972).

Fuchs, C., M. Brasche, U. Donath, H. V. Henning and D. Knoll. *Verh. Dtsch. Ges. Inn. Med.* 79:683 (1973).

Fuller, C. W. *Perkin Elmer Symp.,* London, November 1972a.

Fuller, C. W. *Anal. Chim. Acta* 62:261 (1972b).

Fuller, C. W. *Proc. Soc. Anal. Chem.* 9:279 (1972c).

Fuller, C. W. and J. Whitehead. *Anal. Chim. Acta* 68:407 (1974).

Galanova, A. P., A. K. Kudryavina, V. A. Pronin, I. G. Yudelevitch and G. A. Vall. *Izv. Sib. Otdel. Akad. Nauk. SSR. Ser. Khim. Nauk.* 1:58 (1974).

Galanova, A. P., V. A. Pronin, G. A. Vall, I. G. Yudelevich and E. N. Gilbert. *Zavod. Lab.* 38:646 (1972).

Gallorini, M., N. Genova, S. Meloni and V. Maxia. *Inquinamento* 15:13 (1973).

Gardner, D. and J. P. Riley. *Nature* 241:526 (1973).

Gelman, A. L. *J. Sci. Fd. Agric.* 23:299 (1972).

Gerasimova, L. I. and V. A. Ponomareva. *Zavod. Lab.* 40:360 (1974).

Gilbert, T. R. and D. N. Hume. *Anal. Chim. Acta* 66:5 (1973).

Gomez, M. I. and P. Markakis. *J. Food. Sci.* 39:673 (1974).

Gomiscek, S., V. Hudnik and M. Slamnik. *4th Yugoslav. Conf. Applied Spectroscopy,* Zagreb, October 1972.

Goto, T. *Japan Analyst* 23:1165 (1974).

Goulden, P. D., P. Brooksbank and J. F. Ryan. *Amer. Lab.* 11 (1973).

Govindaraju, K., G. Mevelle, and C. Chouard. *Bull. Soc. Fr. Ceram.* 96:47 (1972).

Govindaraju, K. *4th ICAS,* Toronto, October 1973.

Gross, S. B. and E. S. Parkinson. *Atom. Abs. Newsl.* 13:107 (1974).

Guzeev, I. D., E. S. Blinova, I. A. Mairov and V. V. Nedler. *Ind. Lab.* 39:165 (1973).

Hageman, L. R., L. Torma and B. E. Ginther. *88th Ann. Meeting Assoc. Off. Chem.* Washington, D.C., October 1974.

Hafez, A. A. R., J. R. Brownell and P. R. Stout. *Commun. Soil Sci. Plant Anal.* 4:333 (1973).

Hall, R. J. and T. Farber. *J. Assoc. Off. Anal. Chem.* 55:639 (1972a).

Hall, G., M. P. Bratzel, and C. L. Chakrabarti. *55th Chem. Conf. and Exhib.,* Quebec, June 1972b.

Hall, G., M. P. Bratzel, and C. L. Chakrabarti. *Talanta* 20:755 (1973).

Harrison, T. S., W. W. Foster and W. D. Cobb. *Metallurg. Metal Form.* 41:27 (1974).

Harth, M., D. S. M. Haines and D. C. Bondy. *Amer. J. Clin. Pathol.* 59:423 (1973).

Hauck, G. *Zh. Anal. Chem.* 267:337 (1973).

Hauser, T. R. and T. A. Hinners. *Pittsburg Conf. on Anal. Chem.*, Paper no. 82, Cleveland 1972a.

Hauser, T. R., T. A. Hinners, and J. A. Kent. *Anal. Chem.* 44:1819 (1972b).

Head, P. C. and R. A. Nicholson. *Analyst* 98:53 (1973).

Heistand, R. N. and W. C. Shaner. *Atom. Abs. Newsl.* 13:65 (1974).

Hendel, Y., A. Ehrenthal and B. Bernas. *Atom. Abs. Newsl.* 12:130 (1973).

Henning, S. and T. L. Jackson. *Atom. Abs. Newsl.* 12:100 (1973).

Hermann, P. *17th CSI*, Florence, September 1973.

Hicks, J. E., R. T. Mc Pherson and J. W. Salyer. *Anal. Chim. Acta* 61:441 (1972).

Hodkova, M. and B. Holle. *Rapa Uhlie.* 15:207 (1973).

Hohn, R. and F. Umland. *Zh. Anal. Chem.* 258:100 (1972).

Hollerer, G. and J. Hoffmann. *Z. Lebensm. Unters. Forsch.* 150:277 (1973).

Hoover, W. L. *J. Assoc. Off. Anal. Chem.* 55:737 (1972).

Hoschler, M. E., E. L. Kanabrocki, C. E. Moore and D. M. Hattori. *Appl. Spectrosc.* 27:185 (1973).

Hsu, D. Y. and W. O. Pipes. *Environ. Sci. Technol.* 6:645 (1972).

Hudnik, V. and S. Gomiscek. *2nd Inter.|Symp. Anal. Chem.*, Paper No. S-40, Ljubliana, Yugoslavia, June 1972.

Huffman, C., R. L. Rahill, V. E. Shaw and D. R. Norton. *Prof. Paper USGS 800-C*, 203 (1972).

Hwang, J. Y. and Y. Hashimoto. *Inter. Cong. Anal. Chem. IUPAC*, Kyoto, April 1972.

Ihnat, M. and R. J. Westerby. *Anal. Letter.* 7:257 (1974).

Iskandar, I. K., J. K. Syers, L. W. Jacobs, D. R. Keeny and J. T. Gilmour. *Analyst.* 97:388 (1972).

Jackson, A. J., L. M. Michael and H. J. Schumacher. *Anal. Chem.* 44:1064 (1972).

Jacob, R. A. and L. M. Klevay. *TARC Conf. II*, Halifax, Canada, August 1974.

Janouskova, J., Z. Sulcek and V. Sychra. *Chem. Listy.* 68:969 (1974).

Janssens, M. and R. Dams. *Anal. Chim. Acta* 65:41 (1973).

Janssens, M. and R. Dams. *Anal. Chim. Acta* 70:25 (1974).

Jeanroy, E. *Analysis* 2:703 (1973-74).

Jensen, F. O., J. Dolezal, F. J. Langmyhr. *Anal. Chim. Acta* 72:245. (1974).

John, M. K., H. H. Chuah, and C. J. Van Laerhoven. *Environ. Sci. Technol.* 6:555 (1972).

Johnson, F. J., T. C. Woodis, and J. M. Cummings. *Atom. Abs. Newsl.* 11:118 (1972).

Jones, M., G. F. Kirkbright, L. Ranson and T. S. West. *Anal. Chim. Acta* 63:210 (1973).

Jung, P. D. and D. Clarke. *J. Assoc. Off. Anal. Chem.* 57:379 (1974).
Kahn,H.L.,F. J. Fernandez and S. Slavin. *Atom. Abs. Newsl.* 11:37 (1972a).
Kahn, H. L., F. J. Fernandez and S. Slavin. *Atom. Abs. Newsl.* 11:42 (1972b).
Kamada, T., Y. Hayashi, T. Kumamaru and Y. Yamamoto. *Japan Analyst* 22:1481 (1973a).
Kamada, T., H. Okuda, T. Kumamaru and Y. Yamamoto. *Eisei Kagaku.* 19:314 (1973b).
Kaneko, Y. *Koku Eisei Gakkai Zasshi.* 21:227 (1971).
Kashiki, M., S. Yamozoe, N. Ikeda and S. Oshima. *Analyt. Letter.* 7:53 (1974).
Kato, K. *Chishitsu Cho Sasho Geppo.* 22 (8):437 (1971).
Kato, K., A. Ando and T. Kishimoto. *Japan Analyst* 21:1057 (1972).
Kerber, J. D., A. Koch and G. E. Peterson. *Atom. Abs. Newsl.* 12:104 (1973).
Kidani, Y., K. Takeda and H. Koike. *Japan Analyst* 22:719 (1973).
Kim, C. H., C. M. Owens, and L. E. Smythe. *Talanta* 21:445 (1974).
Kirkpatrick, D. C. and D. E. Coffin. *J. Sci. Fd. Agric.* 24:1595 (1973).
Kivalo, P., A. Visapaa and R. Backman. *Anal. Chem.* 46:1814 (1974).
Koester, H. M. *Neues Jb. Miner. Abh.* 119:145 (1973).
Koirtyohann, S. R. and J. W. Wen. *Anal. Chem.* 45:1986 (1973).
Kometani, T. Y., J. L. Bove, B. Nathanson, S. Siebenberg and M. Magyar. *Environ. Sci. Technol.* 6:617 (1972).
Korkisch, J. and H. Gross. *Talanta* 20:1153 (1973).
Krichever, M. Y., B. B. Makovoz, N. A. Orlov and S. G. Slavnov. *Probl. Izuch. Osvoeniya Prir. Resur. Sev.* 189 (1973).
Krishnan, S. S. *J. Forensic Sci.* 19:351 (1974).
Kubasik, N. P., M. T. Volosin and M. H. Murray. *Clin. Chem.* 18:410 (1972a).
Kubasik, N. P., M. T. Volosin, and M. H. Murray. *24th Nat. Meet. Amer. Assoc. Clin. Chem.,* Paper no. 167 (1972b).
Kubasik, N. P., H. E. Sine and M. T. Volosin. *Clin. Chem.* 18:1326 (1972c).
Kubasik, N. P. and H. T. Volosin. *Clin. Chem.* 19:954 (1973).
Kubo, Y., N. Nakazawa and M. Sato. *Sekiyu Gakkai Shi.* 16:588 (1973).
Kundu, M. K. and A. Prevot. *Anal. Chem.* 46:1591 (1974).
Kwok-tai, K. *Anal. Letters.* 6:603 (1973).
Langmyhr, F. J. and P. E. Paus. *Anal. Chim. Acta* 43:397 (1968).
Langmyhr, F. J. *Euroanalysis I,* Heidelberg, August 1972.
Langmyhr, F. J. and Y. Thomassen. *Zh. Anal. Chem.* 264:122 (1973).
Langmyhr, F. J., R. Solberg and L. T. Wold. *Anal. Chim Acta* 69:267 (1974a).
Langmyhr, F. J., J. R. Stubergh, Y. Thomassen, J. E. Hanssen and J. Dolezal. *Anal. Chim. Acta* 71:35 (1974b).
Langmyhr, F. J., A. Sundli, and J. Jonsen. *Anal. Chim. Acta* 73:81 (1974c).
Lansford, M., E. M. Mc Pherson and M. J. Fishman. *Atom. Abs. Newsl.* 13:103 (1974).
Least, C. J.,T. A. Rejent and H. Lees. *Atom. Abs. Newsl.* 13:4 (1974).

Lech, J. F., D. D. Siemer and R. Woodriff. *Spectrochim. Acta* 28B:435 (1973).
Lech, J. F. and R. Woodriff. *1st Nat. Meet. Fed. Anal. Chim. Spectro. Soc.,* Atlantic City, November 1974.
Le Meur, J. F. and J. Courtot-Coupez. *Bull. Soc. Chim. Fr.* 3:929 (1973).
Lerner, L. A., L. P. Orlova and D. N. Ivanov. *Soviet Soil Sci.* 3:367 (1971).
Lidums, V. *Chemica Scr.* 2:159 (1973).
Long, S. J., D. R. Scott and R. J. Thompson. *Anal. Chim.* 45:2227 (1973).
Lopez-Escobar, L. and D. N. Hume. *Anal. Letters.* 6:343 (1973).
Lund, W. and B. V. Larsen. *Anal. Chim. Acta* 70:299 (1974).
Lunde, G. *J. Sci. Fd. Agric.* 24:1029 (1973).
Lundgren, G. *4th ICAS,* Toronto, October 1973.
Lungren, G. and G. Johansson. *Talanta* 21:360 (1974).
Machata, G. and R. Binder. *Z. Rechtsmed.* 73:29 (1973).
Macias, F. D. *Soil Sci.* 115:276 (1973).
Maessen, F. J. M. J., F. D. Posma and J. Balke. *Anal. Chem.* 46:1445 (1974).
Magos, L. *Analyst* 96:847 (1971).
Magos, L. *Biochem. J.* 130:63 (1972).
Maines, I. S., C. L. Chakrabarti. *21st Can. Spectros. Symp.,* Ottawa, October 1974.
Malaiyandi, M. and J. P. Barrette. *J. Assoc. Off. Anal. Chem.* 55:951 (1972).
Malissa, H. and E. Schoeffman. *Mikrochim. Acta* 1:187 (1955).
Maman, T. and T. Dwerdt. *Meth. Phys. Anal.* 4:206 (1967).
Matousek, J. P. and K. G. Brodie. *Anal. Chem.* 45:1607 (1973).
Matousek, J. P. *4th Varian Techtron Symp.,* London, September 1973.
Mc Intosh, J. L. and K. E. Varney. *Agronomy J.* 65:629 (1973).
Mc Intyre, N. S., N. G. Cook and D. G. Boase. *Anal. Chem.* 46:1983 (1974).
Melton, J. R., W. L. Hoover, J. P. Ayers and P. A. Howard. *Soil Sci. Soc. Amer. Proc.* 37:558 (1973).
Mitchell, D. G., K. W. Jackson and K. M. Aldous. *Symposium Rec. Adv. Ass. Health Effects Environ. Poll.,* WHO, Paris, June 1974a.
Mitchell, D. G., A. F. Ward and K. M. Aldous. *1st Nat. Meeting Fed. Anal. Chem.,* Atlantic City, November, 1974b.
Miwa, E. and F. Yamazoe. *Nogyo Gijutsu Kenkyusho Hokotu B.* 23:1 (1972).
Mizuno, T. *Nippon Kagaku Kaishi* 1904 (1973).
Montiel, A. *Analysis.* 1:66 (1972).
Montaser, A., S. R. Goode and S. R. Crouch. *Anal. Chem.* 46:599 (1974).
Moody, R. R. and R. B. Taylor. *J. Pharm. Pharmacol.* 24:848 (1972).
Morrison, G. H. and H. Freiser. *Solvent Extraction in Analytical Chemistry* (New York: J. Wiley, 1957).
Munoz-Ribadeneira, F. J., M. L. Nazario and A. Vega. *7th Mat. Res. Symp.,* Gaithersburg, Md., October 1974.
Murthy, L., E. E. Menden, P. M. Eller and H. G. Petering. *Anal. Biochem.* 53:365 (1973).
Muzzarelli, R. A. A. and R. Rochetti. *Anal. Chim. Acta* 64:371 (1973).
Muzzarelli, R. A. A. and R. Rochetti. *Anal. Chim. Acta.* 69:35 (1974a).

Muzzarelli, R. A. A. and R. Rochetti. *Anal. Chim. Acta* 70:283 (1974b).
Nakahara, T., H. Nishino, M. Munemori and S. Musha. *Bull. Chem Soc. Japan* 46:1706 (1973).
Nakamura, Y., H. Nagai, D. Kubota and S. Himeno. *Japan Analyst* 22:1543 (1973).
Ng, W. K. *Malaysia Geol. Surv. Ann. Rep.* 120 (1970).
Nullens, H. and H. Deelstra. *32nd Congress GAMS,* Paris, December 1974.
Oelschlaeger, W. and E. Frankel. *Land Wirtsch. Forsch.* 26:281 (1973).
Okuno, L., R. A. Wilson and R. E. White. *J. Assoc. Off. Anal. Chem.* 55:96 (1972).
Okusu, H., Y. Ueda, K. Ota and K. Kawano. *Japan Analyst* 22:84 (1973).
Olafsson, J. *Anal. Chim. Acta* 68:207 (1974).
Omang, S. H. *Kjemi.* 31 (6):13 (1971).
Owa, T., K. Hiizo and T. Tanaka. *Japan Analyst* 21:878 (1972).
Parker, C. R., J. Sanders and C. R. Parker. *4th ICAS,* Toronto, October 1973.
Patriarca, G. and G. Quaglia. *Med. Lab.* 64:276 (1973).
Paus, P. E. *Atom. Abs. Newsl.* 11:129 (1972).
Paus, P. E. *Z. Anal. Chem.* 264:118 (1973).
Pearton, D. C. G. and R. C. Mallett. *Nat. Inst. Met. Rep. S. Afr. Rep.* No. 1435 (1972).
Pickford, C. J. and G. Rossi. *Analyst* 97:647 (1972).
Pierce, F. D., T. C. Lamoreaux, H. R. Brown and R. S. Fraser. *Appl. Spectrosc.* 30:38 (1976).
Pinta, M. *Olagineux.* 28:87 (1973).
Popova, S. A., L. Bezur and E. Dungor. *Z. Anal. Chem.* 271:269 (1974).
Presley, B. J., Y. Kolodny, A. Nissenbaum and I. R. Kaplan. *Geochim. Cosmochim. Acta* 36:1073 (1972).
Prevot, A. and M. Gente. *Rev. Fr. Corp. Gras.* 20:95 (1973).
Purdue, L. J., R. E. Enrione, R. J. Thompson and B. A. Bonfield. *Anal. Chem.* 45:527 (1973).
Quickert, N., A. Zdrojewski and L. Dubois. *Sci. Total. Environ.* 1:309 (1972).
Quickert, N., A. Zdrojewski and L. Dubois. *Int. J. Environ. Anal. Chem.* 2:331 (1973).
Rains, T. C. and O. Menis. *J. Assoc. Off. Anal. Chem.* 55:1339 (1972).
Ramirez-Munoz, J. *Flame Notes* 6:8 (1974).
Rattonetti, A. *Pittsburg, Anal. Chem. Appl. Spectros.,* Cleveland, 1973.
Rattonetti, A. *Anal. Chem.* 46:739 (1974).
Renshaw, G. D. *Symposium Perkin Elmer and Royal Institution,* Paper No. 652, London, November 1972.
Riandey, C. and M. Pinta. *Analysis.* 2:179 (1973).
Riandey, C., P. Linhares and M. Pinta. *Analysis* 3:303 (1975).
Ribeiro, N. J. *Rev. Form. Bioquim. Univ. Sao Paulo* 9:357 (1971).
Rice, A. J., J. O. Rice, W. C. Shaner and C. C. Cerato. *Symp. Role Trace Metals Petrol. Amer. Chem. Soc.,* Chicago, August 1973, p. 609.
Robert, R. V. D., E. Van Wik, R. Palmer and T. W. Steele. *J. South Afr. Chem. Inst.* 25:179 (1972).

Robinson, J. W., P. J. Slevin, G. D. Heindman and D. K. Wolcott. *Anal. Chim. Acta* 61:431 (1972a).

Robinson, J. W. *Proc. Int. Symp. Environ. Health Aspects Leads* 1099 (1972b).

Robinson, J. W., D. K. Wolcott, P. J. Slevin and G. D. Heindman. *Anal. Chim. Acta* 66:13 (1973).

Rollier, M. A. *Chim. Ind.*, Milan 54:895 (1972).

Ropars, J. *Analysis.* 2:199 (1973).

Roques, Y. and J. Mathieu. *Analysis.* 2:481 (1973).

Rosen, J. F. and E. E. Trinidad. *J. Lab. Clin. Med.* 80:567 (1972).

Rosen, J. F., C. Zarate-Salvador and E. E. Trinidad. *Pediatrics.* 84:45 (1974).

Ross, T. T., J. G. Gonzalez and D. A. Segar. *Anal. Chim. Acta* 63:205 (1973).

Rousselet, F., V. Courtois and M. L. Girard. *4th ICAS,* Toronto, October 1973.

Roy-Chowdhury, A. K., T. F. Mooney and A. L. Reeves. *Archs. Environ. Health* 26:253 (1973).

Rubeska, I. and M. Miksovsky. *Coll. Czech. Chem. Comm.* 37:440 (1972).

Rubeska, I. *Atom. Abs. Newsl.* 12:33 (1973).

Scaringelli, F. P., J. C. Puzak, B. I. Bennett and R. L. Denny. *Anal. Chem.* 46:278 (1974).

Schaller, K. H., H. G. Essing, H. Walentin and G. Schaelke. *Z. Klin. Chem. Klin. Biochem.* 10:434 (1972).

Schmidt, F. J. and J. L. Royer. *Anal. Letter.* 6:17 (1973).

Schumacher, E. and F. Umland. *Z. Anal. Chem.* 270:285 (1974).

Schwab, H. and V. Wedel. *Inst. Energet. Mitt. Dtsch.* 95:321 (1968).

Scullman, J. and G. Widmark. *Int. J. Environ. Anal. Chem.* 2:29 (1972).

Sedykh, E. M. *Glass. Ceram.* 28:265 (1972).

Sedykh, E. M. and E. V. Igoshina. *Glass. Ceram.* 28 (7-8):507 (1972).

Segar, D. A. and J. G. Gonzalez. *Anal Chim. Acta* 58:15 (1972).

Segar, D. A. *Int. J. Environ. Anal. Chem.* 3:107 (1973a).

Segar, D. A. and J. L. Gilio. *Int. J. Environ. Anal. Chem.* 2:291 (1973b).

Severne, B. C. and R. R. Brooks. *Anal. Chim. Acta* 58:216 (1972a).

Severne, B. C. and R. R. Brooks. *Talanta.* 19:1467 (1972b).

Severs, R. K. and L. A. Chambers. *Arch. Environ. Health* 25:139 (1972).

Shigematsu, T., M. Matsui and O. Fujino. *Japan Analyst* 22:1162 (1973a).

Shigematsu, T., M. Matsui, O. Fujino and K. Kinoshita. *Nippon Kagaku Kaishi.* 21:23 (1973b).

Shiraishi, N., T. Hasegawa, T. Hisayuki and H. Takahashi. *Japan Analyst* 21:705 (1972).

Siemer, D. D. and R. Woodriff. *4th ICAS,* Toronto, October 1973a.

Siemer, D. D., J. F. Lech and R. Woodriff. *Spectrochim. Acta* 28B:469 (1973b).

Siemer, D. D., J. F. Lech and R. Woodriff. *Appl. Spectrosc.* 28:68 (1974a).

Siemer, D. D. and R. Woodriff. *Spectrochim. Acta* 29B:269 (1974b).

Sighinolfi, G. P. *Atom. Abs. Newsl.* 11:96 (1972).

Sixta, V., M. Miksovsky and Z. Sulcek. *Coll. Czech. Chem. Comm.* 38:3418 (1973).
Smith, W. H. *Science* 176 (4040):1237 (1972).
Smith, R. V. *Amer. Lab.,* March:27 (1973).
Smith, S. B., P. A. Ullucci and J. Y. Hwang. *4th ICAS,* Toronto, October 1973.
Snodin, D. J. *J. Assoc. Publ. Analysts* 11:47 (1973).
Soerensen, N. R. *Tidsskr. Planteavl.* 78:156 (1974).
Sorenson, J. R. J., E. G. Melby, P. J. Nord and H. G. Petering. *Archs. Environ. Health* 27:36 (1973).
Spielholtz, G. I., G. C. Toralballa and J. J. Willsen. *Mikrochim. Acta.* 649 (1974).
Stupar, J., J. Korosin and A. Jersic. *2nd Symp. Yugoslav. Soc. Clean Air Serajevo,* October 1974.
Stupar, J., F. Dolinsek, M. Spenko and J. Furlan. *Landwirtsch. Forsch.* 27:51 (1974).
Szivos, K., L. Polos, L. Bezur and E. Pungor. *2nd Inter. Symp. Anal. Chem.,* Ljubliana, Yugoslavia, June 1972.
Takeuchi, T., M. Suzuki and M. Yanagisawa. *Inter. Cong. Anal. Chem. IUPAC,* Paper No. C-1605, Kyoto, April 1972.
Tanaka, G., H. Kawamura, and Y. Ohyagi. *Inter. Cong. Anal. Chem,* *IUPAC,* Paper No. A-0614, Kyoto, April 1972.
Terashima, S. *Chishitsu Chosajo Geppo.* 24:469 (1973a).
Terashima, S. *Japan Analyst* 22:1317 (1973b).
Terashima, S. *Japan Analyst* 23:1331 (1974).
Thilliez, G. *Anal. Chem.* 39:427 (1967).
Thilliez, G. *Chim. Anal.* 50:226 (1968).
Toffaletti, J. and J. Savory. *26th Natl. Meet. Amer. Assoc. Clin. Chem.,* Las Vegas, August 1974.
Tomikichi, K. *Sangyo Igaku.* 15:223 (1973).
Tsalev, D. L., I. P. Alimarin and S. I. Neiman. *Zh. Anal. Khim.* USSR, 27:1100 (1972a).
Tsalev, D. L., I. P. Alimarin and S. I. Neiman. *Zh. Anal. Khim.* USSR 27:1223 (1972b).
Tsalev, D. L., N. I. Taraseviych and I. P. Alimarin. *Zh. Anal. Khim.* USSR 28:19 (1973).
Tsuji, S., Y. Suzuki, M. Tobe, M. Tonomura, Y. Emoto, M. Saito, M. Araki and M. Muramatsu. *Eisei Shikenjo Hokoku.* 90:20 (1972).
Ullucci, P. A. and J. Y. Hwang. *19th Spectrocs. Symp.,* Canada, Montreal, October 1972.
Ullucci, P. A. and J. Y. Hwang. *17th CSI,* Florence, September 1973.
Van Loon, J. C. *Atom. Abs. Newsl.*
Vigler, M. S. and V. F. Gaylor. *16th ACS Natl. Meeting,* Chicago, August 1973.
Vigler, M. S. and V. F. Gaylor. *Appl. Spectrosc.* 28:342 (1974).
Vijan, P. N., G. R. Wood. *Atom. Abs. Newsl.* 13:33 (1974).
Vitkun, R. A., Y. V. Zelyukova and N. S. Poluektov. *Zav. Lab.* 40:949 (1974).
Vulfson, E. K., A. V. Karyakin and A. I. Shidlovskii. *Zh. Anal. Khim.* USSR 28:1253 (1973).
Vulfson, E. K., A. V. Karyakin and A. I. Shidlovskii. *Zav. Lab.* 40:945 (1974).

Welz, B. *Cz. Chemie. Tech.* 1:455 (1972a).
Welz, B. *Fortschr. Miner.* 50:106 (1972b).
Welyz, B. and E. Wiedeking. *Fresenius Z. Anal. Chem.* 264:110 (1973).
Willey, B. F., C. M. Duke, A. L. Wojcieszak and C. T. Thomas. *J. Amer. Water Wks. Assoc.* 64:303 (1972).
Willis, J. B. *TARC Conf. II,* Halifax, Canada, August 1974.
Willis, J. B. *Anal. Chem.* 47:1753 (1975).
Woodriff, R. and J. F. Lech. *Anal. Chem.* 44:1323 (1972).
Wozniak, J. and W. Tadeusiak. *Form. Pol.* 28:269 (1972).
Yamamoto, Y., T. Kumamaru, Y. Hayashi, and T. Kamada. *Japan Analyst* 22:876 (1973a).
Yamamoto, T. H. and H. Oshio. *Koku Eisei Gakkai Zasshi.* 23:192 (1973b).
Yamasaki, S. *Nippon Dojo-Hiryogaku Zasshi.* 44:383 (1973).
Yamauchi, R. *Tokyo Jikekai Ika Daigaku Zasshi.* 86:694 (1971).
Yasuda, S. and H. Kakiyama. *Japan Analyst* 23:406 (1974a).
Yasuda, S. and H. Kariyama. *Japan Analyst* 23:620 (1974b).
Yudelevitch, I. G., G. V. Poleva and V. L. Kustatas. *Zh. Anal. Khim.* 27:2432 (1972).
Zdrojewski, A., N. Quickert and L. Dubois. *19th Spectrosc. Symp.,* Montreal, October 1972a.
Zdrojewski, A., N. Quickert, L. Dubois, and J. L. Monkman. *Int. J. Environ. Anal. Chem.* 2:63 (1972b).
Zdrojewski, A., N. Quickert and L. Dubois. *21st Canadian Spectro. Symp.* Ottawa, October 1974.

CHAPTER 7

ATOMIC FLUORESCENCE SPECTROMETRY

1.   GENERAL REMARKS AND PRINCIPLE

1.1   The Birth of a Technique

The phenomenon of atomic absorption as we have seen in Chapter 4 is accompanied by an electron transition and return of the electron on its initial orbit as manifested by the emission of a characteristic atomic radiation. This emission, which carries the name of *fluorescence emission*, can also be used for qualitative and quantitative determination of chemical elements. Wood's experiment (Chapter 4, Section 1.2) demonstrates the reciprocity of the emission and absorption phenomena. These two phenomena have been known for a long time (Wood's work dates back to 1905), but their applications in chemical analysis are relatively recent. Atomic absorption as a method of chemical analysis dates back 15 years. With regard to atomic fluorescence, the question arises whether it should be considered as a genuine applied method at all.

We will attempt to answer this question by defining the area of application of atomic fluorescence spectrometry. While the analytical applications of atomic absorption began in 1955 with the well-known works of Walsh and Alkemade, the idea of using atomic fluorescence in chemical analysis appeared only in 1962 with the studies of Alkemade (1962), followed by those of Winefordner *et al.* (1964a, 1964b), of West (1966), and Dagnall *et al.*(1966). Atomic absorption was in a state of constant progress at that time, but now it seems that atomic fluorescence, which appeared to be a highly novel technique, is burdened by some difficulties.

In an attempt to understand this situation, we must go back to the physical principle of fluoresence as compared with atomic absorption.

Furthermore, since atomic absorption spectrometry is the older method, we should establish whether atomic fluorescence is a new or a supplementary technique.

1.2   Basic Principle

If an atomization source, such as a flame, contains a given number of atoms N° in the ground state, the absorption A of the atomic resonance radiation is given by:

$$A = \log \frac{I_o}{I} = k \, 1 \, N^\circ$$

where I_o and I = the intensities of the incident and transmitted resonance radiation flux

  1 = the atomic source thickness
  k = a coefficient defining the capacity of the atoms to produce electron transitions and their photon capture cross section.

Theoretically, the sensitivity A can be increased for a given number of atoms by increasing the optical path length 1 through the flame, for example, by a multiple path system or by guiding the atoms in a long-path tube (Fuwa *et al.*, 1963). However, in the final analysis, these devices have limited applications because of the spectral background, which is the principal factor in the detection limit, increasing with the path length 1.

In atomic fluorescence, the signal F is also proportional to the number of atoms N° in the ground state in the atomizing source. The relation between F and N° has the following form:

$$F = k \, I_o \, \Phi \, s \, \Omega \, N^\circ$$

where I_o = the excitation radiation intensity

  s = the cross section of the atomizing source
  Ω = the solid angle at which the fluorescence is measured and detected
  Φ = the quantum fluorescence efficiency for the analyzed atom
  k = the coefficient defining the capacity of the atom to produce electron transitions as in the case of atomic absorption.

If the formulas relating A to N° on one hand and F to N° on the other are compared theoretically, we find proportionality between F and I_o in the case of fluorescence, while absorption is independent of I_o. In other words, an increase of I_o increases the fluorescence value of F but not that of absorption. This observation shows that absorption measurements are essentially a function of the number of atoms N° formed as well as of the nature of the atom. In fluorescence, the measurement also depends on the exciting radiation intensity.

Thus, theoretically, if the factors k and Φ are favorable, it is possible to increase the sensitivity by increasing the intensity of the excitation source. This conclusion is valid in many cases. As for the parameter l, it has little influence in fluorescence. The cross section of the atomizing source and the angle Ω are important. This offers an additional but limited method for increasing the sensitivity. The optical geometry of the excitation and atomization sources relative to the spectrophotometric receiver constitutes an important factor in fluorescence. Finally, we should add that the factor Φ is high in the majority of flames for numerous atoms. The number of deexciting collisions produced by a substance in atomic dimensions is 10-1000 during its travel through the flame.

Thus, atomic fluorescence appears to offer a much higher sensitivity than atomic absorption, to the extent that sufficiently intense excitation sources are available. In fact, the first experiments were conducted with a classical hollow-cathode lamp, but because of the weak intensity of these lamps, the resulting fluorescence was also very weak. On the other hand high-frequency electrodeless excitation lamps with signals 100-1000 times more intense than the hollow cathode have proved to be particularly effective. In 1970, a rather spectacular development occurred in atomic fluorescence involving the use of this new technique as a supplement to atomic absorption. Interest focused on certain particularly sensitive elements, such as Cd, Co, Cu, Fe, Hg, Mg, Mn, Ni, Tl and Zn.

Today, atomic fluorescence spectrometry has not been as popular as expected initially because the manufacturers have not found new applications compared to absorption spectrometry. However, applications are not lacking, even though the user would be required to construct his own instrumentation. The following sections will discuss applications, and we will see that fluorescence cannot be ignored today. Interesting publications on cadmium, zinc, arsenic, antimony, selenium, sliver, gold and lead have already appeared.

1.3 Types of Fluorescence

When the atom absorbs a characteristic radiation, it can emit monochromatic radiations called fluorescence. This radiation is the result of defined electron transitions. Several types are distinguished (Figure 1).

Resonance fluorescence: This corresponds to a transition between the ground level E_0 and an upper level E_2 (Figure 1a). Absorption and fluorescence radiations have the same wavelength: $\lambda_A = \lambda_F$. The transition can also originate in a metastable level near the ground level. For example, for thallium, the ground level $^2P_{1/2}$ is joined by a metastable level

$^2P_{3/2}$ at 0.966 eV, also giving rise to a resonance fluorescence that can furthermore be qualified as an excited state (Omenetto *et al.*, 1973).

As an example, we can cite:

Zn 219.9: Ni 232.0: Pb 283.3: Tl 377.6 nm.

Direct line fluorescence: Fluorescence emission occurs between two excited states. The upper level is shared by the exciting and the fluorescence radiation (Figure 1b). The lower level is metastable and differs from the ground state. The excitation energy is higher than the fluorescence energy ($\lambda_F > \lambda_A$), which has been called the Stokes process. Actually, this type of fluorescence is observed when the ground level is a multiplet.

An example of this is lead: fluorescence radiations at 405.8 and 722.9 nm correspond to the resonance radiation (emission-absorption) of Pb at 283.3 nm (Sychra *et al.*, 1970a, 1970b). The same applies to the elements Tl, In, Ga. The fluorescence radiations for Tl at 535.0, In at 451.5 and Ga at 417.2 nm correspond to the exciting radiations of Tl at 377.5, In at 410.2 and Ga at 403.3.

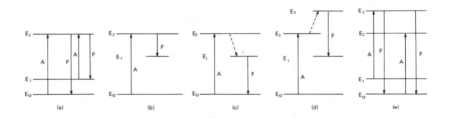

Figure 1. Types of atomic fluorescence. (a) Resonance fluorescence, (b) direct line fluorescence, (c) stepwise line fluorescence, (d) thermally assisted stepwise line fluorescence, (e) thermally assisted anti-Stokes fluorescence.

Stepwise line fluorescence: The upper levels of the exciting line and the fluorescence line are different (Figure 1c). As a result of absorption of a radiation, the excited atom loses a fraction of its energy by deexciting collisions before reemitting its fluorescence radiations. For example, when sodium is excited by the radiation of Na at 330.3 nm, it reemits the Na line at 589.0 nm. We again have $\lambda_F > \lambda_A$.

Thermally assisted stepwise line fluorescence: An atom that is excited under normal conditions from level E_0 to level E_2 (absorption of a characteristic line) can also be excited thermally to a higher electron level (E_3). It is then deexcited with emission of a fluorescence radiation, which corresponds to the transition from level E_3 to a lower level E_1 (different from the ground level) (Figure 1d).

As an example, we may note the case of antimony and bismuth (Dagnall *et al.*, 1967). Thermally assisted line fluorescence is produced when the energy that separates two excited electron states is weaker than the thermal energy responsible for the transition from the lower to the higher electron state.

Thermally assisted anti-Stokes fluorescence: When the exciting radiation (absorption) corresponds to a transition from a neighboring level (E_1) instead of from the ground level E_0, the fluorescence radiation may end at the ground level E_0 (Figure 1e). For example:

$$\text{indium } \lambda_A = 451.1 \text{ nm} \quad \lambda_F = 410.2 \text{ nm.}$$

This is the so-called thermally assisted anti-Stokes fluorescence. Furthermore, when the atom is excited from the ground state to a specific electron state, it can pass through a slightly higher energy state by absorption of the thermal energy from the flame. Fluorescence emission occurs from this state to the ground state. For example:

$$\text{chromium } \lambda_A = 359.3 \text{ nm} \quad \lambda_F = 357.9 \text{ nm.}$$

In these two cases, the energy of the transition giving rise to a fluorescence emission is higher than the intrinsic energy at the absorption radiation. In both cases this involves thermally assisted anti-Stokes fluorescence.

1.4 Practical Examples

When an atom is excited by absorbing a characteristic radiation, the fluorescence spectrum can actually contain lines having different origins. The case of thallium can serve as an example. The resonance radiation defined by the transition from a lower energy is the line λ 377.5 nm, which corresponds to a transition between the zero-energy level 6p $^2P_{1/2}$ and the upper level 7s $^2S_{1/2}$ with the energy of 3.283 eV (Figure 2). Fluorescence is excited from this radiation, and the fluorescence spectrum includes a line group (Table I).

In this series of lines, it is evidently the resonance line at 377.5 nm that has the highest sensitivity in atomic absorption. It is also the most frequently used and the most sensitive in fluorescence spectrometry. The fluorescence at 258.0, 276.8 and 237.9 is also a resonance fluorescence but its sensitivity is low. Radiations at 535.0, 323.0, 292.1 and 352.9 are the result of direct-line fluorescence, since the basic level is the metastable level $^2P_{3/2}$ at 0.966 eV of the ground level, the population of which cannot result from the thermal energy of low-temperature flames (Omenetto *et al.*, 1969). Finally, the fluorescence at 351.9 nm and 291.8 nm is a thermally assisted stepwise line fluorescence (Omenetto *et al.*, 1970). Thus, a thermally assisted excitation to the level $^2D_{5/2}$, followed by a fluorescence emission at 351.9 nm, is observed from the exciting radiation at 276.8 nm (from level $^2D_{3/2}$). The cited authors show that a given fluorescence radiation can be of several types. For example, in high-temperature

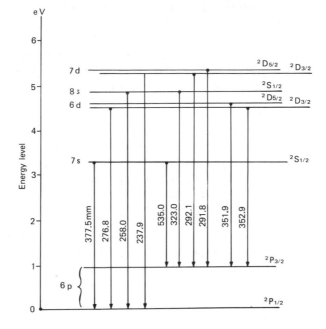

Figure 2. Fluorescence diagram of thallium.

Table I. Fluorescence Spectrum of Thallium

λ nm	Electron Levels	Level Energy (eV)	Type of Fluorescence
377.5	$6p\,^2P_{1/2}$-$7s\,^2S_{1/2}$	0-3.283	Resonance
276.8	$6p\,^2P_{1/2}$-$6d\,^2D_{3/2}$	0-4.478	Resonance
258.0	$6p\,^2P_{1/2}$-$8s\,^2S_{1/2}$	0-4.845	Resonance
237.9	$6p\,^2P_{1/2}$-$7d\,^2D_{3/2}$	0-4.524	Resonance
535.0	$6p\,^2P_{3/2}$-$7s\,^2S_{1/2}$	0.966-3.283	Direct-line fluorescence
323.0	$6p\,^2P_{3/2}$-$8s\,^2S_{1/2}$	0.966-4.845	Direct-line fluorescence
292.1	$6p\,^2P_{3/2}$-$7d\,^2D_{3/2}$	0.966-5.28	Direct-line fluorescence
351.9	$6p\,^2P_{3/2}$-$6d\,^2D_{5/2}$	0.966-4.488	Thermally assisted stepwise line fluorescence
291.8	$6p\,^2P_{3/2}$-$7d\,^2D_{5/2}$	0.966-5.30	

flames ($N_2O\text{-}C_2H_2$), the $^2P_{3/2}$ level is a multiplet of the base level, which can be populated to a significant extent by thermal processes. Omenetto *et al.*, (1970) believe that the fluorescence at 535 nm is a mixture of resonance and direct-line fluorescences.

The fluorescence spectrum of tin also contains several lines. The lines at 224.6, 254.7 and 286.3 nm are due to a resonance fluorescence and those at 326.2, 300.9, 317.5 and 380.1 nm result from a direct-line fluorescence. However, the most important and most sensitive fluorescence at 303.4 nm, which corresponds to a transition $^3P_1$-$^3P_0^0$ (0.210-4.295 eV), is a mixture of resonance and stepwise fluorescence (Sychra *et al.*, 1975).

Thus, the fluorescence spectrum of an element contains several lines of different intensities. Often, a resonance line is the most sensitive, but this is not the general rule. The sensitivity conditions will be discussed later in comparison with those of atomic absorption.

1.5 Fluorescence Quenching

The population of the excited state in atomic fluorescence is large compared to that of thermal equilibrium. The lifetime of the excited atoms is very brief. Deexcitation is accompanied by emission of fluorescence radiation. Collisions between the excited atoms and molecules or atoms present in the source can cause a deexcitation of a part of the excited atom with no fluorescence as a result. The presence of such molecules contributes to a partial reduction or inhibition of fluorescence.

Fluorescence results from the process $M^* \to M + h\nu$. This reaction takes place at a rate A, which is also called the Einstein coefficient or spontaneous emission coefficient for the transition in question.

Fluorescence quenching results from collisions with molecules or atoms X, $M^* + X = M + X$, and the quenching rate is K. From this, the fluorescence yield Y is defined:

$$Y = \frac{A}{\Sigma A + \Sigma K}$$

The sums in the denominator include all possible emission and deexcitation pathways. Fluorescence yields were calculated for a few types of flames (Jenkins, 1970) and are reported in Table II.

The particularly high yields in a hydrogen-oxygen-argon flame will be noted. Unfortunately, for good atomization and for a maximum suppression of interference, it is often necessary to give preference to air-acetylene and nitrous oxide-acetylene flames. A few types of fluorescence quenching are given below.

(a) Collisions with free atoms:

$$M^* + X = M + X$$

M^* = excited atoms, M and X = neutral atoms. This reaction is observed in rare gases.

(b) Collisions with molecules:

$$M^* + AB = M + AB$$

This is the most important cause of quenching. The molecules AB can also originate from combustion of the flame.

(c) Collisions with electrons:

$$M^* + e^- = M + e^-$$

This reaction is observed primarily in plasmas.

(d) Mixing of the excited state by collisions with free atoms:

$$M^* + A = M^X + A$$

where M^* and M^X are different excited states. For example, in the presence of argon, the following is observed with sodium:

$$Na \; (^2P_{3/2}) + Ar = Na \; (^2P_{1/2}) + Ar$$

(e) Mixing of the excited state by collisions with molecules:

$$M^* + AB = M^X + AB$$

(f) Chemical quenching:

$$M^* + AB = M + A + B$$

AB is a molecule or a stable radical at the flame temperature. Studies to be consulted: Kirkbright *et al.*, 1974; Sychra *et al.*, 1975.

Table II. Atomic Fluorescence Power Yields for Some Flames and Various Atoms

Flame	T° K	Fluorescence Power Yield			
		Na	K	Tl	Pb
$2 \, H_2/O_2/4N_2$	2100	0.060	0.047	0.070	0.079
$6 \, H_2/O_2/4N_2$	1800	0.049	0.049	0.099	0.10
$0.4 \, C_2H_2/O_2/4N_2$	2200	0.042	0.028	0.042	0.067
$2 \, H_2/O_2/Ar$	1800	0.75	0.37	0.33	0.22
$H_2/O_2/4N_2$	1600	0.044	0.03	0.051	0.069

2. APPARATUS

2.1 General Properties and Quality Requirements

While the instrument used for atomic fluorescence has some analogies with those of atomic absorption spectrometry, particularly with regard to the dispersion system and the receiver, the primary emission source as well as the atomizing and fluorescence source differ considerably from atomic absorption. The analytical curve (Figure 3) represents the relation between $\log I_F$ and $\log N$, where I_F is the fluorescence intensity and N the number

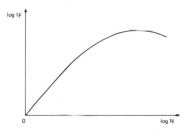

log IF

0   log N

Figure 3. Analytical curve: variation of fluorescence ($\log I_F$) as a function of the atomic concentration ($\log N$) with excitation from a line source.

of atoms in the lowest energy state. This curve should have the greatest possible slope and should be linear in the largest possible concentration range. It will be noted in Figure 3 that the curve of $\log I_F$ as a function of $\log N$ rises linearly, passes through a maximum and then declines.

The fluorescence phenomenon must also be as intense as possible for a given concentration of atoms, and the fluorescence energy emitted must be captured and measured with the best yield. With regard to the exciting source, the fluorescence intensity I_F is the function of the exciting radiation intensity I_E emitted by the emission source. The geometry of the emission and fluorescence sources relative to the spectrophotometer has an important role in the measuring sensitivity (Figure 4).

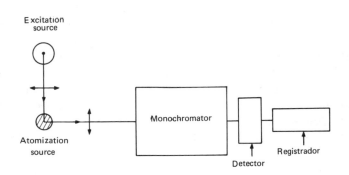

Excitation
source

Atomization
source

Monochromator

Detector

Registrador

Figure 4. Principle of the atomic fluorescence spectrometer.

In addition, we should note that the analytical detection limit is a function of the signal/noise ratio, S/N. Thus, the background N must be reduced to a maximum. For this purpose, the direct emission of the flame (or the atomizing source) is eliminated by modulating the excitation source to the amplifier frequency of the measuring receiver. The light scattered by the flame is also one of the principal sources of background capable of enhancing the detection limit. The solid and liquid particles in the flame are responsible for the "scattering effect" as are defects in the optical surfaces of mirrors and lenses designed to reflect and focus the fluorescence energy on the spectrometer. The best detection limits depend on the following instrumental factors:

- The exciting sources energy must be maximum.
- The scattering effect must be reduced to a minimum.
- The best analysis line should be selected.
- Composition of the flame, temperature and flow rate of gases must be optimal. (The best compromise is represented by premixed flames with a high rare gas concentration having a temperature just sufficient to produce atomization.)
- Efficiency of the nebulizer (in the case of flames) and the transport velocity of the vapor must be high.
- The fluorescence source must be seen by the detector at the largest possible solid angle.
- The optimum slit should be determined by a series of measurements that allow a determination of the highest signal/noise ratio (S/N).

2.2 Excitation Sources

The requirements for high quality excitation sources in fluorescence spectrometry are: radiation intensity, short- and long-term stability, reproducibility, simple handling, applicability to several elements, lifetime and purchase price. In practical terms, the following types of lamps have been tested in atomic fluorescence.

(a) *Gas discharge lamps.* These were tested for the volatile elements Cd, Zn, Hg, Ga, In, Tl, Na and K (Figure 5a). In practice, they have proved to be satisfactory only for Cd, Zn and Tl (Sychra *et al.,* 1975) and especially for Hg. The emitted lines are relatively broad, and sometimes self-absorbed. Detection limits are good for these elements.

(b) *Hollow-cathode lamps.* The classical lamps for atomic absorption (Figure 5b) generally do not give good results in fluorescence because the radiation intensity is too low. On the other hand, the high-intensity hollow-cathode lamps of Sullivan and Walsh (1964) produce a supplementary discharge between two auxiliary electrodes when compared to the classical

hollow cathodes, (Figure 5c) and they have led to good detection limits
with flame atomization as well as nonflame sources. Sychra *et al.* (1970b)
observed better detection limits for Au, Cd and Cu in atomic fluorescence
compared to atomic absorption.

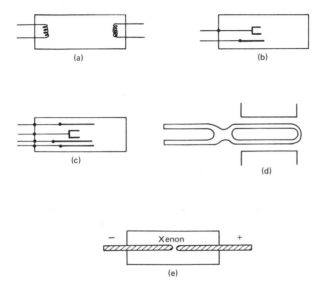

Figure 5. Principal lamps used for fluorescence excitation.

(c) *Electrodeless discharge lamps.* These high-frequency excitation
lamps operate by microwaves in a resonance cavity producing intense radiation. When a volatile compound of a given metal in a sealed ampule under
a low rare gas pressure (Figure 5d) is placed into a high-frequency field
(2450 MHz), the excitation of atoms produces an intense, stable and reproducible line spectrum of the metallic elements. In practice, a few seconds after lighting the spectrum of the filling gas disappears and the spectrum
obtained is purer than that obtained with a hollow cathode. The resonance
lines produced under these conditions are 100-1000 times more intense than
those from hollow cathodes, resulting in a much higher signal/noise ratio
for the emission source than that obtained from a hollow-cathode lamp.
This makes it possible to operate with a lower sensitivity gain of the receiver while reducing the error due to background. Furthermore, at the low
concentrations, the fluorescence signal varies in proportion to the intensity of the line from the exciting source, contributing to a considerable
reduction of the detection limits and explaining why these limits are sometimes 10-100 times lower in fluorescence than in atomic absorption.

In the field of chemical analysis, the development of atomic fluorescence has been subordinated to the advances made in the technology of electrodeless lamps. However, commercial lamps now only correspond to the volatile elements or those giving volatile compounds—As, Sb, Te, Se, Sn, Pb and In—and for the metalloids, such as sulfur, phosphorus and the halogens.

The flux emitted by electrodeless lamps can be easily modulated. Finally, the possibility of fabricating multielemental electrodeless lamps should be noted.

(d) *Continuous-background lamps.* These lamps give a spectral emission extending into a broad UV and visible spectral region (200-700 nm). The value is that the same source can be used for the analysis of several elements. These are mainly high-pressure arc lamps in a xenon atmosphere with a power from 100 to 500 W (Figure 5e), which have been most successful in exciting the fluorescence of Ag, Au, Bi, Cu, Ca, Cd, Co, Cr, Fe, In, Mg, Mn, Pb, Tl and Zn (Browner, 1974).

Continuous background lamps simultaneously excite the fluorescence of several elements present in the atomizing source, causing spectral interference as well as interference due to light scattering (scattering effect). The measuring receiver must record the most narrow possible band corresponding to the fluorescence wavelength, which requires the use of a high-resolution monochromator. The detection limits obtained from continuous-background sources are generally mediocre and far inferior to those obtained with electrodeless lamps.

(e) *Pulsed laser sources.* These have been used successfully to excite atomic fluorescence in flames. Fraser *et al.* (1971) used a colored tunable laser pumped with a nitrogen laser to study the fluorescence of the elements Al, Ca, Cr, Fe, Ga, In, Mn, Sr and Ti, in either air-hydrogen or a nitrous oxide-acetylene flame. The peak energy is higher than 10 kW at all wavelengths, the frequency is 1-25 Hz, the half-width of the spectral band is about 1.0-10Å, and the pulse half-width is 2-8 nsec.

Detection limits attain 0.01-0.3 $\mu g/ml$ depending on the elements (see Table III). It should be noted that the detection limits of laser-excited fluorescence for Al, In and Ti are better than those obtained from classical sources. The use of lasers in atomic fluorescence has two essential advantages: very high signal intensity (kW) and the possibility of modulating the signal in a broad wavelength region. On the other hand, the high price of these sources will limit their use in fluorescence for a long time.

2.3 Atomizing Sources

Atomizing sources must satisfy the following qualities: high atomization efficiency, reduction of chemical and physical interferences, low emission

Table III. Atomic Flame Fluorescence Detection Limits (mg/l) with Different Excitation Sources[a]
(concentrations for S/N = 2)

Element	Wavelength	Vapor Metal Discharge Lamp	Hollow Cathode Lamp	Hollow Cathode High Intensity Lamp	Electrodeless Lamp	Xe-arc Lamp (150 W)	Laser
Ag	328.1	–	0.001	0.0003	0.0001	0.06	–
As	197.3	–	–	1.0	0.07	–	–
Be	234.9	–	–	0.008	0.01	–	–
Bi	306.8	–	0.1	0.25	0.0047	0.3	–
Cd	228.8	0.0001	–	0.00004	0.000001	0.033	–
Co	240.7	–	0.3	0.0015	0.005	0.14	–
Cr	357.9	–	1.0	0.0015	0.05	0.3	0.03
Cu	324.7	–	0.001	0.0003	0.005	0.083	–
Ga	417.2	–	1.0	–	0.007	0.16	0.3
Hg	253.6	0.07	–	–	0.08	100	–
Mn	–	–	0.05	–	0.001	0.03	0.3
Ni	232.0	–	0.5	0.003	0.006	3.0	–
Pb	405.6	–	1.0	0.02	0.01	3.3	–
Sb	231.1	–	–	0.03	0.04	100	–
Se	196.1	–	–	–	0.15	134	–
Te	214.3	–	–	0.05	0.05	4.2	–
Tl	377.6	0.005	0.1	–	0.04	0.13	–
Zn	213.9	0.000001	–	0.0006	0.00004	0.08	–

aAccording to Benetti et al. (1971), Rossi et al. (1969), Larkins et al. (1971), Matousek et al. (1969), Sychra et al. (1975), Zacha et al. (1968), Thompson et al. (1969), Manning et al. (1968), Fraser et al. (1971).

background, stability and reproducibility, low concentration of fluorescence-quenching molecules, and a long residence time of the formed atoms.

2.3.1 Flame Sources

The most common sources still are the classical flames of a premixed laminar type: air-acetylene or nitrous oxide-acetylene. However, the diffusion zone of these flames (plume) has a strong spectral background resulting from the luminescence of unavoidable combustion reactions:

$$2\ CO + O_2 = 2CO_2 + h\nu$$
$$2\ H_2 + O_2 = 2H_2O + h\nu$$

This luminescence extends into the UV and visible regions of the spectrum, thus limiting the sensitivity of the atomic spectra emitted in the flame both by direct emission and fluorescence. This disadvantage can be eliminated by separating the diffusion from the primary zone. The background-free atomizing zone from which fluorescence is excited is located between these two.

In the separated flames, the secondary reaction zone is separated from the primary zone by a quartz separating tube or a concentric laminar flow of argon or nitrogen (Aldous et al., 1970). Figure 6 shows the principle of the separated $N_2O\text{-}C_2H_2$ flame compared to an ordinary flame.

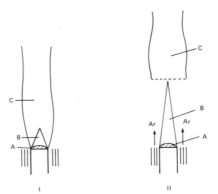

Figure 6. Nitrous oxide-acetylene flame.
I. Standard Flame; II. Separated flame: (A) Primary flame, (B) Pink cone, (C) Secondary reaction zone.

The geometry of the flame and thus of the burner is of great importance in atomic fluorescence, particularly for the shape of the analytical curves. Theoretically, the geometry of the atomizing zone should be rectangular over a height of at least 1 cm above the burner head. In practice, however, the most common burners have a circular head that is simpler to manufacture.

2.3.2 Nonflame Sources (Electrothermal Sources)

The nonflame atomizing sources (see Chapter 5) have also been used successfully in atomic fluorescence (Massmann, 1968). They have numerous advantages over flames, particularly that of suppressing the gases and their combustion reaction, possibly producing fluorescence quenching. Atomization efficiency is generally higher than in flames. On the other hand, electrothermal nebulizers heated by the Joule effect in an inert gas atmosphere have an intense continuous emission background in order to saturate the receiver, making any reading impossible.

The light scattering phenomena (scattering effect) can also be relatively strong in the presence of large amounts of minerals. The background and scattered light emissions are reduced with a suitable optical device, on one hand, and with a background suppressant, on the other (see Chapter 5) (Amos *et al.*, 1971). The principal nonflame nebulizers are graphite or tantalum tubes, ribbons or tantalum, platinum or graphite rods. Heating is generally obtained by a flow of an electrical current of suitable intensity or by induction heating in a high-frequency field. The Massmann (1968) setup is shown in Figure 7.

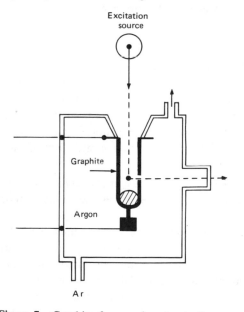

Figure 7. Graphite furnace for atomic fluorescence (according to Massmann, 1968).

Nonflame atomizing sources should be developed in the near future because they allow one to work with very small samples in atomic fluorescence, *i.e.* 10-50 μl or 0.5-5 mg, while permitting very low detection limits.

2.3.3 Atomization from Solids

Attempts are being made to use atomizing sources that accept the samples directly in the solid state. In addition we may note the *pulsed arc* and *cathode sputtering* of relatively classical devices for injecting powder into the flames (Willis, 1975) and furnaces (Langhmyr, 1973, 1974), which are used in atomic absorption. With the pulsed arc, the powder sample is placed into an electrode and vaporized with a pulsed discharge attaining 13 A (Belyaev *et al.*, 1970). The atomic vapor is excited and the fluorescence is measured.

In cathode sputtering, the sample either is placed on the cathode or it constitutes the cathode of a discharge tube filled with argon at 5 torr. A cathode discharge is produced with a current of 35 mA. The region near the cathode (20 mm) contains a quantity of atoms originating from the cathode in which fluorescence can be excited. This principle has been used primarily for the analysis of metal samples (Gough *et al.*, 1973, and Butler *et al.*, 1975).

2.4 Optics, Monochromator, Measurement

2.4.1 Dispersive Spectrometry

The dispersive system and measuring equipment are similar to those used in atomic absorption spectrometry. The optical system, on the other hand, is different. Its object is to make the most efficient use of the fluorescence energy in order to select the analysis radiation and to measure with the highest possible signal/noise ratio (S/N). This results in the best sensitivity and lowest detection limit. Several parameters must be studied to optimize the S/N ratio:

(a) The *number* of solid angles under which the fluorescent source is seen by the detector. This depends on the number of mirrors used in the focusing system.

(b) The *size* of the solid angle under which the fluorescence is seen by the detector. This depends essentially on the ratio of the grating surface to the focal length of the monochromator and, for nondispersive systems, on the ratio of the surface of the focusing lens to its focal length.

(c) The total *transmission* (t) of the optics: lenses, mirrors, monochromator. For a given instrument, t is in the order of 0.4.

(d) Height of the monochromator slit.

(e) Slit width. An optimum width exists, beyond which the S/N ratio decreases.

The principal focusing systems are shown in Figure 8. The best optical gain is obtained with the device using Cassegrain mirrors. Addition of an auxiliary mirror to improve the S/N ratio depends on the arrangement of the assembly as well as on the detector. The secondary source mirror (Figures 8a and 8b) makes it possible to double the S/N ratio at low concentrations (Benetti et al., 1971). In contrast, the field and detector mirrors reflect the atomic fluorescence signal and the spectral background at the same time. Modulation of the emission source to the measuring amplifier frequency eliminates the background of the flame itself but not that of the fluorescence background. The same applies to nonflame cells.

2.4.2 Nondispersive spectrometry

Several authors have replaced the grating monochromator by interference filters (Figure 9). In many cases, fluorescence radiations are sufficiently resolved in the spectrum to allow a selection by a filter. The spectral response of the photomultiplier, the wavelength and band width passing through the filter as well as its transmission must be considered. The solar blind photomultipliers are used in particular. Thus, Zn, Cd and Hg are determined with excitation from an electrodeless lamp in a hydrogen argon-flame with an interference filter for Cd (228.8 nm) and Hg (253.7 nm), and without a filter for Zn. The detection limits are 0.1 ng/ml Cd, 2.5 ng/ml Hg and 0.45 ng/ml Zn (Elser et al., 1971).

Larkins (1971) found a great improvement of the detection limits by using atomization in a separated air-acetylene flame associated with a nondispersive system with or without filter and a solar-blind photomultiplier. For elements with fluorescent radiations between 200 and 300 nm in particular, he obtained the following detection limits in μg/ml: 6 As, 0.1 Au, 0.25 Bi, 0.004 Cd, 0.03 Co, 0.003 Fe, 0.0002 Mg, 0.01 Mn, 0.002 Ni, 1 Pb, 0.04 Sb, and 0.003 Zn.

Nondispersive fluorescence spectrometry is easily applicable to multi-elemental analysis either with the use of several exciting sources and several amplifiers matched to the different frequencies (simultaneous analysis) or by placing filters on a drum so as to bring them successively into the optical beam (sequential analysis). Several exciting sources are necessary. They can be replaced by a multielemental source (Malmstadt et al., 1972). The above-mentioned scattering effect is difficult to correct since the fluorescence measurement is made in a polychromatic spectral region corresponding to the transmission band width of the filter (Larkins et al., 1975).

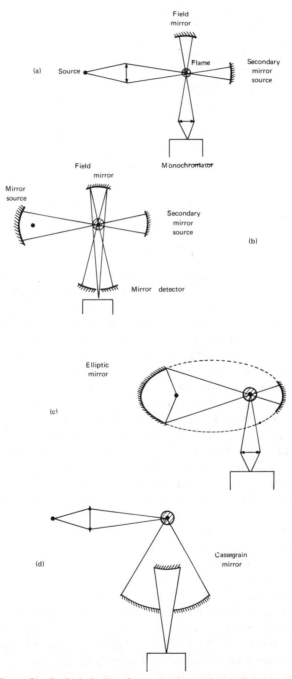

Figure 8. Optical devices for excitation and atomizing sources.

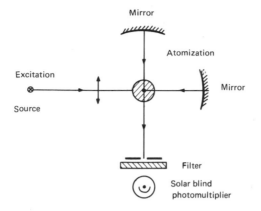

Figure 9. Nondispersive fluorescence spectrometry.

3. PROPERTIES

3.1 Sensitivity and Detection Limit in the Flame

It is apparent from the above that in atomic fluorescense spectrometry, the sensitivity depends essentially on the instrument conditions:
the type of exciting source, the atomizing source, and the dispersive system.

Table III (p. 277) shows the comparative detection limits obtained with
the major exciting sources—metallic vapor discharge lamps, hollow-cathode
lamps, high-intensity hollow-cathode lamps, electrodeless lamps, xenon arc
lamps (150 W) and laser. The best results are obtained with the electrodeless lamps with high-frequency excitation. These are used most frequently.
The detection limits with xenon lamps are mediocre, although the power
of the lamp (450 W) allows a two-to-three-fold lowering of the sensitivity
limit. The classical discharge lamps are used only for mercury. The laser
is used only in a limited number of laboratories because of its high cost.

The type of flame also plays an important role. Table IV shows
the detection limits of some elements in a standard and separated air-
acetylene flame (Larkins, 1971).

A comparison of the values in Table IV with those of Table III
should be made with reservations. The data were obtained by different
authors and under different experimental conditions. We should only
note the improvement of sensitivity by a factor of 2-3 with the use of a
separated flame.

Larkins (1971) made another interesting comparison between dispersive fluorescence spectrometry with a monochromator and nondispersive
spectrometry with interference filters (Table V).

Table IV. Detection Limits Obtained in a Standard and Separated Air-Acetylene Flame (mg/l)

Elements	Wavelength (nm)	Lamps	Normal Flame	Separated Flame
Au	242.8	HIL[a]	4.0	1.0
Bi	306.8	HIL	33.0	5.0
Cd	228.8	HCL[b]	0.06	0.02
Co	240.7	HIL	1.0	0.15
Fe	248.3	HIL	0.30	0.05
Mg	285.2	HIL	0.0006	0.00015
Mn	279.5	HCL	0.20	0.030
Ni	232.0	HIL	0.60	0.30
Pd	340.5	HIL	0.50	0.15
Sb	217.6	HIL	1.5	0.60
Tl	377.6	HCL	20.0	2.0

[a]HIL: High intensity hollow cathode lamp.
[b]HCL: Normal hollow cathode lamp.

Table V. Detection Limits by Dispersive and Nondispersive Spectrometry in a Classical Air-Acetylene Flame (mg/l)

Elements	Wavelength (nm)	Dispersive Spectrometry	Nondispersive Spectrometry
Au	242.8	4	0.9
Cd	228.8	0.06	0.04
Fe	248.3	0.3	0.04
Mg	285.2	0.0006	0.0009
Mn	279.5	0.2	0.09
Ni	232.0	0.6	0.01
Tl	377.6	20	100

3.2 Detection Limit of Nonflame Methods

Table VI lists a few of the absolute detection limits obtained with atomization in a furnace compared to flame atomic absorption. These values, originating from different authors, were reported by Browner (1974). The fluorescence values were obtained with atomization of 0.5-1 μl of sample on a graphite-rod nebulizer. The atomic absorption values were obtained (1) with 5 μl of sample treated with the Varian Model 63 graphite-rod atomizer and (2) with 100 μl of sample treated with the Perkin Elmer HGA 72.

Table VI. Detection Limits in μg Obtained in Fluorescence and Nonflame
Atomic Absorption (concentrations giving a S/N = 2)

	Atomic Fluorescence	Atomic Absorption (1)	Atomic Absorption (2)
Ag	0.4	0.2	0.25
Bi	10	7	10
Cd	0.0015	0.1	0.1
Co	20	6	4
Cu	0.3	7	1
Mg	1	0.06	10
Mn	5	0.5	1
Ni	5	10	10
Pb	10	5	6
Tl	20	3	14
Zn	0.02	0.08	0.06

Massmann (1968) compared the detection limits obtained by atomic and fluorescence absorption when atomization was made in a graphite furnace designed by him. For atomic absorption, he used a horizontal graphite tube of 50 mm length, 6.5 mm diameter and 1.5 mm thickness (see Chapter 5), while a vertical graphite tube of 40 mm length, 6.5 mm diameter and 1.5 mm thickness (Figure 7) served for fluorescence spectrometry. In both cases, the tube was heated to 2600°C in an argon atmosphere. The sample volume amounted to between 5 and 50 μl for fluorescence spectrometry. Table VII lists the comparative values of the detection limits obtained for several elements.

Table VII. Comparative Detection Limits of AAS and AFS with Massmann Cuvette
(absolute values in g)

		AAS	AFS
Ag	328.1	8×10^{-13}	1.5×10^{-12}
Cd	228.8	2×10^{-12}	2.5×10^{-13}
Cu	324.7	1×10^{-11}	4.5×10^{-10}
Fe	248.3·	2×10^{-11}	3×10^{-9}
Mg	285.2	5×10^{-13}	3.5×10^{-12}
Pb	283.3	1×10^{-11}	3.5×10^{-11}
Sb	231.1	1×10^{-10}	2×10^{-10}
Tl	276.8	4×10^{-11}	2×10^{-9}
Zn	213.9	8×10^{-13}	4×10^{-14}

The detection limits in AFS are best for zinc and cadmium, while they are comparable or slightly inferior for the other elements. This also appears from Table VII. With regard to precision, the authors generally agree that AAS and AFS are comparable. When the quantity of dissolved salts in the analysis solution is less than or equal to 2% and the sample volume is between 30 and 100 μl, the standard deviation is between 4 and 12%.

If we accept the above considerations concerning the sensitivity of atomic fluorescence compared to atomic absorption obtained under similar conditions, the question arises as to the real advantages of atomic fluorescence. Other aspects must be considered, particularly the extent of the analyzable concentration range. In atomic absorption, the range is often small (1-10 or 1-100) as shown by the deflection of the analysis curve. Anderson *et al.* (1970) and Alger *et al.* (1971) demonstrated that the analytical range is much wider in atomic fluorescence—a factor of 1000 is classical between the limit of detection and the upper detectable quantity, a concentration beyond which a measurement loses all sensitivity because of the deflection of the analysis curve.

Table VIII lists the values in absolute quantities of the detection limit (column 1) and the maximum determinable quantity in g (column 2).

Table VIII. Analysis Range in AFS

	Limit of Detection (g)	Maximal Determinable Amount (g)
Ag	1×10^{-12}	2×10^{-9}
Bi	1×10^{-11}	1×10^{-8}
Co	2×10^{-11}	6×10^{-9}
Cu	1×10^{-12}	4×10^{-9}
Ga	5×10^{-11}	1×10^{-8}
Mg	1×10^{-12}	1×10^{-9}
Mn	5×10^{-12}	2×10^{-9}
Ni	5×10^{-12}	5×10^{-9}
Pb	1×10^{-11}	1.5×10^{-7}
Sb	1×10^{-9}	3×10^{-8}
Tl	5×10^{-11}	2×10^{-9}
Zn	2×10^{-14}	4×10^{-10}

3.3 Atomic Fluorescence Compared to Flame Emission and Atomic Absorption

The choice of a method depends on the analytical requirements, the composition of the sample to be analyzed, and the possibilities of the method considered. It is certain that the latter should include not only the procedure but also the instrumentation. At the present time, however, it does not seem that an atomic absorption spectrometer is available on the market. The choice will thus depend on other criteria, such as sensitivity. Christian *et al.* (1971) suggested a classification based on the sensitivity of the methods and demonstrated that atomic fluorescence might be recommended for the following elements: Ag, Au, Bi, Cd, Ce, Co, Cu, Ge, Hg, Mn, Se, Te, Tl and Zn. Atomic fluorescence would be equivalent to absorption for As, Fe, Ni, Pb, Sb and Se.

Atomic absorption would be superior to fluorescence for Be, Ho, Mg, Mo, Pd, Si and Sn. Flame emission would be superior to fluorescence and atomic absorption for Al, Ca, Eu, Gd, In, Nd, Pr, Ru, Sr, V and Yb. Finally, flame emission, atomic absorption and fluorescence would be equivalent for Cr, Nb and Tb.

Other criteria to be considered include the possibilities for automation and multielemental analysis. From this prospective, atomic fluorescence can be placed at the top of the list. The analysis range also allows us to classify fluorescence before absorption.

3.4 Interferences in Atomic Fluorescence

Interferences in atomic fluorescence have four origins:
— spectral interferences
— chemical interferences
– light scattering of the exciting source
— quenching by molecules or elements.

Chemical interferences occur not only in fluorescence but are also encountered in atomic absorption (see Chapters 4 and 5). The *spectral interferences* consist of line, band or emission background overlapping on the fluorescence line and are relatively frequent, particularly when the exciting source is a continuous-background or multielemental lamp and when a nondispersive spectrometer is used. They are the result of a fluorescence excitation in several elements of the matrix, the radiations of which are resolved not at all or only poorly by the spectrometer. The spectral interferences can also result from direct flame emissions (continuous background or atomic emission). A direct flame emission is eliminated by modulating the excitation source.

In atomic fluorescence with furnace atomization, the light emitted by the incandescence of the furnace can saturate the photomultiplier if care is not taken to collimate the measuring beam.

The result of the spectral interferences is an apparent enhancement of the recorded signal. The scattering interferences are actually the most annoying. They result from scattering of the light from the excitation source by solid or liquid particles that are not volatilized in the atomizing source when the solution atomized into the flame contains a high mineral salt concentration. In some cases, this interference is eliminated by measuring another fluorescence line (direct-line or stepwise fluorescence) instead of the perturbed resonance line. Unfortunately, few elements emit non-resonance lines with sufficient intensity. In fact, this method is practically applicable only to the elements Pb, Bi and Sb.

Attempts have also been made to excite fluorescence lines from a near line belonging to another element. Dagnall *et al.* (1967) used the iodine line at 206.16 nm with an electrodeless iodine lamp to excite the fluorescence of bismuth, *i.e.,* the lines of Bi at 206.17 and 302.46 nm. The measurement at 302.46 is practically free of a scattering effect. Other line pairs have been proposed (Omenetto *et al.*, 1968).

Tl	377.57	excited by	Hg	377.63 nm
Mg	285.21	excited by	Hg	285.24
Cr	359.35	excited by	Hg	359.35
Fe	248.33	excited by	Hg	248.27
As	228.81	excited by	Cd	228.80

The scattering effect is particularly disturbing and difficult to correct in nondispersive spectrometry.

Finally, quenching interferences are manifested by a reduction of the fluorescence signal occurring when the population in the excited state is reduced. This in turn results from the presence of radical molecules, elements originating from decomposition of the matrix or combustion of the flame (see above, Section 1.5).

4.   PRACTICAL APPLICATIONS

The field of application of atomic fluorescence is still limited because a commercial atomic fluorescence spectrometer is not available on the market. Generally, the users have adapted the necessary accessories for fluorescence excitation to fit an atomic absorption spectrometer. Finally, about 20 elements seem to be validly determinable. These most frequently include mercury, cadmium, zinc, copper and lead. A few applications are listed in Table IX.

Biological media are most commonly analyzed, followed by water, plants, environmental media, sediments, oils and chemical products. Fluorescence is also utilized in multielemental analysis. When compared to atomic absorption this is probably one of its major advantages. In particular, we should note multielemental analysis of natural water (Ag, Cd, Fe, Sb, Se, Te, Zn) (Harsma, 1973), and analysis of trace elements

Table IX. Application of Atomic Fluorescence Spectrometry

Element	Sample	λ nm	Concentration	Atomization	Sample Preparation	References
Hg	Water, rocks, flour, sediments	253.7	ng	Cold vapor	Liberate Hg by reduction + aeration, or by combustion. Collect on Ag wire; heat wire.	Muscat et al., 1972
Hg	Air	253.7	>0.003 mg/l	Cold vapor	–	Thompson 1972
Hg	Water, rocks, sediments	253.7	0-400 ng/g	Cold vapor	Dissolve reduce, aerate.	Muscat et al., 1971
Zn	Water	213.8	0.4-10 μg/l	Flame air-C_2H_2	Heat to 80°C for 2 hr.	Marshall et al., 1972
Cu,Zn	Serum	324.7	1-200 μg/l	Flame air-C_2H_2	Dilute (1:25) with H_2O.	Kolihova et al., 1973
Cd,Ni,Pb	Serum		>10 and 25 μg	Flame N_2O-H_2	Inject μl samples into premixed gases.	Sarbeck et al., 1972
Ag,Au,Cu, Mg,Mn,Zn	Rare earth oxides	Ag 328.1 Au 242.8 Cu 347.8 Mg 285.2 Mn 279.5 Zn 213.9	0.03-1 μg/g 0.1-1 μg/g 0.05-2 μg/g 0.01-10 μg/g 0.03-1 μg/g 0.01-1 μg/g	Impulse-arc (200-400 A) in argon	Mix(1:1) with graphite powder in pack electrode cavity (graphite crucible).	Guzeev et al., 1973
Cd	U and Th compounds		0.01-1 μg/g	Flame air-C_2H_2	Dissolve in HNO_3, add NH_4OH + NH_4 citrate solution, extract (0.1-1 μg Cd with dithizone). Re-extract Cd from organic phase with 0.01 M HCl.	Murucaiyan et al., 1973
Cd,Zn	Ores		>0.02 μg/ml	Flame air-propane	Dissolve in HCl.	Novikov et al., 1971
Hg	Rocks	253.7	>5 ng	Flame	Dissolve in acid.	Vejvodova et al., 1973
Cd,Zn	Orchard leaves		0.1 ppm	Flame	Dissolve in $HClO_4$ 5%.	Rains et al., 1973
Ag,Cd,Fe, Sb,Se,Te, Zn	Natural waters	–	traces	Flame	Dilute sea water (1:1).	Harsma, 1973
Cd,Hg	Air	–	traces	Carbon rod	Signal quenched by N_2 and CO.	Robinson et al., 1973

Table IX, Continued

Element	Sample	λ nm	Concentration	Atomization	Sample Preparation	References
Co,Cr,Cu, Fe,Mn,Zn	Sea water	—	Co,Mn: 0.001-5 μg/ml Zn,Cu,Cr: 0.0004-5 μg/ml Fe: 0.003-5 μg/ml	Flame air-C_2H_2 (N_2 separated)	Extract with Na DDC at pH 5 into butan-1-ol + 4-methylpentan-2-one.	Jones et al., 1973
Various	Orchard leaves, coal, liver	—	μg/g	Flame	Study of application.	Rains et al., 1973
Cu	Blood serum	—	0.06-15 μg/ml	Flame air-H_2, air-C_2H_2, O_2-Ar-H_2	Dilute (1:25) with H_2O.	Kolihova et al. 1973
Zn	Enzymes	—	–	Graphite tube	Automated sampling.	Veillon, 1973
Various	Blood	—	traces	Flame	Use EDL source.	Winefordner et al., 1973
Various	Used oils	—	0-20 μg/g	Flame air-C_2H_2 (Ar-separated)	Dilute (1:24) with n-hexane.	Johnson et al., 1974
Ca,Cu,Fe, K,Mg,Na	Rare earth oxides	—	0.01-0.1 ppm	Graphite crucible	Direct analysis of solid samples.	Guzeev et al., 1973
Cd	Orchard leaves, liver	228.8	>0.1 ppm	Flame Ar-H_2	Digest with HNO_3-$HClO_4$-HF.	Rains et al., 1974
Hg	Soils, water, air	—	traces	Cold vapor	Double Ag amalgamator system.	Corcoran, 1974
Zn	Soils	—	0.0001-20 ppm	Flame air-C_2H_2 (Ar separated)	Nondispersive spectrometer.	Norris et al., 1974
Cd,Zn	Sea water	—	1-100 ng/ml	Flame air-C_2H_2	A study of interference.	Agterdenbos et al., 1974
Hg	Estuary sediments	—	0.02 ng/g	Cold vapor	Study of organic and inorganic Hg compounds.	Andren et al., 1973
Cu	Blood serum	—	traces	Graphite braid	Dilute. Heat atomizer to 2500°C.	Montaser et al., 1974

Element	Sample	Wavelength	Detection limit	Atomization	Notes	Reference
Cu,Fe	Serum	—	traces	Graphite braid	Dilute with H_2O.	Montaser et al., 1974
Pb	Blood	405.8	0.002-40 μg/ml	Flame Ar-H_2-O_2	Dilute (1:20) with water.	Human et al., 1974
Fe	Serum	—	0.5 μg/ml	Flame air-C_2H_2		Rippetoe et al., 1974
Zn	Enzymes	—		Graphite tube	Alimentation with desolvated sample aerosols.	Clyburn et al., 1975
Cu,Fe,Pb	Fuels	—	>0.004, 0.04 0.06 ppm	Flame H_2-O_2-N_2	Direct determination.	Cotton et al., 1970
Ag,Cu,Cr, Fe,Mg,Ni, Pb,Sn	Lubricating oils	—		Flame H_2-Ar-air	Dilute (1:5) with C Cl_4.	Miller et al., 1971
Ni	Gas oil	—	0.04 ppm	Flame air-C_2H_2 (separated)	Direct determination.	Sychra et al., 1970
Ag,Cu	Engine oils	—	Ag 0.1-6 μg/g Cu 0.1-20 μg/g	Graphite rod	Ash on rod at 645°C 18 sec, in Ar-H_2 atmosphere. Atomize at 1800°C, 3 sec.	Patel et al., 1973
Au	Waters	242.8	>0.015 μg/ml	Flame air-C_2H_2 (separated)		Matousek et al., 1970
Mn,Zn,Fe, Ca,Mg	Orchard leaves	279.5	>1.5 ppm	Flame Ar-H_2-air	Dissolve in acid.	Malmstadt et al., 1972
Pb	Blood, urine	405.8	>7.5 pg	Carbon rod	Direct atomization.	Amos et al., 1971
Ag,Sn	Base oil	Ag 328.1 Sn 303.4	>0.4 μg >100 pg	Carbon rod	Direct.	Patel et al., 1973
As,Sb,Se, Te	Animal feed stuffs	As 193.7 Sb 231.1 Se 196.1 Te 214.3	0.1-100 ng/ml 0.1-150 ng/ml 0.06-200 ng/ml 0.08-100 ng/ml	Flame H_2-N_2	Separation of elements as hydrides; dissolve sample in HCl and reduce with 1% $NaBO_2$ solution.	Thompson, 1975

extracted from sea water, such as Co, Cr, Cu, Fe, Mo and Zn (Jones *et al.,* 1973).

Cotton *et al.* (1970) determined copper, iron and lead in kerosene by combusting the latter in a flame with no additional fuel. The oxidant is a mixture of $Ar-O_2$. Excitation was produced with a 450 W xenon lamp. The authors showed that atomic fluorescence is superior to the classical colorimetric methods, with detection limits of 0.04 ppm for Cu, 0.16 for Fe and 0.16 for Pb.

In biology, we may note the determination of Cu, Zn and Mg in urine with a sequential spectrometer (Malmstadt *et al.,* 1972) and of Cu and Zn in blood serum (Kolihova *et al.,* 1972).

Electrothermal atomization on a carbon rod, graphite furnace seems to offer a new development in atomic fluorescence. Ag, Pb and Sn are determined in lubricating oils after atomization on a graphite rod and excitation with an electrodeless lamp (Patel *et al.,* 1973a). The sensitivity is good and the analytical range is sufficiently linear. Some biological applications should also be noted. Lead was determined in urine and serum by Amos *et al.* (1971) with atomization on a carbon rod and excitation by a hollow cathode. The blood sample was diluted by a factor of 2.5 with water and treated directly in the nebulizer after ashing at low temperature. The residue was atomized and the fluorescence read at 405.8 nm. The authors reduced the source intensity to 1% with the use of a filter in order to suppress the scattering effect. The detection limit, particularly for the analysis of urine, depends on the emission of scattered light by the vapors. Analysis of biological media by nonflame atomic fluorescence has also been used by Montaser *et al.* (1974a, 1974b) to determine copper and iron in serum. Veillon (1973) and Clyburn (1975) determined zinc in enzymes.

Atomic fluorescence is also used for the determination of mercury by reducing and exciting cold mercury vapor. The classical atomic absorption method was successfully utilized for fluorescence by numerous authors, such as Thompson (1972), Muscat *et al.* (1971) and Corcoran (1974).

After liberating the mercury and recirculating the vapor into the atomizing cell, Muscat obtained an approximately five-fold improvement of the detection limit, compared to that obtained with atomic absorption. The detection limit is further lowered by depositing the mercury on a silver wire (Muscat *et al.,* 1972) and subsequently liberating it by heating in the atomizing cell. Approximately 0.6 ng of mercury can be quantitatively determined by atomic fluorescence from samples such as rocks and ores. Mercury in air was also determined up to a lower limit of 3 μ/m^3 (Thompson, 1972).

Particular mention is made of the work of Thompson (1975) dealing with the analysis of As, Sb, Se and Te—elements forming volatile hydrides—with an application to complex media, such as foods. The sample is dissolved (5% HCl medium) in a suitable quantity, and 1 ml of a 1% sodium borohydride solution is added to reduce the tested elements into volatile hydrides. These are entrained into a hydrogen flame by a nitrogen stream. The detection limits are very low: 0.1 ng/ml for As, Sb, 0.06 for Se and 0.08 ng/ml for Te. However, because of the impurities of the reagents they are limited to 0.5, 0.1, 0.08 and 0.15 ng/ml for As, Sb, Se and Te, respectively. In the analysis of selenium in animal feeds, the author obtained excellent correlation with fluorometry at a level of 0.07-0.6 ppm. This method might also be extended to the elements Bi, Ge, Sn and Pb.

The applications of atomic fluorescence spectrometry are diverse, and if the technique should be developed commercially, there will be many cases in which fluorescence would have advantages over atomic absorption. In particular, these might involve the sensitivity, detection limits, extended analytical range, multielemental analysis, rapidity of analysis, and simplification of the apparatus if a nondispersive spectrometer is used.

REFERENCES

Agterdenbos, J., J. P. S. Haarsma and J. Vlogtman. *Tenth Spectrometer Conference,* The Hague, Netherlands, May, 1974.

Aldous, K. M., R. F. Browner, R. M. Dagnall and T. S. West. *Anal. Chem.* 42:939 (1970).

Alger, D., R. G. Anderson, I. S. Maines and T. S. West. *Anal. Chim. Acta.* 57:271 (1971).

Alkemade, C. Th. J. *Tenth C.S.I.,* Washington, D.C. (1962).

Amos, J. D., P. A. Bennett, K. G. Brodie, P. W. Y. Lung and J. Matousek. *Anal. Chem.* 43:211 (1971).

Anderson, R. G., J. S. Maines and T. S. West. *Anal. Chim. Acta.* 51: 355 (1970).

Andren, A. W. and R. C. Harriss. *Nature* 245:256 (1973).

Belyaev, Y. L., A. V. Karyakin and A. M. Pchelintsev. *J. Anal. Chem.,* USSR. 25:735 (1970).

Benetti, P., N. Omenetto, and G. Rossi. *Appl. Spectros.* 25:57 (1971).

Browner, R. F. *Analyst* 99:617 (1974).

Butler, L. R. P. and C. D. West. *5th ICAS,* Melbourne, August-September, 1975.

Christian, G. D. and F. J. Feldman. *Appl. Spectros.* 25:660 (1971).

Clyburn, S. A., G. F. Serio, B. R. Bartschmid, J. E. Evans and C. Veillon. *Anal. Biochem.* 63:231 (1975).
Corcoran, F. L. *Amer. Lab.* 6:69 (1974).
Cotton, D. H. and D. K. Jenkins. *Spectrochim. Acta.* 25 B:283 (1970).
Dagnall, R. M., T. S. West, and P. Young. *Talanta* 13:803 (1966).
Dagnall, R. M., K. C. Thompson and T. S. West. *Talanta* 14:1151, and *Talanta* 14:1467 (1967).
Elser, R. C. and J. D. Winefordner. *Appl. Spectros.* 25:345 (1971).
Fraser, L. M. and J. D. Winefordner. *Anal. Chem.* 43:1693 (1971).
Fuwa, K. and B. L. Vallee. *Anal. Chem.* 35:942 (1963).
Gough, D. S., P. Hannaford and A. Walsh. *Spectrochim. Acta.* 28 B: 197 (1973).
Guzeev, I. D., E. S. Blinova, I. A. Mayorov and V. V. Nedler. *Zav. Lab.* 39:165 (1973).
Harsma, J. P. S. *4th Varian Techtron Symp.* London, September, 1973.
Human, H. G. C. and E. Norval. *Anal. Chim. Acta.* 73:73 (1974).
Jenkins, D. R. *Spectrochim. Acta.* 25 B: 47 (1970).
Johnson, D. J., F. W. Plankow and J. D. Winefordner. *Can. J. Spectros.* 19:151 (1974).
Jones, M., G. F. Kirkbright, L. Ranson and T. S. West. *Anal. Chim. Acta.* 63:210 (1973).
Kirkbright, G. F. and M. Sargent. *Atomic Absorption and Fluorescence Spectroscopy.* (London: Academic Press, 1974).
Kolihova, D. and V. Sychra. *Chem. Listy* 66:93 (1972).
Kolihova, D. and V. Sychra. *Anal. Chim. Acta.* 63:479 (1973).
Langmyhr, F. J. and Y. Thomassen. *Z. Anal. Chem.* 264:122 (1973).
Langmyhr, F. J., R. Solberg and L. T. Wold. *Anal. Chim. Acta.* 69: 267 (1974).
Larkins, P. L. *Spectrochim. Acta.* 26 B:477 (1971).
Larkins, P. L. and J. B. Willis. *Spectrochim. Acta.* 26 B:491 (1971).
Larkins, P. L. and J. B. Willis. *5th ICAS,* Melbourne, August-September, 1975.
Malmstadt, H. V. and E. Cordos. *Amer. Lab.* 4:35 (1972).
Manning, D. C. and P. Heneage. *Atom. Abs. Newsletter* 7:80 (1968).
Marshall, G. B. and A. C. Smith. *Analyst* 92:447 (1972).
Massmann, H. *Spectrochim. Acta.* 23 B:215 (1968).
Matousek, J. and V. Sychra. *Anal. Chem.* 41:518 (1969).
Matousek, J. and V. Sychra. *Anal. Chim. Acta.* 49:175 (1970).
Miller, R. L., L. M. Fraser, and J. D. Winefordner. *Appl. Spectros.* 25:477 (1971).
Montaser, A. and S. R. Crouch. *Anal. Chem.* 46:1817 (1974a).
Montaser, A., S. R. Goode and S. R. Crouch. *Anal. Chem.* 46:599 (1974b).
Murucaiyan, P., S. Natarajan and C. H. Venkateswarlu. *Anal. Chim. Acta.* 64:132 (1973).
Muscat, V. I. and T. J. Vickers. *Anal. Chim. Acta.* 57:23 (1971).
Muscat, V. I., T. J. Vickers and A. Angren. *Anal. Chem.* 44:218 (1972).
Norris, J. D. and T. S. West. *Anal. Chim. Acta.* 71:289 (1974).
Novikov, V. M., A. V. Sagulin and S. E. Vorob'eva. *Ezheg. Inst. Geokhim Sib. Otdel. Akad. Nauk.,* SSSR, 401 (1971).

Omenetto, N. and G. Rossi. *Anal. Chim. Acta.* 40:195 (1968).
Omenetto, N. and G. Rossi. *Spectrochim. Acta.* 24 B:95 (1969).
Omenetto, N. and G. Rossi. *Spectrochim. Acta.* 25 B:297 (1970).
Omenetto, N. and J. D. Winefordner. *Appl. Spectros.* 26:555 (1973).
Patel, B. M., R. D. Reeves, R. F. Browner, C. J. Molnar and
 J. D. Winefordner. *Appl. Spectros.* 27:171 (1973a).
Patel, B. M. and J. D. Winefordner. *Anal. Chim. Acta* 64:135 (1973b)
Rains, T. C. , M. S. Epstein and O. Menis. *17th CSI,* Florence, 3:101
 (1973).
Rains, T. C. and O. Menis. *4th ICAS,* Toronto, November, 1973.
Rains, T. C., M. S. Epstein and O. Menis. *Anal. Chem.* 46:207 (1974).
Rippetoe, W. E., V. I. Muscat and T. J. Vickers. *Anal. Chem.* 46:796
 (1974).
Robinson, J. W. and Y. E. Araktingi. *Anal. Chim. Acta.* 63:29 (1973).
Rossi, G. and N. Omenetto. *Talanta* 16:263 (1969).
Sarbeck, J. R., P. A. St. John and J. D. Winefordner. *Mikrochim. Acta.*
 55 (1972).
Smith, R., C. M. Stafford and J. D. Winefordner. *Can. Spectros.*
 14:2 (1969).
Sullivan, J. V. and A. Walsh. *Spectrochim. Acta.* 21:721 (1965).
Sychra, V. and J. Matousek. *Anal. Chim. Acta.* 52:376 (1970a).
Sychra, V. and J. Matousek. *Talanta* 17:363 (1970b).
Sychra, V., V. Svoboda and I. Rubeska. *Atomic Fluorescence Spec-
 troscopy.* (London: Van Nostrand-Reinhold, 1975).
Thompson, K. C. *Lab. Pract.* 21:645 (1972).
Thompson, K. C. *Analyst.* 100:307 (1975).
Thompson, K. C. and P. C. Wildy. *2th ICAS,* Sheffield 1969.
Veillon, C. *4th ICAS,* Toronto, November 1973.
Vejvodova, J. and I. Rubeska. *2nd Czechoslovakian Conf. Flame Spectros.
 Zvikov,* June 1973.
West, T. S. *Analyst* 91:69 (1966).
Willis, J. B. *5th ICAS,* Melbourne, August-September 1975.
Winefordner, J. D. and T. J. Vickers. *Anal. Chem.* 36:161 (1964a).
Winefordner, J. D. and R. A. Staab. *Anal. Chem.* 36:165 (1964b).
Winefordner, J. D. and R. F. Browner. *Pittsburgh Conf. Anal. Chem.*
 Cleveland, March 1973.
Zacha, K. E., M. P. Bratzel, J. D. Winefordner and J. M. Mansfield.
 Anal. Chem. 40:1733 (1968).

CHAPTER 8

X-RAY FLUORESCENCE SPECTROMETRY

1. INTRODUCTION

The analytical application of X-ray spectrometry began with the studies of Moseley in 1913, who used it in qualitative analysis, and with those of W. Bragg and M. de Broglie, who constructed the first rotating-crystal spectrograph. Ten years later, Coster and Hevesy established the basis for quantitative analysis. In fact, X-ray spectrometry, together with optical emission or atomic absorption spectrometry, actually became commercial only in 1955. Today, it takes first place among instrumental techniques. The services rendered by it are not identical to the other two techniques because it has a certain number of special characteristics, especially the possibility of determining practically all but the lightest elements of the periodic table (Z<9). The analytical range is broad since an element can be determined from the ppm level up to percentages. Precision is generally good and affected little by concentration. In contrast to atomic emission or absorption spectrometry, it represents a nondestructive method allowing recovery of the intact sample after spectrometry. Its application to multielemental qualitative and quantitative analysis has made it a method competitive with optical emission spectrometry and of greater interest than atomic absorption spectroplotometry. The structure of the X-ray spectra is much simpler than that of optical spectra and consequently easier to study.

Considerable progress in instrumentation, primarily involving the power of the sources, the precision of the dispersion systems, and the electronic quality of the detection and measuring instruments, allows rapid series analyses with excellent precision. For example, a sequential spectrometer requires a few minutes to determine 10-25 elements in the same sample whereas a multichannel spectrometer requires less than one minute.

The major difficulties reside in the often poor definition, or in an inadequate knowledge of the relation between the measured *X-ray line* and the *concentration* of the element responsible for this line. We will see the importance of the *matrix effect* and attempt to discuss the principle methods to avoid it. The classical methods consist of comparing the analytical sample to reference samples of comparable and similar chemical composition and physical properties. The more modern methods involve mathematical corrections to give consideration to secondary absorption and emission phenomena that can modify the intensity of the measured X-ray line. These corrections depend on the chemical composition of the sample subjected to spectrometric analysis, require careful and detailed adjustments, and necessitate modern computers for their application. At the present time, mathematical methods are being developed for analysis of a matrix medium consisting only of one or two major chemical constituents.

2. PRINCIPLE OF THE METHODS

2.1 Diffraction, Emission, Fluorescence

This method is based on the study and measurement of the X-ray radiation emitted by the constituents of a sample excited in a certain manner. The inner electron shells of the atoms lose one or more electrons, which are then replaced by the outer shell electrons. There is an accompanying loss of energy, which is emitted as radiation. This radiation consists of a more or less intense continuous background and a line spectrum, characteristic of each element. The lines are grouped in K, L, M . . . series that correspond to the different shells or electron orbits (Figure 1).

There are two classic excitation methods: (1) the sample is exposed to cathodic radiation with an energy superior to the excitation threshold of the element under study (direct emission X-ray spectrometry) and (2) the sample is subjected to an X-ray beam of sufficient energy to excite the X-ray spectrum of the element under study (fluorescence spectrometry).

Spectral analysis of the radiation emitted by the excited substance is made with the aid of a crystal diffracting the beam in accordance with Bragg's Law (Figure 2):

$$n \lambda = 2 d \sin \theta$$

where λ is the wavelength diffracted at an indicence angle θ; d is the interplanar spacing of the crystal and n is the order of the diffraction. The various radiations will thus be reflected separately and successively.

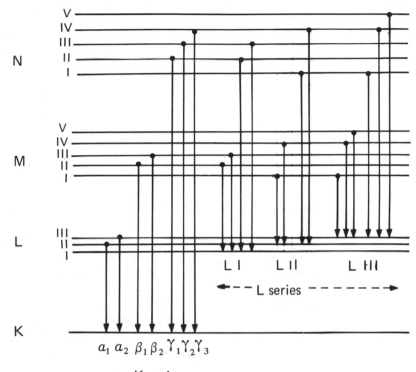

Figure 1.   Energy level diagram and electronic transitions.

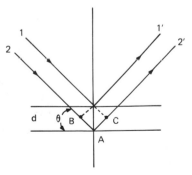

Figure 2.   Bragg's law.   Diffraction of X-rays.
S–sample; 1,2–incident radiation; 1′,2′–fluorescence radiation.

The various radiations will thus be reflected separately and successively. They are recorded on a photographic plate or are directly measured by means of a Geiger-Muller counter or a scintillation counter. In qualitative analysis, the line is identified by its diffraction angle; in quantitative analysis, the intensity of the line is determined.

Of the two methods *direct emission spectrometry* is the older. The sample is placed on the anticathode of an X-ray tube. In this technique an assembly-type X-ray tube must be employed, and there are also certain other technical difficulties. Recent advances in vacuum technique may bring about an improvement in the method, chiefly in its applications to trace analysis. In fact, the intensity of the emitted lines is 100-1000 times as strong in *direct emission* as in *fluorescence emission.*

With the second method, however, the appearance of proportional counters and scintillation counters, as well as the recent advances in electronics, made an important contribution to its development. X-Ray fluorescence spectrometry used with a scintillation counter has now become an analytical method whose sensitivity is comparable with X-ray emission spectrometry using Geiger-Muller recording. The main advantage of the fluorescence method is its simplicity, since the sample is placed outside the X-ray tube. Its further development in the course of recent years may cause the direct emission methods to be gradually abandoned.

Spectrometric X-ray analysis has a number of advantages over other instrumental methods, in particular over optical spectroscopy. (1) The X-ray emission spectrum of an element remains unchanged regardless of the kind of compound in which the element is present in the sample. In fluorescence emission, the sample remains unaltered and may be recovered after analysis. (2) The product being studied may be analyzed directly, without any preliminary chemical operations. (3) The determination is rapid because direct photometry of a line and of the spectral background requires less than a few minutes. (4) The X-ray emission spectrum given by a complex mixture is much simpler than the arc or spark spectrum. In general an X-ray spectrum comprises not more than 20 lines, so direct photometry is much simpler than with an arc or a spark spectrum.

There are disadvantages to this method. (1) Because of the absorption of X-rays by matter, the sample layer subjected to primary radiation must be very homogeneous since only the top layer (0.01-0.1 mm) will emit a fluorescence spectrum. The physical properties of the sample surface must be exactly reproducible so that the perturbations due to absorption effects might be constant. (2) While the experimental technique of X-ray fluorimetric analysis is simple, the apparatus is both complicated and costly. There is no justification in using the method except in laboratories and institutions specializing in trace analysis. (3) Finally, X-ray spectrometry cannot usually be applied to qualitative trace analysis, since the number of elements to which it applicable is small.

2.2   Absorption of Electromagnetic Radiation

X- and γ-Rays are absorbed by matter according to a classical law defined by the relation $I = I_0\ e^{-\mu l}$, where I_0 is the incident radiation intensity, I the transmitted radiation intensity, μ is the mass absorption coefficient of the X-radiation and l is the path length.   For radiations of more than 100 keV energy, μ is mainly related to the density of the absorbent material; for low-energy radiation, μ depends on the atomic number of the absorbent.

For a given element, μ increases when the radiation energy decreases to the excitation value of one of the electron shells.   At these values, μ decreases sharply in a ratio of 2 to 10; these values are called "absorption discontinuities."   The energy corresponding to an absorption discontinuity is characteristic for each element.   Figure 3 shows the variation of the mass absorption coefficient as a function of the atomic number for radiation energies of 1-200 keV.   This property is utilized to filter complex radiations or to determine an element that is highly absorbent for a given radiation in a matrix that is transparent to the same radiation.

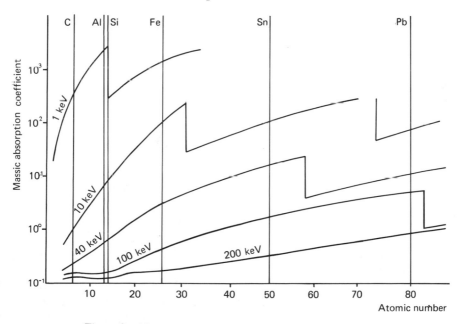

Figure 3.   Mass absorption coefficient as a function of atomic number for different energies.

This property of X-rays, of being absorbed by matter, is one of the major difficulties in X-ray fluorescence spectrometry and is largely responsible for the matrix effect on the measured radiation.

3.   APPARATUS

3.1   X-Ray Spectrometer

The apparatus used in X-ray fluorescence spectrometry consists of an X-ray source, a goniometer and a radiation detector. Various geometrical arrangements of the assembly may be used. Figure 4 represents an assembly with plane crystal diffraction: the sample is placed near the X-ray source in a plane $45°$ to the beam. The polychromatic secondary radiation is collected by a collimator, which sends a parallel beam into the surface of a plane crystal located in the center of the goniometer. The diffracted radiation is then received by the counter. When the crystal is positioned at an angle θ to the parallel beam, the diffraction angle (the angle between the incident and the diffracted beams) is 2θ and the wavelength of the diffracted beam is, in accordance with Bragg's Law, $n\lambda = 2d\sin\theta$.

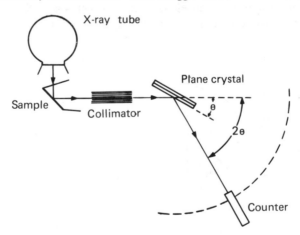

Figure 4. Diffraction by a plane crystal.

In another standard assembly (Figure 5) the beam is reflected on a curved crystal. A slit, placed on the circle of the goniometer, transmits a divergent beam. When the angle between the crystal and the incident beam is θ, the wavelength of the diffracted beam is still in accordance with Bragg's Law. The curvature of the crystal is such that it focuses the diffracted beam on the goniometric circle of motion of the recording counter.

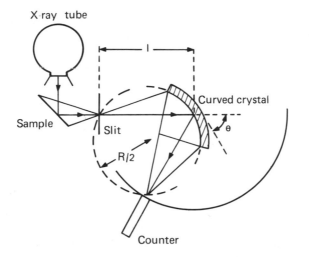

Figure 5. Diffraction by a curved crystal.

The plane crystal assembly is simpler and more easily set up. Moreover, the entire spectrum may be scanned in a continuous manner, but the sensitivity is low. The use of a curved crystal to focus the radiation increases the sensitivity and improves the resolving power. Since a larger part of the surface is utilized, the radiation yield is higher. The crystals most frequently used in fluorescence analysis are quartz, lithium chloride, lithium fluoride and mica.

In order to diminish the atmospheric absorption of the radiation emitted by the sample, especially in the case of light elements, the apparatus should include a system that evacuates the path of X-rays. Air absorbs almost totally X-rays longer than about 3 Å, so only elements with an atomic number higher than 19 may be detected with adequate sensitivity if the determination is made in the atmosphere.

3.2   X-Ray Tubes

An X-ray generating tube consists of a low-pressure chamber (1/100 mm Hg) containing two electrodes: a cathode consisting of a tungsten filament heated by an electrical current with a potential of a few V and a metal anode brought to a high potential of 40-100 kV. The cathode emits electrons by the thermionic effect. They are accelerated in the electrical field produced by the large cathode-anode potential difference and strike the anode to excite the internal shells of the constituent metallic elements. The resulting electron displacements are accompanied by an X-ray emission. The X-radiation is emitted from the tube through a transparent window (beryllium) to be focused on the test sample. The radiation consists primarily of the K and L lines of the cathode metal.

The choice of the incident radiation is an important point in X-ray fluorescence spectrometry. A sealed tube with a tungsten anode is most often used, and the continuous background of tungsten is the exciting radiation. In order that the radiation intensity might be sufficiently high, the X-ray tube must operate at a high power, 50 kV, 50 mA. The fluorescence spectra thus obtained always contain parasite lines of tungsten and of the impurities in the cathode.

Griffoul and Rabillon (1959) improved the procedure by matching the wavelength of the exciting radiation to the discontinuity of the absorption of the element being analyzed. The absorption of X-ray radiation by a given element does not vary smoothly with the wavelength. The absorption rises sharply when the wavelength is somewhat lower than certain discrete values, which are called K, L, M. . .absorption discontinuities and which correspond to the binding energies ($h\nu_K$, $h\nu_L$, $h\nu_M \cdot \cdot \cdot$) of the electrons in shells K, L, M. . . . Thus, in order that the K-series of the fluorescence lines of a given element be emitted with maximum yield, the incident beam must have a wavelength λ slightly shorter than the K-discontinuity of the element, *i.e.*, its energy should be slightly higher than the binding energy of the K-electrons. Between the absorption discontinuities, the absorption drops with decreasing λ, so there is no point in choosing a wavelength much lower than the discontinuity value. Under these conditions, the intensity of the fluorescence spectrum is 2-10 times higher than when the continuous background of tungsten is employed as incident radiation.

3.3 Diffracting Crystal

The choice of the crystal will depend on the wavelength to be studied. Inspecting Bragg's Law shows that, for a crystal with interplanar spacing d, the maximum wavelength that can be diffracted is 2d. Consequently, for the determination of light elements, the crystal employed may be made of quartz (1011 plane) for which d = 3.34 Å, or ethylene diiododitartrate with d = 4.4 Å.

For short wavelength radiations of heavy metals, the interplanar spacing should be sufficiently small for 2 θ to be larger than 10 or 15° and thus provide a sufficient dispersion. Lithium fluoride (200 plane) is most often employed. This spacing d = 2.01 Å is fully suitable for wavelengths between 1 and 3 Å·

Table I lists the more commonly used diffracting crystals and their useful ranges and average reflection efficiencies (Jenkins *et al.*, 1975).

Table I. Diffracting Crystals

Crystal	Reflection Plane	2 d Spacing (Å)	Lowest Atomic Number Detectable			Reflection Efficiency
			K Lines	L Lines		
Topaz	303	2.712	V - 23	Ce - 58		Average
Lithium fluoride	220	2.848	V - 23	Ce - 58		High
Lithium fluoride	200	4.028	K - 19	In - 49		Intense
Sodium chloride	200	5.639	S - 16	Rv- 44		High
Quartz	10$\bar{1}$1	6.686	P - 15	Zr - 40		High
Quartz	10$\bar{1}$0	8.50	Si - 14	Rb- 37		Average
Penta erythritol	002	8.742	Al - 13	Rb- 37		High
Ethylenediamine tartrate	020	8.803	Al - 13	Br - 35		Average
Ammonium dihydrogen phosphate	110	10.65	Mg - 12	As - 33		Low
Gypsum	020	15.19	Na - 11	Cu - 29		Average
Mica	002	19.8	F - 9	Fe - 26		Low
Potassium hydrogen phtalate	1011	26.4	O - 8	V - 23		Average
Lead stearate		100	B - 5	Ca - 20		Average

Reproduced by permission of the Macmillan Press Ltd. from R. Jenkins and J. L. De Vries, *Practical X-Ray Spectrometry*, 2nd ed., 1975.

3.4 Radiation Detectors

The counters most frequently used at present are proportional counters and scintillation counters. The former, which are gas counters, work on the principle of ionization of gases by X-rays and radioactive radiations. A photon with a given energy produces a pulse, the amplitude of which is proportional to the energy of the photon. The latter counters consist of a thallium-activated sodium iodide crystal that scintillates under the action of X-rays or radioactive radiations. An electron photomultiplier cell converts the scintillations to electric pulses, the amplitude of which is a function of the radiation energy. These two kinds of counters have a very short lag time, on the order of one microsecond. In conjunction with a pulse height discriminator, they make it possible to separate different radiations and record pulses only above a certain minimum energy level. In this way the spectral background in the resulting diagram may be eliminated or attenuated.

In the analysis of light elements, however, it would seem preferable to use proportional counters for reasons connected with spectral sensitivity distribution. At wavelengths above 3 Å, the line-to-background intensity ratio obtained with proportional counters is higher than with scintillation counters. Also, the efficiency of proportional counters is better (up to 80%) at longer wavelengths. In sealed counters the window, which is usually made of beryllium, may absorb long wavelength radiations (light metals). For this reason gas flow counters are employed with windows made of Mylar, which is a plastic material of high mechanical strength, transparent to X-rays. Scintillation counters, on the other hand, have the advantage of an unlimited service life and excellent efficiency for short wavelength radiations (0.2-3 Å).

3.5 Special Systems

3.5.1 Excitation with Radioactive Sources

α- and β-particles are also capable of exciting deep shells in the atom. In some cases the use of isotope sources emitting α- or β-particles makes it possible to excite the fluorescence of chemical elements, making X-ray tubes unnecessary.

The radioactive sources have a low particle or photon flux, but have the advantage of being small and exciting a relatively large surface. Furthermore, they are fairly inexpensive. The energy-level of the emitted spectrum is stable. However, the intensity decreases slowly as a function of time by an exponential law, although this decrease is very slow and

and can be neglected. The use of radioactive sources involves no stabilization problem as is the case with X-ray tubes.

A radioactive source must have the following qualities: simple line spectrum, low β- or γ-energy, sufficiently long half-life, and good specific activity. Table II lists some isotopes with their properties (Rhodes, 1971).

The β-sources are used most commonly: certain radionuclides emit pure β-radiation, allowing direct excitation of the X-ray fluorescence spectrum of the analyzed samples. The most common radionuclides used in these sources are strontium + yttrium (^{90}Sr, ^{90}Y), krypton (^{85}Kr) and promethium (^{145}Pm). Electromagnetic radiation sources are also used, which can emit monochromatic X-rays by K-capture, such as tungsten (^{181}W), cadmium (^{109}Cd), and iron (^{55}Fe), or soft γ-radiation such as americium (^{241}Am).

Table II. Properties of X-Ray Radioactive Sources

Isotope	Half-Life Years	Mode of Decay	Energy (keV)	Type of X-Radiation
Fe (55)	2.7	Electron capture	5.9	Mn, K
Co (57)	0.74	Electron capture	6.4	Fe, K
Se (75)	0.33	Electron capture	10.5	As, K
Cd (109)	1.3	Electron capture	22.0	Ag, K
I (125)	0.16	Electron capture	27.0	Te, K
Gd (153)	0.65	Electron capture	41.0	Eu, K
Pb (210)	22.0	β	11-13	Bi, L
Pu (238)	86.4	α	12-17	U, L
Am (241)	458.0	α	14-21	Np, L

Because of the low intensity of fluorescence emission, dispersive crystal spectrometry, which has high resolving power but poor sensitivity, is difficult to use. Nondispersive pulse-height spectrometry (energy resolution) is used, and it has a better yield but lower resolving power. The spectrometric system includes the source, sample detector and measuring electronics (Figure 6).

A pulse-height selector selects the pulses of the detector (counter), the pulse-height of which falls between two limits, constituting a channel. By adjusting the channel to the line of the test elements, its fluorescence is determined directly, followed by its concentration. By sweeping with the channel, the spectrum emitted by the sample is recorded in order to identify its constituents.

Excitation with radioactive sources dates to 1960-1965 (Seibel *et al.,* 1961; Robert, 1964). They are used primarily in portable instruments

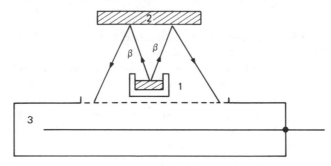

Figure 6. Excitation by radioactive source.
1–radioactive source; 2–sample; 3–counter.

because of their lightness and operating simplicity. However, their use
in trace analysis remains limited, with their analytical applications involving
mainly the major elements (Burkhalter, 1971, Marr *et al.*, 1971). Special
mention is due to Rhodes (1971) who published an objective description
of the properties of the method, the instrumentation requirements and the
fields of practical application.

Finally, we should note charged particle-induced X-ray emission.
When a sample is bombarded with protons or heavy ions, the electrons
from the inner shells of the atoms can be ejected toward higher shells.
This results in X-ray emission. Recently, Duggan *et al.* (1972) and Garcia
et al. (1973) reviewed the applications of this method. Poncet *et al.* (1975)
demonstrated that when excitation is produced with protons of an energy
below 1.5 MeV, the elements Cu, Ag, Sn and Pb can be determined in
thin films with a detection limit of a few ng/cm^2.

For additional information consult also Johansson *et al.* (1970) and
Kraushaar *et al.* (1971).

3.5.2   *Semiconductor Detectors, Nondispersive Spectrometry*

In X-ray spectrometry, the radiations are generally selected by diffraction with an analyzer crystal. This is dispersive spectrometry. For
several years, nondispersive spectrometry by pulse-height discrimination
has been used increasingly. This method was mentioned earlier in Section
3.5.1. Advances in the fabrication of "proportional" detectors delivering
a signal proportional to the incident photon energy and the associated
electronic systems have made it possible to consider the elimination of
the crystal analyzer in classical spectrometers. The present resolution of
semiconductor detectors of the Si(Li) type is attaining 130 eV, which
corresponds to about 0.04 Å for an energy of 5.9 keV (energy corresponding to the MnKα radiation of 2.1 Å). Today, certain spectrometers with

a radioisotope source are equipped with these. If the resolving power of these detectors continues to improve, it may be that in some cases they will replace the classical dispersive systems. However, it is estimated that the ultimate resolution of the Si(Li) detectors cannot decrease below 100 eV for the Kα-line of iron (6.4 keV).

The semiconductor radiation detector operates on the principle of ionization chambers, but the gas is replaced by a semiconductor solid, such as silicon or germanium doped with lithium. The properties of semiconductor detectors have been compared to those of classical detectors by Frankel *et al.* (1970) on the basis of the energy resolution obtained by nondispersive spectrometry with Si(Li) detectors and that obtained with proportional counters and NaI(Tl) scintillation counters. With the MnKα line (obtained from the [55]Fe isotope), the authors obtained the following resolutions:

Si(Li) detector	160 eV
Proportional counter	1000 eV
Scintillation counter	3070 eV

They find that Si(Li) permits resolution of the Kα and Kβ lines and of the Kα lines of nearby elements.

At the present time, the Si(Li) detectors are used mainly in portable nondispersive spectrometric equipment. If the resolution of such an instrument is compared with that of a crystal spectrometer (LiF, plane 200), the resolution of the latter remains better for wavelengths longer than 0.8 Å. Another aspect limiting the use of semiconductor detectors is the maximum measured fluorescence intensity (spectral background + radiation), which should be less than 20,000 counts/sec (CPS) including the potentially high-level primary scattered radiation. Consequently, the fluorescence resulting from elements in the state of traces is difficult to measure.

Semiconductor detectors and nondispersive spectrometry have been used successfully, especially in pollution control (Giauque *et al.*, 1975; Hammerle *et al.*, 1975) and for *in situ* analysis of air and water pollutants. Recently, Price *et al.* (1974) discussed the possibilities of application to S and Pb determinations in fuels.

4. RELATION BETWEEN FLUORESCENCE INTENSITY AND ELEMENTAL CONCENTRATION

4.1 Matrix Effect

When a sample is irradiated by a beam produced by polychromatic X-radiation (most often tubes with a tungsten, gold or chromium anode), it reemits a so-called "primary fluorescence emission." This is a complex emission, the composition of which depends on the elements present. If we consider a given element (A) of concentration C, and if the fundamental fluorescence intensity is I, the relation I = f(C) depends on the matrix of the sample. The matrix effect has been studied extensively by Tertian (1973). Two major factors can modify this relation. First, the elements of the matrix are excited by fluorescence and can produce a secondary fluorescence in the element (A) adding to the primary fluorescence. Also a tertiary fluorescence could result from the combined action of several matrix elements. If the measured element has a higher atomic number than the matrix elements, its fluorescence emission will not be enhanced by the radiations from matrix elements. On the other hand, if elements of higher atomic number are present in the sample, these will produce secondary, and perhaps tertiary, fluorescence of the element that will enhance the primary fluorescence. In practice, the relation I = f(C) will be modified mainly by secondary fluorescence phenomena.

The second perturbing effect on fluorescence results from X-ray absorption (exciting and fluorescence radiation) by the sample constituents. Let us assume a sample S containing an element (A) in concentration C, with I_E the intensity of the exciting radiation of wavelength λ_E and I_F the intensity of the fluorescence radiation of the element (A) at wavelength λ_F. In the path BD, the radiation I_E will be partially absorbed by the matrix elements in the superficial layer of the sample (Figure 7). At point D the fluorescence produced by element (A) is proportional to the absorption of radiation I_E by the atoms A, and on path DF it is inversely proportional to the absorption of radiations I_E and I_F by the constituents of the sample S.

We can write:

$$I_F = K\, I_E\, C\, \frac{\mu_{A(\lambda_E)}}{\mu_{S(\lambda_E)} + \mu_{S(\lambda_F)}} \tag{1}$$

where   K   = a constant depending on the spectrometer geometry and element (A)

$\mu_{A(\lambda_E)}$ = the mass absorption coefficient of the element A at wavelength λ_E

$\mu_{S(\lambda_E)}$ and

$\mu_{S(\lambda_F)}$ = the mass absorption coefficients of all of the constituents of sample S at wavelength λ_E and λ_F, respectively.

Under the given conditions, the product $K \cdot I_E$ and $\mu_{A(\lambda_E)}$ are constant, so the spectrometric measurement of a given element has the form:

$$I_F = \frac{K \cdot C}{\mu_S} \tag{2}$$

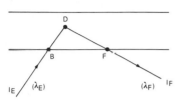

Figure 7. Excitation and fluorescence.

where μ_S is the *resultant* or *effective* absorption coefficient comprising absorption of the primary and fluorescence radiation. This coefficient depends on the chemical composition of the matrix. A knowledge of this factor, which defines the matrix effect, is indispensable. It is determined two ways. First it can be found by relative methods based on a comparison of the unknown sample with reference samples having the most similar matrix as possible. It also may be evaluated by mathematical methods making use of a calculation of the μ_S-coefficient.

4.2 Correction of the Matrix Effect, Calibration

4.2.1 Standards Methods

In the standard methods, an attempt is made to dispense with an exact knowledge of the coefficient μ_S using principle classical optical emission spectroscopy methods.

4.2.1.1 Method with External Standards. The analytical samples are compared to reference samples of the same matrix in the concentration range of the test element. This method is based on the assumption that μ_S is the same or almost identical for all samples (test and reference). If proportionality between I_F and C is verified (Equation 2), a single reference sample may suffice at the limit. However, it is necessary to verify the linearity range of the relation $I_F = K.C/\mu_S$.

4.2.1.2 Method with Internal Standards. This method consists of incorporating into the test sample an element (i) of suitable and known concentration selected to react to the matrix effects the same as the analyzed element. We have:

For element (A): $I_F(A) = K.C(A)/\mu_S$

For the internal standard (i): $I_F(i) = K.C(i)/\mu_S$

A determination of the ratio of the fluorescence radiations $I_{F(A)}$ and $I_{F(i)}$ allows us to write:

$$C(A) = k. \frac{I_{F(A)}}{I_{F(i)}} \tag{3}$$

This method corrects simultaneously for fluorescence enhancement and absorption effects to the extent to which the internal standard meets the required conditions.

Let us assume a chromium matrix in which iron is to be determined. Figure 8 shows the position of the FeK_a line relative to the

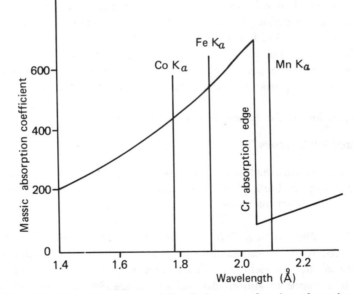

Figure 8. Absorption edges and X-radiations as a function of wavelength.

absorption discontinuity of chromium. The CoK_a radiation can serve as an internal standard, whereas the MnK_a radiation is not suited.

The internal standard must be homogeneously distributed in the entire sample, which is not always easy at the ppm level when a powder pellet is used. On the other hand, addition of the standard is simple with solid and liquid solutions. Recommended element-internal standard line pairs can be found in the study of Muller (1967).

4.2.1.3 Methods with Known Additions. The object of these methods is to add a known quantity of the elements to be determined to the analyzed sample. The sample initially emits a fluorescence:

$$I_F = K.C/\mu_S$$

After adding a quantity ΔC, the fluorescence is I'_F:

$$I'_F = K \frac{C + \Delta C}{\mu_S}$$

from which

$$C = \Delta C \frac{I_F}{I'_F - I_F} \tag{4}$$

The quantity of the element to be added must be of the same order as the concentration of the element in the initial sample and must be introduced homogeneously. The method of known additions is used in passing through the stage of solid or liquid solutions (Vos *et al.*, 1969).

4.2.1.4 Dilution Method. This method consists of diluting the samples with a suitable substance intended to stabilize the variation of the matrix effects of different samples by a buffer effect so as to practically stabilize the coefficient μ_s. Alkali carbonates and borates and cellulose are often used as diluent, but the most effective method consists of diluting with a flux (lithium borate) in a ratio of 1:10.

For a crude sample, we have Equation 2: $I_F = K \cdot C / \mu_S$. If the sample is diluted in a flux (D), in a proportion of x sample and 1-x flux, a similar relation will be obtained with the new fluorescence I_F':

$$I_F' = K \frac{xC}{x\mu_S + (1 - x) \mu_D}$$

which can be written as:

$$I_F' = K \frac{C}{\mu_S + \frac{1-x}{x} \mu_D} \tag{5}$$

In order to approximate the proportionality between I_F' and C, the denominator must be essentially constant, which will be the case if μ_S is small compared to $(1-x)/x \cdot \mu_D$. Two cases are possible for this: (1) x is very small, corresponding to high dilution, (2) μ_F is high, which is realized with a flux having a high mass absorption coefficient. In the first case, a dilution of 50-100 will be used (x = 0.01, flux = 0.99), and sodium tetraborate can be used for the flux ($Na_2B_4O_7$)(Claisse, 1956). Govindaraju (1973) performs the dilution by binding the trace elements to be determined on ion exchange resin.

In the second case, a buffer with a high mass absorption (heavy absorber), such as barium oxide or sulfate, tungstic oxide or lanthanum oxide, is added to the flux (Rose *et al.*, 1963).

Both methods are accompanied by a sensitivity loss. The slope of the calibration line $I_F' = f'(C)$ depends on the denominator of Equation 5, which increases with the dilution x as well as the mass absorption coefficient of the diluent (flux + heavy absorber). Evidently, the dilution methods require the preparation of artificial standards under the same conditions.

4.2.1.5 "Thin-film" Methods. We have seen that the mass absorption coefficient μ_S resulted from absorptions in the inner shells of the atom. It has been demonstrated that fluorescence is independent of the matrix with very thin sample thicknesses, *i.e.,* $I_F \cong KC$, where C is the concentration per cm^2 of test element. Actually, this relation is theoretically valid only for thicknesses of a micron or thin films, in which the element does not exceed 1 mg/cm^2. However, the method has numerous applications, especially for the analysis of filter deposits.

4.2.2 Experimental Determination of the Matrix Mass Absorption Coefficient

A knowledge of the parameter μ_S of the investigated matrix finds the relation

$$I_F = K \frac{C}{\mu_S} \tag{2}$$

Several methods have been proposed.

4.2.2.1 Determination of the Matrix Absorption Coefficient by Emission-Transmission. If the sample thickness is partially transparent to fluorescence radiation I_F at wavelength λ_F, Equation 2 becomes:

$$\mathbf{I_F} = K \frac{C}{\mu_S} (1 - e^{-Ph\mu_S}) \tag{6}$$

where P is the specific gravity of the sample and h its thickness in cm.

If we measure the fluorescence emission of a radiator (Figure 9)

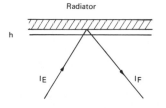

Figure 9. Determination of mass absorption coefficient.

consisting of the pure element through the analyzed sample of thickness h, we can determine the coefficient μ_S (Carr-Brion, 1964, 1965). This method has been used for the analysis of Sr and Rb in rocks (Duchesne *et al.*, 1971).

4.2.2.2 *Determination of the Matrix Absorption Coefficient by the Double-Dilution Method (Tertian, 1968).* If two preparations of dilutions x and nx are made from the sample containing element A in concentration C using quantities 1-x and 1-nx of the same flux, the fluorescences I_F' and I_F'' will be obtained, and Equation 5 will be written as follows:

$$I_F' = K \frac{xC}{x\mu_S + (1 - x)\mu_D}$$

$$I_F'' = K \frac{n \times C}{nx\mu_S + (1 - nx)\mu_D}$$

where μ_S is the matrix absorption coefficient of the sample, which is the unknown, and μ_D is the matrix absorption coefficient of the diluent which is a constant.

By eliminating C, μ_S can be determined from these two equations, and we can also eliminate μ_S and determine C:

$$\mu_S = \frac{K \times C - (1 - x)\mu_D I_F'}{xI_F'}$$

$$\mu_S = \frac{K n \times C - (1 - nx)\mu_D I_F''}{n \times I_F''}$$

From these relations, we derive:

$$C = \frac{(n - 1) \cdot D \cdot I_F' \cdot I_F''}{K \times n (I_F'' - I_F')} \tag{7}$$

Thus, this relation is independent of μ_S. The absorption due to the matrix effect is therefore corrected, and it can even be said that it is corrected in an absolute way. The enhancement effect due to secondary fluorescence is also corrected. Another characteristic of this method is that it is not necessary to know the chemical composition of the analyzed sample. Finally, to obtain a suitable precision, I_F'' and I_F' should be sufficiently different, which implies an adequate value for the number n, which defines the difference between the two dilutions (n = 3,4) (Tertian, 1971).

4.2.3 Mathematical Correction Methods for the Matrix Effect

Several methods have been proposed. We will dwell on the method of influence factors (Shermann, 1955, 1959; Marti, 1962; Claisse *et al.*, 1967; Rasberry *et al.*, 1974; Tertian, 1972). We return to Equation 2 relating fluorescence to the concentration C of element A, then

$$I_F = K \frac{C}{\mu_S}$$

The mass absorption coefficient μ_S can be written in expanded form, taking account of element A and the different matrix constituents in concentrations $C, C_i, C_j, \ldots$ and their characteristic mass absorption coefficients $\mu_A, \mu_i, \mu_j, \ldots$:

$$\mu_S = \mu_A C + \mu_i C_i + \mu_j C_j + \ldots$$

so that we will have:

$$I_F = K \frac{C}{\mu_A C + \mu_i C_i + \mu_j C_j + \ldots}$$

We will consider the intensity of the fluorescence I_{PF} emitted by element A in the pure state under the same conditions ($C = 1$):

$$I_{PF} = \frac{K}{\mu_A}$$

The ratio $R_A = I_F / I_{PF}$ defines the relative fluorescence intensity of element A in concentration C:

$$R_A = \frac{C}{C + \Sigma \frac{\mu_i}{\mu_A} C_i}$$

The ratio $K_{A/i} = \mu_A / \mu_i$ of the mass absorption coefficients of the analyzed element A (μ_A) and the matrix element (μ_i) is called the influence factor. The basic mathematical relation becomes:

$$C = R_A (C + \Sigma K_{A/i} C_i) \tag{8}$$

Since the factors $K_{A/i}$ are known for the different matrix constituents, a series of equations analogous to Equation 8 allows us to calculate the concentrations C_i, C_j, etc. from measured relative intensities $R_i, R_j \ldots$

The method takes a long time, particularly if the matrix of the sample is complex. Despite everything, the method is not absolutely accurate because the incident radiation is not monochromatic, so the theoretical values of coefficients K are actually modified by the chemical composition of the sample. Therefore they must be determined. In the case of a sample containing two elements in concentrations C_A and C_i, Equation 8 becomes:

$$C = R_A (C + K_{A/i} C_i)$$

i.e., $C_i = 1 - C$, and therefore

$$K_{A/i} = \frac{1 - R_A}{R_A} \times \frac{C}{1 - C}$$

The value of R_A (the fluorescence ratio of element A in the sample to the pure element) can be determined easily from a sample of known concentration C. Thus, the effective influence factors are calculated, but these values are valid only for a specific matrix and actually do not give consideration to small matrix variations from one sample to the next. Recently, Tertian (1975a) proposed a more elaborate method involving secondary fluorescence effects due to the different constituents.

The method of influence factors is of interest for routine analysis. The use of computers makes it extremely flexible. Furthermore, it has been used increasingly for the analysis of numerous types of mineral samples (Tertian, 1975b; Vie le Sage, 1975; Tertian, 1975c).

5. PREPARATION OF ANALYTICAL SAMPLE

The preparation of the sample is a very important operation in X-ray fluorescence analysis. The constituents of the medium may seriously interfere due to absorption effects or secondary fluorescence phenomena. Thus, every determination must be preceded by a suitable calibration made with the aid of a synthetic sample whose composition must be as near as possible to that of the sample being studied (see Section 4.2.). The use of an internal standard element to determine the concentration of the sample element by determining the intensity ratio of the lines emitted by the standard and the sample element is not the complete solution to the problem of interfering reactions (see Section 4.2.1.2).

Gunn (1957) recommends that the sample be diluted in a substance that consists of elements of atomic numbers lower than the elements being analyzed so as to reduce absorption effects. A 1:1 mixture of lithium carbonate and starch is an excellent diluent for elements with atomic numbers from 20 to 42. The sample is diluted 20 times with the mixture, and then compressed under a constant pressure of about 5,000 kg/cm^2 to give a pellet 2.5 cm in diameter. The sample pellets should always be prepared under the same pressure in order to have the sample surface the same.

If the sample contains a large proportion of organic matter, it may have to be calcined prior to the determination. Lazzar *et al.* (1958)

showed, however, that molybdenum in plants may be determined directly on the dried material, but the determination of copper is not possible under these conditions. In addition to eliminating organic matter, the calcination of the sample also concentrates the element to be determined. In general it is preferable to destroy the organic matter in vegetable and animal samples; this may also be effected by acid digestion.

Homogeneity of the preparation is very important. In fact, the X-radiations employed for analysis have wavelengths of 0.5-10 Å, so that a powdered sample with a granulometry of 50 μm, for example, constitutes a highly heterogeneous object for these radiations. Jenkins (1975) demonstrated that particle size and microheterogeneity had practically no influence on the fluorescence radiation intensity if the granulometry was smaller than 1/5 of the X-radiation penetration range into the sample (1-50 μm).

Thus, the granulometry depends on the nature of the analyzed elements and the exciting radiation. A very fine granulometry is required (theoretically less than 1 μm) for light elements with X-ray lines having wavelengths between 3 and 10 Å. For the other elements (λ between 0.5 and 3 Å), the theoretical granulometry would be between 1 and 10 μm, which may be difficult to obtain since a defined granulometry is difficult to produce in mills. This is remedied by preparing a solid solution of the analyzed sample. The latter is fused with an alkali borate (lithium tetraborate or metaborate) in a ratio of 1:10 (Claisse, 1974; Hermann, 1973). The mixture is melted in a graphite or gold-platinum-rhodium crucible. The vitreous melting product is poured into a thimble to form a plane and homogeneous surface during cooling. These techniques are widely used for the analysis of mineral media, such as soils, rocks, ores, clays, refractories, cements, ceramics, slags and scoria. The flux, which constitutes a diluent of the sample, consequently allows stabilization of the matrix. Claisse (1956) and Rose et $al.$ (1963) have suggested adding a heavy absorber (BaO, $BaSO_4$, WO_3, La_2O_3 ...) to the flux to act as an absorption buffer, which would allow less dilution and thus improve the sensitivity.

Examples for the preparation of samples by melting and pelletizing are given in the section on "Methods of Analysis and Applications." The precision of the spectrometric reading may attain $\pm$ 0.2% (Tertian et $al.$, 1972).

An attractive application of X-ray fluorescence analysis is direct analysis of the liquid sample or of the dissolved sample. With this method fluorescence analysis has all the advantages of any determination carried out on a sample in solution, such as homogeneity of the sample medium and easy preparation of standards. Kokotailo (1953) determined Br in hydrocarbons by direct irradiation of the liquid material, and measurement of the secondary radiation. Dwiggins *et al.* (1959) determined Ni in lubricating oils with a sensitivity of a few parts per million. The liquid sample was submitted to primary radiation and the intensity of the secondary $K\alpha$ radiation of Ni was measured. Co was used as internal standard.

The sensitivity of the X-ray fluorescence emissions varies from 5 to 500 ppm, depending on the element and the experimental conditions. Chemical concentration of traces is necessary for concentrations lower than 1 ppm, which are frequently encountered in biological materials.

Fagel *et al.* (1957) described fluorescence analysis of compacted organometallic precipitates. After preliminary treatment, a solution of 10 g of sample is prepared, and Cu, Ni, Fe, Mn and Cr are precipitated by 8-hydroxyquinoline with an aluminum carrier. The precipitate is washed, dried, and ground in a mortar. Pellets 2.5 cm in diameter are formed under a pressure of 500 kg/cm^2. The conditions for the preparation of the pellets (including the duration of the application of pressure) must be constant for all the samples of a series and for the respective standards.

6. INSTRUMENTAL TECHNIQUES

6.1 General Instrumental Conditions

Optimum instrumental conditions are defined by the choice of the exciting radiation and the analytical line, the nature of the diffraction crystal, the atmosphere in which the fluorescence radiation is diffracted and the type of detector. The exciting radiation is generally obtained from the tungsten anode of a sealed tube of 2-kW power operating at a voltage of 40-50 kV and a current of 30-50 mA. Chromium, molybdenum or gold anode tubes are also used.

The tube power is important for trace analysis. The intensity of a fluorescence line varies with the square of the voltage applied to the tube anode:

$$I = 1.4 \times 10^{-9} \text{ i } ZV^2$$

where I is the fluorescence radiation intensity, i is the tube current amperes, Z is the atomic number of the element and V the anodic voltage. However, an increase of voltage V also contributes to increasing the spectral fluorescence background. The optimum tube potential in practice is a function of the element and its excitation potential and increases with the atomic number (Table III).

A more judicious choice of the exciting anode improves the fluorescence sensitivity. In particular, we have seen in Section 3.2 that the

Table III. Variation of the Fluorescence Intensity as a Function of the Tube Potential[a]

kV	Al,Z = 13	Cr,Z = 24	Zn,Z = 30	Mo,Z = 42	In,Z = 49
50	100	99	96	67	43.5
60	98	100	100	80	61
70	92	97	100	87	75
80	88	94	98	95	84
90	–	89	95	97	93
100	83	–	–	100	100

[a]According to Jenkins et al., 1975.

exciting radiation must have a slightly shorter wavelength at the absorption discontinuity K of the analyzed element. Table IV shows the type of anode recommended for excitation of the different elements. Table V gives the recommended analytical lines and most favorable instrumental conditions for the elements of atomic number 11 (sodium) to 92 (uranium). The $K\alpha$ lines are used for the elements from sodium No. 11 to silver No. 47 and the $L\alpha$-lines for As No. 33 to U No. 92. It should be noted that from arsenic No. 33 to silver No. 47 a choice is allowed between the K and L lines. This will certainly be made as a function of the matrix.

The nature of the diffraction crystal is a function of the wavelength of the fluorescence radiation. For long wavelengths ($\lambda > 8$ Å), either gypsum (reflection plane 020) or potassium biphthalate (KAP, reflection plane 101) or ammonium dihydrogen phosphate (ADP, reflection plane 110) is used. For radiations between 4 and 8 Å, ethylenediaminetartrate (EDT, reflection plane 020) is recommended.

For wavelengths shorter than 4 Å, lithium fluoride (reflection plane 200, 2d = 4.028 Å) can be used, but for radiations below 2.5 Å, it is preferable to use the reflection plane 220 (2d = 2.848 Å) of lithium fluoride as well as topaz (plane 303, 2d = 2.712 Å) (see Table I). The diffraction angle (2θ) of the fluorescent radiation angles are listed in Table V. The resolving power of a dispersive system depends on the nature of the diffraction crystal but also on the geometry of the goniometer. The use of curved crystals of suitable radius (Figure 5) results in a sensitivity gain and improved resolution. With a plane crystal, the collimator geometry (Figure 4) is an important parameter. In fact, the beam emitted from the tube is normally divergent. A fine and parallel beam must be selected, and this is obtained with a collimator having plates spaced at 0.2-0.5 mm. The collimator thickness depends on the complexity of analyzed spectrum.

Table IV. Type of Anode of the X-Ray Tube Recommended for Excitation
of the Different Elements

	Atomic No.	Line	Anodes		Atomic No.	Line	Anodes
Na	11	Kα	Cr	Br	35	Kα	Mo,W
Mg	12	Kα	Cr	Rb	37	Kα	Mo,W
Al	13	Kα	Cr	Sr	38	Kα	Mo,W
Si	14	Kα	Cr	Zr	40	Kα	W,Au,Mo
P	15	Kα	Cr			Lα	Cr
S	16	Kα	Cr	Mo	42	Kα	W,Au
Ce	17	Kα	Cr			Lα	Cr
K	19	Kα	Cr	Ag	47	Kα	W,Au,Mo
Ca	20	Kα	Cr			Lα	Cr
Sc	21	Kα	Cr	Cd	48	Kα	W,Au,Mo
Ti	22	Kα	Cr			Lα	Cr
V	23	Kα	W	Sn	50	Kα	W,Au,Mo
Cr	24	Kα	W			Lα	Cr
Mn	25	Kα	W	Sb	51	Lα	Cr
Fe	26	Kα	W	I	53	Lα	Cr
Co	27	Kα	W	Cs	55	Lα	Cr
Ni	28	Kα	W	Ba	56	Lα	Cr
Cu	29	Kα	W,Au	La	57	Lα	Cr,W,Au
Zn	30	Kα	Mo,Au,W	Hf	72	Lα	W,Au
Ga	31	Kα	Mo,Au,W	W	74	Lα	Mo,Au
As	33	Kα	Mo,W	Pt	78	Lα	Mo
				Au	79	Lα	Mo
				Pb	82	Lα	Mo,W
				Bi	83	Lα	Mo,W
				U	92	Lα	Mo,W

Radiations of long wavelengths are absorbed by air, so they
must be generated in vacuum, which applies particularly to elements of
$Z = 11$-25 (sodium to manganese). However, in trace analysis requiring
the highest sensitivity, this series can be extended from manganese ($Z = 25$) to zinc ($Z = 30$). When the measurement is made with Lα-lines,
the work should proceed under vacuum except for the elements Pb, Bi and
U. The nature of the counter is also a function of the radiation and its
wavelength. Proportional gas flow counters can be used for sodium
($Z = 11$) to zinc ($Z = 30$), which are measured with the Kα-line, and for
arsenic ($Z = 33$) to tungsten ($Z = 74$), measured with the Lα-line. The
scintillation counter is used from manganese, Kα-line ($Z = 25$) to silver,
Kα-line ($Z = 41$) and from tungsten, Lα-line ($Z = 74$) to uranium, Lα-line
($Z = 92$).

Table V. X-Ray Fluorescence: General Conditions

Element	Atomic No.	Analytical Line (Å)		Diffracting Crystal-2d (A)[a]	$2\theta^o$	Atmosphere	Detector[b]	Collimator
Na	11	$K\alpha$	11.91	Gypsum	15.19	102.76	Vacuum	GF	Coarse
Mg	12	$K\alpha$	9.89	Gypsum	15.19	80.7	Vacuum	GF	Coarse
				ADP	10.65	137			
Al	13	$K\alpha$	8.34	EDT	8.803	142.8	Vacuum	GF	Coarse
Si	14	$K\alpha$	7.12	EDT	8.803	108.0	Vacuum	GF	Coarse
P	15	$K\alpha$	6.16	EDT	8.803	88.8	Vacuum	GF	Coarse
S	16	$K\alpha$	5.37	NaCl	5.639	144.4	Vacuum	GF	Coarse
Cl	17	$K\alpha$	4.73	NaCl	5.639	114	Vacuum	GF	Coarse
K	19	$K\alpha$	3.74	LiF	4.028	136.7	Vacuum	GF	Coarse
Ca	20	$K\alpha$	3.36	LiF	4.028	113.1	Vacuum	GF	Coarse
Sc	21	$K\alpha$	3.03	LiF	4.028	97.6	Vacuum	GF	Coarse
Ti	22	$K\alpha$	2.75	LiF	4.028	86.25	Vacuum	GF	Coarse
V	23	$K\alpha$	2.50	LiF	4.028	76.6	Vacuum	GF	Coarse
Cr	24	$K\alpha$	2.29	LiF	4.028	69.2	Vacuum	GF	Coarse
Mn	25	$K\alpha$	2.10	LiF	4.028	62.8	Vacuum	GF,SC	Coarse
Fe	26	$K\alpha$	1.94	LiF	4.028	57.6	Vacuum	GF,SC	Coarse
Co	27	$K\alpha$	1.79	LiF	4.028	52.8	Vacuum	GF,SC	Coarse
Ni	28	$K\alpha$	1.66	Topaz	2.712	75.4	Vacuum	GF,SC	Coarse
Cu	29	$K\alpha$	1.54	Topaz	2.712	69.3	Vacuum	GF,SC	Coarse
Zn	30	$K\alpha$	1.435	Topaz	2.712	64.0	Vacuum	GF,SC	Coarse
As	33	$K\alpha$	1.175	Topaz	2.712	51.4	Air	SC	Coarse
		$L\alpha$	9.67	ADP	10.65	130.6	Vacuum	GF	Coarse
Br	35	$K\alpha$	1.04	Topaz	2.712	45.1	Air	SC	Coarse
		$L\alpha$	8.375	EDT	8.803	144.1	Vacuum	GF	Coarse
Rb	37	$K\alpha$	0.93	Topaz	2.712	40.0	Air	SC	Fine
		$L\alpha$	7.32	EDT	8.803	112.5	Vacuum	GF	Coarse
Sr	38	$K\alpha$	0.875	Topaz	2.712	37.6	Air	SC	Fine
		$L\alpha$	6.86	EDT	8.803	102.3	Vacuum	GF	Coarse

Element	Z	Line		Crystal[a]				Detector[b]	
Zr	40	Kα	0.786	Topaz	2.712	33.6	Air	SC	Fine
		Lα	6.07	EDT	8.803	87.2	Vacuum	GF	Coarse
Mo	42	Kα	0.71	Topaz	2.712	30.4	Air	SC	Fine
		Lα	5.41	EDT	8.803	78.5	Vacuum	GF	Coarse
Ag	47	Kα	0.55	Topaz	2.712	23.8	Air	SC	Fine
		Lα	4.15	EDT	8.803	56.2	Vacuum	GF	Coarse
Cd	48	Lα	3.96	EDT	8.803	53.5	Vacuum	GF	Fine
Sn	50	Lα	3.60	LiF	4.028	126.9	Vacuum	GF	Coarse
Sb	51	Lα	3.44	LiF	4.028	117.5	Vacuum	GF	Coarse
I	53	Lα	3.15	LiF	4.028	103.0	Vacuum	GF	Coarse
Cs	55	Lα	2.89	LiF	4.028	91.6	Vacuum	GF	Coarse
Ba	56	Lα	2.77	LiF	4.028	87.2	Vacuum	GF	Coarse
La	57	Lα	2.665	LiF	4.028	83.0	Vacuum	GF	Coarse
Ce	58	Lα	2.56	LiF	4.028	79.0	Vacuum	GF	Coarse
W	74	Lα	1.48	LiF	2.848		Vacuum	GF,SC	Coarse
Pt	78	Lα	1.31	LiF	2.848		Vacuum	SC	Coarse
Au	79	Lα	1.28	LiF	2.848		Air	SC	Coarse
Pb	82	Lα	1.175	LiF	2.848		Air	SC	Coarse
Bi	83	Lα	1.144	LiF	2.848		Air	SC	Coarse
U	92	Lα	0.911	LiF	2.848		Air	SC	Fine

[a]Crystal—ADP: ammonium dihydrogen phosphate
EDT: ethylenediamine tartrate
LiF: lithium floride
NaCl: sodium chloride
[b]Detector—GF: gas flow proportional counter
SC: scintillation counter

6.2 Practical Measurement of Fluorescence Radiation

6.2.1 Dispersive Spectrometer and Energy Spectrometry

In practice, dispersive spectrometry with a diffraction crystal is used most often. We have seen the possibilities of spectrometry by energy resolution (Section 3.5.1). A discriminator or pulse-height analyzer selects the pulses recorded by the counter. The height of these pulses is a function of the radiation energy and falls between two limits defining a measuring channel that is characteristic for each radiation.

Modern instruments operating with a dispersive crystal system also benefit from an improvement of spectral resolution by energy discrimination. This is of particular interest for trace analysis, in which the lines of the matrix elements may be very close to the line of the analyzed element and the spectral background may interfere with the real line intensity. For a line with a profile as shown in Figure 10, the pulse height discrimination should be adjusted with a channel width of V'_1-V'_2 and not V_1-V_2 in order to increase the signal:noise ratio to a maximum. If the spectral background is strong, it must be taken into account in the fluorescence reading. If the background is similar on each side of the line, the reading is taken in the vicinity. If it is different, two readings must be made on each side to take the average.

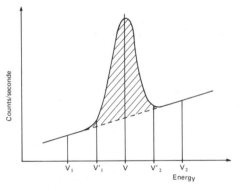

Figure 10. Selection of radiation by pulse height analyzer.

6.2.2 X-Ray Measurement, Counting, Analytical Error

The practical reading of X-radiation is taken by the pulse count of the counter. Actually, the count is made for a given time T. If the reading is repeated several times (n), a certain distribution of the N values is obtained around a mean value N_0. If the number n is sufficiently large, the distribution is Gaussian. It is then characterized by a standard deviation s, which means that the probability for an observed value N to fall between $N_0 - 2s$ and $N_0 + 2s$ is 95.4%.

The standard deviation of the count is approximately equal to $\sqrt{N}$ or $\sqrt{N_0}$ (Jenkins *et al.*, 1975) for a statistical count ($\bar{N}$ being the mean of n readings). We find that this error is essentially equal to the analytical standard deviation, *i.e.*,

$$s = \sqrt{\frac{\Sigma (N - \bar{N})^2}{n - 1}} \cong \sqrt{N}$$

If R is the counting rate (CPS) and T the counting time, we have N = RT. In fact, R is practically proportional to the fluorescence.   For a given value of R, the variation coefficient of R decreases with increasing T. At the same time, the variation coefficient of N decreases with increasing N.

In practice, there are two alternatives for measuring the fluorescence intensity by pulse counting: (1) determination of N for a fixed time T, and (2) determination of T for a given count N. Generally, a reading with constant T is preferred when the element is present in sufficient concentration to enable a limited counting time (for example, 100 sec) that leads to a suitable coefficient of variation. In the case of traces with a low R value, the fixed counting method will give better precision, particularly if the spectral background can be neglected.

However, the spectral background must be taken into account in numerous cases because it influences both the reading precision and the detection limits. We will consider the analytical curve defined by the fluorescence intensity measured by the counting rate as a function of concentration (Figure 11). The equation of the straight line has the form y = mx + C. If the peak measurement is R_p and the spectral background R_b, we have:

$$R_p = mC + R_b \; i.e., \; C = \frac{R_p - R_b}{m}$$

The slope of the line m characterizes the sensitivity. The analytical error (precision) is defined by the standard deviation s ($s \cong \sqrt{N}$) of the line and background measurement.

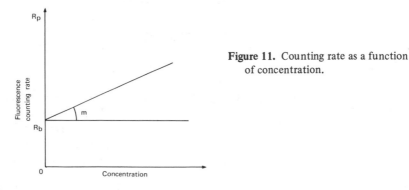

Figure 11. Counting rate as a function of concentration.

The error can be calculated as a function of the measuring method employed. With a count at constant time, if R_p and R_b are the counting rates of the line and the background, respectively, Jenkins *et al.* (1975) has shown that:

$$s_{CT} = \sqrt{2/T} \cdot \sqrt{R_p + R_b}$$

$$(T_p = T_b = T/2)$$

and the relative standard deviation (variation coefficient %) will be:

$$\text{V.C. } \% = \frac{100 \sqrt{2}}{\sqrt{T}} \cdot \frac{\sqrt{R_p + R_b}}{R_p - R_b}$$

This is based on the assumption that the background and line are measured for identical times $T/2$ (T_p and T_b), but the best conditions are obtained with different times T_p and T_b for the line and background measurements. It has been demonstrated that the optimum conditions are obtained for $T_p/T_b = \sqrt{R_p/R_b}$. The standard deviation becomes:

$$s_{CTO} = \frac{1}{\sqrt{T}} (\sqrt{R_p} + \sqrt{R_b})$$

$$(T = T_p + T_b).$$

With a measurement at a constant count, we have:

$$s_{CC} = \frac{1}{\sqrt{T}} \cdot \sqrt{R_p + R_b} \cdot \sqrt{\frac{R_p}{R_b} + \frac{R_b}{R_p}}$$

The concentration giving a fluorescence ($R_p - R_b$) equal to two times the variation of the spectral background determined by s (R_b) can be considered the detection limit. The sensitivity and the detection limit depend on the operating conditions, matrix composition, atomic number of the element, and the counting technique employed. In the next section the analysis ranges for specific cases will be discussed.

7. ANALYTICAL METHODS AND APPLICATIONS

7.1 General Methods

X-Ray fluorescence spectrometry lends itself to the analysis of solids and liquids. Liquid analysis involves samples solubilized after acid or alkaline attack. Its advantage is that the sample is representative and homogeneous, but sample dilution is a limiting factor in trace analysis. Analysis of liquids also includes certain media such as water and organic liquids, especially mineral oils and crude oil products in which fluorescence is utilized to determine certain trace elements.

Solids analysis is more common. Irradiation must be made with a plane and homogeneous surface. The sample is finely ground (see Section 5), mixed with a spectral buffer, if necessary, and with a binder, followed by compacting to form a pellet of 2-4 cm in diameter and 3-5 mm thick. In many cases, the sample is mixed with a flux (lithium tetra- or metaborate) and melted at $1100\text{-}1200°C$. The melting product is cast on a plane surface that will be subjected to X-ray irradiation (Tertian, 1968; Halma, 1973; Gwilliam *et al.*, 1973). The surface must be perfectly smooth and plane, which eliminates perturbations due to the granulometry of the sample. Melting is performed in a graphite or platinum-gold (Au 5%) or platinum-gold-rhodium crucible.

Claisse (1956) has used borate successfully for the determination of light elements (Al, Si). A heavy absorber (La_2O_3) can be added to the flux. The quality of the bead obtained depends on numerous factors, especially the proportion of constituents, heating conditions (time and temperature), and the casting technique. A detailed study was made by Vodinh (1973). Some authors, however, prefer to grind the pellet obtained to a suitable granulometry and subsequently repelletize it with the addition of a heavy absorber, if desired (Rose *et al.*, 1963; Wittmann *et al.*, 1973).

An original method is that of Wittmann *et al.* (1973). For the analysis of metallurgical specimens, the sample (0.125 g) is mixed with the flux (1.35 g) (85% lithium tetraborate + 15% lanthanum oxide) and melted in a graphite crucible in an induction furnace under argon. After cooling, the mixture (melting product + graphite) is finely ground, treated with 0.2 g cellulose and pelletized. The method is not suitable for trace analysis because of the dilution of the final product.

In another method, a small controlled quantity of the solubilized sample (10-100 μl) is deposited on adsorbent paper. Baudin *et al.* (1973) use this technique to determine radioactive uranium and plutonium in solution in a range between 1 and 10 g/l.

Trace analysis often requires chemical enrichment in order to concentrate the test elements in the irradiated product. Organic complex precipitation or solvent extraction are the techniques employed. Fagel *et al.* (1957) precipitated Cu, Ni, Fe, Mn and Cr in the form of oxinates. The precipitate was pelletized and exposed to X-rays.

Knapp *et al.* (1975) formed organic metal complexes from the sample solution (Zn, Cu, Ni, Pb, Hg, Co, Cd, Fe) with sodium diethydithiocarbamade and bound them on a Chromosorb W DMCS column.

They were then eluted with chloroform, the solvent evaporated on filter paper and the residue analyzed by X-ray spectrometry.

Green *et al.* (1970) developed the use of selective ion exchange resins such as Strafion NMRR, which can be used to separate groups of elements. These authors were also able to separate gold from minerals in order to analyze it with a detection limit of 1 ppm.

Ion exchange paper (Campbell *et al.*, 1966) has been used successfully to bind the investigated trace elements. Joly (1969) also uses the Amberlites IR-120, IRC-50, IRA-400, IR-4B in the form of sheets 0.3-mm thick containing 45-50% resin on a cellulose support (manufactured by Reeve Angel). The test solution is filtered repeatedly on the ion exchange paper for a quantitative binding of the investigated elements. Fluorescence is excited from this support. For Ni, Pb, Cr, V and Ca, the author obtained the following detection limits in μg/ml of solution: 0.0012, 0.01, 0.0015, 0.0018, 0.007.

7.2 Analysis of Natural Mineral Media

7.2.1 General Methods

The analysis of geological products is the /principal/ application of X-ray fluorescence spectrometry, which has been used for nearly 20 years to determine the major elements. More recently, methods for trace analysis have been developed. The methods for the determination of the major elements include melting of the sample with borate (see Section 7.1) in a proportion of 1 g of the sample and 9 g of flux. The residue is cast to obtain a plane surface or reground and pelletized.

Generally matrix effects are readily corrected in classical silica rocks. Thus, Hooper (1964) determined Si, Al, Fe, Ca, Mg, K, Ti and Mn in the following concentration ranges:

SiO_2	14.0-80.0%
Al_2O_3	4.0-70.0%
Fe_2O_3	0.10-14.0%
CaO	0.20-44.0%
MgO	2-10.0%
K_2O	0.05-50%
TiO_2	0.01-3.4%
MnO	0.025-0.64% (250-6400 ppm).

For lower concentrations, the author recommends direct pelletizing of the sample powder with the following detection range:

MgO	0.05-2%
P_2O_5	0.05-0.65%
SrO	80-2200 ppm

Table VI. Instrumental Conditions

Element	Atomic No.	Concentration Range (% oxide)	Sample Preparation	Sample Mount	Collimator	Vacuum	Analyzing Crystal	Detector	Peak (Kα) °2θ	Background °2θ	Pulse Height Discriminator
Mg	12	0.05-2.0 2.0-10.0	Pure powder Borax fusion	Pressed disk	Coarse	Yes	ADP	Flow	107.20	106.00	Yes
Al	13	4.0-70.0	Borax fusion	Pressed disk	Coarse	Yes	EDT	Flow	143.28	141.50	Yes
Si	14	14.0-80.0	Borax fusion	Pressed disk	Coarse	Yes	EDT	Flow	108.35		Yes
P	15	0.05-0.65	Pure powder	Pressed disk	Fine	Yes	EDT	Flow	88.60	87.00 90.00	Yes
K	19	0.05-5.50	Borax fusion	Pressed disk	Coarse	Yes	EDT	Flow	116.75	119.00	Yes
Ca	20	0.20-44.0	Borax fusion	Mylar	Fine	Yes	LiF	Flow	113.00	110.50 115.90	Yes
Ti	22	0.01-3.40	Borax fusion	Mylar	Fine	Yes	LiF	Scint.	86.00	84.54	Yes
Mn	25	0.025-0.64	Borax fusion	Mylar	Fine	No	LiF	Scint.	62.90	61.50 64.40	Yes
Fe	26	0.10-14.0	Borax fusion	Mylar	Fine	No	LiF	Scint.	57.41		No
Sr	38	80-2200 ppm	Pure powder	Mylar	Fine	No	Topaz	Scint.	37.30	36.30 38.30	Yes

The instrumental parameters are summarized in Table VI (for a Philips PW 1540 instrument). When the specimen is obtained from natural rock standards, the error is relatively high at low concentrations. For Mg at a level of 0.4% MgO, the error is as high as 32%. For phosphorus it is 40-50% at a concentration level of 0.1-0.15%. This error originates essentially from the matrix effect and not from the fluorescence technique.

Kaye (1965) improved the precision of trace analysis of Cu, Ni, Rb, Sr, Zn, Zr and W. For this purpose, he prepared standards from pure silica (Specpure SiO_2) containing 250 ppm additions of the various elements to be determined. A correction for the matrix effect is made by determining the mass absorption coefficients of the samples (μ_S) relative to SiO_2 ($\mu\ SiO_2$) at a wavelength of 0.9 Å. The absorption coefficients are calculated from the concentration of the various major elements (Jenkins et al., 1975). The trace element concentrations (C) are calculated from the equation:

$$C_{ppm} = \frac{\mu_S}{\mu_{SiO_2}} \times \frac{I_S}{I_{SiO_2}} \times 250$$

where I_S and I_{SiO_2} are the measured fluorescences of the element and of silicon corrected for the spectral background. The operating conditions are summarized in Table VII.

Table VII. Instrumental Conditions for Trace Analysis (Kaye, 1965)

Tube and voltage	W target operated at 48 kV 20 mA						
Crystal	Topaz						
Collimator	480 μ						
Path	Vacuum						
Counter	Scintillation						
Fixed time	3 x 64 sec						
Element	Cu	Ni	Rb	Sr	Zn	Zr	W
Analysis line	K_α	K_α	K_α	K_α	K_α	K_α	L_α
Peak $^\circ 2\theta$	69.32	75.41	39.91	37.65	63.97	29.86	76.50
Background $^\circ 2\theta$	68.30	74.20	38.75	36.60	63.00	28.60	78.00
	70.40		41.30	38.75	64.85	31.12	

A correction must be made for the analysis of Cu and Ni because of contamination of the X-ray tube filament. This is made by measuring the line ratios $NiK\alpha:W_{L_1}$ and $CuK\alpha:W_{L_1}$ on a SiO_2 standard. The tungsten line is measured in all samples and the SiO_2 standards. A suitable correction, which is a function of the Ni and Cu contaminants, is subtracted from the measured fluorescence of the samples and standards.

The errors in precision are approximately a few percent, often less than 1%. The accuracy errors are 5-10% on the average for contents of 100 ppm and more and 10-20% between 10 and 100 ppm. As, Sb, Ni, Rb, Si, V and Zn were determined as trace elements in silica rocks by Fabbi *et al.* (1972). The powdered sample (0.5 g) is mixed with 0.5 cellulose for chromatography and then pelletized under a pressure of 30,000 psi. The pellet is used for the analysis of the different elements. Operating conditions are summarized in Table VIII (General Electric XRD-6 single-channel instrument).

Table VIII. Instrumental Conditions for As, Sb, Ni, Rb, Sc, V and Zn

Element	Target	Voltage kV	Current mA	Crystal-2d	$°2\theta$	Line	Counter	Path	Fixed Time sec
As	W	50	60	LiF-2.85 Å	48.83	$K_{\alpha 1}$	Flow	Air	200
Sb	Cr	50	45	LiF-4.08 Å	106.46	$L_{\alpha 1}$	Flow	Air	200
Ni	W	50	50	LiF-2.85 Å	71.26	$K_{\alpha 1}$	Flow	Air	200
Rb	W	50	60	LiF-2.85 Å	37.99	$K_{\alpha 1}$	Flow	Air	200
Sc	W	50	50	LiF-4.08 Å	97.71	$K_{\alpha 1}$	Flow	Vacuum	200
V	W	50	50	LiF-2.85 Å	123.16	$K_{\alpha 1}$	Flow	Air	200
Zn	Cr	50	50	LiF-2.85 Å	60.58	$K_{\alpha 1}$	Flow	Air	200

The matrix effect gives rise to the following interferences for which a correction must be made:

As, Ni, Rb, Zn—absorption by iron,
Sb—enhancement by calcium,
V—spectral interference by titanium,
As, Ni, Rb, V—absorption by Ca if CaO $>$ 20%.

Furthermore, the $L\beta$-line of antimony is preferred to La_1 because the latter is perturbed spectrally by Ca and K. The following calibration techniques are used:

For Ni, Sc, V, Zn: chemically analyzed external standards
For As, Sb, Zn: synthetic standards
For Rb: addition of Sr as internal standard.

The detection limits are a few ppm: 10 ppm As, 70 ppm Sb, 3 ppm Ni, 10 ppm Rb, 4 ppm Sc, 10 ppm V, 7 ppm Zn.

The thin-film technique was used successfully by Price *et al.* (1968) for the analysis of Sr, Rb, Ba, Ni, Zn, Th, Pb and Zn in concentrations of 10 to 5000 ppm in rocks. The sample (50 mg) is ground to 200 mesh and the powder (10-15 mg) is spread on cellophane adhesive tape, which is then pressed under 5 t/in^2. This preparation is placed on a sample holder and subjected to X-ray irradiation under the conditions indicated in Table IX.

Table IX. Instrumental Conditions for Thin Film X-Ray Analysis

Element	Anode	Voltage (kV)	Current (mA)	Line	Peak (Å)	Scatter Background (Å)	Crystal	Path	Collimator		Counter
Sr	W	55	28	Kα	0.875	± 0.030	LiF (220)	Air	Coarse	480 μ	Scint.
Rb	W	55	28	Kα	0.926	± 0.026	LiF (220)	Air	Coarse	480	Scint.
Ba	W	50	20	Kα	0.385	0.420	Topaz	Air	Fine	160	Scint.
Ni	W	55	28	Kα	1.658	1.618	LiF (220)	Air	Coarse	480	Scint.
Zn	W	50	20	Kα	1.435	1.405	LiF (220)	Air	Fine	160	Scint.
Th	W	60	20	Lα	0.956	1.017	LiF (200)	Air	Fine	160	Scint.
Pb	W	50	20	Lα	1.175	1.192	LiF (200)	Air	Fine	160	Scint.
Zr	W	60	20	Kα	0.786	0.810	LiF (220)	Air	Coarse	480	Scint.

We have seen (Section 4.2.1.4) that the matrix effect becomes negligible if the irradiated sample is in a thin film (1 μ) or distributed over a surface in the amount of 1 mg/cm^2. A layer of 1 μ is difficult to prepare. The authors obtained the best results with 10-15 mg of sample spread on the useful surface and pressed at 5 t/in^2. With consideration of the background, a linear relation is obtained between the concentration and fluorescence. The data scatter obtained from rock standards may originate from the precision of the method as well as the accuracy of the recommended values.

The following studies can also be consulted: Murao (1973) on the determination of Cr, Zn, Rb, Sr, Zr, Ba and Pb; Webber *et al.* (1971) on the correction for matrix effects and use of the computer in trace analysis; Hisano *et al.* (1969) on the determination of Nb, Zr, Mn, Cr, V and Cu in ilmenite; and Labarta Carreno (1968) on the determination of S, Cu, Zn, As and Pb in pyrites.

7.2.2 Special Methods

Studies have been conducted on the determination of one or two given elements. Rabbi *et al.* (1971) determined Ba and Sr in amounts of 10-1000 ppm in geological specimens after pelletizing the sample mixed with cellulose. The spectrometric conditions are classical. Titania causes a spectral interference for barium, and iron gives absorption with the strontium fluorescence. Garavelli *et al.* (1973) determined strontium in limestones (0-500 ppm) and zinc in terra rossa (0-500 ppm). Analysis was made on a powdered sample that was (without pelletizing) placed directly on the sample holder. The detection limit is about 10 ppm. Standardization of strontium is obtained from synthetic samples of $CaCO_3$ + $MgCO_3$, and the slope (m) of the calibration line C = mN depends on the $MgCO_3$ concentration, *i.e.*, m = 0.01069 - 0.00008747 ($MgCO_3$%), leading to the concentration Sr ppm = [0.01069 - 8.747 ($MgCO_3$%) 10^{-5}] N (where N is the Sr fluorescence count). The fluorescence of zinc is not greatly influenced by variations of Si, Al and Fe in terra rossa, so a single calibration may be sufficient.

Willis *et al.* (1969) determined barium in rocks by exciting the $BaL\alpha_1$ line from a chromium anode. Quintin *et al.* (1975) determined traces of U and Th (ppm) in rocks. Mountjoy *et al.* (1968) analyzed cesium after concentration by ion exchange.

Gulacar (1974) determined Cu, Ni and Co as elements by performing a separation of the diethydithiocarbamates in chloroform. The extract is evaporated and pelletized. Detection limits in ppm are 1.5 Co, 1.6 Ni and 2.1 Cu. Fabbi (1971a) determined phosphorus in geological specimens, which requires a mathematical correction for the effects of calcium.

Brown *et al.* (1969) determined sulfur in soils, while Stanton *et al.* (1971) analyzed bismuth in soils, sediments and rocks.

An original application was the analysis of lunar soil on the moon itself by Apollo 15 (Adler *et al.,* 1972) by means of an energy dispersion spectrometer with solar X-ray radiation as the excitation source and with seven channels for the various elements.

7.2.3 Validity and Accuracy

The main concern of the analyst with regard to trace analysis remains the precision and accuracy of the results. Brenner *et al.* (1975) recently discussed the validity of the results obtained with the rock standards of Flanagan (1973) and Abbey (1973): syenite SY-1, diabase W-1, Basalts BR, BCR-1, JB-1, diorite DN-R, granite G-2, andesite AGV-1, tonalite T-1, norite NIM-W. These rocks are used to construct the calibration curves. Two methods of sample preparation have been tested:

(1) Grinding to 300 mesh and introduction of the powdered sample into the sample holder, protecting it with a Mylar film.

(2) Mixing the ground sample (1.5 g) with chromatographic cellulose (1.5 g) by agitation for 30 min and pelletizing at 25 t in vacuum to form pellets of 30 mm diameter.

The samples were analyzed in both of these forms under the classical spectrometric conditions listed in Table X.

Standardization consists of plotting a straight line representing the mean passing through specific points, with the fluorescence of each element on the ordinate and its concentration on the abscissa. This line serves to determine the analytical values deviating more or less from the recommended ones. The authors' results are listed in Table XI and are compared to the standards. The deviations from the standard values are generally permissible and within the range of interlaboratory findings.

Brenner *et al.* (1975) initiated an interlaboratory study to determine elemental Mn, Cr, V, Co, Cu, Pb, Sr, B, Cd, Zr, Ga, Sn and Ba in rocks in order to compare the results obtained by emission, atomic absorption and X-ray fluorescence spectroscopy. The authors published the results in the form of graphs showing the correlations of the paired methods for each element.

Finally, we will note the semiquantitative method proposed by Williams (1976) for the trace analysis of Mn, Fe, Ni, Cu, Zn, As, Br, Rb, Sr, Y, Zr and Pb in soils and Mn, Fe, Ni, Cu, Zn, Br, Rb and Sr in plants. The sample is ground and pelletized in the presence of boric acid and excited with molybdenum and gold anode tubes. The fluorescence diagram is recorded between suitable 2 θ values and under the conditions

Table X. Instrumental Conditions for Trace Elements

Element	Atomic No.	Wavelength K_α (Å)	Anode	Tube kV	mA	Peak °2θ	Background °2θ	Crystal	Collimator	Vacuum	Detector
Ti	22	2.749	Cr	40	40	86.25	84.0	LiF 200	Fine	On	Flow
Ba	56	2.776 (L_α)	Cr	50	50	87.25	89.0	LiF 200	Fine	On	Flow
K	19	2.744	Cr	50	50	136.78		LiF 200	Fine	On	Flow
Zr	40	0.701	W	60	40	20.10	19.50	LiF 200	Fine	On	Flow
Ni	28	1.659	W	50	50	48.72	47.0	LiF 200	Fine	On	Flow, scint.
Sr	38	0.877	W	50	40	25.16	24.0	LiF 200	Fine	On	Flow
Rb	37	0.928	W	60	40	26.63	29.0	LiF 200	Fine	On	Scint.
Zn	30	1.435	W	50	50	41.80	39.50	LiF 200	Fine	On	Scint.
Mn	25	2.103	W	50	50	62.97	60.50	LiF 200	Coarse	On	Flow
Cr	24	2.291	W	50	50	69.35	67.0	LiF 200	Coarse	On	Flow

Table XI. Analysis of Rock Standards by X-Ray Fluorescence

Standard Reference Materials	Mn		Sr		Ba		Zr		Rb		Ni		Cr	Zn
	1	2	1	2	1	2	1	2	1	2	1	2	2	2
BCR-1	1400		310	400	775	720	183	130	50	53	13			113
	1470		*330*	*330*	*680*	*680*	*185*	*185*	*47*	*47*	*13*			*120*
AGV-1	725		690	620	1200		222		70	64	16			
	760		*660*	*660*	*1200*		*220*		*67*	*67*	*17*			
G-2	315		493		1850		295		172	185	9			
	260		*480*		*1850*		*300*		*170*	*170*	*6*			
W-1	1280		160	240	163	200	105	150	23	23	84	120	137	81
	1318		*190*	*190*	*160*	*160*	*105*	*105*	*21*	*21*	*78*	*78*	*120*	*86*
DR-N					367				71	82	15			
					380				*75*	*75*	*22*			
BR	1450		1365	1550	1070	890	246		47	44	265	278	410	135
	1549		*1350*	*1350*	*1050*	*1050*	*240*		*45*	*45*	*270*	*270*	*420*	*160*
SY-1	3000				240	340		2900	163	190	41			290
	3098				*282*	*282*		*3030*	*195*	*195*	*43*			*219*
JB-1									45	45	146			
									40	*40*	*139*			
T-1	880							140						196
	852							*170*						*220*
BCS 269	233			150		580								
	230			*120*		*660*								
NIM-N	1450								8					
	1320								*9*					
DTS-1	1180											2320	3390	41
	969											*2269*	*4000*	*45*
PCC-1	1120											2366	2810	37
	929											*2339*	*2730*	*36*

Column 1:   Analysis on powdered sample;   Column 2:   Analysis on pelletized sample;   In italics: Values recommended by Flanagan and Abbey.

listed in Table XII. The standards consisted of synthetic mixtures (SiO_2, Fe_2O_3, Al_2O_3) containing additions of the trace elements in a suitable range. Detection ranges in ppm were: 5-40 Cu, 20-150 Br, 10-100 Ni, 100-3000 Mn, 100-400 Zr, 10-150 Zn, 20-80 Y, 10-80 Sr, 10-40 As, 10-150 Rb. Quantitative determination requires a consideration of the mass absorptions (calculated according to Norrish *et al.*, 1967). Calibration can also be performed with standard natural rock specimens.

Table XII. Instrumental Conditions for XRF Analysis Semiquantitative Method (Williams, 1976).

Anode	$°2\,\theta$	Crystal	Scintilla-tion EHT (V)	Amplifier-Analyser Lower Level (V)	Window (V)	Elements
Mo	20-31	LiF	600	0.4	0.65	$Zr(K\alpha)$, $Y(K\alpha)$, $Sr(K\alpha)$
	31-35(40)	LiF	630	0.4	0.65	$Rb(K\alpha)$, $Pb(L\beta)$, $Br(K\alpha)$, $As(K\alpha)$, $Ga(K\alpha)$
Au	40-50	LiF	650	0.3	0.85	$Zn(K\alpha)$, $Cu(K\alpha)$, $Ni(K\alpha)$
	50-64	LiF	670	0.3	0.85	$Fe(K\beta)$, $Mn(K\alpha)$

7.3 Water Analysis

Analysis is performed either on water-soluble elements or suspended elements, separated by filtration on a Millipore or Nuclepore filter, for example. Elements in solution can be analyzed either directly in the water placed into a suitable X-ray transparent cuvette or after evaporation of an aliquot. This is done either using a few drops to a few milliliters on a suitable support or after separation of the trace elements by extraction or precipitation in the form of organic complexes. The latter method is used especially for sea water analysis.

7.3.1 Direct Analysis

Direct analysis of water in a suitable cuvette with a Mylar window is of interest because of the maximum reduction of contamination risks. However, the low trace element concentration in natural water (a few ppb) limits the application range.

Deutsch (1974) used this method to determine bromine in water in the range of 0-50 mg/l by subjecting a volume of 8 ml in a Lucite cell with a Mylar window to direct Ka excitation of a tungsten tube (50 kV, 50mA). The Ka-line series of bromine is resolved with a LiF (220) crystal.

7.3.2 Dry Extract Analysis

A suitable volume of water is evaporated on a support, such as aluminum, filter paper, or cellulose, which is then exposed to X-radiation. Herman *et al.* (1973) used this technique to determine elemental Ca, Mn, Fe, Ni, Cu, Zn, Br and Sr in lake water. A drop of sample is evaporated on a carbon support, which is then excited in a proton beam of 2-10 MeV. The authors also used either $^4He^{2+}$ of 6-25 MeV or $^{16}O^{5+}$ of 6-40 MeV. The elements are measured with a nondispersive spectrometer and a Si(Li) detector under 10^{-5} torr. Nonmetallic elements, Br, Cl, I and S, were analyzed in water by Cornil *et al.* (1975) after evaporation of a suitable volume on 2 g powdered cellulose (Ashless Whatman-Standard grade). The residue was dried at 110°C, mixed and pelletized at 200 kg/cm^2

Table XIII summarizes the analytical conditions. The indicated concentration range corresponds to the pelletized sample. These values are referred to the initial water sample as a function of the evaporated volume. Fluorescence is measured by dispersive spectrometry.

Table XIII. Determination of Nonmetals in Water

	Bromine	Chlorine	Iodine	Sulfur
Exciting anode	Mo	Cr	Cr	Cr
Voltage kV	50	45	45	45
Current mA	13	13	13	13
Crystal	LiF	Gypsum	Quartz	Gypsum
Detector	Scintillation	Gas Flow	Gas Flow	Gas Flow
Radiation λ Å	K_α 1.041	$K_{\alpha 1}$ 4.728	$L_{\alpha 1}$ 3.142	$K_{\alpha 1}$ 5.37
Analytical Range (ppm)	5-1000	100-10,000	50-1000	300-9000

Alexander *et al.* (1974) proposed a method to determine K, Ca, Fe, Cu, Br and Sr in sea water. They added yttrium $[Y(NO_3)_2]$ to the sample as an internal standard. A suitable volume of water is evaporated on an aluminum support and then irradiated in a proton beam (3 MeV, 300 mA) for 500 sec. The emitted X-ray lines are measured with an energy dispersion spectrometer and a Si(Li) detector. The authors report that 1 ppm is the detection limit in the irradiated product.

Heres *et al.* (1972) determined Th, V and Zr in concentrations of less than 1 mg/l by evaporating a suitable volume of water on a Millipore filter, which is then exposed to X-rays.

7.3.3 Determination of Trace Elements After Separation

Trace analysis of water actually most often requires chemical separation. Holynska (1974) used separation on cationic exchange resin, Dowex A_1, to bind Pb, Hg, Cu, Zn and Fe from water. Fluorescence is then excited by radioisotopes. Lochmueller et al. (1974) demonstrated that it was possible to bind certain cations (such as Pb, Hg and Cd) from water by immersing cationic exchange membranes (6 mm x 20 mm) in 1 liter of water. The membranes are then dried at 120° and exposed to a proton beam (3 MeV, 7-50 mA). The radiation is measured with an energy dispersion spectrometer.

The authors found that the method is primarily applicable to heavy metals: Pb, Cd and Hg are bound more selectively than the base metals of Ca and Fe. The detection limit is 0.1 ppb for lead. The method is semiquantitative.

Extraction of metal complexes has been used for the analysis of sea water. Morris (1968) separated elemental V, Cr, Mn, Fe, Co, Ni, Cu and Zn by extracting the chelates formed with pyrrolidine and ammonium dithiocarbamate (APDC) in methylisobutylketone (MIBK). The extract is evaporated and then analyzed by X-ray fluorescence. Marcie (1968) also used APDC for the extraction of As, Se, Hg, Tl, Pb and Bi in water (0.01 μg/ml), but he preferred chloroform as the solvent for the complexes. The organic solution is evaporated on filter paper and followed by X-ray analysis. The method is also applicable to elemental Ag, Cd, In, Sn, Cu, Zn, Ni, Co and Mn. Armitage et al. (1974) extracted U, Cu, Ni, Co, Fe and Mn in the form of oxinates in chloroform.

Boiteau et al. (1973) determined Hg, Pb, Zn, Cd, As and Se in water after separation. First, Hg, Zn, Pb and Cd are separated in the form of dithizonates from 1 liter of water: Hg at pH 1.5 Zn at pH 9 and Cd at pH 11. The combined extracts are evaporated on 50 mg cellulose and pelletized at 5 t/cm² in an 8-mm mold on a support of Whatman paper No. 1. Arsenic, after separation of the preceding elements, is precipitated in sulfide form from a volume corresponding 10 100 ml water. After separation of the dithizonates, 10 ml sulfuric acid and 1 ml of 2% thioacetamide are added to the sample, which is then heated to 90° for 15 min, allowed to stand and then filtered on a 20-mm diameter filter. Selenium is also separated from a volume of 100 ml by precipitation after reducing with hydrazine hydrochloride (25 ml of 15% SO_2-saturated solution) in the presence of hydrochloric acid (10 ml). Analysis conditions are given in Table XIV. It will be noted that the authors improved the sensitivity with the use of a curved crystal analyzer.

Table XIV.   Determination of Hg, Zn, Pb, Cd, As and Se as Elements in
Water After Separation

	Hg	Zn	Pb	Cd	As	Se
Atmosphere	Air	Air	Air	Air	Air	Air
Exciting Anode	Mo	W	Mo	W	Mo	Mo
Crystal	LiF	LiF	LiF	LiF	LiF	LiF
Radius of curvature, mm	1100	750	1100	1400	1100	1100
Detector	Scint.	Scint.	Scint.	Scint.	Scint.	Scint.
Measured radiations	$L\beta_1$ 1.04	$K\alpha$ 1.43	$L\beta$ 0.982	$K\alpha$ 0.536	$K\alpha$ 1.17	$K\alpha$ 1.10
Measuring angle (°)	15° 12	20° 90	14° 15	7° 65	17° 04	15° 17
Counting time, sec.	200	200	200	200	200	200
Lower analysis limit, μg/l	10	10	10	10	3	5
Detection limit, μg/l	2	2	2	2	0.5	1

Kato *et al.* (1973) analyzed arsenic in water in concentrations of less than 0.5 μg/ml after separation in the form of a volatile hydride: 100 ml water (containing less than 50 μg As) are treated with 5 ml H_2SO_4, 10 g granulated zinc and stannous chloride. The arsenic hydride evolved is collected on a paper disk (22-mm diameter) impregnated with silver nitrate. The latter is analyzed under classical radiation conditions: tungsten anode (50 kV, 20 mA), measured radiation As$K\alpha$, LiF analyzer crystal, in vacuum. The fluorescence is proportional to the concentrations for 0-10 μg As.

7.3.4 Indirect Methods

The indirect analysis of chlorides after precipitation of silver chloride, of sulfates after precipitation of barium sulfate, and analysis of the excess metal by X-ray fluorescence, is worthy of note. Heres *et al.* (1972) proposed a method with an internal standard using cadmium and cesium. The sample (50 ml) containing 1-5 mg/l of Cl and 2-120 mg/l of SO_4 is treated with an addition of (a) 5 ml cadmium and cesium solution of 500 μg/ml, and (b) a suitable volume of 2.1 g/l of Ag^+, 2.4 g/l of Ba^{2+}, 2 g/l of trinitrophenol and 100 ml/l of HNO_3 (d = 1.38) so that the following are present in excess in solution: 50 to 170 mg/l of Ba and 30 to 150 mg/l of Ag. Trinitrophenol is added to favor the formation of AgCl and $BaSO_4$ crystals. The precipitated salts are separated by centrifuging and the solution is analyzed directly by X-ray fluorescence. The following fluorescence ratios are measured:

$$\frac{F_{Ba}}{F_{Cs}} = f\ (SO_4^{--})\ \text{and}\ \frac{F_{Ag}}{F_{Cd}} = f\ (Cl^-)$$

Cl and SO_4 in concentrations of mg/l are determined with a variation coefficient of 10%.

7.3.5 Determination of Suspended Particles and Sediments in Water

Filtered particles are analyzed in the same way as atmospheric particles (see Burkhalter, 1973) (see Section 7.5). Sufficiently fine Millipore or Nuclepore filters should be used to retain particles of 0.1-0.3 μm. The analysis of suspended particles in sea water is made by filtering 5-20 liters of water on a Nuclepore filter. Hellman (1971) analyzed river sludge by X-ray fluorescence.

7.4 Analysis of Plant and Biological Materials

7.4.1 Plant Materials

Analysis can be performed either directly on the dried, ground and pelletized sample or with a mineral residue after destruction of the organic materials. X-Ray fluorescence is still rarely used for the analysis of plant and biological media, a result of two factors: (1) the basic sensitivity of the method limits its application to elements in concentrations of more than 50 or 100 ppm, and (2) the application does not always justify the high price of an X-ray fluorescence spectrometer. On the other hand, the method does allow nondestructive analysis. The procedures of sample treatment, often consisting of simple pelletizing, consequently reduce the contamination risk. Advances in instrumentation will lead to a reduction of the cost and it is probable that a significant development of X-ray fluorescence spectrometry will occur in the plant and biological sectors in coming years. Furthermore, plant standards are available for P, K, Ca, Mg, Fe, Cu, Mn and Zn (Pinta et al., 1975).

Recently, Williams (1976) published a semiquantitative method for the determination of trace elements in soils and plants (see Section 7.2). The ground sample is simply pelletized, and the fluorescence spectrum is recorded under the conditions listed in Table XII (Section 7.2). The standards consist of synthetic mixtures based on cellulose. Analysis ranges (in ppm) are 5-30 Cu, 20-150 Br, 10-100 Ni, 10-200 Zn, 20-400 Mn, 10-80 Sr and 100-600 Fe.

Quantitative plant analysis involves the elements Cu, Mn, Zn and P. The sample is pelletized (4-cm diameter) under 2 t/cm^2 in the presence of a binder consisting of a Lucite solution in toluene and is directly exposed to X-rays under the following conditions (Cu, Mn, Zn):   tungsten anode, 50 kV, 50 mA, LiF 100 crystal analyzer, gas flow counter as detector, vacuum as atmosphere, electronic pulse height analysis, counting 100 sec. The K$_\alpha$-lines are measured for each element. Detection limits are on the order of 1 ppm. The same elements were also determined in grain under analogous conditions by Wankova (1971).

An important application is the determination of nonmetals, P, Cl, and S (Guennelon *et al.*, 1973). The sample is ground, mixed with magnesia and cellulose (1 g plant material, 1 g MgO, 1 g cellulose) in order to reduce the matrix effects, followed by pelletizing. Classical fluorescence excitation conditions are used. The method is entirely suited for the natural concentrations in plants: 0.5-5‰ P, 0.5-20‰ Cl and 1-10‰ S. A Br analysis (0-200 ppm) in plants was proposed by Ristori *et al.* (1971).

X-Ray fluorescence is used widely not only for trace analysis but also for the determination of major elements (K, Ca, Mg). We should note the publications of Evans (1970) on the direct determination of Mg, Al, Si, P, S and K in plants.

Finally, we will mention the work of Campbell *et al.* (1975) on the analysis of plant and biological media. These authors prefer mineralization of the sample (0.25 g) by a wet method (nitric acid attack). A few drops of the solution obtained are evaporated on a carbon sample holder, which is then exposed to a proton beam (2.2-2.5 MeV) of an accelerator. The X-radiation is detected and measured with a pulse-height analyzer and a Si(Li) detector. Internal standards (Cd and Sr) are added to the analysis solution before evaporation on the sample holder. The method has been used for elemental Mn, Fe, Cu, Zn, Pb, As, Br and Rb in wine, for Mn, Fe, Ni, Cu, Zn, As and Rb in orchard leaves (NBS standards), and for Fe, Zn and Pb in the liver (NBS). The detection limit is between 0.1 and 1 ppm.

7.4.2 Biological Materials

The above methods are evidently applicable, *i.e.* fluorescence excitation either directly with the pelletized sample or after mineralization. An excellent tissue preparation method consists of lyophilization before pelletizing. X-Ray fluorescence has been used particularly for the determination of toxic elements—As, Pb, Br, Sr, Se—as well as for elements normally present in tissue, such as Cu, Zn and Mn.

Today, the US National Bureau of Standards is in the position to furnish biological standards, some of which have also been analyzed by X-ray fluorescence (Reuter, 1975).

At the present time, biological X-ray fluorescence analysis seems to be developing with the use of radioactive excitation sources and energy dispersion spectrometers. Lubozynski *et al.*(1972) studied the application of X-ray fluorescence to blood analysis. They showed that As, Sr, Cs, W, Hg and Pb can be determined in complete blood with detection limits of 2-10 ppm with excitation by means of a Mo anode and an energy dispersion spectrometer. The detection limits can be improved by a factor of 3 with the use of dry blood.

Agarwal *et al.* (1975) determined trace elements such as Cu, Zn, Br, Rb, Sb and Pb in concentrations of less than 1 ppm in urine after extraction on ion exchange resin. For this purpose, the specimen is brought to pH 7 and agitated by slow rotation for 1.5 h with 0.1 g of Chelex 100 resin. The resin is then filtered and dried, and a portion is placed on the sample holder for excitation under classical analysis conditions.

Purdham *et al.* (1975) studied Au, Br and I in biological fluids during their therapeutic application. Br and I can be determined directly in urine and serum. Gold is present in too low a concentration and must be separated by precipitation in sulfide form. Mathies (1974) determined arsenic in urine. The sample was mineralized by sulfonitric acid and then perchloric acid-nitric acid digestion. Arsenic is then reduced into the volatile arsine AsH_3, which is collected on filter paper impregnated with silver nitrate (see Section 7.3).

The arsenic is determined on the paper by excitation under classical conditions. Calibration is made by evaporating an arsenic solution (0 and 10 μg As). Although Sb, Te and Se do not interfere, Sb accompanies the arsenic and collects on the filter paper. Sb can also be determined under similar conditions. The method is also used for the analysis of hair, tissue and gastric fluids.

Selenium is also determined in biological media (Strausz *et al.*, 1975) after sulfonitric acid digestion of the sample. Selenium is reduced by $SnCl_2$ and coprecipitated in the presence of tellurium. The precipitate is filtered and the filter exposed to X-rays. The useful quantity of Se is between 0.5 and 25 μg.

Kneip *et al.* (1972) determined lead in blood. The L_α-lines of lead are excited with ^{238}Pu and measured with a Si(Li) detector. A count of 15 mn allows the determination of 0.1 ppm.

Patti (1969) investigated strontium analysis in biological and plant media. The sample is dissolved and strontium is precipitated with calcium in the form of oxalate. The precipitate is analyzed by X-ray fluorescence, using a tube with a molybdenum anode for excitation.

Nonmetals, such as Br, Cl, I and S, were determined in biological specimens by Cornil *et al.* (1975) (see Section 7.3.2). The sample is dried, lyophilized and pelletized at 200 kg/cm^2. Standards are prepared from cellulose containing the corresponding salt additions (KBr, NaCl, KI, K_2SO_4). The analytical conditions are listed in Table XIII.

7.5   Atmosphere and Atmospheric Particles

This field of application has advanced greatly in recent years. Despite a number of criticisms, interlaboratory studies have demonstrated that X-ray fluorescence is a perfectly valid technique for the analysis of atmospheric particles. A suitable volume of sample is filtered on a Millipore or Nuclepore filter and the filter is examined directly under X-rays. The sampling time may require 1-24 h. The principal problems related with the analysis of aerosols by X-ray fluorescence are sample collection on the filter, the quantity of sample collected, the particle size, product homogeneity, contamination and calibration. Quite often, the problem involves both microanalysis, because only a few hundred micrograms of total product is available, and trace analysis, because the investigated elements are often present in the collected sample at a level of ppm or fractions of ppm, particularly when air pollution is involved.

The matrix effect results from the chemical composition of the medium and the particle size. The latter effect has recently been studied by Criss (1976) who proposed a correction factor $(1 + ba)^2$ giving consideration to the size (a) and composition of the particles (b).

Calibration has been of interest to numerous authors. Cecchetti *et al.* (1969) determined Mn, Fe, Co, Ni, Cu, Zn, Ga, As, Se, Sr, Cd, Sn, Sb and Pb in atmospheric dust filtered on a porous membrane of 0.45 μm. The standards were prepared by depositing suitable quantities of 5-50 μl of titrated solutions on similar membranes, which were then dried by passing a filtered air flow through the filter. In fact, the problem is to prevent penetration of the deposit into the body of the filter membrane.

In order to obtain a homogeneous deposit, Gilfrich *et al.* (1973) preferred evaporation of the titrated solution on the filter. They used this method to prepare their standards for analysis (Pb, Zn and Fe) of incinerating dust and motor vehicle exhausts.

Atmospheric aerosols were studied by Giauque *et al.* (1975). Hammerle *et al.* (1975) developed energy dispersion instruments with detector recording Si(Li) for the control of Ca, Ti, V, Mn, Fe, Ni, Zn, Br, and Pb in the atmosphere.

The nonmetals—Si, P, S and Cl—are determined in atmospheric particles on Whatman and Millipore filters according to Adams *et al.* (1975). Johansson *et al.* (1975) conducted studies to find a method sensitive for K, Cu, Zn and Br. The atmospheric particles are collected on a polystyrene support and irradiated in a proton flux (3.7 MeV) of a Van de Graaf accelerator. The X-radiation is measured with a Si(Li) detector.

O'Connor *et al.* (1975) determined the heavy metals Pb and Br in the atmosphere by collecting 30 m^3 of air on a cellulose filter. The detection limits were 0.05 μg Pb/cm^2 and 0.02 μg Br/cm^2 with a count of 3 sec. Penetration of particles into the filter is a cause of error, remedied by using glass fiber filters.

The spectrometric system developed by Dzubay *et al.* (1975) merits special mention. In order to improve the sensitivity and number of elements to be determined, these authors used a radiation source consisting of the fluorescence spectrum of three elements (Cu, Mo and Tb) obtained from a classical tube with a tungsten anode. These elements correspond to anodic potentials of 35, 50 and 70 kV. Each sample is analyzed with the three excitation sources for 5 min.

With molybdenum, U, Bi, Pb and Tl can be determined with a detection limit of 20-30 ng/m^3; Sr, Rb, Br, Se, As, Zn, Cu, Ni, Co, Fe have a detection limit of 1-10 ng/m^3. With terbium, Cs, I, Sb, Sn, Cd, Ag, Mo, Zn can be determined with a detection limit of 100-200 ng/m^3 With copper for Mn, Cr, V, Ti, the detection limit is 1 ng/m^3, and for Cl, S, P, Si, and Al it is 1-100 ng/m^3. The authors also propose a sample collection device allowing the separation and collection of particles < 2 μm as a homogeneous layer on a first filter and of > 2 μm on a second.

In conclusion, we will note the studies of Cullen *et al.* (1975) with marine aerosols. The study discusses sampling methods for particles and droplets from the ocean atmosphere as well as X-ray fluorescence analysis.

7.6   Industrial and Miscellaneous Products

7.6.1 Oil and Crude-Oil Products Analysis

X-Ray fluorescence has been used successfully for the determination of metals (Fe, Cu, Zn, Pb, V) in oils, crude-oil products and fuels. A direct analysis of lubricating oils is possible. The sample is placed into a cell with a 4-μ thick Mylar window. The operating conditions for Fe, Ni and V are summarized in Table XV.

The elements Cr, Fe, Ni, Sn and Pb have also been determined in aviation lubricating oil by Marangoni *et al.* (1972). A concentration on ion exchange resins has been used successfully by Louis (1970) for the determination of V, Cu, Ni, Fe and Pb with a detection limit of 1 ppm.

Table XV. Determination of Fe, Ni and V in Oils

	Fe	Ni	V
Exciting anode	W	W	W
	50 kV	50 kV	50 kV
	50 mA	50 mA	50 mA
Crystal	LiF	LiF	LiF
Detector	Gas flow	Scint.	Gas flow
Radiation	K_α	K_α	K_α
Counting time	100 sec	100 sec	300 sec
Detection limit	1 ppm	1 ppm	1 ppm

Smith *et al.* (1973) compared fluorescence X-ray and atomic absorption spectrometry in trace analysis of Fe, Ni, Cu and V in crude-oil products. Eschalier (1973) has demonstrated that X-ray fluorescence could be used to observe automotive engine wear by determining metallic Fe, Cu and Zn particles suspended in lubricating oils. The detection limits were 2 ppm Fe and Cu and 5 ppm Zn.

Lead in fuels is analyzed in concentrations of 25-500 ppm by X-ray fluorescence, Larson *et al.*,1974. Determination is made directly with the liquid sample. The L_α-line of Pb is excited with a molybdenum anode. A strong spectral background must be taken into account.

7.6.2 Miscellaneous Products

Cement and glass manufacturers are using X-ray fluorescence widely for the analysis of major elements, such as Si, Ca, Na, Al, Fe, K and Li. Trace analysis has few important applications. However, X-ray fluorescence has numerous industrial uses, particularly for trace analysis of rare earths (Berman *et al.*, 1969). The specimen is melted with a lithium and sodium borate, and the L_{α_1}-lines are measured, but numerous interferences are present. The detection limits are between 1 and 3 ppm.

Rouche (1969) determined rare earths in silica-alumina products with a gold anode as the excitation source, measuring the following interference-free lines: La L_{α_1}, Ce L_{α_1}, Pr L_{β_1}, Nd L_{β_1}, and Sm L_{β_1}.

Lahanier (1973) used X-ray fluorescence and microfluorescence for the analysis of paints and art objects, making use of the nondestructive character of the method. Analyses for court decisions and forensic medicine make use of X-ray fluorescence for the *in situ* determination of trace elements in microsamples: Rayburn *et al.* (1970) have reported methods allowing the identification of fragments of paint, inks and papers in connection with criminal matters.

The possibility of *in situ* analysis, nondestructive analysis, routine analysis, high-speed and sensitive analysis as well as the development of portable instruments will contribute to a definite development of X-ray fluorescence in all fields of instrumental analytical chemistry in the years to come.

REFERENCES

Abbey, S. *Geol. Surv. Canada,* paper 73-36 (1973).

Adams, F. C. and R. E. Van Grieken. *Anal. Chem.* 47:1767 (1975).

Adler, I., J. I. Trombka and P. Gorenstein. *Anal. Chem.* 44:28A (1972).

Agarwal, M., R. B. Bennett, I. G. Stump and J. M. D'Auria. *Anal. Chem.* 47:924 (1975).

Alexander, M. E., E. K. Biegert, J. K. Jones, R. S. Thurston, V. Valkovic and R. M. Wheele. *Inst. J. Appl. Radiation Isotopes* 25:229 (1974).

Armitage, B. and H. Zeitlin. *Anal. Chim. Acta* 53:47 (1971).

Baudin, G., G. Derarue and P. Sarrat. *Coll. Rayons X Matiere,* Siemens, Monaco, May 1973.

Berman, S. S., P. Semeniuk and D. S. Russel. *Can. Spectros.* 14:68 (1969).

Boiteau, H. L. and M. Robin. *Coll. Rayons X Matiere,* Siemens, Monaco, May 1973.

Brenner, I. B., L. Argov, and H. Eldad. *Appl. Spectros.* 29:423 (1975).

Brenner, I. B., L. Gleit and A. Harel. *Appl. Spectros.* 30:335 (1976).

Brown, G. and R. Kanaris-Sotiriou. *Analyst* 94:782 (1969).

Burkhalter, P. G. *Anal. Chem.* 43:10 (1971).

Burkhalter, P. G. *Naval. Res. Lab. Rep.* 7637 (1973).

Campbell, W. J., E. F. Spano and T. E. Green. *Anal. Chem.* 38:987 (1966).

Campbell, J. L., B. H. Orr, A. W. Herman, L. A. McNelles, J. A. Thomson, and W. B. Coor. *Anal. Chem.* 47:1542 (1975).

Carr-Brion, K. G. *Analyst* 89:346 (1964).

Carr-Brion, K. G. *Analyst* 90:9 (1965).

Cecchetti, G., S. Cerquiglini Monterolo, F. Cotta Ramusinc and C. De Sena. *8th Coll. Anal. Mat.,* Florence, September 1969.

Claisse, F. R. P. No. 327. Ministère des Mines. Quebec (1956).

Claisse, F. *3th Coll. Internation. Meth. Anal. Ray. X,* Nice (1974).

Claisse, F. and M. Quintin. *Can. Spectros.* 12:129 (1967).

Cornil, J. and G. Ledent. *Analusis* 3:11 (1975).

Criss, J. W. *Anal. Chem.* 48:179 (1976).

Cullen, W. H., P. E. Wilkniss and D. J. Bressan. *Memo Rep. U.S., Naval. Res. Lab.* n° 3017 (1975).

Deutsch, Y. *Anal. Chem.* 46:437 (1974).

Duchesne, J. C. and I. Roelandts. *2th Coll. Internation. Meth. Anal. Ray. X,* Toulouse (1971).

Duggan, J. L., W. L. Beck, L. Albrecht, L. Munz and J. D. Spaulding. *Adv. X-Ray. Anal.* 15:407 (1972).

Dwiggins, C. W. and H. N. Dunning. *Anal. Chem.* 31:1040 (1959).
Dzubay, T. G. and R. K. Stevens. *Environ. Sci. Tech.* 9:663 (1975).
Escahalier, G. *Coll. Rayons X Matiere*, Siemens, Monaco (May 1973).
Evans, C. C. *Analyst* 95:919 (1970).
Fabbi, B. P. *Appl. Spectros.* 25:41 (1971a).
Fabbi, B. P. *Appl. Spectros.* 25:315 (1971b).
Fabbi, B. P. and L. F. Espos. *U.S. Geol. Survey Prof. Paper* 800-B, B 147 (1972).
Fagel, J. E., E. W. Balis and L. B. Bronk. *Anal. Chem.* 29:1287 (1957).
Fagel, J. E., H. A. Ir. Liebhafsky and P. D. Zemany. *Anal. Chem.* 30:1918 (1958).
Flanagan, F. J. *Geochim. Cosmochim. Acta.* 37:1189 (1973).
Frankel, R. S. and D. W. Aitken. *Appl. Spectros.* 24:557 (1970).
Garavelli, C. L. and M. Moresi. *8th Coll. Anal. Mat.* Florence, September 1973.
Garcia, J. D., R. J. Fortner and T. M. Kavanagh. *Rev. Mod. Phys.* 45:111 (1973).
Giauque, R. D., L. Y. Goda and R. B. Garrett. *Univ. California,* L.B.L. Rep., 4414 (1975).
Gilfrich, J. V., P. G. Burkhalter and L. S. Birks. *Anal. Chem.* 45:2002 (1973).
Govindaraju, K. *X-Ray Spectrom.* 2:57 (1973).
Green, T. E., S. L. Law and W. J. Campbell. *Anal. Chem.* 42:1749 (1970).
Griffoul, R. and R. Rabillon. *Rev. Univ. Min. Met. Meca.* 15:533 (1959).
Guennelon, R. and N. Souty. *Analusis* 2:130 (1973).
Gulacar, O. F. *Anal. Chim. Acta.* 73:255 (1974).
Gunn, E. L. *Anal. Chem.* 29:184 (1957).
Gwilliam, L. and A. Coee. *17th C.S.I.,* Florence, September 1973.
Halma, G. *17th C.S.I.,* Florence, September 1973.
Hammerle, R. H. and W. R. Pierson. *Environ. Sci. Technol.* 9:1058 (1975).
Hellmann, H. *Z. Anal. Chem. Dtsch.* 254:192 (1971).
Heres, A., O. Girard-Devasson, J. Gaudet and J. C. Spuig. *Analusis* 1:408 (1972).
Hermann, H. *Coll. Rayons X Matiere,* Siemens, Monaco 1972, C. R. 415 (1973).
Herman, A. W., L. A. McNelles and J. L. Campbell. *Inst. J. Appl. Radiat. Isotopes* 24:677 (1973).
Hisano, K. and K. Oyama. *Jap. Analyst* 16:1508 (1969).
Holynska, B. *Radiochem. Radioanal. Letter* 17:5 (1974).
Hooper, P. R. *Anal. Chem.* 36:1271 (1964).
Jenkins, R. and J. L. DeVries. *Practical X-Ray Spectrometry* (London: Macmillan, 1975), 2nd Ed.
Johansson, T. B., R. Akselsson and S. A. E. Johansson. *Nucl. Instr. Meth.* 84:142 (1970).
Johansson, T. B., R. E. Van Grieken, J. W. Nelson and J. M. Winchester. *Anal. Chem.* 47:855 (1975).
Joly, D. *8eme Coll. Anal. Matiere,* Florence, September 1969.
Kato, K. and M. Murano. *Japan Analyst.* 22:1312 (1973).
Kaye, M. J. *Geochim. Cosmochim. Acta.* 29:139 (1965).
Knapp, G., B. Schreiber, and R. W. Frei. *Anal. Chim. Acta.* 77:293 (1975).
Kneip, T. J. and G. R. Laurer. *Anal. Chem.* 44:57A (1972).
Kokotailo, G. T. and G. F. Damon. *Anal. Chem.* 25:1185 (1953).

Kraushaar, J. J., R. A. Ristinen and W. R. Smytae. *Bull. Am. Phys. Soc.* 16:545 (1971).
Labarta Carreno, C. E. *Inf. Quim. Anal. Esp.* 22:53 (1968).
Lahanier, C. *Coll. Rayons X Matiere,* Siemens, Monaco, (May 1973).
Larson, J. A., M. A. Short, S. Bonfiglio and W. Allie. *X-Ray Spectrom.* 3:125 (1974).
Lazzar, V. A. and K. C. Beeson. *J. A. O. A. C.* 41:416 (1958).
Lochmueller, C. H., J. W. Galbraith and R. L. Walter. *Anal. Chem.* 46:440 (1974).
Louis, R. *Erdöl Kohle Erdgas Pet. Dtsch.* 28:347 (1970).
Lubozynski, M. F., R. J. Baglan, G. R. Dyer and A. B. Brill. *Int. J. Appl. Radiat. Isotopes* 23:487 (1972).
Marangoni, C., P. Lutrario and A. Tronca. *Metallurg. Ital.* 64:381 (1972).
Marcie, F. J. *Norelco Rep. USA* 15:3 (1968).
Marr, H. E. and W. J. Campbell. *US. Bur. Mines Rep. Invest.* (1971).
Marti, W. *Spectrochim. Acta.* 18:1499 (1962).
Mathies, J. C. *Appl. Spectros.* 28:165 (1974).
Morris, A. W. *Anal. Chim. Acta.* 42:397 (1968).
Mountjoy, W. and J. S. Wahlberg. *Geol. Surv. Prof. Paper* 600B:119 (1968).
Müller, R. *Spektrochemische Analysen mit Röntgenfluorescenz,* R. Oldenbourg (1967).
Murao, E. *Anal. Chim. Acta.* 67:37 (1973).
Norrish, K. and B. N. Chapell, in *Physical Methods in Determinative Mineralogy,* J. Zussman, Ed. (London: Academic Press, 1967).
O'Connor, B. H., G. C. Kerrigan, W. W. Thomas and R. Gasseng. *X-Ray Spectrom.* 14:190 (1975).
Patti, F. *8th Coll. Anal. Mat.,* Florence, September 1969.
Pinta, M., Comité Inter-Instituts *Analusis* 3:345 (1975).
Poncet, M. and C. Engelmann. *Analusis* 3:283 (1975).
Price, N. B. and G. R. Angell. *Anal. Chem.* 40: 660 (1968).
Price, J. B. and K. M. Field. *Amer. Lab.* 6:62 (1974).
Purdham, J. T., O. P. Strausz and K. I. Strausz. *Anal. Chem.* 47:2031 (1975).
Quintin, M. and A. Martin. *18th C.S.I.,* Grenoble, September 1975.
Rasberry, S. B. and K. F. J. Heinrich. *Anal. Chem.* 46:81 (1974).
Rayburn, K. *Eastern Anal. Symp. N.Y.,* December 1970.
Reuter, F. W. *Anal. Chem.* 47:1763 (1975).
Rhodes, J. R. *Amer. Soc. Test. Mat. Spec. Tech. Publ.* 485:243 (1971).
Ristori, G. G., P. Fusi and V. Bruno. *Agrochim. Ital.* 15:427 (1971).
Robert, A. *Contribution à l'étude des éléments légers par fluorescence X excitée au moyen de radioelements.* Thèse CNAM, Paris (1964).
Rose, H. J., L. Adler and F. J. Flanagan. *Appl. Spectros.* 17:81 (1963).
Rouche, J. P. *8th Coll. Anal. Mat.,* Florence, September 1969.
Seibel, G., J. Y. Le Traon and P. Martinelli. *Rapport IRSID,* ser. A, 246 (1961).
Shermann, J. *Spectrochim. Acta.* 7:283 (1955).
Shermann, J. *Spectrochim. Acta.* 15:466 (1959).
Smith, A. J., J. O. Rice, W. C. Shaner and C. C. Cerato. *Amer. Chem. Soc. Div. Pet. Chem. Prep. USA* 18:609 (1973).

Stanton, R. E. *Aust. Inst. Min. Metal. Proc. Austral.* 240:113 (1971).

Strausz, K. L., J. T. Purdham and O. P. Strausz. *Anal. Chem.* 47:2032 (1975).

Tertian, R. *Anal. Chim. Acta* 41:554 (1968).

Tertian, R. *Spectrochim. Acta* 26 B:71 (1971).

Tertian, R. *X-Ray Spectrom.* 1:83 (1972).

Tertian, R. *X-Ray Spectrom.* 2:95 (1973).

Tertian, R., R. Geminasca and A. Vidal. *3th Coll. Internation. Meth. Anal. Ray. X,* Nice (1974).

Tertian, R. *X-Ray Spectrom.* 4:52 (1975a).

Tertian, R. *24th Ann. Denver Conf. Appl. X-Ray Analysis.* Denver (1975b).

Tertian, R. *Symp. Adv. Geol. Geochem and Treat. Bauxites,* Dubrovnik (1975c).

Vie-Le-Sage, R. *X-Ray Spectrom.* 4:171 (1975).

Vodinh, K. *Coll. Rayons X. Matiere,* Siemens, Monaco, May 1973.

Vos, G. and A. Hubeaux. *Spectrochim. Acta* 24B:545 (1969).

Wankova, J. *16th C.S.I.,* Heidelberg, October 1971.

Webber, G. R. and M. L. Newbury. *Can. Spectrosc.* 16:90 (1971).

Williams, C. *J. Sci. Food Agric.* 27:561 (1976).

Willis, J. P., H. W. Fesq, E. J. D. Kable and G. W. Berg. *Can. Spectros.* 14:150 (1969).

Wittmann, A. and J. Chmeleff. *Analusis* 2:271 (1973).

ACTIVATION ANALYSIS

1. INTRODUCTION

Neutrons, gamma rays and energetic charged particles can react with isotopes of various elements and produce radioactive nuclides. The characteristic radiation emitted by the nuclides produced can be used for qualitative detection and quantitative determination of various elements. Often elements in parts per million or parts per billion level can be analyzed by this technique.

1.1 Nature of Radiation

When a radioactive isotope decays, it gives rise to another stable or radioactive nucleus with a significant release of energy emitted in the form of radiation, the measurement of which represents a value proportional to the quantity of irradiated element. In practice, α-, β-, γ- and X-radiations are measured.

α-radiation consists of helium particles ^4_2He (2 protons + 2 neutrons) with kinetic energies of 4-8 million electron volts (MeV), having the property of strongly ionizing the atoms, yielding a positive ion M^+ and an electron e^-. α-emitters are found primarily among heavy nuclei ($Z > 82$). The element formed as a result has an atomic mass lower by 4 units and an atomic number lower by two units:

$$^A_Z\text{X} \rightarrow {}^4_2\text{He} + {}^{A-4}_{Z-2}\text{X}'$$

$$^{226}_{88}\text{Ra} \rightarrow {}^4_2\text{He} + {}^{222}_{86}\text{Rn}.$$

An α-decay event may be accompanied by emission of γ-radiation.

β-Radiation is constituted of electrons or positrons of different energies, *i.e.*, of greater or lesser velocity, resulting from conversions within the nucleus, such as:

$$\begin{array}{l}{}_0^1 n \rightarrow {}_1^1 p + \beta\text{- (electron) or } {}_{-1}^0 e \\[2mm] {}_1^1 p \rightarrow {}_0^1 n + \beta\text{+ (positron) or } {}_{+1}^0 e\end{array}$$

The atomic mass remains unchanged, while the atomic number becomes one unit larger or smaller:

$$\begin{array}{l}{}_Z^A X \rightarrow {}_{Z+1}^{A} X + {}_{-1}^0 e \text{ (electron)} \\[2mm] {}_Z^A X \rightarrow {}_{Z-1}^{A} X + {}_{+1}^0 e \text{ (positron)}\end{array}$$

The β^--emitters are nuclei with an excess of neutrons, while the β^+-emitters have an excess of protons:

$$\begin{array}{l}{}_{27}^{60} Co \rightarrow \beta\text{- } + {}_{28}^{60} Ni \\[2mm] {}_9^{16} F \rightarrow \beta\text{+ } + {}_8^{16} O\end{array}$$

In general, β-decay (+ or -) is accompanied by a γ-photon emission ($h\nu$). For example, with radioactive sodium isotopes ^{24}Na and ^{22}Na, we have:

$$\begin{array}{l}{}_{11}^{24} Na \rightarrow {}_{12}^{24} Mg + \beta^- + \gamma \\[2mm] {}_{11}^{22} Na \rightarrow {}_{10}^{22} Ne + \beta^+ + \gamma\end{array}$$

The β^--particles lose their energy either by ionization of the penetrated material or by emitting electromagnetic radiation analogous to x-radiation and called "γ-bremsstrahlung." The most important phenomenon is ionization.

γ-Rays are energy released during nuclear stabilization. The emission of either X-rays or γ-rays does not alter either the atomic mass or the atomic number of the elements. This radiation is absorbed by matter, which becomes ionized in the process. γ-Rays are the most penetrating form of radiation; their energies run from a few KeV to a few MeV.

Gamma radiation may be emitted following α-, β^- and β^+-decay; it can also occur as a result of a nuclear reaction with neutrons [(n, γ) reaction]; *e.g.*,

$$^{23}\text{Na} + \text{n} \rightarrow [^{24}\text{Na}]^* \rightarrow {}^{24}\text{Na}$$

However, cases of pure β^--radioactivity exist, *i.e.*, without being accompanied by γ-photons: $^{14}_{6}\text{C}$, $^{35}_{16}\text{S}$, $^{32}_{15}\text{P}$, $^{45}_{20}\text{Ca}$.

X-Rays are energy released as a result of internal electron transitions. Numerous radioisotopes decay with capture of an orbital electron (K-shell), resulting in a nucleus of atomic number Z-1. Then, the electron reacts with a proton from the nucleus:

$$^{1}_{1}\text{p} + {}^{0}_{-1}\text{e} = {}^{1}_{0}\text{n}.$$

The vacancy left by the captured K-electron is filled by an electron from a higher shell, with the emission of x-radiation from the element with atomic number Z-1. Measurement of this radioactivity constitutes a measurement of the activity of the radioisotope. A radioisotope decays at a statistically predictable rate. The half-life (T) of a radioisotope is the time necessary for it to lose half of its activity (see Section 5). Figures 1, 2 and 3 show some β- and γ-emission processes from a radioactive isotope leading to a stable isotope. Figures 2 and 3 show the decay of ^{58m}Co and ^{66}Cu isotopes. We have:

$$^{58m}_{27}\text{Co} \rightarrow {}^{58}_{27}\text{Co} \text{ (isomers) and}$$

$$^{58}_{27}\text{Co} + {}^{0}_{-1}\text{e} \rightarrow {}^{58}_{26}\text{Fe} + \text{X} + \gamma$$

$({}^{0}_{-1}\text{e} = \text{K-capture electron}).$

Moreover,

$$^{64}_{29}\text{Cu} + {}^{0}_{-1}\text{e} \rightarrow {}^{64}_{28}\text{Ni} + \text{X} + \gamma$$

Copper $^{64}_{29}\text{Cu}$ also decays with β-emission:

$$^{64}_{29}\text{Cu} \rightarrow {}^{64}_{28}\text{Ni} + \beta^+$$

2.   NUCLEAR REACTIONS

Radionuclides can be prepared with the aid of neutrons or high-energy charged particles or gamma rays. These nuclei are:

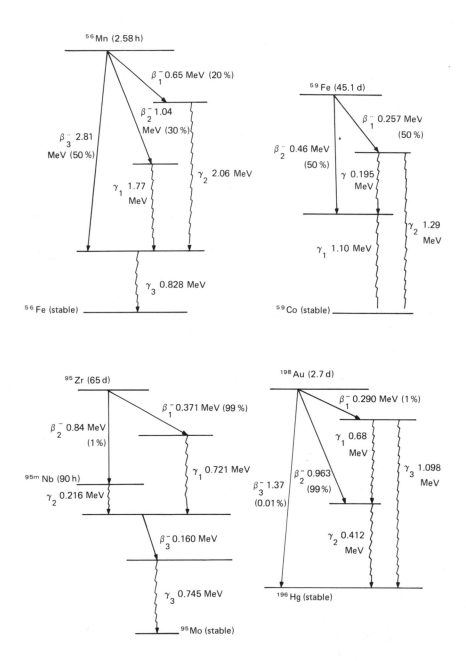

Figure 1. Typical patterns of β- and γ-radiation emitted by a few radioisotopes.

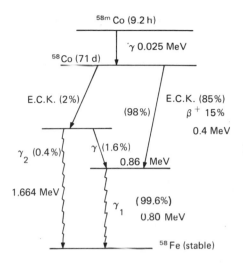

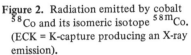

Figure 2. Radiation emitted by cobalt ^{58}Co and its isomeric isotope ^{58m}Co. (ECK = K-capture producing an X-ray emission).

2.1. Particles

Charged particles, including protons $^1_1$H, deutrons $^2_1$H, tritons $^3_1$H and helions $^4_2$He, are involved only in artificial radiations. The protons are designated p, the deutrons d, the tritons t and the helions h.

Neutrons (n), or neutral particles, are produced artificially except in the case of californium which undergoes spontaneous fission. They are called thermal neutrons when they are in equilibrium with matter at room temperature. They have a velocity on the order of 2200 m/sec, and their energy is about 0.025 eV. They are called "fast" if their energies and velocities are much higher (energy above 0.5 MeV). The following are the energies of these particles:

protons, deutons:	a few keV to a few billion electron-volts
thermal neutrons:	about 0.025 eV
epithermal neutrons:	1-10 eV
fast neutrons:	higher than 0.5 MeV.

2.2 Nuclear Changes

Nuclear changes that result from such bombardment occur according to one of the following reactions:

(a)   α-particles and proton production:

$$^A_ZX + {}^4_2He \rightarrow {}^{A+3}_{Z+1}X' + {}^1_1H$$

(α, p) reaction

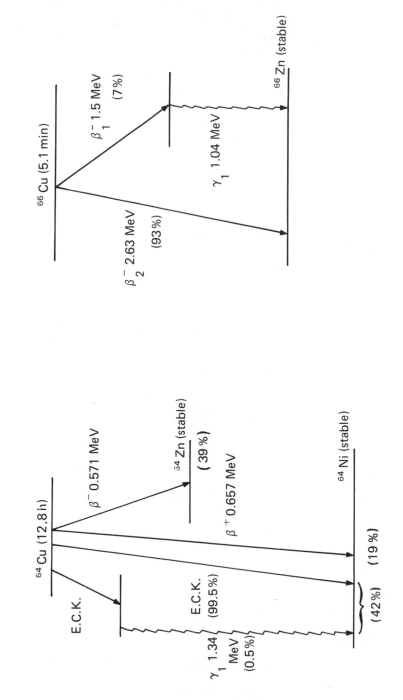

Figure 3. Radiation emitted by copper radioisotopes, ^{64}Cu and ^{66}Cu.

(b) α-rays and neutron production:

$$^A_Z X + ^4_2 He \rightarrow ^{A+3}_{Z+2} X' + ^1_0 n$$

(α, n) reaction

(c) γ-photons and neutron production

$$^A_Z X + \gamma \rightarrow ^{A-1}_2 X + ^1_0 n$$

(γ, n) reaction

(d) γ-photons and production of an isomeric isotope:

$$^A_Z X + \gamma \rightarrow ^{Am}_Z \alpha + \gamma'$$

(γ, γ) photoexcitation reaction

(e)   deutrons and α-ray production:

$$^A_Z X + ^2_1 H \rightarrow ^{A-2}_{Z-1} X' + ^4_2 He$$

(d, α) reaction

(f)   deutrons and protons production:

$$^A_Z X + ^2_1 H \rightarrow ^{A+1}_Z X' + ^1_1 H$$

(d, p) reaction

(g)   deutrons and neutrons production:

$$^A_Z X + ^2_1 H \rightarrow ^{A+1}_{Z+1} X' + ^1_0 n$$

(d, n) reaction

(h)   neutrons and production of a higher isotope:

$$^A_Z X + ^1_0 n \rightarrow ^{A+1}_Z X'$$

(n, γ) reaction

(i)   neutrons and protons production:

$$^A_Z X + ^1_0 n \rightarrow ^A_{Z-1} X' + ^1_1 H$$

(n, p) reaction

(j)   neutrons and production fission elements:

$$^{A+A'+A''}_{Z+Z'}X + ^1_0n \rightarrow ^A_ZX' + ^{A'}_ZX'' + (A'' + 1)^1_0n$$

The most typical reaction of this type is the fission of uranium:

$$^{235}_{92}U + ^1_0n \rightarrow ^{84}_{36}Kr + ^{138}_{56}Ba + 14^1_0n.$$

2.3   Specific Examples of a Few Principal Reactions

1. (d, n): Reactions produced in particle accelerators of the Van de Graaff type or in cyclotrons.

$$^9_4Be + ^2_1H = ^{10}_5B + ^1_0n$$

or: ^9Be (d, n) ^{10}B

$$^7_3Li + ^2_1H = ^8_4Be + ^1_0n$$

or: ^7Li (d. n) ^8Be.

The target and isotope produced differ in atomic number and mass number.

2. (a, n):   Reactions produced in the same particle accelerators.

$$^9_4Be + ^4_2He = ^{12}_6C + ^1_0n$$

or ^9Be (α, n) ^{12}C.

The target and isotope again differ in atomic number and mass number.

3. (γ, n):   Reactions produced in linear accelerators, such as betatron and synchrotron.

$$^9_4Be + \gamma = ^8_4Be + ^1_0n$$

or ^9Be (γ, n) ^8Be.

The target and isotope are of the same atomic number but the isotope loses one unit of mass number.

4. (d, n):   Reactions obtained in deuteron accelerators.

$$^2_1H + ^2_1H = ^3_2He + ^1_0n$$

or ^2H (d, n) ^3He.

This reaction allows the production of 2.6 MeV neutrons.   Furthermore, the $^3H(d, n)^4He$ reaction yields 14 MeV neutrons.

5. (n, p): Bombardments from 14 MeV neutrons.

$$^{16}_{8}O + {}^{1}_{0}n \longrightarrow {}^{16}_{7}N + {}^{1}_{1}H.$$

or $^{16}O(n, p) \ ^{16}N$

The target and isotope are of different atomic number but of the same mass number.

6. (n, γ): The most common reactions from thermal neutrons (0.025 eV).

$$^{63}_{29}Cu + {}^{1}_{0}n = {}^{64}_{29}Cu + \gamma$$

or $^{63}Cu \ (n, \gamma) \ ^{64}Cu$

$$^{59}_{27}Co + {}^{1}_{0}n = {}^{60}_{27}Co + \gamma$$

or $^{59}Co \ (n, \gamma) \ ^{60}Co.$

The target and isotope are of the same atomic number but the mass number differs by one unit.

7. (n, p): Fast neutron bombardment ($>$ 1 MeV).

$$^{32}_{16}S + {}^{1}_{0}n = {}^{32}_{15}P + {}^{1}_{1}H$$

or $^{32}S \ (n, p) \ ^{32}P.$

The target and isotope produced are of different atomic number but have the same mass number.

8. (n, a): Fast neutron bombardment ($>$ 1 MeV).

$$^{27}_{13}Al + {}^{1}_{0}n = {}^{24}_{11}Na + {}^{4}_{2}He$$

or $^{27}Al \ (n, \alpha) \ ^{24}Na.$

The general diagram of activation analysis is shown in Figure 4.

3. ENERGIES INVOLVED IN NUCLEAR REACTIONS

The particles or nucleons (neutron, proton) constituting the atomic nucleus are bound by energies between 7.5 and 8.5 MeV. In addition energy is added to the nucleus by the colliding particle or gamma radiation. Thus, a radioactive disintegration event that is accompanied by a- β- or γ-radiation releases energy in the order of millions of electron-volts. In other words, a nuclear reaction accompanied by a modification of the

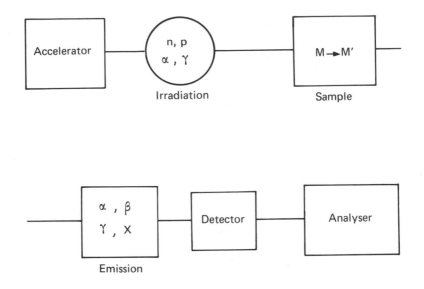

Figure 4. General diagram of activation analysis (M = analyzed element, M$'$ = isotope
produced).

number of nuclear nucleons will require or release an energy approxi-
mating MeV.

The orbital electrons are bound to the nucleus by weaker energies:
a few eV for external electrons (valence electrons), a few thousand eV
for internal electrons (115,000 eV for the K-electrons of uranium). The
energy resulting from a nuclear reaction is emitted in the form of corpuscular (a, p, n) or electromagnetic radiation (γ-rays).

In fact, a nuclear reaction occurs with a very small probability
known as the "cross section." For a particle flux   per cm^2 incident per
unit of time on a target of thickness x containing N atoms of target
element per cm^2, of which N_t are radioisotopes, the cross section is
given by:

$$\sigma = \frac{N_t}{.N.x} \ cm^2$$

σ is expressed in barns; 1 barn = 10^{-24} cm^2.

The activation cross section is a nuclear characteristic of the bombarded isotope, and varies according to the reaction. The cross section
is often about a barn. It may attain several thousand barns (2200 for
^{164}Dy).

The probability of a nuclear reaction (σ) varies with the energy of
the incident particle. The nuclear reactions produced by fast neutrons

resulting in knock-out of (a, p, n) particles require a minimum energy or excitation threshold supplied by the kinetic energy of the neutron. More generally, a nuclear reaction produced by a flux of neutrons or charged particles (a, p, d) or photons (γ) will occur only if their kinetic energy is higher than the above-defined excitation threshold. In fact, the incident energy must be much higher than this threshold for the reaction probability to reach its maximum value.

4. IRRADIATION SOURCE

The principal irradiation sources provide thermal neutrons, fast neutrons, high-energy γ-photons, and charged particles, mainly protons. We will not present a detailed description of these radiation sources, but will summarize only the principal ones.

4.1 Neutron Sources

The most powerful neutron source is a nuclear reactor. The neutron fluxes range from 10^{12} n/cm^2/sec to 10^{14} n/cm^2/sec. Reactors provide an abundant flux of thermal neutrons. Higher energy neutrons are available at special irradiation positions. Other neutron sources are available. Neutrons of specific energies are obtained as a result of various nuclear reactions. The following paragraphs give an outline of these reactions.

Nuclear Reactions

Fission of radioisotopes (^{235}U, ^{239}Pu) is accompanied by a radiation containing γ-rays, neutrons, electrons and α-particles. In modern reactors, thermal neutron fluxes of 10^{11}-10^{13} n/cm^2/sec are obtained, also accompanied by fast neutrons.

Accelerators

Particle accelerators of the Van de Graaff or cyclotron type yield neutrons resulting from bombardment of a suitable target (^{9}Be, ^{7}Li, ^{2}H, ^{3}H) by deuterons, protons, α-particles, etc. With linear accelerators, a betatron or synchrotron, it is possible to produce the (γ, n) -reactions: ^{9}Be(γ, n)^{8}Be. Fast neutrons are generated by 100-200 kV deuteron accelerators according to the (d, n) -reaction:

$$^2\text{H (d, n) }^3\text{He} \rightarrow 2.6 \text{ MeV neutrons}$$

$$^3\text{H (d, n) }^4\text{He} \rightarrow 14 \text{ MeV neutrons.}$$

(p, n)- and (d, n)-Reactions. For low-energy neutrons, the following reactions are used:

$$^7\text{Li}(p, n)^7\text{Be:} \quad \text{energy higher than 1.88 MeV}$$
$$^2\text{H}(d, n)^3\text{He:} \quad \text{energy between 2.5 MeV and 7.0 MeV.}$$

Neutrons of 14 MeV are obtained with the $^3\text{H}(d, n)^4\text{He}$-reaction. Higher energy neutrons are produced by bombarding targets such as Be, U, Pb and W, with a proton or deuteron beam generated in a cyclotron.

Isotopic Neutron Sources

Portable isotope sources exist which produce a stable neutron flux, but this is much weaker than that of nuclear reactions. They are obtained from the (γ, n)- and (α, n)-reactions as above. The neutron flux is from 1.3×10^7 n/cm^2/sec to 2.5×10^6 n/cm^2/sec with energies of 3.6-4 MeV.

(γ, n)-Reaction. The target employed is beryllium or deuterium because of their low energy threshold in the (γ, n)-reaction (Table I). The γ-radiation is obtained from a γ-ray source. The resulting neutrons have a low energy of less than 1 MeV.

(α, n)-Reaction. Typical α-emitters are ^{226}Ra, ^{239}Pu and ^{210}Po (Table I). The most common target is beryllium, which yields neutrons according to the reaction $^9\text{Be}(\alpha, n)^{12}\text{C}$. The energy is 5-6 MeV.

^{252}Cf Source. Californium is an element that decays by spontaneous fission of the ^{252}Cf nucleus with neutron emission. The yield obtained is 2.34×10^{12} n/g/sec with an energy of 2.3 MeV. ^{252}Cf has a half-life of 2.73 years. This source is now finding wider use in analytical laboratories.

4.2 Charged Particle Sources

The charged particles used in activation analysis are essential protons, deuterons and helions. The helions ^4_2He (α-particles) can be obtained from isotope sources of certain heavy elements (Table II). Their energies are lower than 8 MeV, which limits the activation possibilities to light elements. For example, a polonium source ^{210}Po of 160 mCi intensity, emitting a-rays of 5.31 MeV with a half-life of 138.4 days allows the detection of boron, fluorine, sodium, magnesium and aluminum, with detection limits of 10^{-3}-10^{-4} g (Engelmann, 1976).

Table I. Sources Used as Neutron Emitters

Emitting γ- and α- isotope	Reaction	Target	Half-Life	Neutron Flux $(n/cm^2/sec)$	Radiation Dose (in $\square$R/h at 1 m for 1 Ci)
^{124}Sb	(γ, n)	Be	60.4 days	1.4×10^6	980
^{227}Ac	(α, n)	Be	21.8 years	1.5×10^7	120
^{226}Ra	(α, n)	Be	1602 years	1.3×10^7	780
^{228}Th	(α, n)	Be	1.9 years	2×10^7	600
^{238}Pu	(α, n)	Be	87.7 years	2.2×10^6	<1.5
^{241}Am	(α, n)	Be	458 years	2.2×10^6	<2.65
^{242}Cm	(α, n)	Be	163 days	2.5×10^6	2.5

Table II. Sources Used as α-Emitters

Isotope	Half-Life	α-Radiations Energy (MeV)	Emission/ 100 decay events	Other Radiations
^{210}Po	138.4 days	5.305	100	γ 803 keV (10^{-3}%)
^{210}Pb	22.8 years	5.305	100	X LBi; γ 47 keV (4%)
^{226}Ra	1602 years	4.78		X KRn; γ
^{241}Am	458 years	5.49	85	X LNp; γ 59,6 keV (36%)

4.3 Gamma Ray Sources

Cobalt-60 and antimony-124 are the principal isotope sources used, and they have the following properties:

^{60}Co half-life 5.26 yr energies 1.17-1.33 MeV
^{124}Sb half-life 60.2 days energies 1.69-2.09 MeV

An isomeric state of the interacting nuclide may be produced by the (γ, γ')-reaction. The gamma ray emitted in the deexcitation of the isomeric state is used for the measurement.

Particle accelerators allow the production of γ-ray beams by bombarding a metal target (Pt, Au) with the accelerated electrons. The γ-photon energy ranges from a few MeV to several dozen MeV. The linear electron accelerators are presently the most common. They are followed by the Van de Graaff accelerators with energies of a few MeV and several mA, the cyclotrons and finally the betatrons.

Studies to be consulted: Albert, 1964; May, 1975; Engelman, 1976.

5. PRINCIPLE OF ACTIVATION ANALYSIS

When a sample is bombarded with neutrons, charged particles or photons radioactive nuclei are often produced. The number of radionuclei dN_t produced at a given time dt is:

$$dN_t = -\lambda N_t \, dt \qquad (1)$$

where λ is the decay constant and N_t the number of atoms of the radio-isotope existing at time t. Integration yields:

$$N_t = N_0 \, e^{-\lambda t} \qquad (2)$$

where N_O is the number of radioisotope atoms at time t = o, the start of the measurements.

The half-life of a radioisotope (T) is the time necessary for it to lose half of its activity. Equation 2 is written as follows:

$$\frac{N_T}{N_O} = \frac{1}{2} = e^{-\lambda T}$$

$$Ln\ 2 = \lambda T = 0.693$$

$$\lambda = \frac{0.693}{T} \tag{3}$$

The half-life of a radioisotope is an important property in activation analysis. Tables III and IV list the half-life of some radionuclides.

Table III. γ-Ray Spectrometry of Elements Activated in a Flux of 10^{-8} neutron/sec/cm^2

Element	Radioisotope	Half-Life	γ-Ray Energy (MeV)	Detection Limit (ppm)
Aluminum	^{28}Al	2.3 min	1.78	2.4
Antimony	^{122m}Sb	3.5 min	0.061,0.075	8.9
Arsenic	^{76}As	27 hr	0.56	5.2
Bromine	^{80}Br	18 min	0.51,0.62	1.4
Bromine	^{82}Br	36 hr	0.55,0.83	1.4
Chlorine	^{28}Cl	37 min	1.59,2.16	6.8
Cobalt	^{60m}Co	10.5 min	0.059	0.45
Copper	^{69}Cu	13.5 hr	0.51	4
Copper	^{66}Cu	1 min	0.04	4
Gold	^{198}Au	2.7 days	0.41	1.3
Indium	^{116}In	54 min	0.41,1.09,1.27	0.08
Iodine	^{128}I	25 min	0.45	1.0
Magnesium	^{27}Mg	9.5 min	0.84	82
Manganese	^{56}Mn	2.6 hr	0.84	0.12
Molybdenum	^{101}Mo	15 min	0.19	14
Nickel	^{65}Ni	2.6 hr	0.39,1.10	180
Platinum	^{199}Pt	30 min	0.074,0.20,0.54	17
Potassium	^{42}K	12.5 hr	1.53	150
Rhenium	^{188}Re	17 hr	0.06,0.16	0.5
Silver	^{108}Ag	2.3 min	0.63	1.2
Sodium	^{24}Na	15 hr	1.35,2.75	6.4
Tin	^{123}Sn	40 min	0.15	40
Tin	^{125}Sn	9.5 min	0.33	40
Titanium	^{51}Ti	5.8 min	0.32	40
Tungsten	^{187}W	24 hr	0.07,0.13,0.48, 0.69	3.5
Vanadium	^{52}V	3.8 min	1.44	0.2

^{a} m = Isomeric transition isotope.

Table IV. Detection Limits Obtained in Neutron Activation in a Flux of 10^{12} n/cm^2/sec for 1000 cpm[a]

Element	A Isotope	Isotope Abundance	Nuclear Reaction and Activation Cross Section (barns)	Isotope Produced	Half-Life	Detection Limit (μg)
Ag	107	0.513	n,γ 45 ± 4	Ag 108	2.3 min	1×10^{-4}
Al	27	1	n,γ 0.21 ± 0.02	Al 28	2.37 min	4×10^{-3}
As	75	1	n,γ 5.4 ± 1	As 76	26.7 hr	4×10^{-4}
Au	197	1	n,γ 96 ± 10	Au 198	2.7 days	5×10^{-5}
Ba	138	0.7	n,γ 0.5 ± 0.1	Ba 139	1.41 hr	1×10^{-2}
Br	79	0.50	n,γ 10.4 ± 1	Br 80	4.4 hr	5×10^{-4}
Cd	114	0.288	n,γ 1.1 ± 0.3	Cd 115	54 hr	1×10^{-2}
Cl	37	0.246	n,γ 0.56 ± 0.12	Cl 38	37.3 min	7×10^{-3}
Co	59	1	n,γ 20 ± 3	Co 60	5.2 yr	8×10^{-5}
Cr	50	0.04	n,γ 13.5	Cr 51	27.8 day	5×10^{-3}
Cu	65	0.309	n,γ 18 ± 0.4	Cu 66	5.10 min	3×10^{-3}
Cs	133	1	n,γ 30 ± 1	Cs 134	2 yr	5×10^{-3}
Fe	56	0.917	n,p 0.11 (13.5 MeV)	Mn 56	2.58 hr	1×10^{-2}
Ga	69	0.602	n,γ 1.4 ± 0.3	Ga 70	21 min	2×10^{-3}
Hg	196	0.001	n,γ 3100 ± 1000	Hg 197	25 hr	2×10^{-3}
I	127	1	n,γ 5.6 ± 0.3	I 128	24.99 min	6×10^{-4}
La	139	0.999	n,γ 8.2 ± 0.8	La 140	40.2 hr	5×10^{-4}
Li	6	0.075	n,α 945	H 3	12.2 yr	6×10^{-4}
Mg	24	0.786	n,p 0.19 (14 MeV)[b]	Na 24	15 hr	4×10^{-3}
Mn	55	1	n,γ 13.3 ± 0.2	Mn 56	2.58 hr	1×10^{-4}

Mo	98	0.237	n, γ 0.45 ± 0.1	Mo 99	67 hr	3×10^{-2}
Na	23	1	n, γ 0.53 ± 0.02	Na 24	15 hr	2×10^{-3}
Ni	58	0.677	n,p 0.56 (14 MeV)	Co 58	71 day	2×10^{-2}
P	31	1	n, γ 0.19 ± 0.01	P 32	14.3 day	6×10^{-3}
Pb	206	0.236	n, γ 25 ± 5 mb	Pb 207	0.85 sec	1×10^{-3}
Pt	192	0.0078	n, γ 90 ± 40	Pt 193	4.4 day	8×10^{-3}
Rb	85	0.721	n, γ 0.80 ± 0.08	Pb 86	1.02 min	6×10^{-3}
S	32	0.950	n, p 0.3 (14 MeV)	P 32	14.2 day	4×10^{-3}
Sb	121	0.572	n, γ 6.8 ± 1.5	Sb 122	2.8 day	9×10^{-4}
Se	76	0.090	n, γ 7 ± 3	Se 77	17 sec	3×10^{-3}
	80	0.498	n, γ 30 ± 10 mb	Se 81	18 min	9×10^{-3}
Sn	120	0.329	n, γ 0.14 ± 0.03	Sn 121	27 hr	7×10^{-2}
Sr	86	0.098	n, γ 1.3 ± 0.4	Sr 87	2.9 hr	2×10^{-2}
Te	126	0.187	n, γ 0.8 ± 0.2	Te 127	9.4 hr	2×10^{-2}
Ti	48	0.734	n, p 92.7 mb (14 MeV)	Sc 48	44 hr	2×10^{-2}
V	51	0.998	n, γ 4.5 ± 0.9	V 52	3.77 min	3×10^{-4}
W	186	0.284	n, γ 34 ± 7	W 187	24 hr	5×10^{-4}
Y	89	1	n, γ 1.2 ± 0.3	Y 90	64.8 hr	2×10^{-3}
Zn	68	0.185	n, γ 1 ± 0.2	Zn 69	59 min	1×10^{-2}
Zr	90	0.514	n, p 0.247 (14 MeV)	Y 90	64.8 hr	2×10^{-2}

aMay 1974

bThe cross sections followed by the indication (14 MeV) correspond to irradiations by monoenergetic fast neutrons of 14 MeV.

According to Equation 1, the activity A of a number of atoms N of a radioisotope, *i.e.*, the number of nuclear decays per second, will be:

$$A = \lambda N = \frac{0.693}{T} N_t \qquad (4)$$

Actually, the change in the number of nuclei dN existing during time dt during irradiation of the studied element results from the difference between the rate of radionuclide production ($N_A \sigma \Phi$) and the decay constant of the nuclei (λN):

$$\frac{dN}{dt} = N_A \sigma \Phi - \lambda N \qquad (5)$$

where N_A = number of atoms present in the sample which can give rise to the product activity

   N   = number of radioactive nuclei of the studied element after exposure time t

   t   = exposure time (sec)

   σ = probability of a specific nuclear reaction or cross section of the investigated reaction (in cm^2)

   Φ = incident particle flux (cm^2/sec)

   λ = decay constant of the isotope produced

Integration of Equation 5 yields:

$$N = \frac{N_A \sigma \Phi}{\lambda} \, [1 - e^{-\lambda t}] \qquad (6)$$

and if we replace λ by 0.693/T (3), we will have:

$$N = \frac{N_A \sigma \Phi}{\lambda} \, [1 - e^{-\frac{0.693t}{T}}]$$

This equation shows that if σ, Φ, t and T are known, N_A is proportional to N, *i.e.*, to the produced activity. This is the case when all atoms N_A are bombarded by particles or γ-photons of the same energies and when the absorption of the latter in the sample is negligible.

The radioactivity is written as follows:

$$A = \lambda N = N_A \cdot \sigma \cdot \Phi (1 - e^{-\lambda t})$$

and is expressed in curies/g of target. With the use of suitable units (see Section 6) we have:

$$A \, (\text{curie/g}) = \frac{0.6023 \, \Phi \sigma C a \, (1 - e^{-0.693 \, t/T})}{(3.7 \, 10^{10} \, M)}$$

where   T = half-life, sec (see Section 6 for definition)

σ = cross section in barns

a = activatable isotope content of the element (in at %)

M = atomic mass of the bombarded element

C = wt element per gram sample

In this formula, two factors must be considered:

1. The product $\Phi\,\sigma$: since σ is a constant of the element, the activity A depends on Φ. An increase of Φ improves the sensitivity.
2. The factor $(1 - e^{-0.693t/T})$, as we have seen earlier, expresses that the activity results from two phenomena: the production of radioactive atoms that are a linear function of time (since Φ is constant) and the radioactive decay of the element that follows an exponential law.

In practice, irradiation is performed for a time t equal to one or two half-lives (T) if T is shorter than a few days. For the short half-lives, t will be much longer than T. The factor assumes a value of about 1 and we say that we have saturation. The expression $(0.6023\ \Phi\ \sigma\ aC)/(3.7\ 10^{10}M)$ is called the saturation activity. This is the activity acquired by a sample irradiated for an infinitely long time compared to the half-life of the radioisotope. This value depends only on σ and Φ. The activity produced during a nuclear reaction is measured by counting the α-, β-, γ- or X-radiation produced.

6. DEFINITION OF UNITS USED IN ACTIVATION PROCEDURES

The curie is the unit of radioactivity. It is the quantity of radioisotopes undergoing 3.7×10^{10} disintegrations per second. Use is also made of the millicurie (mCi) and the microcurie (μCi). The number of atoms corresponding to 1 μCi is

$$N = \frac{3.7 \times 10^4}{0.693}\ T = 5.3 \times 10^4 T$$

and the weight

$$P = \frac{N}{6.02 \times 10^{23}}\ M = 8.8 \times 10^{-20}\ T.M$$

(T = half-life of the radionuclide in sec and M = atomic mass).

The energy of radioactive particles is expressed in eV (with its multiples, keV and MeV). An electron-volt represents a quantity of energy equal to the kinetic energy gained by an electron subjected to a potential difference of 1 V.

$$1\ eV\quad = 1.602 \times 10^{-12}\ erg$$

$$1\ eV/mol\ = 1.602 \times 10^{-12}\ erg \times 6.024 \times 10^{23} = 9.65 \times 10^{11}$$
$$erg/mol = 23.07\ kcal/mol$$

The γ- and X-radiations are measured in roentgens: one roentgen is the quantity of γ- or X-radiation, the associated corpuscular emission of which produces a quantity of ions of either sign equivalent to 1 ESU of electricity in 1 cm^3 of air under standard temperature and pressure (0.001293 g). The energy corresponding to 1 roentgen is 0.110 erg.

7.   PRACTICAL RADIOACTIVITY MEASUREMENT

Today, the measurement of radioactivity is based on three phenomena—ionization of gases by particles, fluorescence excitation of crystals (scintillation), and modification of the electrical properties of semiconductors.

An ionization chamber is an argon chamber containing two electrodes (Figure 5). When α-rays pass through this chamber, strong ionization is produced. The intensity of the ionization between the electrodes is proportional to the number of α-rays penetrating the chamber per unit of time. Ionization by β- and γ-rays is much weaker than that produced by α-radiation. The neutron activity is measured by using the reaction of a neutron with the ^{10}B atom: $^1_0 n + ^{10}_5 B \rightarrow ^7_3 Li + \alpha$ and by measuring the α-radiation.

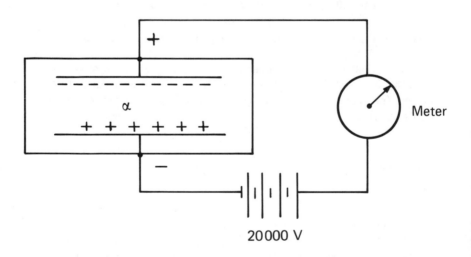

Figure 5. Ionization chamber.

Geiger-Muller counters or proportional counters are special ionization chambers with high sensitivity for β-particles and low sensitivity for γ-radiation. A Geiger counter is a metal tube filled with argon at low pressure containing a small amount of alcohol or chlorine (Figure 6). The tube is the

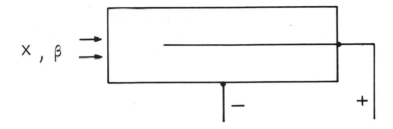

Figure 6. Geiger counter.

cathode and a wire in its axis is the anode. The electrons released by ionization are accelerated by a very high electrical field near the anode (20,000 V/cm)
and produce new ionizations in a cascade reaction. This increase in number
of electrons is about 10^9.

Proportional counters operate with argon-methane gas mixtures at
atmospheric pressure. They permit the detection of β-particles and X-rays of
low penetrating power. The number of ion pairs produced by a single emission
is proportional to its energy. Geiger counters are used primarily for the
measurement of pure β^--emitters.

In scintillation counters, the fluorescence excitation of certain solids or
liquids by α-, β-, γ- and X-radiations is utilized. The fluorescence is proportional to the radiation intensity and is measured by a photomultiplier
(Figure 7). The most common substances for γ-counting are sodium iodide
crystals containing traces of thallium: NaI/Tl. For low-energy β-radiation
detection, liquid scintillators are used. Those consist of a solution of an
organic compound in an aromatic solvent. The most common mixture is a
solution of 2,5-diphenyloxazole in toluene. Scintillation detectors have a
good counting efficiency but poor resolution (half-width of the photoelectric
peak). For this reason, solid semiconductor detectors are used increasingly
today. Ge(Li) and Si(Li) have a very good resolution.

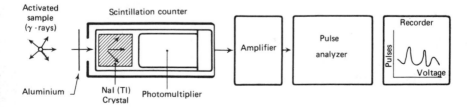

Figure 7. Scintillation counter and γ-spectrometer.

In γ-spectrometry the output signals of scintillation counters and semiconductor detectors are proportional to the incident γ-photon energy. The electrical pulse height measured by amplitude is proportional to the measured γ-photon energy. The γ-spectrometer allows a discrimination and measurement of the different pulse heights, corresponding to γ-photons of different energies. Their measurement amounts to a potential measurement in volts. The analyzer selects the pulses in a specific "channel," *e.g.,* 10 V. By shifting this channel from 0 to 100 V, the energy spectrum can be recorded at the output of the receiver and the pulses of the different γ-photons can be counted successively. Therefore the most energetic ones produce 100 V pulses at the output of the linear amplifier (Figure 8). Scanning γ-spectrometers and multichannel spectrometers (100-400 channels) exist, allowing a simultaneous count of the different photons of the studied spectrum.

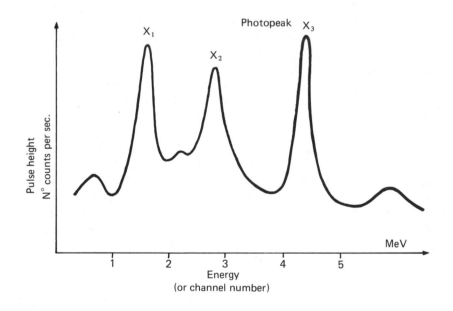

Figure 8. γ-spectrum.

8. ANALYTICAL TECHNIQUE

8.1 Principles

As a rule, more than one of the following techniques are employed in one determination.

1. By a suitable chemical separation the activity can be measured on a "radiochemically pure" product.
2. If the half-lives of the activated elements are sufficiently different, it is possible to use the radioactive decay curve of the sample to determine the individual radioactivities of each element with the aid of the counter alone.
3. Several radionuclides may be determined in the presence of each other by β or γ spectrometry.

8.2 Activity Decay

We will first consider the decay of a mixture of several isotopes. If they have sufficiently different half-lives, they can be identified.

Figure 9 shows the radioactive decay of two elements, X and Y. The decay of the mixture is represented by a resultant curve comprising two parts, AB and BD, corresponding to the two elements X and Y. Element X has the longest half-life. If segment AB is extended, it intersects the ordinate at a point C. The value $\overline{OC}$ represents the radioactivity of element X. The activity of element Y is $\overline{OD} - \overline{OC} = CD$. Three elements can be determined simultaneously under the same conditions if they have sufficiently different half-lives.

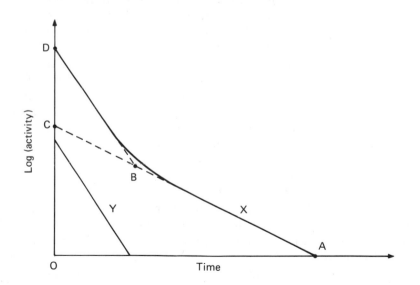

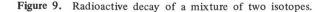

Figure 9. Radioactive decay of a mixture of two isotopes.

8.3 γ- and X-Ray Spectrometry

Spectrometry consists of the separation and determination of gamma-ray energies emitted by a mixture of radioisotopes, just as the emission spectrometry in the ultraviolet, visible or infrared consists of the dispersion and measurement of radiations of different wavelengths. The selective recording of the energies of each radionuclide present in the sample is an important simplification of activation analysis, dispensing with certain chemical separations.

Figure 7 is a schematic diagram illustrating the principle of the γ-spectrometer. The apparatus consists of a detector, an amplifier, pulse height discriminator/multichannel analyzer, and a printer/recorder. The detector is usually a scintillation counter consisting of a thallium-activated sodium iodide crystal and a photomultiplier, similar to the one used in X-ray spectrography. Certain counters are equipped with wells to permit the study of solutions.

The pulses received by the counter are converted into electric signals that are amplified before being resolved in a pulse height analyzer, which discriminates pulse heights. The incident γ radiations are thus resolved according to their energies. The analyzer also resolves different energy radiations originating from the different radionuclides in the sample.

Figure 8 shows the γ-spectrum of a mixture of radionuclides (X_1, X_2, X_3) defined on the ordinate by the measured γ-activity and on the abscissa by the energy of the different radiations, in MeV (photoelectric peaks), that can originate either from several nuclear reactions of the same element or from nuclear reactions of different elements. The thallium-activated sodium iodide crystal detector has very good efficiency but mediocre resolution. The number of detectable photoelectric peaks in the spectrum is very limited. In contrast, in spectrometry with Si(Li) or Ge(Li) detectors, the spectrum of a mixture of radioisotopes generally has a large number of peaks that must then be identified and measured. Several radioisotopes may have identical or very similar γ-energies. In this case, the decay is measured.

Recently, Engelman *et al.* (1973) showed that it is possible to determine elements in the microtrace state from their X-ray spectrum when excitation is produced with 0.4-1.5 MeV protons or 13.5 MeV charged particles. Elements with an atomic number of up to 50 (Sn) can be determined by the K-lines and beyond this, with the L-lines. For elements having an atomic number of about 30, the detection limit is about 10^{-10} g with proton excitation of 1.5 MeV.

8.4 General Method

The utilization of nuclear reactions in activation analysis can be based on direct observation of nuclear reactions during irradiation of the analyzed sample, with emission of a secondary excitation radiation characteristic of the element to be determined and differing from the primary nuclear radiation. Alternatively it could be based on the detection and measurement of the radionuclides produced in the sample after exposure to a radiation flux or to primary nuclear particles. In the latter case, the new characteristic nuclear radiation is analyzed. The majority of methods are based on this principle.

The most general experimental procedure comprises the following operations: irradiation of sample, separation of trace elements and the determination of the energy associated with each element by a γ-spectrometer.

In certain cases it is possible to make the γ-spectrometric determination of each element immediately after irradiation without any chemical separation. Guinn *et al.* (1960) studied the application of activation to routine analysis in this manner. A 25-gram sample was irradiated by a flux of 10^8 neutrons per cm^2 per second, emitted by a 3 MeV Van de Graaff accelerator. Spectrometric analysis of the sample immediately after irradiation determines with a satisfactory sensitivity more than some 20 elements. Table III lists the elements that can be so determined, the isotopes that are formed, the respective energies and the detection sensitivity of each element.

The method seems to be suitable for trace analysis. A low-power accelerator is used, and no chemical treatment of samples is required. The method has been employed in the analysis of organic products, cracking catalysts, and new and used lubricating oils. It may, on the other hand, be too insensitive for certain elements, and is not suited to the analysis of complex materials.

The general procedure for determining a given element E by the activation method is as follows:

(a) Irradiation of a known amount of the sample under well-defined conditions.

(b) Solubilization of the irradiated product and the addition to the solution of a known amount of inactive element E in the same chemical form; this will then serve as carrier of the radioactive isotope.

(c) Chemical separation of the total amount of element E (initial amount plus added amount) in the form of a known chemical compound.

> (d)   Determination of the yield of the compound obtained in (c) if the yield is not quantitative.
>
> (e)   Preparation of a series of radioactive standards under the same conditions.
>
> (f)   Determination and comparison of the radioactivities of the unknown sample with the standards.

A known amount of element E is added during operation (b) to be able subsequently to separate during operation (c) a chemical compound of E in an amount large enough to avoid the large analytical error usually involved in separating microgram amounts of trace elements. It is also necessary that the sample be large enough to measure the radioactivity of the element. If the element being determined is present in large amounts, the addition may be dispensed with. This separation is readily carried out by precipitation or solvent extraction. Paper chromatography is also an excellent method of isolating the elements whose radioactivity is then measured directly on the chromatogram.

The measurement of the radioactivity (f) is carried out with a scintillation counter if the separated product is "radiochemically pure," *i.e.*, if it contains no other isotopes that have become radioactive as a result of the irradiation. If the values of the half-lives are suitable, it is sometimes possible to measure the activity of an element after enough time has elapsed for the other irradiated elements to have become practically inactive.

On the other hand, γ-ray spectrometry permits the determination of several radioactive elements in the presence of each other in the irradiated sample. This often simplifies certain fractionations, since it may be enough to concentrate all trace elements in one single fraction. Even though activation analysis is one of the most recently introduced analytical methods, it has numerous applications, particularly in the determination of trace elements. Activation analysis is now one of the most sensitive methods and is applied to an increasing extent, both in trace analysis and in microanalysis. For additional information, consult Fedoroff, 1973.

8.5 Precision, Accuracy, Sensitivity and Detection Limit

The irradiation conditions, particularly the exposure time and particle flux, allow an increase of the sensitivity and detection limit in a wide range. For example, with a thermal neutron flux of 10^{13} $n/cm^2/sec$ of a nuclear reactor, an exposure time of 1 hr and a standard counting time of the activity produced, the detection limits of 70 elements are approximately 10 ng in absolute value (Guinn, 1966). The detection limits depend on the spectral background near the photoelectric peak. The spectral background, consisting of the activity recorded by the detector in the absence of a radioactive source, results from the activity input from cosmic radiation and natural radioactive material in the detector environment. Also there is a component caused by higher energy gamma rays emitted by the sample.

The detection limits obtained in activation analysis are low and are rarely attained with other physical analysis methods. Table IV lists the detection limits of various elements. Expressed in μg for irradiation in a flux of 10^{12} n/cm^2/sec and for a registration of 1000 counts/min, these values correspond to an activity double that of the random deflection.

The precision of activation analysis is often better than that of classical methods (chemical or physical). In fact, one of the principle causes of error in precision (reliability) in analytical chemistry originates from contaminations of all kinds, often accidental, not predictable and difficult to evaluate even in a blank test. In activation analysis, this risk is minimized since it exists only during sampling prior to irradiation. All phases of analysis which follow irradiation, including chemical separations, are not subject to contamination risks, so the measurement involves only the radioisotope. In practice, the reproducibility (coefficient of variation of a series of analyses) generally ranges between 5 and 10%.

With regard to accuracy, the analytical error can reside in calibration and in spectral interference. Calibration is generally made with a primary standard, the matrix of which is similar to that of the analytical samples and which is placed side by side with the samples during irradiation so as to receive exactly the same neutron flux. The calibration errors therefore can be easily reduced.

Spectral interference refers to the influence of foreign radionuclides producing an energy peak close to that of the measured elements. This cause of error can often be remedied by selecting the radioisotope serving for analysis in such a way that the photoelectric peak measured is not perturbed by the activity of other matrix elements. It may also be that the same radioisotope may originate from two different nuclear reactions. For example, $^{35}_{16}S$ may originate from the reactions:

$$^{31}_{16}S \ (n, \ \gamma) \ ^{35}_{16}S \ \text{and} \ ^{35}_{17}Cl \ (n, \ p) \ ^{35}_{16}S \qquad (7)$$

and $^{32}_{15}P$ from the reactions:

$$^{32}_{15}P \ (n, \ \gamma) \ ^{32}_{15}P \ \text{and} \ ^{32}_{16}S \ (n, \ p) \ ^{32}_{15}P \qquad (8)$$

$$\text{and} \ ^{24}_{11}Na \ \text{from} \ ^{23}_{11}Na \ (n, \ \gamma) \ ^{24}_{11}Na \ \text{and} \ ^{27}_{13}Al \ (n, \ \alpha) \ ^{24}_{11}Na \qquad (9)$$

In a determination of phosphorus (Equation 8) this interference can be eliminated by calculating the quantity of isotope $^{32}_{15}P$ obtained in the (n, p)-reaction by irradiating a sample containing the same amount of sulfur but no phosphorus. The sulfur activity will then be subtracted from the total activity (S + P).

Another cause of error in accuracy is caused by the production of the isotope to be measured by a (n, γ)-reaction, followed by a β^-- or β^+-decay. An example is the analysis of arsenic in germanium. Arsenic is determined by the reaction $^{75}_{33}As(n, \gamma)^{76}_{33}As$. The germanium isotope is produced in a germanium matrix:

$$^{74}_{32}Ge \ (n, \ \gamma) \ ^{75}_{32}Ge.$$

The isotope produced decays with β^--emission:

$$^{75}_{32}Ge \rightarrow \beta^- + {}^{75}_{33}As$$

and then with

$$^{75}_{33}As \ (n, \ \gamma) \ ^{76}_{33}As.$$

This interference is corrected as before. Finally, it must be noted that calibration errors can occur with samples having a very large cross section (>10 barns) because of the strong neutron absorption in the bulk of the sample.

9.   PRACTICAL ANALYSIS METHODS AND APPLICATIONS

The practical analysis methods are customarily distinguished as a function of the type of activation: thermal neutron activation, fast neutron activation, charged particle activation and γ-photon activation.

9.1   Thermal Neutron Activation Analysis

9.1.1   Field of Application and Experimental Procedure

This, the most common method, results from a neutron capture by the investigated atom. It is the simplest and also the most probable nuclear reaction.   (n, γ)-Reactions exist for the majority of elements. For example, the determination of copper, which has two stable isotopes, results from the reactions:

$$^{63}_{29}Cu \ (n, \ \gamma) \ ^{64}_{29}Cu$$

and
$$^{65}_{29}Cu \ (n, \ \gamma) \ ^{66}_{29}Cu$$

The radioisotopes have the following properties:

$^{64}_{29}Cu$ half-life 12.8 h, β - and β^+ - emission, K-capture

$^{66}_{29}Cu$ half-life 5.1 min, β - and γ emission of 1.04 MeV and 0.83 MeV

The two radioisotopes are obtained by bombardment with thermal neutrons. Trace analysis is possible by the following procedures:

· irradiation of sample,
· solubilization and addition of a known amount of entraining agent (the same inactive element),
· chemical separation of the element and determination of the yield of this extraction,
· comparison of the activities obtained in the studied sample and standards treated under the same conditions.

The last step involves a control of the radiochemical purity of this measurement, or, in other words, a verification that the measured γ-photons are characteristic of the desired element and not perturbed by stray activity. Generally, the chemical separation of the above element associated with γ-spectrometry eliminates such interferences. Quite often, γ-spectrometry, especially with the Ge (Li) detector, eliminates a preliminary chemical separation. In any case, the radiochemical purity can be established by verifying the half-life of the measured element.

9.1.2 Properties

The principal advantages of thermal neutron activation analysis are:

· high sensitivity and low detection limit (Table IV),
· absence of risks of contamination,
· specificity of the determination,
· precision, particularly when the analysis does not require chemical separation,
· number of elements that can be determined simultaneously.

However, certain elements are not detected by neutron activation analysis. This applies to those forming radioisotopes with too short a half-life. Thus, it is difficult to determine an element with good sensitivity when the radioisotope has a half-life similar to or less than the time required for chemical separation. Any chemical separation becomes difficult if not impossible. Extremely rapid separation techniques must be used, such as paper electrophoresis and binding in the form of a mercury amalgam. For certain elements (Ag, Br, In), however, the sensitivity obtained with short-lived isotopes may be higher than that obtained from the long-lived ones of the same element.

9.1.3 Nondestructive Analysis

In some cases, a radiochemical separation of the isotopes is not necessary. The sample is suitably irradiated in a thermal neutron flux and the characteristic activity of an isotope of the analyzed element is measured directly by γ-spectrometry.

In practice, a γ-spectrum is always complex. Today, the advances in electronics allow the construction of very high-performance γ-spectrometers, capable of resolving and measuring the characteristic line of the analyzed element among numerous γ lines of different energies in a spectrum. However, nondestructive analysis is possible with high sensitivity if the analyzed sample does not contain large amounts of elements which can be highly activated. This applies to C, Si, O, N and H which constitute the matrix of organic, biological and plant compounds. These elements become only weakly active and emit only weak γ-radiation.

A second possibility for nondestructive analysis concerns compounds in which the matrix consists of elements such as Al, Mg, V, Pb and Ni, forming short-lived radioisotopes (less than 30 or 60 min). After waiting a few hours, the matrix spectrum will no longer interfere.

Finally, it is possible to analyze a matrix containing elements of low activity after short irradiation: Be, C, Fe and Zr. These elements have long half-lives (1-2 months and more), but their maximum activity is obtained only after a relatively long irradiation time. For example, the saturation activity of an iron sample is reached only after one year (the half-life of ^{59}Fe is 45 days) and irradiation for 1.5 months is about half of this. The activity of ^{59}Fe gained in 2.3 min of exposure (138 sec) is calculated to be $1/10^5$ of the activity gained in one month (Albert, 1964).

The essential advantages of nondestructive or instrumental analysis are rapidity and simplicity. However, the sensitivity as well as the detection limits are not as high as in methods making use of radiochemical separation.

9.1.4 Practical Applications

These are numerous practical applications today involving all media. only a few examples will be cited in the following tables, indicating the analyzed elements, the analyzed medium and possibly, the radiochemical pretreatment performed after irradiation of the sample. The field of application concerns primarily natural mineral media, such as rocks, minerals, soils (Table V), water, atmospheric dust (Table VI), biological and plant media (Table VII), industrial products (Table VIII), and petroleum products and their derivatives (Table IX). These examples were selected from a large number of publications to illustrate the possibilities of activation analysis of trace elements.

As an example of the entire procedure, the practical conditions for the nondestructive analysis of atmospheric dust samples and suspended particles in sea water will be described. First, sample is collected on a filter. The filter fractions (corresponding to 0.1 to a few mg of sample) are irradiated in a thermal neutron flux of 10^{12}-10^{14} n/cm^2/sec. Exposure time depends on the neutron flux and the properties of the measured isotope. Most counts are made in a high-resolution γ-spectrometer with a Ge(Li) detector, with counting times depending on the half-lives of the analyzed isotopes. The physical and instrumental conditions are summarized in Table X (Darcourt-Rieg, 1973).

It will be noted that the isotopes employed are not always those of highest sensitivity (see Table IV). They are selected as a function of the concentration of the element, the irradiated isotope abundance and the possible matrix effects. In fact, despite the very broad applications of neutron activation analysis for the chemical determination of trace elements, the method was used for a long time only in radiochemistry laboratories especially equipped for activation analysis, i.e., with irradiation sources at their disposal. In the initial stage of development, neutron activation demonstrated its considerable possibilities for application to the analysis of traces in levels much lower $(10^{-9}$-$10^{-6})$ than those accessible to other physiochemical methods (colorimetry, polarography, atomic absorption spectrometry). In fact, no competition exists between activation analysis, on one hand, and the other methods, on the other. However, the development of small portable neutron generators in recent years should contribute to a spectacular advance of the method. The progress in electronics and γ-spectrometry has also facilitated the development of new applications, particularly in direct and nondestructive analysis of trace elements at a part per million level. At the same time, progress in instrumentation, such as spark mass spectrometry, plasma emission spectrometry and flameless atomic absorption spectrometry, allow us to anticipate interesting comparisons of analytical methods in which activation analysis, because of its characteristic qualities, might play the role of a control and reference. For additional information, consult May, 1975 and Albert, 1964.

Table V. Application of Neutron Activation Analysis to Natural Mineral Media

Elements	Media	Remarks	References
Pt	Ores	Nondestructive analysis	Turkstra et al., 1970
Au, Ga, Ge, In, Ir	Lunar terrestrial, meteoritic basalts	—	Wasson et al., 1970
Noble Metals	Geological samples	—	Millard et al., 1970
Pt, Pd	Geological samples	With extraction	Rowe et al., 1971
I, Hg	Forest soils	—	Lag et al., 1971
Ta, Cs	Rocks	—	Lobanov et al., 1970
Bi	Minerals	—	Babaev et al., 1970
Cu, Zn	Chondritic meteorites	—	Greenland et al., 1965
Na, Mn, Cr, Sc, Co	Ultramafic rocks	—	Stueber et al., 1967
Ce, Br, I	Sedimentary rocks	Simultaneous determination	Tajima et al., 1971
V, Th	Geological materials	Nondestructive analysis	Meyer, 1971
Sc	Clays	—	Craft, 1974
Au, Pt, Noble Metals	Geological materials	With ion exchange separation	Nadkarni et al., 1974
Trace Elements	Standard rocks	Nondestructive analysis with high resolution γ-spectrometry	Turkstra et al., 1971
Trace Elements	Soils profiles		Bate, 1973
Au	Rocks	With extraction	Anastase et al., 1969
Au	Rocks	With fire-assay extraction	Simon et al., 1968
La	Rocks	With extraction	Nadkarni et al., 1971, 1972
B	Minerals	Reaction n, α	Berzina et al., 1969
Sb	Rocks	After extraction	Alian et al., 1968
Mn	Rocks	γ-spectrometry	Johansen et al., 1968
Co, Cu, Fe, Ga, W, Zn	Rocks	With anion exchange separation	Johansen et al., 1970
V, Cr, Hf, Zr, Tl, Rb, Cs	Clays, zircons, feldspars	Nondestructive analysis for Cs, Rb, Zr, Hf; extraction for Tl, V	Jaffrezic et al., 1973
Cu	Ores	Nondestructive analysis; detection: 0.2 μg	Doctor et al., 1969

Element(s)	Material	Method	Reference
Ta	Rocks	With extraction	Vados et al., 1969
Ta, Hg	Silicates	With extraction	Greenland, 1968
Zr, Hf	Rocks	With extraction	Brooks, 1968
Trace Elements	Rocks	Destructive and nondestructive methods	Aubouin et al., 1968
Trace Elements	Soils	Determination of 31 elements	Buenafama, 1973
Dy	Rocks	With extraction; detection: 0.005 µg	Tomura et al., 1968a
Sr, Ba	Rocks	—	Hamaguchi, 1960
Cu	Geological materials	Nondestructive analysis for Cu; 100-1000 ppm	Michelsen et al., 1968
Mo	Soil extracts	With radiochemical extraction; detection limit: 1 ng	Zmijewska et al., 1969
Sb	Rocks, minerals	With radiochemical extraction; level: ppm	Brunfelt et al., 1968b
Ao, Sb	Soils	With extraction, detection: As-0.005 ppm, Sb-0.01 ppm	Ohno et al., 1970
Se	Standard rocks	With extraction; level: submicrogram	Lavrakas et al., 1974
Hg, Se	Coal	Nondestructive analysis	Weaver, 1973
Trace Elements	Coal	Nondestructive analysis	Sheibley, 1973
Trace Elements	Diamonds	Direct determination for 30 elements	Randa et al., 1973
W	Rocks	—	Khajdarow et al., 1974
Hg	Rocks	With separation by volatilisation	Rosenberg et al., 1975
Cu	Ores	With separation	Akbarov et al., 1973
Sb	Rocks	—	Zajtsev et al., 1972
Sb	Standard rocks	Nondestructive analysis	Lombarc et al., 1971
Cu	Rocks	With extraction	Glukhov et al., 1970a
Ta	Rocks and ores	With extraction	Glukhov et al., 1970b
V	Silicate rocks	With extraction	Das et al., 1970
Rare Earths	Rocks	With hydroxides precipitation	Tomura et al., 1968b
Rare Earths	Rocks	With extraction	Treuil et al., 1973
Hf, Sc, Cr, Mn, Co, Cu, Zn,Rb,Cs,Ra,Rare Earths	Rocks	With extraction	Brunfelt et al., 1974
Trace Elements	Marine sediments	In situ analysis with ^{252}Cf	Wogman et al., 1973
Trace Elements	Coal	Nondestructive method for 40 elements	Block et al., 1974
Ni	Rocks	Nondestructive analysis	Steinnes, 1974
P	Rocks	With phosphate precipitation, β activity determination	Brunfel, 1968a
Sn	Standard rocks	With selective extraction	Alian et al., 1974
Ni,Rb,Br,Co,Cs,Ga,Se,Te,V,Zn Ag,Bi,Cd,Ge,In,Sb,Tl,Au,Ir,Re	Terrestrial lunar, meteoritic material	Detection limits: $10^{-6} - 10^{-12}$	Keays et al., 1974
Trace Elements	Meteorites	With wet-chemical separation	Seitner et al., 1971

Table VI. Application of Neutron Activation Analysis to the Environment, Water and Air

Elements	Media	Remarks	References
Sc, Cr, Fe, Co, Zn, Sb	Airborne particulates	Nondestructive method	Bando et al., 1970
Se, Te, Au	Natural waters	—	Abdullaev et al., 1971
Hg	Rhine river	—	Heidinga et al., 1971
Au	Natural waters	With preconcentration before	Schiller et al., 1971
Al, Sb, Br, Ga, Na	Air particulates	Nondestructive method	Dams et al., 1972
Sc, Cr, Fe, Co, Cu, Zn, Ag, Sb, Cs, Hf	Seawater	Irradiation, evaporation of samples and determination by γ-spectrometry	Robertson, 1968
Co, Cr, Cs, Fe, Rb, Sc, Sr, Zn	Seawater and suspension in seawater	Evaporation, irradiation, γ-spectrometry	Piper et al., 1969
Trace elements	Air particulates	Nondestructive method	Health Lab. Sci., 1973
V	Waters	Detection 0.1 μg	Linstedt et al., 1969
S	Pure water	Nondestructive method	Glazov, 1968
Mo, Cr	Natural waters and sediments	—	Turekian, 1968
Au	Natural waters	With chromatographic concentration	Abdullaev et al., 1968b
Rare earths	Water	With separation	Shigematsu, 1968
Hg	Water and air	Hg adsorption on charcoal	Van Der Sloot et al., 1974
As, Cd, Hg	River waters	With As, Cd, Hg coprecipitation	Yung-Yang Wu et al., 1974
As, Cd, Cu, Hg, Sb, Zn	Natural waters		Schneider et al., 1973
U	Sea and natural waters	With adsorption on charcoal	Van Der Sloot et al., 1975
Trace elements	Marine environment	Nondestructive method	Bressan et al., 1974
W, Co, Ta, Hg	Environmental contamination	Nondestructive method	Pillay et al., 1973

Elements	Sample	Method	Reference
Trace elements	Atmospheric dusts	Nondestructive method	Vogg et al., 1973
In	Seawater	With anionic exchange separation, detection: 0.006 ng/l	Matthews et al., 1970
Na, Al, Cl, V, Mn, Ca, Br,	Atmospheric aerosols	Nondestructive analysis	Heindryckx et al., 1974
Hg	Environment	—	Filby et al., 1970
Trace elements	Environment	—	Lukens et al., 1972
Hg, Cd, Cu, Cr, Zn, As	Seawater	With separation by precipitation or ion exchange	May, 1974
Trace elements	Water	—	Brune, 1973
Sc, Cr, Fe, Co, Ni, Zn, As, Se, Br, Ag, Sb, Cs, W, Hg	Waters	Direct determination	Clemente et al., 1974
Cd	Air particulates	Nondestructive analysis	Das et al., 1971
Trace elements	Waters	General survey	Thatcher et al., 1971
Lanthanides	Seawater	With precipitation from 10 l	Hogdahl et al., 1968
Trace elements	Air particulates	—	Gilfrich et al., 1973
Trace elements	Air particulates	Nondestructive analysis	Salmon et al., 1971
Trace elements	Air particulates	Nondestructive analysis	Dams et al., 1971
Trace elements	Environmental research	A review	Filby et al., 1974
Al, As, Br, Ca, Cl, Cu, Mg, Mn, Hg, K, Na, V	Air particulates	Nondestructive analysis	Navarrete et al., 1974

Table VII. Application of Neutron Activation Analysis to Biological and Plant Media

Elements	Media	Remarks	References
Pb	Milk powder	—	Dutilh et al., 1971
Mo	Biological materials	—	Nadkarni et al., 1971
Ag, Se	Biological materials	—	Lucas et al., 1970a
Th, U	Blood, urine, cigarettes	—	Mahler et al., 1970
Sn, Zn	Fingernails, hair	Nondestructive analysis	Ohno, 1971
I, Br	Biological materials	—	Das et al., 1971
Cd	Foodstuffs, hair	Nondestructive analysis	Lucas et al., 1970b
Se, Cd, Co, Th, Re	Fishes	—	Henke et al., 1971
Ba, La, Rb, Zn	Human brain	—	Schmolzer et al., 1971
I	Thyroxine, blood	—	Morgan et al., 1968
I	Urine	—	Blotcky et al., 1976
Al	Urine	With cation exchange separation	Abdullaev et al., 1968a
U	Plants	—	Cohn et al., 1971
Na, Ca, Cl, N, P	Man	In vivo analysis	Bate, 1971
Hg	Biological materials	—	Brune, 1973
Trace elements	Biological materials, foods	—	Lukens et al., 1972
Trace elements	Biomedical studies	—	Filby et al., 1970
Hg	Biological materials	Nondestructive analysis	

Elements	Material	Method	Reference
Cu	Plants	Nondestructive analysis	Grimanis, 1968
Cs	Marine organisms	Nondestructive analysis	Merlini, 1968
Cu, Mn	Fingernails	Nondestructive analysis	Kanabrocki et al., 1968
Fe, Co, Ni	Vegetal oils	With separation	Shinbori, 1968
Cu, Zn	Vegetal oils	With separation	Shinbori et al., 1968
Trace elements, Hg	Marine environment	With separation	Bressan et al., 1974
Ag, Co, Cr, Cs, Fe, Hg, Rb, Sb, Se, Zn	Serum and biological materials	Nondestructive analysis	Cornelis et al., 1973
As, Br, Mo, Cd, Sb, W, Au	Biological materials	Irradiation with epithermal neutrons, nondestructive analysis	Bagdavadze et al., 1975
Fe, Co, Zn, Se	Biological materials	Nondestructive analysis of ash	Maziere et al., 1975
Mn, Cu	Biological materials	With ion exchange separation	Maziere et al., 1975
Trace elements	Plant, animal	35 elements	Nadkarni, 1971
Mo, Se, Cr, Fe, Zn, Co, As, Br, Sb, Hg, Rb, Cs, Sc, Eu	Foods	—	Becker et al., 1974
Cu, Hg, Zn, Cd, As, Se, Cr	Fish	With extraction	Pillay et al., 1974
As, Cd, Co, Cu, Ga, Hf, Hg, Mn, Na, Sb, Sc, Se, Zn, lanthanides	Biological materials	With extraction	Henzler et al., 1974

Table VIII. Application of Neutron Activation Analysis to
Chemical, Industrial and Miscellaneous Products

Elements	Media	Remarks	References
As, Sb, Ba, Eu, Hf, Th	Glass fragments, forensic science	Nondestructive analysis	Goode et al., 1971
Ba, Cu, Ga, Sr	Ruby laser crystals	Nondestructive	Thompson et al., 1971
F, Sr	Phosphates	Fast neutrons activation	Lisovskii et al., 1971
Cl, As, Cu, Mn, Na	Pure reagents	–	Kosta et al., 1971
Fe, Cr, Co, Ni, Sc	Asbestos	–	Holmes et al., 1971
Br	Drugs	–	Margosis et al., 1971
Br	Drugs	–	Peisach et al., 1971
Sn, In, Te	Semiconductors	–	Artyukhin et al., 1968
Trace elements	Forensic science	Gunpowder residues	Albu-Yaron et al., 1969
Trace elements	Forensic science	Comparison with other methods	Krishnan, 1973
Trace elements	Forensic science	Gunshot residues	Renfro et al., 1973
Trace elements	Forensic science	Bullet lead and jacket material	Guy et al., 1973
Trace elements	Ancient ceramics	Nondestructive analysis	Delcroix et al., 1973
Trace elements	Roman silver coins	Nondestructive analysis	Van Dalen et al., 1973
Ni, Mn, Co, Fe	Zn sulfide phosphor	With extraction	Persiani et al., 1968

Table IX. Application of Neutron Activation
Analysis to Oils and Fuels

Elements	Media	Remarks	References
Na, S, Cl, K, Ca, V, Mn, Cu, Ga, Br	Petroleums	Nondestructive analysis	Shah et al., 1970
S	Petroleum products	Irradiation with fast neutrons	Tamura, 1970
V	Mineral oils	Nondestructive analysis	Mayer, 1971
Sr, As, Cr, Se, Fe	Petroleums		Filby et al., 1971
V	Petroleum products	Detection: ppm	Passaglia et al., 1973
Trace elements	Crude oils	29 elements	Al-Shahristani, 1973
Co, As	Petroleums		Berkutova et al., 1973
V, S, Na, Br	Crude oils	Nondestructive analysis	Zaghloul et al., 1973
Trace elements	Crude and fuel oils	Nondestructive analysis	Guinn et al., 1971

Table X. Neutron Activation Analysis of Suspended Particles in Seawater; Physical and Instrumental Conditions

Elements	Nuclear Reactions	Half-life[a]	Cross Section (Barns)	Irradiation Time (min)	Measured Peak Energy (keV)	Percent Peak Emission	Recommended Cooling Time[b]	Counting Time (min)
Ag	^{109}Ag (n, γ) ^{110m}Ag	253 day	3.2	480-1020	884	74	a few months	1200
Al	^{27}Al (n, γ) ^{28}Al	2.37 min	0.210	0.5-1	1780	100	10 min	c
Au	^{197}Au (n, γ) ^{198}Au	2.7 day	96	480-1020	412	99	1-6 d	600
Ba	^{130}Ba (n, γ) ^{131}Ba	12 day	10	480-1020	373	13	2-4 weeks	1200
Br	^{81}Br (n, γ) ^{82}Br	35.9 hr	3.1	480-1020	776-1317	83-28	7 d	1200
Ce	^{140}Ce (n, γ) ^{141}Ce	32 day	0.31	60-1020	145	49	2-4 weeks	1200
Co	^{59}Co (n, γ) ^{60}Co	5.2 yr	20	60-1020	1332	100	a few months	1200
Cr	^{50}Cr (n, γ) ^{51}Cr	27.8 day	13.5	60-1020	320	9	1 month	1200
Fe	^{58}Fe (n, γ) ^{59}Fe	45 day	0.98	60-1020	1099	56	a few months	1200
Hf	^{180}Hf (n, γ) ^{181}Hf	45 day	10	60-1020	482	81	1 month	1200
Hg	^{202}Hg (n, γ) ^{203}Hg	47 day	3.8	480-1020	279	77	a few months	1200
K	^{41}K (n, γ) ^{42}K	12.5 hr	1.1	1-60	1524	18	1 d	60
La	^{139}La (n, γ) ^{140}La	40.2 hr	8.2	60-400	1596	96	1-5 d	240
Mn	^{55}Mn (n, γ) ^{56}Mn	2.58 hr	13.3	1-60	847	99	a few hours	30
Na	^{23}Na (n, γ) ^{24}Na	15 hr	0.53	1-60	1369	100	1 d	60
Sb	^{121}Sb (n, γ) ^{122}Sb	2.8 day	6.8	480-1020	564	66.3	7 d	1200
Th	^{232}Th (n, γ) ^{233}Th	4.4 yr	7.4	60-1020	312	100	2-4 weeks	1200
W	^{186}W (n, γ) ^{187}W	24 hr	34	1020	479-686	23-27	a few days	1200
Zn	^{64}Zn (n, γ) ^{65}Zn	245 day	0.44	480-1020	1115	50.6	a few months	1200

[a]The half-lives and cross sections are those given by May (1975). The irradiation and counting times are a function of the concentration of the element.

[b]The cooling times are the recommended values that can be reduced with small amounts of irradiated sample (1 mg).

[c]For Al, the decay is measured every minute.

9.2   Fast Neutron Activation Analysis

Activation with thermal neutrons is a method theoretically applicable to all elements, but it does not always allow sensitive determinations. A typical example is the analysis of nitrogen, which is impossible by the (n, γ)-reaction. In fact, the isotope obtained, ^{16}N, has a half-life of 7.4 sec, a cross section of 0.00002 b and an isotope abundance of 0.37%. Lithium forms the ^{8}Li isotope with a half-life of 0.8 sec in the (n, γ)-reaction. When the isotope produced has a very short half-life as well as a small cross section, the detection limits are poor or mediocre. The nuclear reactions produced by irradiation with high-energy neutrons (14 MeV) offered new prospects. Thus, for example, the reaction $^{14}_{7}N$ $(n, 2n)$ $^{13}_{7}N$ is obtained in a 14 MeV-neutron flux of 10^{10} and the ^{13}N isotope has a half-life of 10 min, with a cross section of 5.67 mb. The detection limit is 7×10^{-2} μg N.

A characteristic reaction obtained with fast neutron irradiation of oxygen is $^{16}_{8}O(n, p)$ $^{16}_{7}N$. Although the half-life of ^{16}N is very short (7.4 sec), the emitted γ-radiation has a very high energy (6.13-6.9-7.12 MeV) and allows direct analysis without chemical separation.

Table IV lists the detection limits obtained for some elements by neutron activation. In the majority of cases, the values result from (n, γ)-reactions, and for those with poor sensitivity, from fast-neutron (n, a), (n, p), $(n, 2n)$-reactions.

Table XI lists the detection limits of elements that are more readily analyzed by fast neutron activation analysis than by irradiation with thermal neutrons. While the table values indicate a desirable choice of irradiation technique, other analytical conditions, especially the matrix and the concentration level of the unknown, are important. For example, it may happen that the same radioisotope is produced by two different elements, such as:

$$^{16}_{8}O \ (n, p) \ ^{16}_{7}N \qquad \text{cross section 50 mb at 14.5 MeV} \qquad (9)$$

$$^{19}_{8}F \ (n, \alpha) \ ^{16}_{7}N \qquad \text{cross section 140 mb at 14.5 MeV.} \quad (10)$$

For the analysis of oxygen in the presence of a small amount of fluorine it is advisable to determine fluorine separately. Table XII lists some examples of analysis with the use of 14 MeV fast-neutron irradiation with the minimum necessary neutron energy, the half-life of the radioisotope produced and the interferences.

In practice, fast-neutron activation has recently been applied for the determination of light elements, particularly O, N, Be, Fe because of the

Table XI. Comparative Detection Limits of Irradiations with Thermal and with Fast Neutrons (n, α), (n, 2n), (n, p)[a]

Element	Isotope	Isotope Abundance	Nuclear Reaction Cross Section (barns)	Isotope Produced	Half-life	Detection Limit (μg)
Al	27	1	(n, γ) 0.210	Al 28	2.37 min	4×10^{-3}
			(n, α) 116 mb (14 MeV)	Na 24	15 h	6×10^{-3}
Au	197	1	(n, γ) 96	Au 198	2.7 d	5×10^{-5}
			(n, 2n) 1.72 (14 MeV)	Au 196	14 h	3×10^{-3}
Br	79	0.505	(n, γ) 10.4	Br 80	4.4 h	5×10^{-4}
			(n, 2n) 1.14 (14 MeV)	Br 78	6.4 min	4×10^{-3}
	81	0.494	(n, 2n) 0.83 (14 MeV)	Br 80	17.6 min	5×10^{-3}
Cl	35	0.754	(n, α) 0.19 (14 MeV)	P 32	14.2 d	9×10^{-3}
	37	0.246	(n, γ) 0.56	Cl 38	37.2 min	7×10^{-3}
Cu	63	0.691	(n, 2n) 0.51 (14 MeV)	Cu 62	9.9 min	5×10^{-3}
	65	0.309	(n, γ) 1.8	Cu 66	5.10 min	3×10^{-3}
F	19	1	(n, γ) 9 mb	F 20	11 sec	6×10^{-2}
			(n, p) 0.135 (14 MeV)	O 19	29 sec	4×10^{-3}
			(n, 2n) 61 mb (14 MeV)	F 18	1.87 h	9×10^{-3}
Fe	54	0.0584	(n, γ) 2.5	Fe 55	2.6 y	5
	56	0.917	(n, p) 0.11 (13.5 MeV)	Mn 56	2.58 h	1×10^{-2}
Ga	69	0.602	(n, γ) 1.4	Ga 70	21 min	2×10^{-3}
			(n, 2n) 0.55 (14 MeV)	Ga 68	1.13 h	6×10^{-3}
Ge	70	0.02	(n, γ) 3.9	Ge 71	11 d	3×10^{-3}
	76	0.077	(n, 2n) 1.82 (14 MeV)	Ge 75	1.36 h	1×10^{-2}
I	127	1	(n, γ) 5.6	I 128	25 min	6×10^{-4}
			(n, 2n) 1.12 (14 MeV)	I 126	2.6 h	3×10^{-3}
Mg	24	0.786	(n, p) 0.19 (14 MeV)	Na 24	15 h	4×10^{-3}
	26	0.113	(n, γ) 27 mb	Mg 27	9.5 s	2
Mo	98	0.237	(n, γ) 0.45	Mo 99	67 h	3×10^{-2}
	100	0.096	(n, 2n) 3.79 (14 MeV)	Mo 99	67 h	7×10^{-3}

Table XI, continued

Element	Isotope	Isotope Abundance	Nuclear Reaction Cross Section (barns)		Isotope Produced	Half-life		Detection Limit (µg)
N	14	0.996	(n, 2n) 5.67 mb	(14 MeV)	N 13	10	min	7×10^{-2}
	15	0.0037	(n, γ) 24 µb		N 16	7.4	s	4
Na	23	1	(n, γ) 0.53		Na 24	15	h	2×10^{-3}
			(n, p) 33.9 mb	(14 MeV)	Na 23	38	sec	2×10^{-2}
Ni	58	0.677	(n, p) 0.56	(14 MeV)	Co 58	71	d	2×10^{-2}
	64	0.0116	(n, γ) 1.6		Ni 65	2.65	h	9×10^{-2}
O	16	0.9959	(n, p) 49 mb	(14 MeV)	N 16	7.4	sec	9×10^{-3}
S	32	0.950	(n, p) 0.30	(14 MeV)	P 32	14.2	d	4×10^{-3}
	34	0.042	(n, γ) 0.26		S 35	87	d	0.4
Sb	121	0.572	(n, γ) 6.8		Sb 122	2.8	d	9×10^{-4}
	123	0.427	(n, 2n) 1.25	(14 MeV)	Sb 122	2.8	d	6×10^{-3}
Se	76	0.090	(n, γ) 7		Se 77	17	sec	3×10^{-3}
	82	0.092	(n, 2n) 1.5	(14 MeV)	Se 81	18.4	min	2×10^{-2}
Si	28	0.923	(n, p) 0.22	(14 MeV)	Al 28	2.3	min	4×10^{-3}
	30	0.030	(n, γ) 0.11		Si 31	2.62	h	0.20
Sr	86	0.099	(n, γ) 1.3		Sr 87	2.9	h	2×10^{-2}
	88	0.825	(n, α) 64 mb	(14 MeV)	Kr 85	4.4	h	5×10^{-2}
Te	126	0.187	(n, γ) 0.8		Te 127	9.4	h	2×10^{-2}
	128	0.318	(n, 2n) 0.78	(14 MeV)	Te 127	9.4	h	2×10^{-2}
Ti	48	0.734	(n, p) 92.7 mb	(14 MeV)	Sc 48	44	h	2×10^{-2}
	50	0.053	(n, γ) 0.14		Ti 51	5.8	min	0.20
V	51	0.998	(n, γ) 4.5		V 52	3.77	min	3×10^{-4}
			(n, α) 28.6 mb	(14 MeV)	Sc 48	44	h	5×10^{-2}
			(n, p) 27 mb	(14 MeV)	Ti 51	5.8	min	5×10^{-2}
Zn	64	0.489	(n, p) 0.39	(14 MeV)	Cu 64	12.8	h	1×10^{-2}
	68	0.186	(n, γ) 1		Zn 69	59	min	1×10^{-2}
	90	0.514	(n, p) 0.247	(14 MeV)	Y 90	64.8	h	2×10^{-2}
	96	0.028	(n, γ) 0.10		Zr 97	17	min	0.9

Table XII. Fast Neutron Activation

Element	Nuclear Reaction		Necessary Neutron Energy (MeV)	Radioisotope Half-life		Interference; Element Capable of Producing the Same Isotope
Al	$^{27}_{13}$Al (n, α)	$^{24}_{11}$Na	3.3	15	h	Na, Mg[a]
Cl	$^{35}_{17}$Cl (n, α)	$^{32}_{15}$P	1	14.3	d	P, S[a]
Fe	$^{54}_{26}$Fe (n, p)	$^{54}_{25}$Mn	2	310	d	Mn, Cr[c]
	$^{56}_{26}$Fe (n, p)	$^{56}_{25}$Mn	2.9	2.6	h	Mn, Co[a]
Mg	$^{24}_{12}$Mg (n, p)	$^{24}_{11}$Na	4.9	15	h	Na, Al[a]
Ni	$^{58}_{29}$Ni (n, p)	$^{58}_{27}$Co	2	71	d	Fe[b]
S	$^{32}_{16}$S (n, p)	$^{32}_{15}$P	1	14.3	d	P, Cl[a]
Ti	$^{46}_{22}$Ti (n, p)	$^{46}_{21}$Sc	1	85	d	Sc,[a] Ca[b]
Zn	$^{64}_{30}$Zn (n, p)	$^{64}_{29}$Cu	2	12.8	h	Cu,[a] Ni[b]

[a] Isotopes produced by low-energy thermal or fast neutrons ($<$10 MeV).

[b] Isotopes produced by high-energy protons, α-particles and neutrons ($\geqslant$ 10 MeV).

high sensitivity of the method. A review of applications was published by Csikai (1973). The majority concern the determination of oxygen in metals, especially steels. Silicon is also determined according to the reaction $^{28}_{14}Si$ (n, p) $^{28}_{13}Al$ obtained in a 14 MeV neutron flux. The determination is based on a measurement of the γ-radiation of ^{28}Al at 1.78 MeV, which has a half-life of 2.3 min. Santos et al. (1968) used this method to determine silicon in rocks. Anderson et al. (1964) used 14 MeV neutrons for *in vivo* analyses related with medical and biological research. Lisovskii et al. (1971) determined fluorine and strontium in phosphates after fast-neutron activation.

9.3 Activation With High-Energy Charged Particles

9.3.1 General Remarks

Charged-particle bombardment is a complementary neutron technique allowing analyses that are often more sensitive and more specific than would be possible with neutron activation. The particles employed are the protons ^1_1H, α-particles ^4_2He, deuterons ^2_1H and tritons ^3_1H. The technique applies to the light elements Li, Be, B, C, N, O, F, S, which can be determined in various materials with detection limits approximately 10^{-7}-10^{-9}

As with neutron activation analysis, the characteristics of activation analysis with charged particles are the cross section σ and energy threshold. The determination of an element A bombarded by particles x, yielding an isotope B emitting a radiation y is represented by the equation A(x, y)B. The same interference problem exists as in neutron activation. Isotope B can also originate from the irradiation of an element C in the matrix C(x, y')B.

For example, lithium can be determined from the ^7Be isotope obtained by the reaction $^7Li(p, n)^7Be$, but the ^7Be isotope can also be obtained from proton reactions: $^9Be(p, t)^7Be$, $^{10}B(p, a)^7Be$ and $^{11}B(p, an)^7Be$. In other words boron and beryllium can interfere with the analysis of lithium. However, the interference also depends on the bombarding particle energy.

Nitrogen is determined by the reaction $^{14}N(p, n)^{14}O$ produced with an energy threshold of 6.36 MeV. In practice, bombardment is performed with particles of 10 MeV energy. The ^{14}O isotope can also originate from fluorine, $^{19}F(p, a2n)^{14}O$ and $^{19}F(p, {}^6He)^{14}O$, which have thresholds of 21.8 and 20.8 MeV. Consequently, fluorine will not interfere with the nitrogen determination nor will oxygen which undergoes the reaction $^{16}O(p, t)^{14}O$ (threshold of 21.7 MeV). In practice, it is almost always possible to find activation conditions (type of particles and energy) so that the effect of interfering elements will be almost negligible.

9.3.2 Practical Applications

Table XIII lists the principal nuclear reactions performed by bombardment of light nuclei by protons (p), deuterons (d), helions (a) and tritons (t) (Engelman, 1976). The analysis of light elements was the first application of proton activation.

MacGinley et al. (1973) demonstrated that the majority of elements with an atomic number between 34 and 82 were determinable by proton or deuteron activation, with detection limits of 10^{-3} $\mu g/g$. Table XIV also indicates proton bombardment methods for Ti, Fe, Ni and Zn (Schweikert et al., 1966a and b; Swindle et al., 1973) The sensitivity and detection limit were investigated by Flocchini et al. (1972) and by Umbarger et al. (1972). Lead, which shows little sensitivity for neutron activation, can be determined by proton bombardment with a detection limit of 10^{-2} $\mu g/g$ (Riddle et al., 1974). In addition Tl and Bi are determinable with a limit of 10^{-3} $\mu g/g$, provided the matrix contains no interfering elements (Riddle et al., 1973).

Nondestructive analysis is widely used. Bankert et al. (1973) used 14 MeV proton activation to determine B, N, Na, Cr, Se, Br and Cd in water without chemical separation. Delmas et al. (1976) used 10 MeV proton activation to determine trace elements nondestructively in rocks and silica minerals. The specimens are bombarded for 30 min and then measured in the γ-spectrometer with a Ge(Li) detector and a multichannel analyzer. The authors have obtained good results for the trace elements Li, V, Cr, Fe, Ga, Ge, As, Br, Rb, Sr, Y and Zr. After comparing proton and neutron activation, they recommend proton activation for the elements Li, Ti, V, Fe, Ga, Ge, Sr, Y and Zr and neutron activation for Na, Sc, Cr, Co, Rb, Cs, Ba, Eu and Ta. The following detection limits were obtained for the types of minerals investigated (feldspar, magnesite, quartz):

> Li 0.2-4 ppm; Ti 0.7-25; V 0.7; Fe 0.6-70; Ga 1.4-6;
> Ge 0.35-3; Sr 0.7-20; Y 0.25-0.5; Zr 20.

The glass standards of the National Bureau of Standards, SRM 610, 612 and 614, which contain 500, 50 and 1 ppm of various trace elements, were analyzed by MacGinley et al. (1976) by activation in a 20 MeV proton flux and direct spectrometry of the X-ray lines (generally K_a) of the radionuclides formed. In the SRM 610 sample, which was

Table XIII. Nuclear Reactions Used for the Analysis of Light Elements by Charged-Particle Activation

Element	Nuclear Reaction	Energy Threshold (MeV)	Energy Employed in Practice (MeV)	Radioisotope Half-life Produced	Interferences— Element Capable of Producing the Same Isotope	Detection Limit ($\mu g/g$)
Li	$^7Li\,(p, n)\,^7Be$	1.9	14	53.4 d	B	100
	$^6Li\,(d, n)\,^7Be$	$>0^a$	25	53.4 d	Be, B	25
	$^7Li\,(d, 2n)\,^7Be$	5		53.4 d	Be, B	
Be	$^9Be\,(\alpha, 2n)\,^{11}C$	18.8	40	20.3 min	B, C, N, O	3
B	$^{11}B\,(p, n)\,^{11}C$	3	10	20.3 min	N	0.25
	$^{10}B\,(d, n)\,^{11}C$	>0	15	20.3 min	N	0.75
	$^{11}B\,(d, 2n)\,^{11}C$	5.9	11	20.3 min	N	
N	$^{14}N\,(p, n)\,^{14}O$	6.35	10	71 sec	B	10
	$^{14}N\,(p, \alpha)\,^{11}C$	3.1	10	20.3 min	B	0.5
	$^{14}N\,(d, n)\,^{15}O$	>0	15	2.03min	O	0.5
O	$^{16}O\,(d, \alpha)\,^{13}N$	5.5	10	9.96min	C, N	3
	$^{16}O\,(d, n)\,^{17}F$	1.8	15	66 sec	F, Ne	0.1 - 1
	$^{16}O\,(t, n)\,^{18}F$	>0	10	109.7 min	Ne	0.1 - 1
F	$^{19}F\,(p, d)\,^{18}F$	8.6	15	109.7 min	O, Ne	2
	$^{19}F\,(p, pm)\,^{18}F$	11	15	109.7 min	O, Ne	
	$^{19}F\,(d, t)\,^{18}F$	4.6	15	109.7 min	O, Ne	2 - 5
	$^{19}F\,(d, dn)\,^{18}F$	11.5	15	109.7 min	O, Ne	
S	$^{33}S\,(p, \gamma)\,^{34m}Cl$	>0	15	32 min	Cl, Ar	30
	$^{34}S\,(p, n)\,^{34m}Cl$	6.3	15			

[a] > 0 Means that the reaction occurs with approximately zero energy.

Table XIV.

Element	Reactions	Radioisotope Half-life	Irradiation Energy	Detection Limit (μg/g)
Fe	^{56}Fe (p, n) ^{56}Co	77.3 d	17 MeV	10^{-2}
Ni	^{58}Ni(p, pn) ^{57}Ni	36 h	30 MeV	10^{-3}
Ti	^{48}Ti(p, n) ^{48}V	16.2 d	17 MeV	10^{-3}
Zn	^{66}Zn(p, n) ^{66}Ga	9.4 h	30 MeV	10^{-3}

activated for 15 min (0.5 μA), 26 elements were easily determined. The SRM 612 sample (50 ppm) was irradiated for 30 min at 0.5 μA and the elements were determined by 300 min of counting. Twenty-seven elements were easily determined. The SRM 614 sample (1 ppm) was bombarded for 1-2 h at 1-2 μA. Only seven elements could be detected and analyzed: Er, Fe, Gd, Hf, Pb, Sr and W. After 2-3 days of irradiation, a high spectral background remained, which prevented the detection of numerous elements at a concentration level of 1 to a few ppm.

Parsa *et al.* (1974) determined lead in atmospheric dust. The sample was irradiated with ^{3}He. The isotope formed (^{207}Po) was measured after radiochemical separation. The detection limit was 0.5 ppb in the solid matrix.

Another interesting application is surface analysis. In fact, the 8 MeV helions ($^3_2$He) practically do not penetrate the irradiated material. Aluminum irradiation penetrates only a surface layer of 50 μm. The penetration of 6.7 MeV deuterons is 100 μm and that of 40 MeV α-particles is 0.3 mm. These properties are utilized for the determination of light elements, such as C, O, B, N and Cl, in thin films or at the surface of samples consisting of intermediate or heavy elements.

Wolicki (1972) recently reviewed the principal analytical applications of charged-particle activation.

9.4 High-Energy γ-Photon Activation Analysis

We have seen that γ-photons can produce nuclear reactions. Two processes must be distinguished, *i.e.,* nuclear photoexcitation and photonuclear reactions.

9.4.1 Photoexcitation

In nuclear photoexcitation, the irradiated nuclei do not undergo transmutation but are excited, yielding a radioisotope isomer of the analyzed element. Return to the stable state (deexcitation) takes place

with γ-photon emission. The nuclear reaction is written as $A(\gamma, \gamma')A'$.
The energy thresholds are generally less than 1 MeV.

While the method is not always highly sensitive, it is characterized
by simplicity and applicability to routine analysis.

The elements that can be determined by nuclear photoexcitation
are listed in Table XV. The detection limits in this table (Engelman, 1976)
have orders of magnitude corresponding to the bombardment of 10 g
in a 7-8 MeV photon beam of 100 μA intensity for 10 min. Abrams
et al. (1967) have determined Br, Ag and In, and Kodiri et al. (1969)
have measured Se, Y, Ag, Ba, Hf and A . The same authors (1972)
determined hafnium in ores.

9.4.2 Photonuclear Reactions

Photonuclear reactions can be produced by bombarding with high-
energy γ-photons (more than 8 MeV), resulting in isotopes of the bombarded element according to the reactions (γ, n) and $(\gamma, 2n)$ or in chemically different elements according to the reactions (γ, p), $(\gamma, 2p)$, (γ, pn),
(γ, a), (γ, an). These reactions permit the determination of nearly all
elements with a higher sensitivity than does photoexcitation. Bremsstrahlung of 30 MeV, electron accelerators and γ-photon generators permit
utilization of the reactions (γ, n) and (γ, p) for the determination of a
large number of trace elements. Combination with a high-resolution γ-
spectrometer allows nondestructive analysis with high sensitivity.

The technique was used first for light elements, C, N, O and F,
according to the reactions given in Table XVI. The determination of
these elements is very specific and generally free of interference. However,
high-energy γ-photon activation is also applicable to other elements and
has certain advantages over the other nuclear methods. First of all, it is
possible to select a nuclear reaction by the choice of the incident γ-
photon energy. The γ-rays penetrate the bulk of the material. Consequently, samples of several grams can be activated.

Engelmann et al. (1967), Lutz (1969), Debrun et al. (1969) and
Kato (1973) reported the detection limits for a few elements listed in
Table XVII as obtained by bombardment with a 30-40 MeV electron beam
of 100 μA mean intensity for 1 hr. The detection limits can be further
improved by bombarding the sample in the electron flux from bremsstrahlung of 72 MeV (Kato et al., 1972) and 110 MeV (Sastri 1971).

Applications are constantly developing. They have been reviewed by
Baker (1967), Debrun et al. (1969, 1972) and Lutz (1971). Only a few
examples will be described. Wilkniss et al. (1968) determined fluorine in
seawater with 5 g of sample, obtaining a detection limit of 0.001 ppm.

Table XV. Elements Determined by Nuclear Photoexcitation [A (γ, γ') A$'$]

Element	Isotope A (stable)	Isotope Abundance	Active Isomer Half-life A$'$	γ'-Ray Energy (MeV)	Detection Limit ($\mu g/g$)
Ag	^{107}Ag	0.513	44.3 s	0.093	10
Au	^{197}Au	1	7.8 s	0.279	1
Br	^{79}Br	0.505	4.8 s	0.208	5
Ba	^{135}Ba	0.0659	28.7 h	0.268	100
	^{137}Ba	0.112	2.55 min	0.661	
Cd	^{111}Cd	0.127	48.6 min	0.150-0.245	5
Hg	^{199}Hg	0.168	42.6 min	0.158-0.374	200
Hf	^{180}Hf	0.352	5.5 h	0.057-0.093	1
In	^{115}In	0.957	4.5 h	0.336	5
Ir	^{191}Ir	0.373	4.9 s	0.129	20
Pt	^{195}Pt	0.338	4.1 d	0.099-0.129	50
Se	^{77}Se	0.076	18.1 s	0.161	10
Sr	^{87}Sr	0.070	2.8 h	0.388	100
Y	^{89}Y	1	16.1 s	0.909	50
W	^{183}W	0.144	5.3 s	0.099-0.102	200

Table XVI. Nuclear Reactions (γ, n) for the Analysis of Light Elements

Element	Nuclear Reactions	Energy Threshold (MeV)	Radionuclide Half-life	Beam Energy	Detection Limit (μg)
C	^{12}C (γ, n) ^{11}C	18.7	20.3 min	35 MeV	0.05
N	^{14}N (γ, n) ^{13}N	10.6	9.96 min	35 MeV	0.15
O	^{16}O (γ, n) ^{15}O	15.7	2.03 min	35 MeV	0.05
F	^{19}F (γ, n) ^{18}F	10.4	109.7 min	35 MeV	0.04

Table XVII. γ-Photon Activation (30-40 MeV)[a]

Element	Detection Limit (ng)	Element	Detection Limit (ng)
Au	1	Sc	20
Ag	10	Sr	0.8
As	10	Ta	10
Cu	1	Ti	2
Ga	5	Tl	10
I	10	Zn	5
Ni	10	Zr	10
Pb	100		
Sb	20		

[a]These detection limits are absolute values.

Kato *et al.* (1973) utilized the (γ, n) and (γ, p) reactions obtained from 30 MeV generators (bremsstrahlung) to determine 16 trace elements in standard rocks by nondestructive analysis. The same was done by Das *et al.* (1973), who analyzed Mg, Ca, Ti, Mn, Sr and Nb in rocks. Ratynski *et al.* (1973) determined trace elements in copper ores. A review of applications in biology and environmental pollution, dealing particularly with the elements F, I, Pb, Sr and Be, was published by Hislop (1973). Olmez *et al.* (1974) analyzed atmospheric dust samples and determined Si, Rb and Y from 2500 m³ sample of air. The filter was pelletized and bombarded in the electron flux of 35-40 MeV bremsstrahlung at 50 μA.

The recent study of Kato *et al.* (1976) on the nondestructive analysis of atmospheric dust constitutes a typical practical example of photon activation. The samples were filtered on a millipore filter, and a piece of the filter (20 x 12.5 cm) was compressed in a press to form a pellet of 13 mm diameter and 5 mm thickness. This was bombarded with a bremsstrahlung source in a 30 MeV electron beam of 70 μA intensity for 6 hr. The counts were made in a multichannel γ-spectrometer with a Ge(Li) detector.

Table XVIII summarizes the analytical conditions for some elements. The detection limits correspond to a 500 m³ sample of air. They are expressed in ng of element per m³ producing a photoelectric peak equal to three times the standard deviation of the background. The sensitivity and precision of the method are adequate for more than 20 elements analyzed in the same sample. In particular, this table contains information on interfering elements and the nuclear reactions leading to the isotope of the analyzed element. Such interference must be anticipated with elements present in large amounts. For example, the presence of a

Table XVIII. Photon Activation Analysis of Atmospheric Dust

Element	Nuclear Reaction	Stable Isotope Abundance	Radioisotope Half-life	Photoelectric Peak γ-Energy (keV)	Time Interval for Measurement after Irradiation	Practical Detection Limit (ng/m^3)	Interfering Element and Nuclear Reaction
As	^{75}As (γ, n) ^{74}As	1	17.9 d	596	10-15 d	0.4	^{76}Se (γ, pn) ^{74}As
Ce	^{140}Ce (γ, n) ^{139}Ce	0.885	140 d	166	30-40 d	0.5	^{141}Pr (γ, pn) ^{139}Ce
Cl	^{35}Cl (γ, n) ^{34m}Cl	0.754	32.0 min	2130	30-60 min	90	^{39}K $(\gamma, \alpha n)$ ^{34m}Cl
Co	^{59}Co (γ, n) ^{58}Co	1	71.3 d	811	30-40 d	0.6	^{60}Ni (γ, pn) ^{58}Co
Cr	^{52}Cr (γ, n) ^{51}Cr	0.837	27.8 d	319	10-15 d	3.6	^{56}Fe $(\gamma, \alpha n)$ ^{51}Cr
Fe	^{57}Fe (γ, p) ^{56}Mn	0.022	2.576 h	847	2- 5 hr	85	^{55}Mn (n, γ) ^{56}Mn
I	^{127}I (γ, n) ^{126}I	1	13 d	388	10-15 d	0.5	^{128}Xe (γ, pn) ^{126}I
Mg	^{25}Mg (γ, p) ^{24}Na	0.101	15.0 d	1368	1- 2 d	13	^{27}Al (n, α) ^{24}Na
Mn	^{55}Mn (γ, n) ^{54}Mn	1	303 d	835	10-15 d	2.5	^{56}Fe (γ, pn) ^{54}Mn
Ni	^{58}Ni (γ, n) ^{57}Ni	0.677	36.0 h	1378	1- 2 d	2.9	None
Pb	^{204}Pb (γ, n) ^{203}Pb	0.013	52 h	279	1- 2 d	12	None
Rb	^{85}Rb (γ, n) ^{84}Rb	0.721	33.0 d	881	10-15 d	0.6	^{86}Se (γ, pn) ^{84}Rb
Sb	^{123}Sb (γ, n) ^{122}Sb	0.427	2.80 d	564	1- 2 d	1.1	^{123}Te (γ, p) ^{122}Sb
Sr	^{88}Sr (γ, n) ^{87m}Sr	0.826	2.83 h	388	2- 5 hr	0.04	^{89}Y (γ, pn) ^{87m}Sr ^{92}Zr $(\gamma, \alpha n)$ ^{87m}Sr
Ti	^{48}Ti (γ, p) ^{47}Sc.	0.734	3.43 d	160	2- 5 hr	5.6.	^{51}V (γ, α) ^{47}Sc ^{48}Ca $(\gamma, n\beta)$ ^{47}Sc
Y	^{89}Y (γ, n) ^{88}Y	1	108 d	1836	10-15 d	0.3	^{90}Zr (γ, pn) ^{88}Y
Zn	^{68}Zn (γ, p) ^{67}Cu	0.186	59 h	185	2- 3 d	20	^{71}Ga (γ, α) ^{67}Cu
Zr	^{90}Zr (γ, n) ^{89}Zr	0.514	78.4 h	910	1- 2 d	1.4	^{94}Mo $(\gamma, \alpha n)$ ^{89}Zr

large amount of nickel interferes with the analysis of cobalt (^{58}Co) because of the reaction ^{60}Ni$(\gamma, pn)^{58}$Co. Another example is the determination of magnesium by formation of the ^{24}Na isotope. This isotope can also be produced simultaneously from aluminum by the reaction ^{27}Al$(n,a)^{24}$Na. It is necessary to calibrate this interference and take it into account in the analysis of Mg as a function of the Al content. Furthermore, the ^{54}Mn isotope, which serves for the analysis of manganese, is also obtained from iron: ^{56}Fe$(\gamma, pn)^{54}$Mn.

In addition spectral interferences must always be anticipated. Calcium forms a peak at 808 keV near that of cobalt (^{58}Co, 811 keV), but if the cobalt reading is taken 30-40 days after activation, the calcium peak has become negligible. Furthermore, cesium (^{139}Ce, 166 keV) may be perturbed by a scandium peak (^{47}Sc, 160 keV). Since the half-life of ^{47}Sc is 3.4 days, it will not interfere with the cesium reading taken 30 days later.

10. TRACE ANALYSIS BY ACTIVATION AND OTHER ANALYTICAL TECHNIQUES

In order to place activation analysis into perspective compared to the other analytical techniques, we might consider two different aspects concerning the comparative properties of the method. One aspect includes characteristics such as precision, sensitivity, sample size, analysis time and the possibility of multielemental analysis.

The intrinsic comparative properties of the methods are summarized in Table XIX. Its data are of a very general nature, not involving chemical pretreatment for enriching the trace elements. The values listed refer to samples analyzed directly after being subjected to a simple physical treatment (for example, pelletizing), or upon being dissolved. This table does not include the use of methods for routine analysis. In fact, the advances made in electronic technology and instrumentation allow an extensive automation of the sample pretreatment procedures (chemical and physical treatment) as well as of the analysis itself.

Another criterion not mentioned in Table XIX is the basic cost of an analysis. Activation analysis is still very expensive compared to other techniques, essentially because of the necessary investment for the equipment. However, the development and multiplicity of applications as well as the appearance of commercial activation sources seen in the last few years should contribute to a signficiant reduction of activation analysis costs.

The second aspect that may be considered in evaluating activation analysis is its position as a function of applications. Several years ago,

Table XIX. Activation Analysis Compared to Other Analytical Techniques for Trace Elements[a]

Methods	Precision	Detection: Necessary Quantity of Element	Sample Size: Necessary Quantity of Sample	Possibility of Nondestructive Analysis	Possibility of Multielemental Analysis	Analysis Time
Activation Analysis	2-10%	$>10^{-12}$ g	a few mg - a few g	yes	yes	a few min to a few weeks
Spark Mass Spectrometry	20-30%	$10^{-3} - 10^{-9}$ g	a few mg	no	yes	a few hours
Molecular Absorption Spectrometry	4-8%	$10^{-5} - 10^{-7}$ g	10 mg - 1 g	no	no	a few hours
Emission Spectrometry (arc, spark, plasma)	5-20%	$10^{-5} - 10^{-6}$ g	10-100 mg	no	yes	1-2 h
Flame Emission and Atomic Absorption Spectrometry	2-5%	$10^{-6} - 10^{-7}$ g	0.1 - 1 g	no	no	1-2 h
Flameless Atomic Absorption Spectrometry	8-12%	$10^{-11} - 10^{-12}$ g	0.1 - 2 mg	no	no	1-2 h
X-Ray Fluorescence	1-2%	$10^{-3} - 10^{-5}$ g	0.1 - 2 g	yes	yes	1 h
Polarography	4-8%	$10^{-6} - 10^{-7}$ g	0.010 - 1 g	no	no	1-2 h

[a]The *precision* applies to the total analysis.

The *detection limit* is the absolute quantity of element necessary to obtain a measurable signal under classical conditions.

The *sample size* is the useful quantity of sample to perform the analysis (one or more elements).

Nondestructive analysis is defined as the possibility of reusing the sample.

The *analysis time* represents the complete analysis, including dissolution. In fact, analyses by atomic and molecular absorption, emission and polarography require only a few minutes.

the International Union of Pure and Applied Chemistry (IUPAC) sponsored a survey on "Trace Analysis Applicable to the Determination of Minor Amounts of Impurities in Chemicals: General Survey" (Pinta, 1974). The objective of this study was to make an inventory of the principal analytical techniques applied to the determination of trace elements. A questionnaire addressed to researchers known for their interest in trace element problems produced about 200 responses. They allowed a classification of the methods in order of numerical importance. Table XX summarizes the results obtained for the principal methods employed. A percentage of laboratories is using each method, and numerous laboratories are employing several methods.

Table XX. Analytical Techniques Used in Trace Analysis, Percentage of Laboratories Carrying Out Trace Analysis

Technique	Percentage
Absorption Spectrophotometry and Colorimetry	41.5
Atomic Absorption Spectrometry and Flame Emission	48.5
Emission Spectrometry (arc and spark)	23
Polarography	18.5
Radioactivation	14
X-Ray Fluorescence	13
Mass Spectrometry	9
Other Methods (electric, chromatographic)	20

The survey also shows the methods preferred for each element. The resulting classification is presented in Table XXI. Thus, for example, in 100 laboratories where arsenic is determined, activation analysis is preferred by 23, absorption spectrometry by 41, emission spectrometry by 15.4, polarography by 12.8 and X-ray fluorescence by 7.8.

The results of this survey, though of interest, correspond to the situation in 1972-1973. Since then, the development of certain instrumental techniques has modified these data. Molecular absorption as well as emission spectrometry have been losing favor to atomic absorption spectrometry with a flame and especially of the flameless type. Activation analysis, especially with the development of charged-particle and γ-photon activation, continues to show advances. However, activation analysis, particularly neutron activation, is retaining its individual characteristics, which confer upon its own field of application in which other methods are often inefficient. We might note the analysis of microsamples requiring the determination of traces at the ppm level and lower. Until the appearance of flameless atomic absorption spectrometry, only spark mass spectrometry could rival activation analysis (Roth, 1968).

Table XXI. Techniques Used for Each Particular Element
(Percentage of Laboratories)

	AA[a]	AS	AAS EF	ES	P	XRF
As	23	41	---	15.4	12.8	7.8
Au	50	19	31	---	---	---
Bi	20	---	---	50	30	---
Co	10	22	30	27	11	---
Cr	8	21	23	29	11	8
Cs	25	---	---	25	---	50
Cu	9	19	33	15.5	15.5	8
Fe	7.5	32.5	32.5	16	4	7.5
K	16	---	84	---	---	---
Mn	5.5	20	35	19	11	9.5
Mo	8	32	13	30	11	6
Na	22	---	78	---	---	---
Ni	10	20	29	27	10	4
Sb	25	30	---	25	20	---
Si	13	45	---	42	---	---
Sn	8.5	23	---	46	14	8.5
W	31	38	---	31	---	---
Zn	9	8	41	15	18	9

[a]AA: activation analysis
  AS: absorption spectrometry and colorimetry
  AAS-EF: atomic absorption spectrometry and flame emission
  ES: emission spectrometry (and spectrography)
  P: polarography
  XRF: X-ray fluorescence

Finally, it may be said that activation analysis is certainly the method in which the risks of contamination can be minimized most easily, and chemists involved in trace analysis unanimously recognize that contamination, with all of its possible sources, is the first cause of analytical errors. At the trace level, this error is often of random nature and unpredictable. It requires precautions that must be more rigid the lower the concentration level of the analyzed trace elements. In activation analysis, in which activation is performed in the first step, the subsequent procedures are no longer disturbed by this cause of error.

In conclusion, among physical analysis methods, activation analysis should not become a competing method but, to the contrary, a complementary technique, utilized as a function of its specific properties. In fact, it is being considered in this light among the available physical or chemical analysis methods and sometimes even as a reference technique.

REFERENCES

Abdullaev, K., B. B. Zakhvataev and V. P. Perelygin. *Radiobiolog.* 8:168 (1968a).

Abdullaev, A. A., E. S. Gureev, V. A. Grakhov, L. I. Zhuk and A. Sh. Zakhidov. *Izvest. Akad. Nauk Uzbek. SSR., fiz.-mat. Nauk* 12:59 (1968b).

Abdullaev, A. A., E. B. Sharipov, U. Khudajbergenov and A. S. Khasanov. *Izvest. Akad. Nauk uzbek. SSR., Fiz.-mat. Nauk* 15:71 (1971).

Abrams, I. A. and L. L. Pelekis. *Latv. PSR Zinat. Akad. Vestis. Fiz. Tech. Zinat. Ser.* 5:45 (1967).

Akbarov, U., B. R. Allabergenov, P. Kh. Nishanov, U. Uzakova and K. Umirbekov. *Zavodsk. Lab. SSSR.* 39:551 (1973).

Albert, P. *Activation Analysis* (Paris: Gauthier Villars, 1964).

Albu Yaron, A., A. Nitzan, and S. Amiel. *Israel Atomic Energy Comm.* IA 1190:117 (1969).

Alian, A., R. Shabana, W. Sanad, B. Allam and K. Khalifa. *Talanta* 15:263 (1968).

Alian, A. and W. Sanad. *Radiochem. Radioanal. Letter* 17:155 (1974).

Al-Shahristani, H. and M. J. Al-Atyia. *J. Radioanal. Chem.* 14:401 (1973).

Anastase, S. and V. Cercasov. *An. Univ. Bucuresti, Chim. Roman* 18:111 (1969).

Anderson, J. *Lancet* 1201 (1964).

Artyukhin, P. I., E. N. Gilbert and V. A. Pronin. *Tr. Kom. Anal. Khim. Akad. Nauk. SSSR, Inst. Geokhim. Anal.* 16:169 (1968).

Aubouin, G., H. Dabrowski, J. Laverlochere and J. Vial. *Trace Element Analysis in Rocks and Other Natural Mineral Substances* (Paris: CNRS, 1970), p. 331.

Babaev, A. and A. Khaidarov. *Dolk. Akad. Nauk Tadzh. SSR* 13:21 (1970).

Bagdavadze, N. V. and L. M. Mosulishvili. *J. Radioanal. Chem.* 24:65 (1975).

Baker, C. A. *Analyst* 92:601 (1967).

Bando, S. and T. Imahashi. *Bunseki Kagaku* 20:49 (1970).

Bankert, S. F., S. D. Bloom and G. D. Sauter. *Anal. Chem.* 45:692 (1973).

Bate, L. C. in *Nuclear Methods in Environmental Research, Proc. American Nuclear Society Topical,* Columbia, Missouri (1971), p. 226.

Bate, L. C. *J. Radioanal. Chem.* 15:193 (1973).

Becker, R. R., A. Veglia and E. R. Schmid. *Radiochem. Radioanal. Letter* 19:343 (1974).

Berkutova, I. D., E. S. Viunova, T. I. Jalnina, I. M. Zlotova and K. I. Yakubson. *J. Radioanal. Chem.* 18:119 (1973).

Berzina, I. G. and S. V. Malinko. *Dolk. Akad. Nauk SSSR.* 189:849 (1969).

Block, C. and R. Dams. *Anal. Chim. Acta* 68:11 (1974).

Blotcky, A. J., D. Hobson, J. A. Leffler, E. P. Rack and R. R. Recker. *Anal. Chem.* 48:1084 (1976).

Bressan, D. J., R. A. Carr, P. J. Hannan and P. E. Wilkniss. *J. Radioanal. Chem.* 19:373 (1974).

Brooks, C. K. *Radiochim. Acta* 9:157 (1968).
Brune, D. *Sci. Total Environment* 2:111 (1973).
Brunfelt, A. O. and E. Steinnes. *Anal. Chim. Acta* 41:155 (1968a).
Brunfelt, A. O. and E. Steinnes. *Analyst* 93:286 (1968b).
Brunfelt, A. O., I. Roelandts and E. Steinnes. *Analyst* 99:277 (1974).
Buenafama, H. D. *J. Radioanal. Chem.* 18:111 (1973).
Clemente, G. F. and G. G. Mastinu. *J. Radioanal. Chem.* 20:707 (1974).
Cohn, S. H. and C. S. Dombrowski. *J. Nucl. Med.* 12:499 (1971).
Cornelis, R., A. Speecke and J. Hoste. *Anal. Chim. Acta* 68:1 (1973).
Craft, T. F. *Radiochem. Radioanal. Letter* 8:199 (1971).
Csikay, J. *Atomic Energy Review,* Vienna, IAEA 11 (1973).
Dams, R., K. A. Rahn and J. W. Winchester. *Environ. Sci. Technol.*
 6:441 (1972).
Dams, R., K. A. Rahn, G. D. Nifong, J. A. Robbins and J. W. Winchester.
 in *Nuclear Methods in Environmental Research, Proc. American Nuclear
 Society Topical*, Columbia, Missouri (August 1971).
Darcourt-Rieg, C. *Study of Suspended Matter in the Deep Water of the
 Atlantic Ocean: Trace Element Concentrations Measured by Neutron
 Activation; Comparison With Sediment,*, Thesis, Faculty of Science,
 Paris (1973).
Das, H. A., N. de Graaff, D. Hoede and J. Zonderhuis. *Radiochem.
 Radioanal. Letter* 4:307 (1970).
Das, H. A. and H. H. DeVries. *Reactor Centrum Nederland* 136:10
 (1971).
Das, H. A., G. A. V. Gerritsen, D. Hoede and J. Zonderhuis. *J. Radioanal.
 Chem.* 14:415 (1973).
Debrun, J. L. and P. Albert. *Bull. Soc. Chim. Fr.* 3:1020 (1969).
Debrun, J. L., D. C. Riddle and E. A. Schweikert. *Anal. Chem.* 44:1386
 (1972).
Delcroix, G. and J. C. Philippot. *J. Radioanal. Chem.* 15:87 (1973).
Delmas, R., J. N. Barrandon and J. L. Debrun. *Analusis* 4:339 (1976).
Doctor, Z. K. K. and B. C. Haldar. *J. Indian Chem. Soc.* 46:295 (1969).
Dutilh, C. E. and H. A. Das. *Radiochem. Radioanal. Letter* 6:195
 (1971).
Engelmann, C. and D. Y. Jerome. *Proceedings 2nd Conference Practical
 Aspects Activation Analysis Charged Particles*, Liege, Publication EUR
 3896 d-f-e 1968, 119, Comm. Communautes Eur. Bruxelles (1967).
Engelmann, C. and M. Poncet. DGRST-72 70 857, Fr., 1 (1973).
Engelmann, C. "Activation Analysis with Charged Particles and γ-Photons,
 in *Engineering Techniques, Measurements and Analyses* (Paris, 1976),
 p. 2575.
Fedoroff, M. *J. Radioanal. Chem.* 15:435 (1973).
Filby, R. H., A. I. Davis, K. R. Shah and W. A. Haller. *Mikrochim. Acta*
 6:1130 (1970).
Filby, R. H. and K. R. Shah. *Proc. American Nuclear Society Topical*
 Columbia, Missouri (August 1971), p. 86.
Filby, R. H. and K. R. Shah. *Toxic. Environ. Chem. Rev.* 2:1 (1974).
Flocchini, R. G., P. J. Feeney, R. J. Sommerville and F. A. Chahill.
 Nucl. Instrum. Methods 100:397 (1972).
Gilfrich, J. V., P. G. Burkhalter and L. S. Birks. *Anal. Chem.* 45:2002
 (1973).

Glazov, V. M.   *Trudy Khim. Tekhnol.* SSSR 3:85 (1968).

Glukhov, G. G. and E. N. Gil'bert   *Radiokhimija,* SSSR 12:533 (1970a).

Glukhov, G. G. and E. N. Gil'bert   *Radiokhimija,* SSSR 12:534 (1970b).

Goode, G. C., G. A. Vood, N. M. Brooke and R. F. Coleman.   *AWRE-O-24/71,* Atom. Weapons Research Establishment (1971), p. 32.

Greenland, L. P. and G. G. Goles.   *Geochim. Cosmochim. Acta* 29:1285 (1965).

Greenland, L. P.   *Anal. Chim. Acta* 42:365 (1968).

Grimanis, A. P.   *Talanta* 15:279 (1968).

Guinn, V. P. and C. D. Wagner.   *Anal. Chem.* 32:317 (1960).

Guinn, V. P.   in *The Encyclopedia of Physics,* R. M. Besancon, Ed. (New York: Reinhold Pub. Corp., 1966).

Guinn, V. P., D. E. Bryan and H. R. Lukens.   *Nuclear Techniques in Environmental Pollution,* Vienna, IAEA (1971).

Guy, R. D. and B. D. Pate.   *J. Radioanal. Chem.* 15:135 (1973).

Hamaguchi, H.   *Bunseki Kiki* 703 (1969).

Health Lab. Sci. 10:251 (1973).

Heidinga, M. C., J. H. Koeman, J. J. M. De Goeij, C. Zegers, J. H. P. Verwey, W. Van Driel and A. J. De Groot.   Reprint from *TNO-Nieuws* 26:382 (1971).

Heindryckx, R. and R. Dams.   *Radiochem. Radioanal. Letter* 16:209 (1974).

Henke, G., H. Mollmann and H. Alfer.   *Z. Neurol.* 199:283 (1971).

Henzler, T. E., R. J. Korda, P. A. Helmke, M. R. Anderson, M. M. Jimenez and L. A. Haskin.   *J. Radioanal. Chem.* 20:649 (1974).

Hislop, J. S.   in *Int. Conf. Photonucl. React. Appl. Proc.* (Livermore, Calif.: Lawrence Livermore Lab., 1973), p. 1159.

Hogdahl, O. T., S. Melson and V. T. Bowen.   *Trace Inorganics in Water, Advances in Chemistry* Ser. 73 (Washington, D.C.: American Chemical Society, 1968).

Holmes, A., A. Morgan and F. J. Sandalls.   *Amer. Ind. Hyg. Assoc. J.* 32:281 (1971).

Jaffrezic, H., A. Decarreau, J. P. Carbonnel and N. Deschamps.   *J. Radioanal. Chem.* 18:49 (1973).

Johansen, O. and E. Steinnes.   *Anal. Chim. Acta* 40:201 (1968).

Johansen, O. and E. Steinnes.   *Talanta* 17:407 (1970).

Kanabrocki, E. L., L. F. Case, L. Graham, T. Fields, Y. T. Oester and E. Kaplan.   *J. Nucl. Med.* 9:478 (1968).

Kato, T. and Y. Oka.   *Talanta* 19:515 (1972).

Kato, T.   *J. Radioanal. Chem.* 16:307 (1973).

Kato, T., I. Morita and N. Sato.   *J. Radioanal. Chem.* 18:97 (1973).

Kato, T., N. Sato and W. Suzuki.   *Talanta* 23:517 (1976).

Keays, R. R., R. Ganapathy, J. C. Laul, U. Kraehenbuehl and J. W. Morgan.   *Anal. Chim. Acta* 72:1 (1974).

Khajdarov, A. and S. M. Babakhodzhaev.   *Dokl. Akad. Nauk. Tadzhik. SSR* 17:28 (1974).

Kodiri, S. and L. P. Starchik.   *Dokl. Akad. Nauk. Tadzhik. SSR* 12:17 (1969).

Kodiri, S., I. A. Abrams, L. L. Pelekis. and L. P. Starchik.   *Atom. Energ. SSR* 32:428 (1972).

Kosta, L. and V. Ravnik. *Radiochem. Radioanal. Letter* 7:295 (1971).
Krishnan, S. S. *J. Radioanal. Chem.* 15:165 (1973).
Lag, J. and E. Steinnes. in *Nuclear Technique in Environmental Pollution*, International Atomic Energy Agency (1971), p. 429.
Lavrakas, V., T. J. Golembeski, G. Pappas, J. E. Gregory and H. L. Wedlick. *Anal. Chem.* 46:952 (1974).
Linstedt, K. D. and P. Kruger. *J. Amer. Water Works Assoc.* 61:85 (1969).
Lisovskii, I. P., L. A. Smakhtin, M. M. Vinnik and S. N. Zakharova. *Khim. Prom.* 47:235 (1971).
Lobanov, E. M. and D. Nurmatov. *Dokl. Akad. Nauk Uzb. SSR* 3:17 (1970).
Lombard, S. M., K. W. Marlow and J. T. Tanner. *Anal. Chim. Acta* 55:13 (1971).
Lucas, H. F. and F. Markun. *ANL*-7760, 47, Argonne Nat. Lab (1970a).
Lucas, H. F., D. N. Edgington and P. J. Colby. *J. Fish. Res. Board Can.* 27:667 (1970b).
Lukens, H. R. and J. John. *Adv. X-Ray Anal.* 16:10 (1972).
Lutz, G. J. *Anal. Chem.* 41:424 (1969).
Lutz, G. J. *Anal. Chem.* 43:92 (1971).
MacGinley, J. R. and E. A. Schweikert. *J. Radioanal. Chem.* 16:385 (1973).
MacGinley, J. R. and E. A. Schweikert. *Anal. Chem.* 48:429 (1976).
Mahler, D. J., A. E. Scott, J. R. Walsh and G. Haynie. *J. Nucl. Med.* 739 (1970).
Margosis, M., J. T. Tanner and J. P. F. Lambert. *J. Pharm. Sci.* 60:1550 (1971).
Matthews, A. D. and J. P. Riley. *Anal. Chim. Acta* 51:287 (1970).
May, S. "Neutron Activation Analysis," in *Engineering Techniques, Measurements and Analyses* (Paris, 1975), p. 2565.
May, S. "Development of the Systematic Determination of Trace Elements in Seawater (Specifically Mercury, Cadmium, Copper, Chromium, Zinc and Arsenic) by Neutron Activation Analysis," DGRST-7371501, Fr. 1 (1974).
Mayer, W. A. *Erdoel Kohle Erdgas Petrochem.* 24:416 (1971).
Maziere, B., J. Gros and D. Comar. *J. Radioanal. Chem.* 24:279 (1975).
Merlini, M. "Internation. Symposium Application Neutron Activation Analysis in Oceanography," Bruxelles, June, Inst. Royal Sci. Naturelles, Belgique (1968).
Meyer, H. G. *J. Radioanal. Chem.* 7:67 (1971).
Michelsen, O. B. and E. Steinnes. *Talanta* 15:574 (1968).
Millard, H. T. and A. J. Bartel. in *Activation Analysis in Geochemistry and Cosmochemistry*, A. O. Brunfelt and E. Steinnes, Eds. Proc. NATO Advanced Study Institute Kjeller, Norway (1970), p. 353.
Morgan, D. J. and A. Black. *AERE-PR/HPM-12* 44, U. K. Atomic Energy Authority Res. Group, Atomic Energy Res. Establ. (1968).
Nadkarni, R. A. and W. D. Ehmann. in *Trace Substances in Environmental Health, IV, Proc. 4th Annual Conference Trace Substances Environmental Health*, Univ. of Missouri (1971), p. 407.
Nadkarni, R. A. and B. C. Haldar. *Radiochem. Radioanal. Letter* 9:205 (1972).

Nadkarni, R. A. *Radiochem. Radioanal. Letter* 19:17 (1974).
Nadkarni, R. A. and G. H. Morrison. *Anal. Chem.* 46:232 (1974).
Navarrete, M., L. Galvez, E. Tzontlimatzin and C. Aguilar. *Radiochem. Radioanal. Letter* 19:163 (1974).
Ohno, S. and M. Yatazawa. *Radioisotopes, Jap.* 19:565 (1970).
Ohno, S. *Analyst* 96:423 (1971).
Olmez, I., N. K. Aras, G. E. Gordon and W. H. Zoller. *Anal. Chem.* 46:935 (1974).
Parsa, B. and S. S. Markowitz. *Anal. Chem.* 46:186 (1974).
Passaglia, A. M. and F. W. Lima. *Rev. Bras. Technol., Bras.* 4:31 (1973).
Peisach, M., D. Comar and C. Kellershohn. *Radiochem. Radioanal. Letter* 8:267 (1971).
Persiani, C. and J. F. Cosgrove. *Electrochem. Technol.* 6:205 (1968).
Pillay, K. K. S., C. C. Thomas, Jr. and G. F. Mahoney. *J. Radioanal. Chem.* 15:33 (1973).
Pillay, K. K. S., C. C. Thomas, Jr. and C. M. Hyche. *J. Radioanal. Chem.* 20:597 (1974).
Pinta, M. *Pure Appl. Chem.* 37:483 (1974).
Piper, D. Z. and G. G. Goles. *Anal. Chim. Acta* 47:560 (1969).
Randa, Z., J. Benada, J. Kuncir and J. Kourimsky. *J. Radioanal. Chem.* 14:437 (1973).
Ratynski, W., A. Stegner and Z. Sujkowski. in *Nucl. Data Sci. Technol. Proc. Symp.*, Paris, Vienna, International Atomic Energy Agency, 2:411 (1973).
Renfro, W. B. and W. A. Jester. *J. Radioanal. Chem.* 15:79 (1973).
Riddle, D. C. and E. A. Schweikert. *Anal. Chem.* 46:395 (1974).
Riddle, D. C. and E. A. Schweikert. *J. Radioanal. Chem.* 16:413 (1973).
Robertson, D. E. *Anal. Chem.* 40:1067 (1968).
Rosenberg, J. and S. Salmela. *Radiochem. Radioanal. Letter* 21:23 (1975).
Roth, E. *Applied Nuclear Chemistry* (Paris: Masson Cie, 1968).
Rowe, J. J. and F. O. Simon. *Talanta* 18:121 (1971).
Salmon, L. and M. G. Creevy. in *Nuclear Techniques in Environmental Pollution* (Vienna: IAEA, 1971).
Santos, G. G. and R. E. Wainerdi. *J. Radioanal. Chem.* 1:509 (1968).
Sastri, M. N. in *Vistas in Analytical Chemistry* (New Delhi: S. Chand and Co., Ltd., 1971)
Schiller, P. and G. B. Cook. in *Nuclear Techniques for Mineral Exploration and Exploitation*, International Atomic Energy Agency (1971), p. 137.
Schmolzer, G. and K. Mullet. *J. Appl. Radiadion Isotopes* 22:509 (1971).
Schneider, J. and R. Geisler. *Z. Anal. Chem.* 267:270 (1973).
Schweikert, E. A. and P. Albert. *C. R. Acad. Sci.* 262:87 (1966a).
Schweikert, E. A. and P. Albert. *C. R. Acad. Sci.* 262:342 (1966b).
Seitner, H., W. Kiesl, F. Kluger and F. Hecht. *J. Radioanal. Chem.* 7:235 (1971).
Shah, K. R., R. H. Filby and W. A. Haller. *J. Radioanal. Chem.* 6:185 (1970).
Sheibley, D. W. *Amer. Chem. Soc. Div. Fuel Chem., Prep.* 18:59 (1973).
Shigematsu, T. *Bull. Soc. Sea Water Sci.* 21:241 (1968).
Shinbori, Y. *Yukagaku* 17:430 (1968).

Shinbori, Y. and T. Tamachi. *Yukagaku* 17:606 (1968).
Simon, F. O. and H. T. Millard. *Anal. Chem.* 40:1150 (1968).
Steinnes, E. *Anal. Chim. Acta* 68:25 (1974).
Stueber, A. M. and G. G. Goles. *Geochim. Cosmochim. Acta* 31:75 (1967).
Swindle, D. L. and E. A. Schweikert. *Anal. Chem.* 45:2111 (1973).
Tajima, E. and H. Akaiwa. *Radioisotopes* 20:165 (1971).
Tamura, M. *Genshiryodu Kogyo* 16:68 (1970).
Thatcher, L. L. and J. O. Johnson. in *Nuclear Techniques in Environmental Pollution* (Vienna: IAEA, 1971).
Thompson, B. A. and E. C. Miller. *J. Research Nat. Bureau Standards* 75A:429 (1971).
Tomura, K., H. Higuchi, N. Miyaji, N. Onuma and H. Hamaguchi. *Anal. Chim. Acta* 41:217 (1968a).
Tomura, K., H. Higuchi, N. Onuma and H. Hamaguchi. *Anal. Chim. Acta* 42:389 (1968b).
Treuil, M., H. Jaffrezic, N. Deschamps, C. Derre, F. Guichard, J. L. Joron, B. Pelletier, S. Novotny and C. Courtois. *J. Radioanal. Chem.* 18:55 (1973).
Turekian, K. K. *Yale*-2912-20:31 (1968).
Turkstra, J., P. J. Pretorius and W. J. De Wet. *Anal. Chem.* 42:835 (1970).
Turkstra, J., H. J. Smit and W. J. De Wet. *J. S. Afr. Chem.* 24:113 (1971).
Umbarger, G. J., R. C. Bearse, D. Close and A. Malanify. *Sensitivity and Detectability Limits for Elemental Analysis by Proton-Induced X-Ray Fluorescence with a 3 MeV Van de Graaff*, La–DC-72-1069. Nat. Tech. Inf. Ser. Springfield (1972).
Vados, I., M. Mohai and E. Upor. *Acta Chim. Acad. Sci.* 59:171 (1969).
Van Dalen, A., H. A. Das and J. Zonderhuis. *J. Radioanal. Chem.* 15:143 (1973).
Van der Sloot, H. A. and H. A. Das. *Anal. Chim. Acta* 73:235 (1974).
Van der Sloot, H. A., R. Massee and H. A. Das. *J. Radioanal. Chem.* 25:99 (1975).
Vogg, H. and R. Hartel. *Z. Anal. Chem.* 267:257 (1973).
Wasson, J. T. and P. A. Baedecker. *Proc. Apollo 11 Lunar Science Conference 2* p. 1741 (Pergamon, 1970).
Weaver, J. N. *Anal. Chem.* 45:1950 (1973).
Wilkniss, P. K. and V. J. Linnenbom. *Limnol. Oceanog.* 13:530 (1968).
Wogman, N. A., H. G. Rieck, Jr., J. R. Kosorok and R. W. Perkins. *J. Radioanal. Chem.* 15:591 (1973).
Wolicki, E. A. "Nuclear and Ion Beam Techniques for Surface and Near-Surface Analysis," NRL-7477 (1972).
Yung-Yang Wu, Tong-Chuin Pung, Hui-Tuh Tsai and Shaw-Chii Wu. *Nucl. Sci. J.* 11:105 (1974).
Zaghloul, R., M. Obeid and H. Staerk. *Radiochem. Radioanal. Letter* 15:363 (1973).
Zajtsev, E. I., V. A. Skakodub and Yu. P. Sotskov. *Zh. Anal. Khim. SSSR* 27:2055 (1972).
Zmijewska, W. and J. Minczewski. *Chem. Anal.* 14:23 (1969).

CHAPTER 10

TRACE ANALYSIS AND THE ENVIRONMENT

The notion of trace elements was born in the field of biology. The studies of Claude Bernard published in 1857 led to the discovery of metalloenzymes and the catalytic role of trace elements in the organism. Dutoit and Zbinden (1929) detected the systematic presence of silver, aluminum, copper, iron, potassium, manganese, phosphorus, silicon, titanium, zinc and the occasional presence of cobalt, chromium, lead, nickel and strontium in human blood. At about the same time, Gabriel Bertrand distinguished the *plastic elements* from the *oligo-elements*. The first group includes carbon, oxygen, hydrogen, nitrogen, chlorine, sulfur, phosphorus, calcium, magnesium, potassium and sodium. The second group is manganese, iron, copper, zinc, boron, molybdenum, cobalt, arsenic, fluorine, bromine and iodine. These elements, often in very small amounts, were said to play a dynamic role and act as catalysts controlling the function of the plastic elements.

In 1935, research on trace elements in the earth's crust began with Goldschmidt, permitting a true definition of these elements as a function of their concentrations in a natural medium. In fact, the earth's crust is the very source of trace elements. Table I lists the concentration of trace elements in the principal natural mineral media: eruptive rocks, sedimentary rocks, terrestrial and lunar soils. These are mean values, particularly for terrestrial soils where the coefficient of variation is much larger than in the rocks from which the soils originated.

The concentration dispersion increases from soils to plants and finally to animal media. Table II shows the variation of the mean, the variability of some of the trace elements from the earth's crust up to the animal medium. We must note that some elements (oligo-elements) are indispensable for the growth of plants (Fe, Cu, Mn, Zn, B, etc.) and animals (Fe, Cu, Zn, Mo, Co, etc.). However, while plants just like animals require certain elements in relatively well-defined concentrations, their presence can exceed the useful

413

Table I.   Trace Elements in Rocks and Soils
(Goldschmidt, 1935, 1936; Turekian *et al.*, 1961; Vinogradov, 1959;
Aubert *et al.*, 1971, 1977; Taylor, 1971, 1972)

Elements	Eruptive Basalt	Rocks– Granite	Sedimentary Rocks	Earth Soils	Lunar Soils
Ag	0.1	0.04	0.9	0.1	0.01-0.06
As	2	1.5	6	5	0.03-0.09
Au	0.005	0.004	0.001	0.001	-
B	4	10	12	10	1-4
Ba	300	600	80	500	185-1030
Be	1	2	3	6	1.6-6.6
Bi	0.007	0.001	1	1	-
Cd	0.2	0.1	0.3	0.6	0.3-1
Ce	60	60	50	50	50-200
Co	50	2	20	8	24-58
Cr	200	8	150	200	$1\text{-}3.10^3$
Cs	1	2	12	5	0.1-0.7
Cu	80	10	60	20	4.5-9.0
Ga	15	15	30	30	3.3-5.5
Hg	0.1	0.05	0.4	0.01	0.04
La	15	50	40	40	16-74
Li	15	30	60	30	9.2-23
Mn	1,500	400	700	800	6,000
Mo	1.5	1.2	1	2	0.3-0.7
Ni	200	5	100	40	185-420
Pb	6	20	20	10	2.2-10
Rb	30	170	200	100	2.7-13
Se	0.05	0.05	0.6	0.01	0.2-0.9
Sn	1.5	3	30	10	0.3-1.2
Sr	450	200	400	300	180
Ti	10,000	4,000	4,000	5,000	10,000
Tl	0.1	2	1	-	-
U	1	3	3	1	0.6-1.2
V	250	40	130	100	27-125
W	0.7	2	1	-	0.66
Y	30	30	30	50	49-276
Zn	100	40	80	50	6.7-24.9
Zr	140	170	200	300	273-815

limits and the element can become toxic. Moreover, an excess of certain elements (Ni, Al, V, Ba, etc.) in the soil can be transferred to the cultivated plant without biological function.

While the earth's crust remains the direct source of trace elements supplying the soil and then plant and animal media, today numerous external sources

Table II. Distribution of Trace Elements in Natural Media (ppm) (Pinta, 1961)

Elements	Rocks	Soils	Plants	Animals
Mn	400-2000 (500)[a]	200-3000 (300)	8-700 (80)	0.1-10 (1)
F	300-600	-	1-100	2-1000
Cu	10-100 (60)	2-100 (20)	1-200 (10)	0.5-100 (10)
Sr	200-500 (120)	50-1000 (120)	2-800 (80)	0.01-0.8 (0.1)
Li	10-100 (60)	8-300 (70)	0.02-8 (0.5)	0.1-10 (1.2)
B	2-60 (10)	1-100 (15)	1-100 (20)	0.08-1 (0.3)
Mo	2-20 (5)	0.2-8 (0.5)	0.03-12 (1)	0.03-10 (0.8)
Hg	0.02-0.5 (0.06)	- (0.01)	0.001-0.07	0.001-0.1
Sc	0.1-0.01 (0.07)	0.1-2 (0.05)	0.05-5 (0.6)	0.08-8 (0.8)

[a]( ) average.

are modifying the natural cycle of trace elements in nature. Industrial, domestic and agricultural activities sometimes make use of large quantities of elements such as iron, copper, lead, nickel, cadmium or mercury, a part of which returns to the environment via discharges into the atmosphere or waters.

The development of physical methods such as activation or flameless atomic absorption has increased our knowledge of trace elements in the atmosphere. Aerosols of terrestrial origin contain silica and the silicates of iron, aluminum, calcium, magnesium, etc., while marine aerosols are the sources of carbonates and chlorides of sodium, magnesium and other metals. Both contain considerable quantities of such trace elements as Cu, Mn, V, Ti, Cd, Zn, Pb, Co, Ni and Cr (Table III). The concentration of particulate atmospheric aerosols is in the order of 20-150 $\mu g/m^3$ and can reach 1000 $\mu g/m^3$ in the case of high air pollution.

The chemical composition of aerosols varies with the geographical conditions, the altitude and the weather. Air currents constitute the principal vector of trace elements. A significant change of this composition can be

Table III. Trace Elements in the Atmosphere (ng/m^3)
(Valkovic, 1975; Duce *et al.*, 1975; Zoller *et al.*, 1974)

Element	U.S.A.–Chicago	North Atlantic	South Pole
Al	2175	8-370	0.57
Fe	13800	3.4-220	0.84
V	18.1	0.06-14	0.0015
Cr	·113	0.07-1.1	0.0053
Pb	35-380 (U.K.)	0.10-64	0.63
Se	3.8	0.09-0.40	0.0056

observed during transport from terrestrial to oceanic zones, for example. Considerable enrichments in trace elements can be observed in marine aerosols compared to terrestrial aerosols. Taking aluminum, a major element of the earth's crust, as a reference, one can compare the ratio of trace elements to aluminum in oceanic aerosols (X/Al air) to the same ratio in the earth's crust (X/Al crust). We often find high enrichment factors [X/Al air / X/Al crust] attaining 2,000 for lead and 10,000 for selenium (Table IV). The composition of atmospheric particles is very important, since the enriched substances include particularly toxic elements (Pb, Cd, Se and As). Fallout of these particles results in environmental pollution (land or water). The vegetation as well as surface water can be seriously modified.

Table IV. Concentration Range and Enrichment Factors for Any Trace Elements (X) in the Atmospheric Particles Relative to the Crust: [X/Al (air) / X/Al (crust)] (Duce *et al.*, 1975)

Element	North Atlantic Ocean (ng/m^3)	X/Al Air / X/Al Crust
Al	8-370	1
Fe	3.4-220	1.4
Mn	0.05-5.4	2.6
Cr	0.07-1.1	11
Zn	0.3-27	110
Cd	0.003-0.62	730
Pb	0.10-64	2,200
Se	0.09-0.40	10,000

Recently, we have become conscious of these problems which are likely to have serious effects on human health. Solid suspended particles in the air which we breathe can penetrate the organism through the respiratory tract. Particles larger than 1 μm will be retained, while the finest ones will penetrate the lungs with additional risk of accumulation. It has been demonstrated that certain toxic metals undergo preferential concentration in submicrometric particles. Certain substances can pass through the lungs and bronchi into the gastrointestinal system, blood and other tissues and produce their undesirable effects there.

Until now, interest essentially has been turned to the effects of toxic agents absorbed from the gastrointestinal tract, although the absorption of toxins by the airways is at least equally important. In the U.S., Dulka and Risby (1976) have reported the quantities of some important elements discharged annually into the atmosphere (Table V). Although these values are highly instructive, we should distinguish natural discharges resulting, for example, from soil erosion on land or from "bubbling" in oceanic zones, and from the particulate emissions resulting from human domestic, agricultural

Table V. Annual Emission of Trace Elements in the Atmosphere (Dulka *et al.*, 1976)

Elements	Emission (tons/yr)	Emission (%)
Arsenic	10,600	1.60
Barium	15,420	2.33
Beryllium	172	0.03
Cadmium	2,160	0.33
Chromium	18,136	2.73
Copper	13,680	2.07
Lead	16,000	
Highway	197,437	34.72
Off highway	16,563	
Manganese	17,900	2.70
Magnesium	75,293	11.38
Mercury	857	0.13
Molybdenum	990	0.14
Nickel	7,310	1.10
Selenium	986	0.14
Silver	417	0.06
Titanium	88,351	13.33
Vanadium	20,300	3.07
Zinc	159,922	24.14
Total	662,494	100

or industrial activities. Furthermore, some of the elements listed in Table V
are not especially toxic. The problem of air pollution, not a matter of serious
concern a few years ago, is presently of considerable importance. In addition
to direct effects resulting from the inhalation of polluted air, the fallout of
atmospheric dust can also pollute the soil, vegetation and surface waters.
Today the risk exists that discharges of any nature, either directly into the
river network or into the atmosphere will increasingly disturb the natural biological balance. With regard to the problem of trace elements, physical and
physicochemical methods are available today which should permit an effective
monitoring of the environment and thus permit action to eliminate or limit
the development of pollution where it exists.

The role of trace elements in nature should be considered not only because
of the undesirable effect on the environment. A number of elements normally
present in soils play a necessary and active role in plant growth (Aubert and
Pinta, 1971). It is now well established that B, Co, Cu, Fe, Mn, Mo and Zn
are indispensable for plant growth. They must exist in soil in a form assimilable
by plants, for otherwise the yields of agricultural crops may be greatly reduced.
Fertilizers must contain suitable quantities of these elements.

The same is true for animal nutrition in which the trace elements Cu, Zn,
Mo, Fe, Co and Se are recognized as having an essential role in the organism
particularly in the composition of numerous enzymes.

In considering this aspect of chemical elements, however, it is difficult to
propose a classification of essential and nonessential elements. Some are
essential for plants, without being so for animals. Furthermore, all plant
species, like all animal species, do not necessarily require all of the above
elements. While the presence of a given element is necessary, an excess can
become toxic even on a trace scale. Such cases have been reported for B, Cu
and Mo. Here again, we recognize the importance of a chemical control of
trace elements in the environment, soils, plants and animals. Furthermore,
the list of essential elements will likely increase with expanded research and
improved analytical methods. For example, in addition to the elements
traditionally listed as essential for animals, tin, titanium, vanadium, chromium
and nickel have recently been reported as essential (Dulka and Risby, 1976).

In recent years, instrumental and analytical methods have made it possible
to learn more about the biochemical role of essential or toxic trace elements
in the organism. We will limit ourselves to a few examples. *Copper* was
demonstrated in numerous enzymes: hematocuproine, cytochrome C oxidase
and ceruloplasmin. However, in excess, it may play a hemolytic role. *Zinc*
is also present in many proteins and enzymes and is used in the treatment of
skin diseases. An excess can cause numerous disorders, such as dizziness,
vomiting and diarrhea. *Selenium* is essential in very small amounts. It forms
a complex in the plasma proteins and is thus distributed to the tissues. It can

replace sulfur in cysteine and methionine. A selenium excess causes irritation of the eyes, nose, throat and lungs as well as degeneration of tissues and organs such as the liver and kidneys. It has been reported to induce cancer. *Chromium* is found in the β-globulins. It plays an essential role in glucose metabolism. However it is also highly toxic, particularly in its hexavalent state, causing various problems such as congestion, emphysema, tracheitis, bronchitis, bronchopneumonia, cancer of the lungs and dermatoses.

In contrast, there are elements which are highly toxic. *Beryllium* is toxic for the skin and mucosa and is a possible carcinogen for the lungs and bones. The role of *cadmium* has been the subject of numerous studies. It depresses growth and inhibits protein and fat digestion. It is bound in the liver, kidneys and genital organs and can cause disorders such as hypertension, cardiovascular disturbances, proteinuria, glycosuria, cancer and edema. *Mercury* is an important toxin. The mercurous salts (Hg^+) are oxidized by tissues and erythrocytes, forming highly toxic mercuric salts (Hg^{2+}). Mercury damages the nerve centers; it is also bound in the liver, kidneys, brain, heart, lungs and muscles.

Many other examples could be cited. In fact, this brief review demonstrates the importance of the problem. Continuous progress may be anticipated.

REFERENCES

Aubert, H. and M. Pinta. *Trace Elements in Soils* (Paris: Orstom, 1971; Amsterdam: Elsevier, 1977).

Bernard C. *Lectures on the Effects of Medicinal Substances* (Paris: Bailliere, 1857).

Bertrand, G. *Ann. Inst. Pasteur* 26:852 (1912).

Bertrand, G. *Scientific Jubilee of Gabriel Bertrand* (Paris: Gauthier-Villars, 1938).

Duce, R. A., G. L. Hoffman and W. H. Zoller. *Science* 187:59 (1975).

Dulka, J. J. and T. H. Risby. *Anal. Chem.* 48:640A (1976).

Dutoit, P. and C. Zbinden. *C. R. Acad. Sci.* (Paris) 188:16 (1929).

Goldschmidt, V. M. *Ind. Eng. Chem. Anal. Ed.* 7:1100 (1935).

Goldschmidt, V. W. *J. Chem. Soc.* 655 (1937).

Pinta, M. *Detection and Determination of Trace Elements* (Paris: Dunod, 1961; Ann Arbor, Michigan: Ann Arbor Science Publishers, 1971).

Taylor, S. R., P. Muir and M. Kaye. *Geochim. Cosmochim. Acta* 35:975 (1971).

Taylor, S. R. *et al. The Apollo 15 Lunar Samples*. J. W. Chamberlain and C. Watkins, Eds. Lunar Science Institute (1972).

Turekian, K. K. and K. Wedepohl. "Distribution of the Elements in Some Major Units of the Earth's Crust," *Geol. Soc. Am. Bull.* 72:175 (1961).

Valkovic, V. *Trace Elements Analysis* (London: Taylor, 1975).

Vinogradov, A. P. *Geochemistry of Rare and Dispersed Chemical Elements in Soils*, translated from Russian (New York: Consultants Bureau Inc., 1959).

Zoller, W. H., E. S. Gladney and R. A. Duce. *Science* 183:198 (1974).

APPENDICES

APPENDIX 1. PERIODIC CHART OF THE ATOMS

1a	2a	3b	4b	5b	6b	7b		8	
1 +1 H −1 1.00797 1									
3 +1 Li 6.939 2 1	**4** +2 Be 9.0122 2–2				Transition Elements				
11 +1 Na 22.9898 2-8 1	**12** +2 Mg 24.312 2-8-2							Group 8	
19 +1 K 39.102 ·8–8–1	**20** +2 Ca 40.08 -8-8-2	**21** +3 Sc 44.956 –8–9–2	**22** +2 +3 Ti +4 47.90 -8-10-2	**23** +2 +3 V +4 +5 50.942 -8-11-2	**24** +2 +3 Cr +3 +6 51.996 -8-13-1	**25** +2 Mn+3 +4 +7 54.9380 -8-13-2	**26** +2 Fe +3 55.847 -8-14-2	**27** +2 Co +3 58.9332 -8-15-2	**28** Ni 58.7 -8-
37 +1 Rb 85.47 18-8-1	**38** +2 Sr 87.62 -18-8-2	**39** +3 Y 88.905 -18-9-2	**40** +4 Zr 91.22 -18-10-2	**41** +3 Nb+5 92.906 -18-12-1	**42** +6 Mo 95.94 -18-13-1	**43** +4 Tc +6 +7 (97) -18-13-2	**44** +3 Ru 101.07 -18-15-1	**45** +3 Rh 102.905 -18-16-1	**46** Pd 106 18–
55 +1 Cs 132.905 -18-8 1	**56** +2 Ba 137.34 -18-8-2	**57*** +3 La 138.91 -18-9-2	**·72** +4 Hf 178.49 -32-10-2	**73** +5 Ta 180.948 -32-11-2	**74** +6 W 183.85 -32-12-2	**75** +4 Re +6 +7 186.2 -32-13-2	**76** +3 Os +4 190.2 -32-14-2	**77** +3 Ir +4 192.2 -32-15-2	**78** Pt 195 -32
87 +1 Fr (223) 18 8 1	**88** +2 Ra (226) -18-8-2	**89**** Ac +3 (227) -18-9-2							

		58 +3 Ce +4 140.12 -20-8-2	**59** +3 Pr 140.907 -21-8-2	**60** +3 Nd 144.24 -22· 8-2	**61** +3 Pm (145) -23-8-2	**62** +2 Sm+3 150.35 -24-8-2	**63** +2 Eu +3 151.96 -25-8-2	**64** +3 Gd 157.25 -25-9-2	**65** Tl 158 -2
*Lanthanides									
**Actinides		**90** +4 Th (232) -18-10-2	**91** +5 Pa +4 (231) -20-9-2	**92** +3 U +4 +5 +6 (238) -21-9-2	**93** +3 Np+4 +5 +6 (237) -22-9-2	**94** +3 Pu +4 +5 +6 (244) -24-8-2	**95** +3 Am+4 +5 +6 (243) -25-8-2	**96** +3 Cm (247) -25-9-2	**97** Bl (24 -2

b	2b	3a	4a	5a	6a	7a	0	Orbit
							2 0 He 4.0026 2	K
		5 +3 B 10.811 2–3	6 +2 C +4 –4 12.01115 2–4	7 +1 N +2 +3 +4 +5 –1 14.0067 –2 2–5 –3	8 –2 O 15.9994 2–6	9 –1 F 18.9984 2–7	10 0 Ne 20.183 2–8	K–L
		13 +3 Al 26.9815 2–8–3	14 +2 Si +4 –4 28.086 2–8–4	15 +3 P +5 –3 30.9738 2–8–5	16 +4 S +6 –2 32.064 2–8–6	17 +1 Cl +5 +7 –1 35.453 2–8–7	18 0 Ar 39.948 2–8–8	K–L–M
+1 +2	30 +2 Zn 65.37 –8–18–2	31 +3 Ga 69.72 –8–18–3	32 +2 Ge +4 72.59 –8–18–4	33 +3 As +5 –3 74.9216 –8–18–5	34 +4 Se +6 –2 78.96 –8–18–6	35 +1 Br +5 –1 79.904 –8–18–7	36 0 Kr 83.80 –8–18–8	–L–M–N
+1	48 +2 Cd 112.40 8–1 –18–18–2	49 +3 In 114.82 –18–18–3	50 +2 Sn +4 118.69 –18–18–4	51 +3 Sb +5 –3 121.75 –18–18–5	52 +4 Te +6 –2 127.60 –18–18–6	53 +1 I +5 +7 –1 126.9044 –18–18–7	54 0 Xe 131.30 –18–18–8	–M–N–O
+1 +3	80 +1 Hg +2 200.59 8–1 –32–18–2	81 +1 Tl +3 204.37 –32–18–3	82 +2 Pb +4 207.19 –32–18–4	83 +3 Bi +5 208.980 –32–18–5	84 +2 Po +4 (209) –32–18–6	85 At (210) –32–18–7	86 0 Rn (222) –32–18–8	–N–O–P
								–O–P–Q

| +3 | 67   +3
Ho

164.930
29–8–2 | 68   +3
Er

167.26
–30–8–2 | 69   +3
Tm

168.934
–31–8–2 | 70   +2
Yb   +3
173.04
–32–8–2 | 71   +3
Lu

174.97
–32–9–2 | | | –N–O–P |
| +3 | 99
Es

(254)
–29–8–2 | 100
Fm

(257)
–30–8–2 | 101
Md

(256)
–31–8–2 | 102
No

(254)
–32–8–2 | 103
Lw

–32–9–2 | 104
—

–32–10–2 | | –O–P–Q |

Legend:

Atomic number ⟶ | 19   +1 | ⟵ Atomic valence
K
Atomic weight ⟶ | 39.102

2-8-1 ⟵ Electron configuration

APPENDIX 2. ABUNDANCE AND PHYSICAL CONSTANTS OF ELEMENTS

Symbol	Element	Atomic Weight (g/mol)	Abundance[1] (percent)	Melting Point (°K)	Boiling Point (°K)	Ionization Energy (eV)
Ag	Silver	107.87	$1 \cdot 10^{-5}$	1234	2483	7.57
Al	Aluminum	26.98	8.80	933	2723	5.98
Ar	Argon	39.95	$4 \cdot 10^{-5}$	84	87	15.76
As	Arsenic	74.92	$5 \cdot 10^{-4}$	1090 (36 atm)sub.	885	9.81
Au	Gold	196.97	$5 \cdot 10^{-7}$	1336	3243	9.22
B	Boron	10.81	$3 \cdot 10^{-4}$	2300	4200	8.30
Ba	Barium	137.34	0.05	987	1913	5.21
Be	Beryllium	9.01	$6 \cdot 10^{-4}$	1550	3043	9.32
Bi	Bismuth	208.98	$2 \cdot 10^{-5}$	545	1833	7.28
Br	Bromine	79.91	$1.6 \cdot 10^{-4}$	266	331	11.84
C	Carbon	12.01	0.1		5103	11.26
Ca	Calcium	40.08	3.6	1111	1713	6.11
Cd	Cadmium	112.40	$5 \cdot 10^{-5}$	594	1038	8.99
Ce	Cerium	140.12	$4.5 \cdot 10^{-3}$	1077	3743	6.54
Cl	Chlorine	35.45	0.045	172	239	13.01
Co	Cobalt	58.93	$3 \cdot 10^{-3}$	1768	3173	7.86
Cr	Chromium	52.00	0.02	2148	2938	6.76
Cs	Caesium	132.91	$7 \cdot 10^{-4}$	302	963	3.89
Cu	Copper	63.54	0.01	1356	2868	7.72
Dy	Dysprosium	162.50	$4.5 \cdot 10^{-4}$	1680	2873	(6.82)
Er	Erbium	167.26	$4 \cdot 10^{-4}$	1770	3173	(6.7)
Eu	Europium	151.96	$1.2 \cdot 10^{-4}$	1100	1703	5.67
F	Fluorine	19.00	0.027	53	85	17.42
Fe	Iron	55.85	$5 \cdot 10$	1812	3273	7.90
Ga	Gallium	69.72	$1.5 \cdot 10^{-3}$	303	2516	6.00
Gd	Gadolinium	157.25	$1 \cdot 10^{-3}$	1585	3273	6.16
Ge	Germanium	72.59	$7 \cdot 10^{-4}$	1210	3103	7.88
H	Hydrogen	1.01	0.15	14	20	13.60.
He	Helium	4.00	$3 \cdot 10^{-7}$	3	4	24.58
Hf	Hafnium	178.49	$3.2 \cdot 10^{-4}$	2495	>5613	5.5
Hg	Mercury	200.59	7.10^{-6}	234	630	10.43
Ho	Holmium	164.93	$1.3 \cdot 10^{-4}$	1734	2873	(6.9)
In	Indium	114.82	$1 \cdot 10^{-5}$	429	2273	5.79
Ir	Iridium	192.20	$1 \cdot 10^{-7}$	2727	5573	9.2
I	Iodine	126.90	$3 \cdot 10^{-5}$	386	456	10.44
K	Potassium	39.10	$2 \cdot 60$	337	1033	4.34
Kr	Krypton	83.80	$2 \cdot 10^{-8}$	116	121	14.00
La	Lanthanum	138.91	$1.8 \cdot 10^{-3}$	1193	3743	5.61
Li	Lithium	6.94	$6.5 \cdot 10^{-3}$	382	1603	5.39
Lu	Lutetium	174.97	$1 \cdot 10^{-4}$	1928	3600	6.15
Mg	Magnesium	24.31	$2 \cdot 10$	923	1380	7.64
Mn	Manganese	54.94	0.09	1517	2423	7.43
Mo	Molybdenum	95.94	$3 \cdot 10^{-4}$	2890	5833	7.13
N	Nitrogen	14.01	0.01	63	77	14.54
Na	Sodium	22.99	2.64	372	1165	5.14
Nb	Niobium	92.91	$1 \cdot 10^{-3}$	2770	4573	6.88

Appendix 2 (continued)

Symbol	Element	Atomic Weight (g/mol)	Abundance[1] (percent)	Melting Point (°K)	Boiling Point (°K)	Ionization Energy (eV)
Nd	Neodymium	144.24	$2.5 \cdot 10^{-3}$	1297	3300	(6.31)
Ne	Neon	20.18	$7 \cdot 10^{-9}$	24	27	21.56
Ni	Nickel	58.71	$8 \cdot 10^{-3}$	1728	3003	7.63
O	Oxygen	16.00	47.2	54	90	13.61
Os	Osmium	190.20	$5 \cdot 10^{-6}$	2973	5773	8.7
P	Phosphorus	30.97	0.08	870 (red)		10.55
				317 (white)	553	
Pb	Lead	207.19	$1.6 \cdot 10^{-3}$	600	1998	7.42
Pd	Palladium	106.4	$1 \cdot 10^{-6}$	1823	4253	8.33
Pr	Praseodymium	140.91	$7 \cdot 10^{-4}$	1208	3400	(5.76)
Pt	Platinum	195.09	$5 \cdot 10^{-7}$	2043	4803	8.96
Rb	Rubidium	85.47	0.03	312	961	4.18
Re	Rhenium	186.2	$(1 \cdot 10^{-7})$	3453	6173	7.87
Rh	Rhodium	102.91	$1 \cdot 10^{-7}$	2239	4773	7.46
Ru	Ruthenium	101.07	$6 \cdot 10^{-7}$	2773	5173	7.36
S	Sulphur	32.06	0.05	392	718	10.36
Sb	Antimony	121.75	$4 \cdot 10^{-5}$	903	1653	8.64
Sc	Scandium	44.96	$6 \cdot 10^{-4}$	1811	3003	6.56
Se	Selenium	78.96	$6 \cdot 10^{-5}$	490	958	9.75
Si	Silicon	28.09	27.6	1683	2953	8.15
Sm	Samarium	150.35	$7 \cdot 10^{-4}$	1345	2173	6.6
Sn	Tin	118.69	$4 \cdot 10^{-3}$	505	2543	7.33
Sr	Strontium	87.62	0.04	1043	1653	5.69
Ta	Tantalum	180.95	$2 \cdot 10^{-4}$	3270	5698	7.88
Tb	Terbium	158.92	$1.5 \cdot 10^{-4}$	1629	3073	(6.74)
Te	Tellurium	127.60	$1 \cdot 10^{-6}$	723	1263	9.01
Th	Thorium	232.04	$8 \cdot 10^{-4}$	2023	4123	6.95
Ti	Titanium	47.90	0.6	1941	3533	6.83
Tl	Thallium	204.37	$3 \cdot 10^{-4}$	577	1730	6.11
Tu	Thulium	168.93	$8 \cdot 10^{-5}$	1818	1993	(6.6)
U	Uranium	238.03	$3 \cdot 10^{-4}$	1406	4091	(4)
V	Vanadium	50.94	0.02	2173	3723	6.74
W	Tungsten	183.85	$1 \cdot 10^{-4}$	3683	6203	7.98
Xe	Xenon	131.30	$3 \cdot 10^{-9}$	161	165	12.13
Y	Yttrium	88.91	$2.8 \cdot 10^{-3}$	1782	3200	6.38
Yb	Ytterbium	173.04	$3 \cdot 10^{-4}$	1097	1700	6.22
Zn	Zinc	65.37	$5 \cdot 10^{-2}$	693	1180	9.39
Zr	Zirconium	91.22	0.02	2128	3853	6.84

[1] Mean concentration in Earth Crust, from VINOGRADOV, A.P., Geokhim, 1956, 1, 52.

Others constants from L'VOV. Atomic absorption spectrochemical analysis, 1970, Hilger, London.

APPENDIX 3. EMISSION SPECTROSCOPY

Persistent Lines of Elements (Classified According to Wavelength)
(From Pinta, M. *Detection and Determination of Trace Elements*, Paris: Dunod, 1961,
Ann Arbor, Michigan: Ann Arbor Science Publishers, Inc., 1973).

Wavelength Å	Element	Relative Intensity	Wavelength Å	Element	Relative Intensity
2,138.56	Zn I	800 R	3,302.32	Na I	600 R
2,142.75	Te I	600	3,302.99	Na I	300 R
2,288.02	Cd I	1,500 R	3,311.16	Ta	300
2,348.61	Be I	2,000 R	3,321.34	Be I	1,000 R
2,383.25	Te I	600	3,341.87*	Ti I, II	100
2,385.76	Te I	600	3,345.02	Zn I	800
2,427.95	Au I	400 R	3,349.41*	Ti II	100
2,496.78	B I	300	3,382.89	Ag	1,000 R
2,497.73	B I	500	3,391.97	Zr II	300
2,516.12	Si I	500	3,404.58	Pd I	2,000 R
2,536.52	Hg I	2,000 R	3,405.12	Co I	2,000 R
2,543.97	Ir I	200	3,414.76	Ni I	1,000 R
2,576.10	Mn II	300 R	3,421.24	Pd I	2,000 R
2,651.18	Ge I	400	3,434.89	Rh	1,000 r
2,659.45	Pt I	2,000 R	3,436.74	Ru I	300 R
2,675.95	Au I	250 R	3,438.23	Zr II	250
2,714.67*	Ta	200	3,453.50	Co I	3,000 R
2,833.07	Pb I	500 R	3,460.47	Re I	1,000
2,837.30*	Th	15	3,466.20	Cd I	1,000 R
2,839.99	Sn I	300 R	3,492.96	Ni I	1,000 R
2,852.13	Mg I	300 R	3,498.94	Ru I	500 R
2,881.58	Si I	500	3,519.24	Tl I	2,000 R
2,897.97	Bi I	500 R	3,524.54	Ni I	1,000 R
2,909.06	Os I	500 R	3,581.19	Fe I	1,000 R
2,943.64	Ga I	10	3,653.50	Ti I	500
3,020.64*	Fe I	1,000 R	3,683.47	Pb I	300
3,039.06	Ge I	1,000	3,692.36	Rh I	500
3,058.66	Os I	500 R	3,710.29	Y II	80
3,064.71	Pt I	2,000 R	3,719.93	Fe I	1,000 R
3,067.72	Bi I	3,000 R	3,775.72	Tl I	3,000
3,082.15	Al I	800	3,798.25	Mo I	1,000 R
3,092.71	Al I	1,000	3,838.26	Mg I	300
3,130.41	Be II	200	3,902.96	Mo I	1,000 R
3,131.07	Be II	200	3,911.81	Sc I	150
3,133.32*	Ir I	40	3,933.66	Ca II	600 R
3,170.35*	Mo	1,000 R	3,944.03	Al I	2,000
3,175.04	Sn I	500	3,961.52	Al I	3,000
3,185.39	V I	500 R	3,968.46	Ca I	500 R
3,220.78	Ir I	100	3,998.64*	Ti I	150
3,232.61	Li I	1,000 R	4,008.75	W I	45
3,242.28	Y II	60	4,019.14	Th II	8
3,247.54	Cu I	5,000 R	4,023.69	Sc I	100
3,256.09	In I	1,500 R	4,030.75	Mn I	500 r
3,261.06	Cd I	300	4,032.98	Ga I	1,000 R
3,262.33	Sn I	400	4,033.07	Mn I	400 r
3,269.49	Ge I	300 R	4,034.49	Mn I	250 r
3,273.96	Cu I	3,000 R	4,044.14	K I	800
3,280.68	Ag	2,000 R	4,047.20	K I	400
			4,057.82	Pb I	2,000 R
			4,077.71	Sr II	400 r

Appendix 3 (continued)

Wavelength Å	Element	Relative Intensity	Wavelength Å	Element	Relative Intensity
4,172.06	Ga I	2,000 R	4,810.53	Zn I	400
			4,889.17	Re I	2,000
4,201.85	Rb I	2,000 R			
4,215.56	Rb I	1,000 R	4,934.08	Ba II	400
4,226.73	Ca I	500 R	4,981.73	Ti I	300
4,241.67	U	40	5,183.62	Mg I	500
4,246.83*	Sc II	80			
4,254.34	Cr I	5,000 R	5,350.46	Ti I	5,000 R
4,274.80	Cr I	4,000 R			
			5,535.55	Ba I	1,000 R
4,289.72	Cr I	3,000 R			
4,294.61	W I	50	4,889.95	Na I	9,000 R
			5,895.93	Na I	5,000 R
4,302.11	W I	60			
4,358.35	Hg I	3,000	6,103.64	Li I	2,000 R
4,374.93*	Y II	150	6,707.84	Li I	3,000 R
4,379.24	V I	200 R			
			7,664.91	K I	9,000 R
4,511.32	In I	5,000 R	7,698.98	K I	5,000 R
4,554.04	Ba II	1,000 R			
4,555.35	Cs I	2,000 R	7,800.23	Rb I	9,000 R
4,593.18	Cs I	1,000 R	7,947.60	Rb I	5,000 R
4,602.86	Li I	800			
4,607.33	Sr I	1,000 R	8,521.10	Cs I	5,000 R
4,687.80	Zr I	125	8,943.50	Cs I	2,000 R
4,722.55	Bi I	1,000			

Note: I = atom
II = ion
R = wide self-reversal line
r = narrow self-reversal line

APPENDIX 4. EMISSION SPECTROSCOPY
Persistent Lines of Elements (Classified According to Elements)

The elements are arranged in alphabetical order according to their chemical symbols. Only the most sensitive lines are given in this table; for a more complete list of the persistent lines a table of wavelengths should be consulted.

The first column gives the symbol of the element and the type of line; I indicates the line of the normal atom and II of the once-ionized atom. The second column gives the wavelengths. The third column gives relative line intensities and indicates the resonance of inversion lines: R for a wide line and r for a narrow line. The fourth column gives the principal lines which may interfere with each persistent line because of their sensitivity and wavelength (to within ± 0.5 Å); the most sensitive interfering lines for each persistent line are noted in italics. (From Pinta, M, *Detection and Determination of Trace Elements*, Paris: Dunod, 1961; Ann Arbor, Michigan: Ann Arbor Science Publishers, Inc., 1973.

Element	Wavelength	Relative Intensity	Interferences
Ag I	3,382.89	1,000 R	Nd : 3,382.81
Ag I	3,280.68	2,000 R	Mn : 3,280.76 − Rh : 3,280.55 − Fe : 3,280.26
(Silver)			
Al I	3,961.52	3,000	Zr : 3,961.58
Al I	3,944.03	2,000	
Al I	3,092.71	1,000	V : 3,093.11 − 3,092.72 − Mg : 3,092.99 − Fe : 3,092.78 − Na : 3,092.73
Al I	3,082.15	800	Co : 3,082.62 − Re : 3,082.43 − N : 3,082.11 − Mn : 3,082.05
(Aluminum)			
Au I	2,675.95	250 R	Ta : 2,675.9
Au I	2,427.95	400 R	Pt : 2,428.20 − 2,428.03 − Ta : 2,427.64
(Gold)			
B I	2,497.73	500	
B I	2,496.78	300	Cr : 2,496.31
(Boron)			
Ba I	5,535.55	1,000 R	Fe : 5,535.41
Ba II	4,934.08	400	Yb : 4,935.50 − La : 4,934.82
Ba II	4,554.04	1,000 R	Ru : *4,554.51* − Ta : 4,553.69
(Barium)			
Be I	3,321.34	1,000 r	
Be II	3,131.07	200	Os : 3,131.11 − Eu : 3,130.74
Be II	3,130.41	200	Ta : 3,130.58 − V : 3,130.27
Be I	2,348.61	2,000 R	Ru : 2,348.33
(Beryllium)			
Bi I	4,722.55	1,000	Ta : 4,722.88 − Ti : 4,722.62 − Zn : 4,722.16
Bi I	3,067.71	3,000 R	Fe : 3,068.18 − 3,067.24
Bi I	2,897.97	500 R	Fe : 2,898.35 − Pt : 2,897.87
(Bismuth)			
Ca I	4,226.73	500 R	Cr : 4,226.76 − Ge : 4,226.57 − Fe : 4,226.43
Ca II	3,968.46	500 R	Zr : 3,968.26 − Ag : 3,968.22
Ca II	3,933.66	600 R	V : 3,934.01 − Co : 3,933.91 − 3,933.65 − Ag : 3,933.62 − Fe : 3,933.60
(Calcium)			
Cd I	3,466.20	1,000	Fe : *3,465.86* − Co : *3,465.80*
Cd I	3,261.06	300	Co : 3,260.82
Cd I	2,288.02	1,500 R	As : 2,288.12
(Cadmium)			
Co I	3,453.50	3,000 R	
Co I	3,405.12	2,000 R	
(Cobalt)			
Cr I	4,289.72	3,000 R	Ti : 4,289.07
Cr I	4,274.80	4,000 R	Cu : 4,275.13 − Ti : 4,274.58
Cr I	4,254.34	5,000 R	
(Chromium)			

Appendix 4 (continued)

Element	Wavelength	Relative Intensity	Interferences
Cs I	8,943.50	2,000 R	
Cs I	8,521.10	5,000 R	Ti : 8,518.32
Cs I	4,593.18	1,000 R	Fe : 4,592.65
Cs I	4,555.35	2,000 R	Ti : 4,555.49
(Cesium)			
Cu I	3,273.96	3,000 R	Fe : 3,274.45
Cu I	3,247.54	5,000 R	Mn : 3,247.54 − Co : 3,247.18
(Copper)			
Fe I	3,719.93	1,000 R	Ti : 3,720.38 − Os : 3,720.13 − 3,719.52
Fe I	3,581.19	1,000 R	Re : 3,580.96 − V : 3,580.82
Fe I	3,020.64*	1,000 R	Fe : 3,021.07 − Cr : 3,020.67 − Co : 3,029,64
(Iron)			Fe : 3,020.49
Ga I	4,172.06	2,000 R	Ir : 4,172.56 − Fe : 4,172.13 − Cr : 4,171.67
Ga I	4,032.98	1,000 R	Mn : 4,033.07 − Ta : 4,033.07 − Fe : 4,032.63
Ga I	2,943.64	110	Ru : 2,943.92 − Ni : 2,943.91
(Gallium)			
Ge I	3,269.49	300 R	Os : 3,269.21 − Ta : 3,269.14
Ge I	3,039.06	1,000	In : *3,039.36*
Ge I	2,651.18	400	Fe : 2,651.71 − Ta : 2,651.48 − Ru : 2,651.29
(Germanium)			Pt : *2,650.86*
Hg I	4,358.35	3,000	Re : 4,358.69 − Fe : 4,358.50
Hg I	2,536.52	2,000 R	Pt : 2,536.49 − Ta : 2,536.23
(Mercury)			
In I	4,511.32	5,000 R	Ta : 4,511.50 − Sn : 4,511.30 − Ta : 4,510.98
In I	3,256.09	1,500 R	Mn : 3,256.14
(Indium)			
Ir I	3,220.78	100	Pb : 3,220.54
Ir I	3,133.32*	40	V : 3,133.33 − Cd : 3,133.17 − Ru : 3,132.88
Ir I	2,543.97	200	Co : 2,544.25 − Ru : 2,544.22 − Fe : 2,543.65
(Iridium)			
K I	7,698.98	5,000 R	Yb : 7,699.49
K I	7,664.91	9,000 R	
K I	4,047.20	400	
K I	4,044.14	800	Fe : 4,044.61
(Potassium)			
Li I	6,707.84	3,000 R	Co : 6,707.86 − Mo : 6,707.85
Li I	6,103.64	2,000 R	
Li I	4,602.86	800	Fe : 4,602.94
Li I	3,232.61	1,000 R	Fe : 3,233.05 − Ni : 3,232.96 − Co : 3,232.87
(Lithium)			Ru : 3,232.75 − Os : 3,232.54 − Sb : 3,232.50
Mg I	5,183.62	500	La : 5,183.42
Mg I	3,838.26	300	
Mg I	2,852.13	300 R	Na : 2,852.83 − Fe : 2,852.13 − 2,851.80
(Magnesium)			
Mn I	4,034.49	250 r	Ti : 4,033.91
Mn I	4,033.07	400 r	Ta : 4,033.07 − Ga : *4,032.98* − Fe : 4,032.63
Mn I	4,030.75	500 r	Cr : 4,030.68 − Ti : 4,030.51 − Fe : 4,030.49
Mn II	2.576.10	300 R	Fe : 2,575.74 − Mn : 2,575.51
(Manganese)			
Mo I	3,902.96	1,000 R	Fe : 3,902.95 − Cr : 3,902.91
Mo I	3,798.25	1,000 R	Fe : 3,798.51
Mo I	3,170.35*	1,000 R	Ta : 3,170.29 − Dy : 3,169.98
(Molybdenum)			

Appendix 4 (continued)

Element	Wavelength	Relative Intensity	Interferences
Na I	5,895.93	5,000 R	
Na I	5,889.95	9,000 R	Mo : 5,889.98
Na I	3,302.99	300 R	La : 3,303.11 – Zn : *3,302.94 – 3,302.59*
			Bi : 3,302.55
Na I	3,302.32	600 R	Cr : 3,302.19 – Pd : 3,302.13 – Pt : 3,301.86
(Sodium)			Sr : 3,301.73
Ni I	3,524.54	1,000 R	Fe : 3,524.24 – Cu : 3,524.24 – Fe : 3,524.07
Ni I	3,492.96	1,000 R	Fe : 3,493.47
Ni I	3,414.76	1,000 R	Co : 3,414.74 – Ru : 3,414.64
(Nickel)			
Os I	3,058.66	500 R	Fe : *3,059.09* – Re : 3,058.78 – Ta : 3,058.64
Os I	2,909.06	500 R	Fe : 2,909.50 – Cr : 2,909.05 – Ta : 2,908.91
(Osmium)			Fe : 2,908.86 – V : 2,908.82
Pb I	4,057.82	2,000 R	Fe : 4,058.23 – Co : 4,058.19 – Ti : 4,058.14
			Mn : 4,057.95 – In : 4,057.87 – Zn : 4,057.71
			Ti : 4,057.62
Pb I	3,683.47	300	Fe : 3,684.11 – V : 3,683.13 – Fe : 3,683.06
			Co : 3,683.05
Pb I	2,833.07	500 R	Ta : 2,833.64
(Lead)			
Pd I	3,421.24	2,000 R	Cr : 3,421.21 – Co : 3,420.79
Pd I	3,404.58	2,000 R	Re : 3,404.72 – Fe : 3,404.36
(Palladium)			
Pt I	3,064.71	2,000 R	Ru : 3,064.83 – Ni : 3,064.62 – Co : 3,064.37
			Mo : 3,064.28
Pt I	2,659.45	2,000 R	Ru : 2,659.61
(Platinum)			
Rb I	7,947.60	5,000 R	Sm : 7,948.12
Rb I	7,800.23	9,000 R	
Rb I	4,215.56	1,000 R	Sr : 4,215.52 – Fe : 4,215.42 – Gd : 4,215.02
Rb I	4,201.85	2,000 R	Os : 4,202.06 – Fe : *4,202.03* – Mn : 4,201.76
(Rubidium)			Zr : 4,201.45
Re I	4,889.17	2,000	Cr : 4,888.53
Re I	3,460.47	1,000	Pd : 3,460.77 – Cr : 3,460.43 – Mn : 3,460.33
(Rhenium)			
Rh I	3,692.36	500	Mn : 3,692.81 – V : *3,692.22* – Sm : 3,692.22
Rh	3,434.89	1,000 r	Ru : 3,435.19 – Mo : 3,434.79
(Rhodium)			
Ru I	3,498.94	500 R	Rh : *3,498.73* – Os : 3,498.54
Ru I	3,436.74	300 R	Ni : *3,437.28* – Fe : 3,437.05
(Ruthenium)			
Sc II	4,246.83*	80	Fe : 4,247.43 – Gd : 4,246.55
Sc I	4,023.69	100	La : 4,023.59 – Co : 4,023.40
Sc I	3,911.81	150	V : 3,912.21 – Cr : 3,912.00
(Scandium)			
Si I	2,881.58	500	Cd : 2,881.23
Si I	2,516.12	500	Re : 2,516.12 – Zn : 2,515.81 – Rh : 2,515.75
(Silicon)			Bi : 2,515.69
Sn I	3,262.33	400	Os : 3,262.75 – *3,262.29* – Fe : 3,262.28
Sn I	3,175.02	500	Fe : 3,175.45 – Co : 3,174.90
Sn I	2,839.99	300 R	Fe : 2,840.42
(Tin)			

Appendix 4 (continued)

Element	Wavelength	Relative Intensity	Interferences
Sr I Sr II (Strontium)	4,607.33 4,077.71	1,000 R 400 r	Fe : 4,607.65 – Mn : 4,607.62 Dy : 4,077.97 – Hg : 4,077.81 – Co : 4,077.41 La : *4,077.34*
Ta Ta (Tantalum)	3,311.16 2,714.67*	300 200	Os : 3,310.91 Rh : 2,715.04 – Fe : 2,714.87 – Os : 2,714.64 Fe : *2,714.41* Rh : 2,714.41 – V : 2,714.20
Te I Te I Te I (Tellurium)	2,385.76 2,383.25 2,142.75	600 500 600	Rh : 2,386.14 Sb : 2,383.63 – Rh : 2,383.40 – 2,382.89
Th II Th (Thorium)	4,019.14 2,837.30*	8 15	Co : 4,019.30 – Tb : 4,019.12 Ce : 2,837.29 – Zr : 2,837.23 – Co : 2,837.15 In : 2,836.92 – Cd : 2,836.91
Ti I Ti I Ti I Ti II Ti I, II (Titanium)	4,981.73 3,998.64* 3,653.50 3,349.41* 3,341.87*	300 150 500 100 100	Fe : *4,982.51* Os : 3,998.93 – V : 3,998.73 Cr : 3,653.91 Cr : 3,349.32 – Cu : 3,349.29 – Cr : 3,349.07 Ti : 3,349.03 Re : 3,342.26 – La : 3,342.22 – Fe : 3,342.21 Cb : 3,341.97 Fe : 3,341.90 – Ce : 3,341.87 – Ru : 3,341.66 Hg : 3,341.48
Tl I Tl I Tl I (Thallium)	5,350.46 3,775.72 3,519.24	5,000 R 3,000 2,000 R	Ti : 5,351.08 – Cb : 5,350.74 V : 3,776.16 – 3,775.72 – Ni : *3,775.57* Ni : *3,519.77* – Ru : 3,519.63 – Zr : 3,519.60
U (Uranium)	4,241.67	40	Zr : 4,241.69 – 4,241.20
V I V I (Vanadium)	4,379.24 3,185.39	200 R 500 R	Pr : 4,379.33 – Ta : 4,378.82 Rh : 3,185.59 – Re : 3,185.56 – Os : 3,185.33 Fe : 3,184.90
W I W I W I (Tungsten)	4,302.11 4,294.61 4,008.75	60 50 45	Ca : 4,302.53 – Fe : 4,302.19 Fe : *4,294.13* – Ti : 4,294.12 Ti : 4,008.93 – Pr : 4,008.71
Y II Y II Y II (Yttrium)	4,374.93* 3,710.29 3,242.28	150 80 60	Sm : 4,374.97 – Mn : 4,374.95 – Er : 4,374.93 Rh : *4,374.80* – Sc : 4,374.45 Ti : 3,709.96 Pd : *3,242.70* – Ru : 3,242.16 – Ta : 3,242.05 Ti : 3,241.99
Zn I Zn I Zn I (Zinc)	4,810.53 3,345.02 2,138.56	400 800 800 R	Mo : 4,811.06 – Cb : 4,810.60 Zn : *3,345.57* – Ru : 3,345.32 – Ce : 3,344.76 Mo : 3,344.75 La : 3,344.56 – Ca : 3,344.51
Zr I Zr II Zr II (Zirconium)	4,687.80 3,438.23 3,391.97	125 250 300	Eu : 4,688.2 Co : 3,438.71 – Ru : 3,438.37 Fe : 3,392.31 – Ru : 3,391.89

APPENDIX 5. EMISSION SPECTROSCOPY
Optical Density Determination
Table of $\log (i_0/i)$ for $i_0 = 50$

i	0	1	2	3	4	5	6	7	8	9
0	∞	2.699	2.398	2.222	2.097	2.000	1.921	1.854	1.796	1.745
1	1.699	1.659	1.620	1.585	1.553	1.523	1.495	1.469	1.444	1.420
2	1.398	1.377	1.357	1.337	1.319	1.301	1.284	1.268	1.251	1.236
3	1.222	1.208	1.194	1.181	1.168	1.155	1.143	1.131	1.119	1.108
4	1.097	1.086	1.076	1.066	1.056	1.046	1.036	1.027	1.018	1.009
5	1.000	0.991	0.982	0.975	0.967	0.959	0.951	0.943	0.936	0.928
6	0.921	0.914	0.907	0.900	0.890	0.886	0.880	0.873	0.867	0.860
7	0.854	0.848	0.842	0.836	0.830	0.824	0.818	0.813	0.807	0.801
8	0.796	0.791	0.785	0.780	0.775	0.770	0.765	0.760	0.755	0.750
9	0.745	0.740	0.735	0.730	0.726	0.721	0.717	0.713	0.708	0.703
10	0.699	0.695	0.690	0.686	0.682	0.678	0.674	0.670	0.666	0.662
11	0.659	0.654	0.650	0.646	0.642	0.638	0.635	0.631	0.627	0.624
12	0.620	0.616	0.613	0.609	0.606	0.602	0.599	0.595	0.592	0.588
13	0.585	0.582	0.579	0.575	0.572	0.569	0.566	0.562	0.559	0.556
14	0.553	0.550	0.547	0.544	0.541	0.538	0.535	0.532	0.529	0.526
15	0.523	0.520	0.517	0.514	0.512	0.509	0.506	0.503	0.500	0.498
16	0.495	0.492	0.490	0.487	0.484	0.482	0.479	0.476	0.474	0.471
17	0.469	0.466	0.463	0.461	0.459	0.456	0.454	0.451	0.449	0.447
18	0.444	0.441	0.439	0.437	0.434	0.432	0.430	0.428	0.424	0.423
19	0.420	0.418	0.416	0.413	0.411	0.409	0.407	0.405	0.403	0.400
20	0.398	0.396	0.394	0.392	0.389	0.387	0.385	0.383	0.381	0.379
21	0.377	0.375	0.373	0.371	0.369	0.367	0.365	0.363	0.361	0.359
22	0.357	0.355	0.353	0.351	0.349	0.347	0.345	0.343	0.341	0.339
23	0.337	0.335	0.334	0.332	0.330	0.328	0.326	0.324	0.322	0.321
24	0.319	0.317	0.315	0.313	0.312	0.310	0.308	0.306	0.305	0.303
25	0.301	0.299	0.298	0.296	0.294	0.293	0.291	0.289	0.287	0.286
26	0.284	0.282	0.281	0.279	0.277	0.276	0.274	0.273	0.271	0.269
27	0.268	0.266	0.264	0.263	0.261	0.260	0.258	0.257	0.255	0.253
28	0.251	0.250	0.249	0.247	0.246	0.244	0.243	0.241	0.240	0.238
29	0.236	0.235	0.234	0.232	0.231	0.229	0.228	0.226	0.225	0.223
30	0.222	0.220	0.219	0.218	0.216	0.215	0.213	0.212	0.210	0.209
31	0.208	0.206	0.205	0.204	0.202	0.201	0.199	0.198	0.197	0.196
32	0.194	0.193	0.191	0.190	0.189	0.187	0.186	0.185	0.183	0.182
33	0.181	0.179	0.178	0.177	0.175	0.174	0.173	0.171	0.170	0.169
34	0.168	0.166	0.165	0.164	0.162	0.161	0.160	0.159	0.157	0.156
35	0.155	0.154	0.153	0.151	0.150	0.149	0.148	0.146	0.145	0.144
36	0.143	0.142	0.140	0.139	0.138	0.137	0.136	0.134	0.133	0.132
37	0.131	0.130	0.129	0.127	0.126	0.125	0.124	0.123	0.122	0.120
38	0.119	0.118	0.117	0.115	0.114	0.113	0.112	0.111	0.110	0.109
39	0.108	0.107	0.106	0.105	0.103	0.102	0.101	0.100	0.099	0.098
40	0.097	0.096	0.095	0.094	0.093	0.091	0.090	0.089	0.088	0.087
41	0.086	0.085	0.084	0.083	0.082	0.081	0.080	0.079	0.078	0.077
42	0.076	0.075	0.074	0.073	0.072	0.071	0.070	0.069	0.068	0.067
43	0.066	0.064	0.063	0.062	0.061	0.060	0.059	0.058	0.057	0.057
44	0.056	0.055	0.054	0.053	0.052	0.051	0.050	0.049	0.048	0.047
45	0.046	0.045	0.044	0.043	0.042	0.041	0.040	0.039	0.038	0.037
46	0.036	0.035	0.034	0.033	0.032	0.031	0.031	0.030	0.029	0.028
47	0.027	0.026	0.025	0.024	0.023	0.022	0.021	0.020	0.020	0.019
48	0.018	0.017	0.016	0.015	0.014	0.013	0.012	0.011	0.011	0.010
49	0.009	0.008	0.007	0.006	0.005	0.004	0.003	0.003	0.002	0.001

APPENDIX 6. ATOMIC ABSORPTION SPECTROMETRY. PHYSICAL CONSTANTS AND GENERAL CONDITIONS
(from general data of literature).

Element	Wave-length (Å)	Intensity (arc)	Line Atom I / Ion II	Energy Levels (eV)	Resonance Line	Self-Reverse Line (arc)	Oscillator Strength	Flame	Flame Quality	Sensitivity (µ g/ml)	Analytical Range (µ g/ml)
Ag	328.1	2000	I	0-3.778	+	+	0.51	A.P.	lean	0.05	0.2-10
	358.3	1000	I	0-3.664	+	+	0.25	A.P.	lean	0.1	0.5-20
Al	237.3	—	I	0.014-5.235	−	+	—	N.A.	rich	2	20-500
	308.2	800	I	0-4.020	+	−	0.22	N.A.	rich	0.5	5-100
	309.3	1000	I	0.014-4.020	−	−	0.23	N.A.	rich	0.3	1-20
	394.4	2000	I	0-3.143	+	−	0.15	N.A.	rich	0.6	5-100
	396.1	3000	I	0.014-3.143	−	−				0.4	4-50
As	193.7	400	I	0-6.398	+	+	0.095	Ar-H	−	0.5	5-100
	197.2	250	I	0-6.285	+	+	0.07	Ar-H	−	1	10-100
Au	242.8	1000	I	0-5.105	+	+	0.3	A.A.	lean	0.15	0.5-20
	267.6	1000	I	0-4.632	+	+	0.19	A.A.	lean	0.5	2-50
Ba	455.4	2000	I	0-2.710	+	+	—	N.A.	rich	1	5-100
	553.6	1000	II	0-2.230	+	+	1.4	N.A.	rich	0.4	1-25
Be	234.9	2000	I	0-5.277	+	+	0.24	N.A.	rich	0.05	0.2-20
Bi	222.8	100	I	0-5.562	+	+	0.002	A.A.	lean	0.5	5-50
	223.1	100	I	0-5.556	+	+	0.012	A.P.	lean	0.8	10-100
Ca	239.9	500	I	0-5.170	+	+	0.037	A.A.	lean	20	−
	396.9	500	II	0-3.110	+	+	1.49	N.A.	−	−	−
	422.7	1500	I	0-2.920	+	+	1.2	N.A.−A.A.	lean	0.03	0.1-10
Cd	228.8	300	I	0-5.390	+	+	0.0018	A.P.	lean	0.01	0.1-50
	326.1	100	I	0-3.780	+	−	0.22	A.A.	lean	20	−
Co	240.7	250	I	0-5.149	+	+	0.19	A.A.	lean	0.06	0.2-10
	242.5	75	I	0-5.111	+	+	0.19	A.A.	lean	0.2	1-20
	252.1	500	I	0-4.916	+	+	0.34	A.A.	lean	0.5	−
Cr	357.9	500	I	0-3.460	+	+	0.27	A.A.	rich	0.04	0.1-10
	359.3	500	I	0-3.440	+	+	0.10	N.A.	rich	0.10	1-50
	425.4	5000	I	0-2.900	+	+	—	A.A.	rich	0.5	−
Cs	455.4	2000	I	0-2.720	+	+	0.8	A.A.	lean	20	−
	852.1	1000	I	0-1.450	+	+	0.8	A.A.	lean	0.5	5-100

Element	Wave-length (Å)	Intensity (arc)	Line Atom I / Ion II	Energy Levels (eV)	Resonance Line	Self-Reverse Line (arc)	Oscillator Strength	Flame	Flame Quality	Sensitivity (μg/ml)	Analytical Range (μg/ml)
Cu	217.9	30	I	0-5.688	+	+	0.011	A.A.	lean	0.6	5-100
	222.6	40	I	0-5.569	+	+	0.004	A.A.	lean	2.0	5-100
	324.8	5000	I	0-3.817	+	+	0.74	A.A.	lean	0.02	0.1-10
	327.4	3000	I	0-3.786	+	+	0.38	A.A.	lean	0.08	2-50
Fe	248.3	500	I	0-4.991	+	+	0.34	A.A.	lean	0.1	0.5-25
	248.8	600	I	0.052-5.033	-	+	-	A.A.	lean	0.08	0.2-20
	252.3	300	I	0-4.913	+	+	0.30	A.A.	lean	0.2	1-50
	271.9	500	I	0-4.558	+	+	0.15	A.A.	lean	0.4	2-50
Ga	287.4	10	I	0-4.290	+	+	0.32	A.A.	lean	2	10-100
Hg	253.7	2000	I	0-4.870	+	+	0.03	cold vapour		0.5 ng	1-10 ng
K	404.4	800	I	0-3.050	+	+	0.11	A.P.	lean	10	-
	766.5	9000	I	0-1.610	+	-	0.69	A.P.	lean	0.1	0.5-10
La	550.1	100	I	0	+	-	0.15	N.A.	rich	40	-
Li	323.3	1000	I	0-3.830	+	+	0.026	A.A.-A.P.	-	20	-
	670.7	3000	I	0-1.850	+	+	0.71	A.A.-A.P.	-	0.02	0.1-10
Mg	279.6	150	II	0-4.410	+	+	1.65	N.A.	-	0.2	0.05-2
	285.2	300	I	0-4.330	+	+	1.2	A.A.	-	0.002	0.1-10
Mn	279.5	1000	I	0-4.428	+	+	0.58	A.A.	lean	0.02	2-25
	403.1	500	I	0-3.073	+	+	-	A.A.	lean	0.5	2-20
Mo	313.3	1000	I	0-3.957	+	+	0.2	N.A.-A.A.	rich	0.2	5-50
	319.4	1000	I	0-3.881	+	+	-	A.A.	rich	1	-
	379.8	1000	I	0-3.263	+	+	0.13	A.A.	rich	1	-
Na	330.2	600	I	0-3.740	+	+	0.055	A.P.	lean	5	-
	589.0	9000	I	0-2.090	+	+	0.76	A.A.	lean	0.01	0.1-10
Ni	231.1	50	I	0-5.363	+	+	0.095	A.A.	lean	0.5	5-100
	232.0	30	I	0-5.342	+	+	0.30	A.A.	lean	0.1	0.25-20
	341.5	1000	I	0.025-3.655	-	+	0.12	A.A.	lean	0.2	0.5-50
	352.4	1000	I	0.025-3.542	-	+	0.39	A.A.	lean	2.5	-
Pb	217.0	1000	I	0-5.712	+	+	-	A.A.	lean	0.15	0.5-25
	261.4	200	I	0.969-5.712	-	+	0.21	A.A.	lean	5.0	-
	283.3	500	I	0-4.375	+	+	0.12	A.A.	lean	0.2	1-50
Pt	265.9	2000	I	0-4.660	+	+	-	-	-	2.5	10-100
	306.5	2000	I	0-4.044	+	+	-	A.A.	-	5.0	-

Element	λ (Å)	No.	I/II	Working range	±	±	Value	Method	lean/rich	Sens.	Range
Rb	780.0	9000	I	0–1.590	+	+	0.8	A.A.	lean	0.5	2–25
Rh	343.5	1000	I	0–3.590	+	+	0.073	A.A.	lean	0.15	1–20
Sb	206.8	300	I	0–5.992	+	+	0.1	A.A.	–	0.5	2–20
	217.6	300		0–5.696	+	+	0.045	A.A.	–	1.1	–
Se	196.1		I	0–6.323	–	–	0.12	A.A.	–	0.8	5–50
Si	251.6	500	I	0.028–4.953	+	+	0.26	N.A.	rich	2	10–100
Sn	224.6	100	I	0–5.518	+	–	0.41	N.A.	rich	0.8	4–50
	286.3	300	II	0–4.329	+	+	0.23		rich	2.5	–
Sr	407.8	400	I	0–3.040	+	+	0.76	A.A.	–	3	0.2–10
	460.7	1000	I	0–2.680	+	+	1.54	A.A.–N.A.	–	0.05	2–50
Te	214.3	600	I	0–5.783	+	+	0.08	A.A.	lean	0.5	–
Ti	363.5	200	I	0–3.409	+	–		N.A.	rich	1.5	10–100
Tl	364.3	300	I	0.021–3.424	–	–	0.25		rich	0.2	1–20
	276.8	400	I	0–4.478	+	+	0.27	A.A.	–		–
	377.6	3000	I	0–3.283	–	–	0.13	N.A.	–		–
U	358.5	30	I		–	–		N.A.	–	100	–
V	306.6	400	I	0.68–4.110	+	+	0.6	N.A.	–	7	2.5–25
	318.4	500	I	0.040–3.920	+	–		N.A.	–	1	–
	370.4	200	I	3.00–3.645	+	–	1.2	N.A.	–	7	0.05–2
Zn	213.9	800	I	0–5.792	+	+	0.00017	A.A.	–	0.01	–
	307.6	150	I	0–4.010	–	–		A.A.	–	150	–
Zr	301.2	100	I	1.537	–	–	0.22	N.A.	rich	20	–
	360.1	400	I		–	–		N.A.	rich	30	–

APPENDIX 7. ABSOLUTE SENSITIVITIES IN NONFLAME ATOMIC ABSORPTION

(From l'Vov, *Atomic Absorption Spectrochemical Analysis*, London: A. Hilger, 1970).

Elements	Line (Å)	Curvette Diameter r (mm)	Argon Pressure P (atm)	Temperature T (°C)	Absorbance A	Quantity c (g)	Sensitivities M (g)
Ag	3281	2.5	2	1800	0.020	$5.0 \cdot 10^{-13}$	$1 \cdot 10^{-13}$
Al	3093	4.5	1	2100	0.03	$2.5 \cdot 10^{-11}$	$1 \cdot 10^{-12}$
As	1973	2.5	2	1700	0.30	$5.6 \cdot 10^{-10}$	$8 \cdot 10^{-12}$
Au	2428	2.5	2	1700	0.28	$7.0 \cdot 10^{-11}$	$1 \cdot 10^{-12}$
B	2498	2.5	2	2400	0.105	$5.0 \cdot 10^{-9}$	$2 \cdot 10^{-10}$
Ba	5535	3.0	1	2200	0.05	$1.0 \cdot 10^{-10}$	$6 \cdot 10^{-12}$
Be	2349	4.5	6	2800	0.17	$2.6 \cdot 10^{-12}$	$3 \cdot 10^{-14}$
Bi	3068	2.5	2	1800	0.031	$2.5 \cdot 10^{-11}$	$4 \cdot 10^{-12}$
Ca	4227	2.5	2	2300	0.244	$2.5 \cdot 10^{-11}$	$4 \cdot 10^{-13}$
Cd	2288	1.2	1	1500	0.014	$6.0 \cdot 10^{-14}$	$8 \cdot 10^{-14}$
Co	2407	2.5	2	2200	0.020	$7.5 \cdot 10^{-12}$	$2 \cdot 10^{-12}$
Cr	3579	2.5	2	2200	0.125	$5.0 \cdot 10^{-11}$	$2 \cdot 10^{-12}$
Cs	8521	2.5	2	1900	0.072	$6.6 \cdot 10^{-12}$	$4 \cdot 10^{-13}$
Cu	3248	2.5	2	2100	0.044	$6.3 \cdot 10^{-12}$	$6 \cdot 10^{-13}$
Eu	4594	4.0	2.5	2450	0.13	$3.9 \cdot 10^{-10}$	$5 \cdot 10^{-12}$
Fe	2483	2.5	2	2100	0.11	$2.5 \cdot 10^{-11}$	$1 \cdot 10^{-11}$
Ga	2874	2.5	2	2100	0.12	$2.5 \cdot 10^{-11}$	$1 \cdot 10^{-12}$
Ge	2652	7.0	5	2400	0.19	$8.7 \cdot 10^{-10}$	$3 \cdot 10^{-12}$
Hg	2537	2.5	2	700	0.014	$5.0 \cdot 10^{-11}$	$2 \cdot 10^{-11}$
I	1830	4.0	1	1900	0.11	$2.0 \cdot 10^{-9}$	$3 \cdot 10^{-11}$
I	2062	2.5	2	2400	0.02	$5.0 \cdot 10^{-9}$	$1 \cdot 10^{-9}$
In	3039	2.5	1	1900	0.08	$8.0 \cdot 10^{-12}$	$4 \cdot 10^{-13}$
K	4044	2.5	2	1800	0.074	$6.3 \cdot 10^{-10}$	$4 \cdot 10^{-11}$
Li	6708	3.0	1	1900	0.05	$5.0 \cdot 10^{-11}$	$3 \cdot 10^{-12}$
Mg	2852	4.5	2	1800	0.10	$3.0 \cdot 10^{-12}$	$4 \cdot 10^{-14}$
Mn	2795	2.5	2	2000	0.048	$2.5 \cdot 10^{-12}$	$2 \cdot 10^{-13}$
Mo	3133	2.5	2	2500	0.08	$5.0 \cdot 10^{-11}$	$3 \cdot 10^{-12}$
Ni	2320	2.5	2	2200	0.012	$2.5 \cdot 10^{-11}$	$9 \cdot 10^{-12}$
P	1775	2.5	1	1900	0.10	$8.0 \cdot 10^{-11}$	$3 \cdot 10^{-12}$
P	2136	2.5	3	2400	0.01	$5.0 \cdot 10^{-10}$	$2 \cdot 10^{-10}$
Pb	2833	2.5	2	1900	0.063	$3.0 \cdot 10^{-11}$	$2 \cdot 10^{-12}$
Pd	2476	2.5	2	2100	0.05	$5.0 \cdot 10^{-11}$	$4 \cdot 10^{-12}$
Pt	2659	2.5	2	2300	0.086	$2.5 \cdot 10^{-10}$	$1 \cdot 10^{-11}$
Rb	7800	2.5	2	1900	0.030	$7.5 \cdot 10^{-12}$	$1 \cdot 10^{-12}$
Rh	3435	2.5	2	2300	0.034	$6.3 \cdot 10^{-11}$	$8 \cdot 10^{-12}$
S	1807	4.0	1	1900	0.12	$1.0 \cdot 10^{-8}$	$1 \cdot 10^{-10}$
Sb	2311	2.5	2	2000	0.040	$5.0 \cdot 10^{-11}$	$5 \cdot 10^{-12}$
Sc	3912	4.0	2.5	2350	0.08	$2.8 \cdot 10^{-9}$	$6 \cdot 10^{-11}$
Se	1961	2.5	2	1600	0.092	$2.0 \cdot 10^{-10}$	$9 \cdot 10^{-12}$
Si	2516	2.5	2	2250	0.23	$2.7 \cdot 10^{-12}$	$5 \cdot 10^{-14}$
Sn	2863	2.5	2	2000	0.022	$1.0 \cdot 10^{-11}$	$2 \cdot 10^{-12}$
Sr	4607	3.0	1	2200	0.05	$2.0 \cdot 10^{-11}$	$1 \cdot 10^{-12}$
Te	2143	2.5	2	2000	0.030	$7.6 \cdot 10^{-12}$	$1 \cdot 10^{-12}$
Ti	3653	2.5	2	2500	0.054	$5.0 \cdot 10^{-10}$	$4 \cdot 10^{-11}$
Tl	2768	2.5	2	1800	0.01	$2.5 \cdot 10^{-12}$	$1 \cdot 10^{-12}$
V	3184	4.5	2	2800	0.10	$2.5 \cdot 10^{-10}$	$3 \cdot 10^{-12}$
Yb	3988	4.0	2.5	2150	0.125	$5.0 \cdot 10^{-11}$	$7 \cdot 10^{-13}$
Zn	2138	4.5	4	1500	0.04	$1.0 \cdot 10^{-12}$	$3 \cdot 10^{-14}$

A : absorbance for concentration c in experimental conditions. M : absolute detection limit in g for standard conditions : P = 2 atm., r = 2.5 mm, absorbance A = 0.004 (1% absorption).

APPENDIX 8. ATOMIC FLUORESCENCE SPECTROMETRY

Physical and Instrumental Conditions: Flame Atomizer
(From literature data and V. Sychra, V. Svoboda and I. Rubeska: *Atomic Fluorescence
Spectroscopy*, London: Van Nostrand Reinhold Co., 1975; G. F. Kirkbright and
M. Sargent. *Atomic Absorption and Fluorescence Spectroscopy*, London:
Academic Press, 1974)

Element	Fluorescence Line (nm)	Energy Levels (eV)	Excitation Source	Atomizer	Detection Limit (g/ml)
Ag	328.1	0-3.78	EDL, HIL	A.H.	0.002
Al	396.15	0.014-3.143	HIL	S.N.A.	0.1
As	197.2	0-6.28	EDL	Ar-H	0.1
Au	242.8	0-5.105	HCL, HIL	A.A.-S.A.A.	0.005
Ba	553.5		HCL	A.A.	
Be	234.9		EDL	N.A.	0.04
			HIL	S.N.A.	0.008
Bi	302.5	1.91-6.01	EDL (Ir)	Ar-H	0.1
			EDL	S.A.A.	0.04
Ca	422.7		EDL	A.A.	0.02
			HCL	A.H.	0.2
Cd	228.8		EDL	A.H.	0.001
Co	240.7	0-5.15	EDL, HIL	A.A.	0.05
Cr	357.8		EDL, HIL	A.H.	0.01
				A.P.	0.005
Cu	324.7	0-3.82	EDL	A.H.	0.01
Fe	248.3	0-4.99	EDL	A.A.	0.05
			HIL	S.A.A.	0.003
Ga	417.2		EDL, MVL	A.A.-A.H.	0.1
Ge	265.16	0-4.67	HIL	N.A.-A.A.	0.2
Hg	253.6		EDL	A.H.	0.1
In	410.2		EDL, MVL	A.H.-A.A.	0.1
Mg	285.2		EDL, HIL	A.A.-A.H.	0.01
Mn	279.5	0-4.43	EDL	A.H.	0.05
Mo	313.3	0-3.96	HIL	S.N.A.	0.5
Ni	232.0	0-5.34	EDL	A.H.	0.1
			HIL	S.A.A.	0.002
Pb	405.8	1.32-4.375	EDL	A.H.	0.1
Pt	265.9	0-4.66	HIL	S.A.A.	0.5
Se	196.0	0-6.32	EDL	A.A.-A.H.	0.2
Sb	217.6	0-5.696	EDL	A.P.	0.05
				S.A.A.	0.04
Si	251.6	0.028-4.95	EDL	S.N.A.	0.5
Sn	303.4	0.210-4.295	EDL	A.H.	0.3
Sr	460.7		EDL	Ar-H	0.03
Te	214.3	0-5.78	EDL	A.H.	0.2
Ti	319.99		HIL	S.N.A.	4.0
Tl	377.57	0-3.28	EDL	A.H.	0.1
V	318.4		EDL	S.N.A.	0.1
Zn	213.9		EDL	A.H.	0.0001

HCL :	hallow cathode lamp
EDL :	electrodeless discharge lamp
HIL :	high intensity hallow cathode lamp
MVL :	metal vapor discharge lamp
AP :	air-propane flame
AH :	air-hydrogen flame
Ar-H :	argon-hydrogen flame
AA :	air-acetylene flame
SAA :	air-acetylene separated flame
NA :	nitrous oxide-acetylene flame
SNA :	nitrous oxide separated flame

Appendix 8, Continued

Electrothermal Atomizer

Element	Wavelength (nm)	Excitation Source	Atomizer	Absolute Detection (ng)	Remarks
Ag	328.1	EDL	MFA	2.0	Solution
		HIL	CFA	0.001	Solution
		EDL	GEA	0.0012	Solid sample
		HCL	CRA	0.0005	Solution
As	231.1	HCL	CRA	0.5	Solution
Au	242.8	HIL	CFA	0.005	Solution
	267.6	HIL	CFA	0.004	
Be	234.9	HCL	CRA	0.030	Solution
Bi	306.8	EDL	CFA	0.010	Solution
		EDL	MFA	0.040	Solution
Ca	422.7	HCL	CRA	0.0001	Solution
Cd	228.8	EDL	MFA	0.00002	Solution
		EDL	CFA	0.00002	Solution
		EDL	GEA	0.0010	Solid sample
Cu	324.7	HIL	CFA	0.04	Solution
		HCL	CRA	0.005	Solution
			GEA	0.10	Solid sample
Fe	242.8	HCL	CRA	0.002	Solution
Ga	417.2	EDL	CFA	0.050	Solution
Hg	253.6	Hg V	Q C	0.6	Cold vapor
Mg	285.2	HCL	GFA	0.0035	Solution
		HIL	GFA	0.001	Solution
Mn	279.5		GEA	0.15	Solid sample
Pb	283.3	HCL	GFA	0.035	Solution
	405.7	HIL	CFA	0.010	Solution
Tl	377.6	EDL	CFA	0.050	Solution
Zn	213.9	HCL	GFA	0.00004	Solution
		EDL	MFA	0.020	Solution
		EDL	CFA	0.00002	Solution
			GEA	0.010	Solid sample

EDL	:	electrodeless discharge lamp
HIL	:	high intensity hallow cathode lamp
HCL	:	hallow cathode lamp
Hg V	:	mercury vapor lamp
MFA	:	metal filament atomizer
CFA	:	carbon filament atomizer
CRA	:	carbon rod atomizer
GEA	:	graphite electrode atomizer
GFA	:	graphite furnace atomizer
Q C	:	quartz cell

APPENDIX 9. X-RAY SPECTROMETRY
K. Spectra of Elements (Angstroms) Na (11) to U (93)
(From Liebhafsky, H. A., *et al.* *X-Ray Absorption and Emission in Analytical Chemistry,* New York: J. Wiley, 1960.)

Atomic Number	Element	$K\alpha1$	$K\alpha2$	$K\beta1$	$K\beta2$	Atomic Number	Element	$K\alpha1$	$K\alpha2$	$K\beta1$	$K\beta2$
11	Na	11.909	11.909	11.617	— —	48	Cd	0.535	0.539	0.475	0.465
12	Mg	9.889	9.889	9.558	— —	49	In	0.512	0.517	0.455	0.445
13	Al	8.337	8.339	7.981	— —	50	Sn	0.491	0.495	0.435	0.426
14	Si	7.125	7.128	6.768	— —	51	Sb	0.470	0.475	0.417	0.408
15	P	6.155	6.155	5.804	— —	52	Te	0.451	0.456	0.400	0.391
16	S	5.372	5.375	5.032	— —	53	I	0.433	0.438	0.384	0.375
17	Cl	4.728	4.730	4.403	— —	54	Xe	0.416	0.420	0.368	0.360
18	A	4.192	4.195	— —	— —	55	Cs	0.400	0.405	0.354	0.346
19	K	3.741	3.745	3.454	— —	56	Ba	0.385	0.390	0.341	0.333
20	Ca	3.358	3.362	3.090	— —	57	La	0.371	0.375	0.328	0.320
21	Sc	3.031	3.035	2.780	— —	58	Ce	0.357	0.362	0.316	0.308
22	Ti	2.748	2.752	2.514	— —	59	Pr	0.344	0.349	0.304	0.297
23	V	2.503	2.507	2.284	— —	60	Nd	0.332	0.336	0.293	0.286
24	Cr	2.290	2.294	2.085	— —	61	Pm	0.321	0.325	0.282	— —
25	Mn	2.102	2.106	1.910	— —	62	Sm	0.309	0.314	0.273	0.266
26	Fe	1.936	1.940	1.757	— —	63	Eu	0.299	0.303	0.264	0.257
27	Co	1.789	1.793	1.621	— —	64	Gd	0.288	0.293	0.254	0.248
28	Ni	1.658	1.662	1.500	1.489	65	Tb	0.279	0.283	0.246	0.240
29	Cu	1.541	1.544	1.392	1.381	66	Dy	0.270	0.274	0.238	0.232
30	Zn	1.435	1.439	1.295	1.284	67	Ho	0.261	0.266	— —	— —
31	Ga	1.340	1.344	1.208	1.196	68	Er	0.252	0.257	0.223	0.217
32	Ge	1.254	1.258	1.129	1.117	69	Tm	0.244	0.249	0.215	— —
33	As	1.176	1.180	1.057	1.045	70	Yb	0.237	0.241	0.209	0.204
34	Se	1.105	1.109	0.992	0.980	71	Lu	0.229	0.234	0.202	0.197
35	Br	1.040	1.044	0.933	0.921	72	Hf	0.222	0.227	0.196	0.191
36	Kr	0.980	0.984	0.878	0.866	73	Ta	0.215	0.220	0.190	0.185
37	Rb	0.926	0.930	0.829	0.816	74	W	0.209	0.214	0.184	0.179
38	Sr	0.875	0.879	0.783	0.771	75	Re	0.203	0.208	0.179	0.174
39	Y	0.829	0.833	0.741	0.729	76	Os	0.197	0.202	0.174	0.169
40	Zr	0.786	0.790	0.702	0.690	77	Ir	0.191	0.196	0.169	0.164
41	Nb	0.746	0.750	0.666	0.654	78	Pt	0.186	0.190	0.164	0.159
42	Mo	0.709	0.714	0.632	0.621	79	Au	0.180	0.185	0.159	0.155
43	Tc	0.673	0.676	0.602	— —	81	Tl	0.170	0.175	0.150	— —
44	Ru	0.643	0.647	0.572	0.562	82	Pb	0.165	0.170	0.146	0.142
45	Rh	0.613	0.618	0.546	0.535	83	Bi	0.161	0.166	0.142	0.138
46	Pd	0.585	0.590	0.521	0.510	90	Th	0.133	0.138	0.117	0.114
47	Ag	0.559	0.564	0.497	0.487	92	U	0.126	0.131	0.111	0.109

Appendix 9, Continued

L Spectra of Elements (Angstroms) Cr (24) to Am (95)

Atomic Number	Element	L_I						L_II			L_III		
		β3	β4	γ3	γ2	β1	γ1	η	α1	β2	α2	ℓ	β6
24	Cr	19.429	---	---	---	21.323	---	24.339	21.714	---	---	24.840	---
25	Mn	17.575	---	---	---	19.159	---	21.864	19.489	---	---	22.315	---
26	Fe	15.742	---	---	---	17.290	---	19.770	17.602	---	---	20.202	---
27	Co	14.269	---	---	---	15.699	---	17.896	16.000	---	---	18.358	---
28	Ni	13.167	---	---	---	14.308	---	16.304	14.595	---	---	16.694	---
29	Cu	12.115	---	---	---	13.079	---	14.940	13.357	---	---	15.297	---
30	Zn	11.225	---	---	---	12.009	---	13.720	12.282	---	---	14.081	---
31	Ga	---	---	---	---	11.045	---	12.620	11.313	---	---	12.976	---
32	Ge	---	---	---	---	10.195	---	11.608	10.456	---	---	11.944	---
33	As	8.930	---	---	---	9.414	---	10.733	9.671	---	---	11.069	---
34	Se	---	---	---	---	8.736	---	9.959	8.990	---	---	10.293	---
35	Br	---	---	---	---	8.125	---	9.254	8.375	---	---	9.583	---
37	Rb	6.787	6.820	---	6.046	7.076	---	8.041	7.318	---	7.325	8.363	6.984
38	Sr	6.367	6.403	---	5.644	6.624	---	7.517	6.863	---	6.869	7.836	6.519
39	Y	5.983	6.018	---	5.283	6.212	---	7.040	6.449	---	6.456	7.356	6.094
40	Zr	5.633	5.668	---	4.953	5.836	5.384	6.607	6.070	5.586	6.078	6.918	5.710
41	Nb	5.310	5.345	---	4.654	5.492	5.036	6.211	5.724	5.238	5.732	6.517	5.361
42	Mo	5.013	5.049	---	4.380	5.177	4.726	5.847	5.406	4.923	5.414	6.150	5.049
44	Ru	4.487	4.523	---	3.897	4.620	4.182	5.205	4.846	4.372	4.853	5.503	4.487
45	Rh	4.252	4.289	---	3.685	4.374	3.943	4.922	4.597	4.131	4.605	5.217	4.242
46	Pd	4.034	4.071	---	3.489	4.146	3.724	4.660	4.368	3.909	4.376	4.952	4.016
47	Ag	3.833	3.870	---	3.306	3.934	3.523	4.418	4.154	3.703	4.163	4.707	3.808
48	Cd	3.645	3.682	---	3.138	3.738	3.336	4.193	3.956	3.514	3.965	4.480	3.614
49	In	3.470	3.507	---	2.980	3.555	3.162	3.983	3.772	3.338	3.781	4.269	3.436
50	Sn	3.306	3.343	---	2.833	3.385	3.001	3.789	3.600	3.175	3.609	4.071	3.269
51	Sb	3.152	3.190	---	2.695	3.226	2.851	3.608	3.439	3.023	3.448	3.888	3.115
52	Te	3.009	3.046	---	2.567	3.077	2.712	3.438	3.289	2.882	3.298	3.717	2.971
53	I	2.874	2.911	---	2.447	2.937	2.582	3.280	3.148	2.750	3.158	3.557	2.837
55	Cs	2.628	2.666	2.233	2.237	2.683	2.348	2.993	2.892	2.511	2.902	3.267	2.593
56	Ba	2.516	2.555	2.134	2.138	2.567	2.241	2.862	2.775	2.404	2.785	3.135	2.482

Appendix 9, Continued

Atomic Number	Element	L_I					L_II				L_III		
		β3	β4	γ3	γ2	β1	γ1	η	α1	β2	α2	ι	β6
57	La	2.410	2.449	2.041	2.046	2.458	2.142	2.740	2.665	2.303	2.674	3.006	2.379
58	Ce	2.311	2.349	1.955	1.960	2.356	2.048	2.620	2.561	2.209	2.570	2.892	2.281
59	Pr	2.217	2.255	1.874	1.879	2.258	1.961	2.512	2.463	2.119	2.473	2.784	2.190
60	Nd	2.126	2.167	1.796	1.801	2.167	1.878	2.409	2.370	2.036	2.380	2.676	2.104
61	Pm	—	—	—	—	2.081	—	—	2.283	—	—	—	—
62	Sm	1.962	2.000	1.655	1.659	1.998	1.727	2.218	2.199	1.882	2.210	2.482	1.946
63	Eu	1.887	1.926	1.591	1.597	1.920	1.658	—	2.121	1.812	2.132	2.395	1.874
64	Gd	1.815	1.853	1.529	1.534	1.846	1.592	2.049	2.046	1.745	2.057	2.312	1.807
65	Tb	1.746	1.785	1.471	1.477	1.776	1.530	—	1.975	1.682	1.986	2.234	1.741
66	Dy	1.681	1.720	1.417	1.423	1.710	1.473	1.897	1.909	1.623	1.920	2.158	1.681
67	Ho	1.619	1.659	1.364	1.370	1.647	1.417	1.826	1.845	1.567	1.856	2.086	1.622
68	Er	1.561	1.601	1.315	1.321	1.587	1.364	1.756	1.784	1.514	1.796	2.019	1.567
69	Tm	1.505	1.544	1.268	1.274	1.530	1.315	1.696	1.726	1.463	1.737	1.955	1.515
70	Yb	1.452	1.491	1.222	1.229	1.476	1.268	1.636	1.672	1.415	1.683	1.894	1.466
71	Lu	1.401	1.440	1.179	1.185	1.424	1.222	1.578	1.619	1.370	1.630	1.836	1.419
72	Hf	1.353	1.392	1.138	1.144	1.374	1.179	1.523	1.570	1.326	1.580	1.781	1.374
73	Ta	1.307	1.346	1.099	1.105	1.327	1.138	1.471	1.522	1.284	1.533	1.728	1.331
74	W	1.263	1.302	1.062	1.068	1.282	1.099	1.421	1.476	1.245	1.487	1.678	1.290
75	Re	1.220	1.259	1.026	1.032	1.239	1.061	1.373	1.433	1.207	1.444	1.630	1.251
76	Os	1.180	1.218	0.992	0.998	1.197	1.025	1.328	1.391	1.170	1.402	1.585	1.213
77	Ir	1.141	1.180	0.959	0.965	1.158	0.991	1.284	1.351	1.135	1.362	1.541	1.178
78	Pt	1.104	1.142	0.928	0.934	1.120	0.958	1.243	1.313	1.102	1.324	1.499	1.144
79	Au	1.068	1.107	0.898	0.904	1.084	0.926	1.203	1.276	1.070	1.288	1.460	1.111
80	Hg	1.034	1.072	0.869	0.875	1.049	0.896	1.164	1.241	1.040	1.253	1.422	1.080
81	Tl	1.001	1.039	0.841	0.848	1.015	0.868	1.128	1.207	1.010	1.219	1.385	1.050
82	Pb	0.969	1.007	0.815	0.821	0.982	0.840	1.092	1.175	0.983	1.186	1.350	1.021
83	Bi	0.939	0.977	0.789	0.796	0.952	0.813	1.059	1.144	0.955	1.155	1.316	0.993
84	Po	0.909	0.947	0.764	—	0.922	0.787	—	1.114	0.929	1.126	—	0.967
87	Fr	—	—	—	—	0.840	0.716	—	1.030	0.858	—	—	—
88	Ra	0.803	0.841	0.675	0.682	0.814	0.695	0.907	1.005	0.835	1.016	1.167	0.871
90	Th	0.755	0.793	0.636	0.642	0.765	0.653	0.854	0.956	0.794	0.968	1.115	0.828
91	Pa	0.732	0.770	0.617	0.624	0.742	0.634	0.829	0.933	0.774	0.945	1.091	0.808
92	U	0.710	0.748	0.599	0.605	0.720	0.615	0.805	0.911	0.755	0.922	1.067	0.788
93	Np	—	—	—	—	0.698	0.597	—	0.889	0.736	—	—	—
95	Am	—	—	—	—	0.658	0.562	—	0.849	0.701	0.860	—	—

APPENDIX 10. MASS ABSORPTION COEFFICIENTS IN THE REGION 0.1–10 ANGSTROMS. ELEMENTS H (1) to Sn (50)

From Liebhafsky, H. A. et al. X-Ray Absorption and Emission in Analytical Chemistry, New York: J. Wiley, 1960)

Atomic No.	Element	0.1	0.15	0.2	0.25	0.3	0.4	0.5	0.6	0.7	0.8	0.9	1.0	1.5	2.0	2.5	3	4	5	6	7	8	9	10
1	H	0.29	0.32	0.34	0.36	0.37	0.38	0.40	0.42	0.43	0.44	0.44	0.45	0.49	0.52	0.62	0.75	1.25	2.12	3.28	4.85	7.1	10.0	13.7
2	He	0.114	0.124	0.132	0.140	0.146	0.159	0.173	0.186	0.203	0.222	0.241	0.255	0.355	0.715	1.04	1.48	3.55	6.9	11.6	18.1	26.6	37.7	51
3	Li	0.124	0.132	0.143	0.153	0.163	0.180	0.198	0.223	0.254	0.302	0.358	0.428	1.02	2.18	3.98	6.6	15.2	28.8	48.8	76	113	157	213
4	Be	0.131	0.142	0.153	0.162	0.171	0.185	0.210	0.240	0.292	0.362	0.445	0.57	1.55	3.38	6.2	10.3	22.7	43.7	74	118	174	245	333
5	B	0.138	0.152	0.164	0.173	0.182	0.198	0.222	0.277	0.355	0.470	0.61	0.76	2.31	5.05	9.6	15.8	36.0	69	116	187	285	405	560
6	C	0.142	0.155	0.170	0.186	0.204	0.240	0.305	0.41	0.55	0.75	1.05	1.40	4.2	9.7	14.0	32	74	145	250	390	570	810	1100
7	N	0.144	0.159	0.175	0.195	0.216	0.288	0.395	0.62	0.89	1.25	1.73	2.20	6.9	16.0	30.5	52	123	235	400	620	910	1290	1800
8	O	0.145	0.162	0.181	0.206	0.236	0.345	0.508	0.87	1.25	1.80	2.40	3.20	10.5	24.0	45.5	78	180	350	580	910	1350	1900	2600
9	F	0.147	0.165	0.192	0.228	0.270	0.417	0.675	1.20	1.85	2.60	3.40	4.40	14.2	32	62	102	240	450	760	1190	1750	2450	3300
11	Na	0.150	0.175	0.228	0.287	0.380	0.630	1.18	2.05	3.25	4.75	6.70	8.80	32.5	53.5	132	250	520	920	1450	2200	3100	5200	5300
12	Mg	0.152	0.190	0.251	0.330	0.445	0.78	1.47	2.70	4.30	6.50	9.10	11.7	41.5	68	185	280	610	1100	1800	2700	3900	4300	340
13	Al	0.155	0.205	0.277	0.380	0.525	0.97	1.82	3.70	5.75	8.80	11.8	15.2	52	87	235	360	780	1400	2250	3300	280	390	520
14	Si	0.159	0.215	0.310	0.442	0.615	1.22	2.25	4.65	8.30	11.1	14.2	18.2	66	140	290	440	950	1700	2800	270	400	550	740
15	P	0.165	0.228	0.346	0.510	0.725	1.48	2.80	5.45	9.6	13.3	16.5	21.7	78.5	170	330	490	1050	1900	245	380	550	770	1070
16	S	0.170	0.241	0.392	0.592	0.855	1.78	3.45	6.40	11.3	15.5	19.3	26.0	95	202	375	590	1250	2200	320	500	730	1040	1400
17	Cl	0.176	0.260	0.433	0.667	1.02	2.13	4.25	7.45	13.0	18.5	22.6	29.7	130	280	430	680	1530	250	410	620	900	1250	1700
19	K	0.191	0.310	0.542	0.86	1.39	3.02	5.75	10.2	18.0	24	30.7	44	150	288	540	860	205	370	600	910	1300	1800	2350
20	Ca	0.200	0.327	0.601	0.98	1.63	3.45	6.50	11.5	20.2	26.5	36	48	150	328	625	950	255	460	730	1100	1550	2100	2800
21	Sc	0.210	0.358	0.667	1.12	1.83	3.95	7.4	13.1	22.7	30	41.0	55	172	375	720	145	310	550	880	1320	1850	2520	3300
22	Ti	0.221	0.395	0.740	1.26	2.10	4.50	8.4	14.8	25.5	34	47	62.5	195	425	113	175	360	630	1000	1450	2050	2750	3550
23	V	0.231	0.431	0.830	1.42	2.37	5.15	9.6	16.8	28.7	38	54	71.5	222	485	125	195	400	710	1120	1620	2250	3000	3700
24	Cr	0.241	0.480	0.925	1.60	2.68	5.88	11.0	18.9	32.2	42.5	60.3	80.5	245	540	136	210	440	760	1200	1720	2350	3200	4100
25	Mn	0.253	0.528	1.03	1.79	3.05	6.65	12.3	21.4	36.0	47.5	66.6	90	277	73	148	232	480	820	1260	1850	2550	3400	4350
26	Fe	0.265	0.58	1.16	2.02	3.45	7.6	14.1	23.8	39.8	53.2	74	100	312	80	162	255	510	880	1400	2050	2800	3700	4700
27	Co	0.285	0.64	1.28	2.25	3.80	8.35	15.5	26.6	43.6	58.2	80.5	110	345	87	177	275	550	960	1500	2200	3000	4000	5200
28	Ni	0.303	0.71	1.42	2.51	4.15	9.20	17.0	29.6	48.5	64	87.5	121	44	96	193	295	600	1050	1570	2350	3300	4300	5600
29	Cu	0.328	0.785	1.57	2.79	4.55	10.1	18.6	32.4	53.3	70	95.5	130	48.5	101	212	320	660	1150	1750	2550	3500	4600	6000
30	Zn	0.350	0.818	1.76	3.07	5.00	11.1	20.6	35.3	59	76.5	106	141	53.5	116	227	350	700	1200	1870	2700	3750	5000	6400
31	Ga	0.378	0.94	1.91	3.36	5.50	12.2	22.3	38.7	63.5	82	112	152	59	126	248	385	760	1300	2020	2950	4050	5050	5500
32	Ge	0.404	1.02	2.06	3.67	6.15	13.3	24.4	42.1	68.5	88	116	163	63	138	267	410	820	1400	2150	3100	4300	1650	4000
33	As	0.427	1.12	2.22	4.01	6.45	14.6	26.3	45.1	79.5	94	120	175	69	150	287	440	870	1500	2320	3330	4600	1750	2000
34	Se	0.472	1.21	2.41	4.33	7.00	15.8	28.5	48.4	93	101	129	43	74	165	311	470	940	1600	2500	3600	4900	2000	2150
35	Br	0.502	1.31	2.60	4.70	7.60	17.0	30.8	52.0	100	108	138	46	80	180	335	510	1020	1720	2700	3900	5200	2150	2300
37	Rb	0.572	1.50	3.03	5.45	9.05	19.7	35.9	59.5	108	113	28.0	49	94	207	386	590	1170	2000	3100	4200	5500	2350	2620
38	Sr	0.61	1.61	3.28	5.82	9.80	21.3	38.7	63	113	22	29.8	51.8	102	223	412	640	1260	2150	3300	4500	1250	2450	2800
39	Y	0.65	1.72	3.52	6.20	10.3	22.6	41.3	67.5	118	23.2	32	54.8	109	238	442	690	1380	2300	3450	4700	1440	2650	3000
40	Zr	0.69	1.83	3.73	6.60	11.0	24.0	43.5	72	16.8	24.5	33.8	61.2	118	255	478	740	1500	1840	3550	4900	1540	2800	3200
41	Nb	0.74	1.93	4.00	7.02	11.7	25.4	46.0	76	17.8	26	36.0	65	127	271	507	800	1600	1900	800	1300	1650	2950	3400
42	Mo	0.79	2.04	4.29	7.50	12.4	26.9	49.0	81	19.0	27.7	38.4	69	136	290	544	860	1700	600	910	1400	1770	3000	3650
44	Ru	0.89	2.27	4.70	8.4	13.8	30.5	53	15.0	21.6	31.4	43.7	72.7	158	342	625	960	1840	700	980	1650	1900	3400	4150
45	Rh	0.94	2.40	5.01	8.9	14.6	32.3	55.5	15.8	23.0	33.5	46.5	76.2	170	355	670	1050	1900	750	1090	1770	2050	3650	4450
46	Pd	1.00	2.53	5.25	9.4	15.5	34.3	58.2	16.6	24.5	35.5	49.5	80	183	380	720	1140	440	800	1160	1900	2500	3900	4750
47	Ag	1.05	2.67	5.50	9.9	16.3	36.5	10.0	17.5	26.0	38.0	52.8	84	196	405	770	1250	480	850	1230	2050	2700	4100	5100
48	Cd	1.10	2.80	5.74	10.3	17.2	38.5	10.7	18.5	27.7	40.2	56.2	88	210	432	820	1340	510	900	1350	2150	2900	4300	5400
49	In	1.15	2.94	5.98	10.8	17.8	40.6	11.4	19.5	29.5	42.7	60.0	92	222	460	880	1120	560	940	1450	2250	3100	4500	5700
50	Sn	1.20	3.07	6.22	11.3	18.7	43.1	12.0	20.5	31	45.5	63.5	96	236	495	940	775	620	990	1520	2360	3350	5400	5900

The stepped lines through the table mark the K and L (L_I, L_II, L_III) absorption edges.

Appendix 10, Continued. Elements Sb (51) to U (92)

Atomic No.	Element	0.1	0.15	0.2	0.25	0.3	0.4	0.5	0.6	0.7	0.8	0.9	1.0	1.5	2.0	2.5	3	4	5	6	7	8	9	10
51	Sb	1.25	3.22	6.48	11.8	19.5	45.5	12.8	21.6	33.1	48.5	67	88	250	525	1000	805	590	1050	1660	2480	3500	4700	6200
52	Te	1.30	3.37	6.76	12.3	20.3	7.4	13.6	22.7	35.2	51.5	70.6	93	265	557	1070	295	620	1100	1730	2600	3650	4900	6400
53	I	1.36	3.52	7.05	12.9	21.2	7.8	14.3	24	37.5	54.5	74.8	97	280	590	880	310	650	1160	1810	2700	3700	5100	6700
55	Cs	1.48	3.85	7.65	13.9	22.7	8.6	16.2	26.5	42.0	61.3	82.5	106	312	656	215	360	740	1300	2050	3000	4200	5600	7200
56	Ba	1.53	4.03	7.98	14.5	23.1	9.05	17.2	27.9	44.2	64	87	111	347	690	227	390	790	1370	2150	3150	4400	5800	5400
57	La	1.60	4.22	8.27	15.0	24.5	9.5	18.1	29.3	46.7	67.2	91	116	365	650	240	415	840	1450	2250	3300	4600	5800	5600
58	Ce	1.66	4.37	8.55	15.6	25.2	10.0	19.1	30.9	49.2	70.5	95	127	385	685	252	440	920	1570	2360	3400	4800	4750	5800
59	Pr	1.72	4.53	8.88	16.1	4.7	10.5	20.1	32.5	51.6	73.8	100	132	405	540	267	465	980	1600	2470	3650	5000	4950	6100
60	Nd	1.80	4.69	9.20	16.7	4.95	11.0	21.1	34.2	54.0	77.2	104	139	425	179	283	485	1040	1670	2580	3750	3900	5100	5800
61	Pm	1.86	4.86	9.57	17.3	5.20	11.5	22.1	36	56.9	80.7	108	144	446	187	295	515	1080	1750	2700	3900	3500	5300	6000
62	Sm	1.93	5.03	9.95	17.7	5.45	12.0	23.2	37.7	59.5	84	114	150	470	196	313	535	1120	1830	2810	4100	3600	3900	4500
63	Eu	2.02	5.20	10.3	18.3	5.71	12.5	24.2	39.8	62.5	87.8	119	157	470	206	330	555	1170	1900	2910	3300	3700	2900	3300
64	Gd	2.09	5.37	10.5	3.95	6.00	13.2	25.3	41.8	65.5	91.5	125	163	440	215	348	575	1230	1980	3020	3370	2700	3000	3400
65	Tb	2.18	5.52	10.8	4.15	6.28	13.9	26.6	44.0	69.5	95.5	130	170	465	225	368	610	1300	2090	3200	3440	2800	3100	3500
66	Dy	2.26	5.70	11.2	4.36	6.57	14.4	27.2	46.2	72	99.5	135	177	365	236	388	635	1350	2200	3400	3530	2200	3200	2800
67	Ho	2.33	5.87	11.6	4.55	6.90	15.1	28.2	48.6	75	103	141	185	365	247	409	665	1410	2280	2500	2800	2260	1890	2170
68	Er	2.42	6.03	11.9	4.75	7.2	15.8	30.2	50.8	79	107	147	193	136	258	432	690	1480	2350	2570	2300	2330	1970	2270
69	Tm	2.50	6.23	12.3	5.02	7.6	16.5	31.4	53.2	82.8	112	152	201	142	272	456	720	1540	2500	2650	2390	2400	2050	2360
70	Yb	2.58	6.41	12.7	5.27	7.95	17.3	32.7	55.8	87	115	158	210	148	285	482	750	1600	2600	2740	2480	1950	2130	2490
71	Lu	2.66	6.61	3.07	5.50	8.4	18.2	34.2	58.4	90.5	120	165	219	155	298	510	780	1660	1980	2060	2570	1580	2210	2500
72	Hf	2.75	6.80	3.22	5.80	8.8	19.0	35.5	61	94.2	124	172	229	162	313	540	825	1720	2050	2100	2650	1660	2290	2600
73	Ta	2.82	7.02	3.36	6.05	9.2	19.9	37.2	64	99	129	178	239	176	327	570	855	1800	2210	1520	1900	1740	2370	2700
74	W	2.90	7.24	3.50	6.27	9.7	20.8	38.7	66.5	103	134	185	163	184	342	600	890	1900	2280	1660	1350	1820	2450	2800
75	Re	2.96	7.45	3.63	6.60	10.2	21.8	40.5	69	106	139	192	170	193	358	636	925	1980	1750	1700	1400	1900	2560	2930
76	Os	3.03	7.65	3.78	6.9	10.6	22.8	42.2	72	110	144	200	178	200	376	672	970	1450	1300	1740	1460	1980	2670	3020
77	Ir	3.09	7.86	3.92	7.3	11.1	23.8	44.0	74.5	113	150	208	185	208	392	710	1010	1500	1340	1400	1530	2060	2780	3130
78	Pt	3.17	8.06	4.08	7.6	11.6	25.0	46.0	77.6	117	155	170	192	217	410	750	1070	1550	1380	1160	1590	2150	2900	3250
79	Au	3.23	8.28	4.22	8.0	12.2	26.2	47.5	80.5	122	160	170	200	225	430	795	1130	1150	1430	1200	1650	2250	2950	3370
80	Hg	3.30	2.17	4.38	8.3	12.7	27.4	49.5	83.7	126	165	135	208	233	445	825	1190	1200	1480	1250	1700	2300	3110	3490
81	Tl	3.36	2.24	4.53	8.7	13.3	28.6	51.5	86.6	130	170	145	72	242	460	865	1250	1250	1120	1300	1790	2370	3270	3600
82	Pb	3.41	2.32	4.67	9.2	13.9	30.0	53.5	89.2	135	147	150	75	252	478	905	1310	920	1150	1360	1880	2430	3430	3700
83	Bi	3.45	2.37	4.81	9.6	14.5	31.3	55.5	92	140	109	135	77	262	495	950	1380	730	910	1410	1970	2560	3600	3500
90	Th	3.81	2.76	5.45	12.4	19.5	41.2	69	110	109	52	70	95	328	638	930	920	730	1200	1700	2600	3460	4500	4300
92	U	3.91	2.85	5.61	13.1	20.9	44.0	83.1	79	103	55	74.4	101	354	683	975	750	790	1300	1890	2790	3690	3600	4820

Absorption-edge labels appearing along the staircase boundaries of the table: K; L_I, L_II, L_III; M_I, M_II, M_III, M_IV, M_V; N_I.

APPENDIX 11. ACTIVATION ANALYSIS.
PRINCIPAL STABLE AND RADIOACTIVE ISOTOPES OF ELEMENTS

Atomic N°	Element Total Absorption Cross Section (barn)	Stable Isotope (in italic): %, Thermal Neutrons Cross Section (barn). Radioisotope: Period, Emissions and Energy (MeV)
1	H $\sigma_t = 0.33$	^1H 99.985% $\sigma = 0.33 - {}^2H$ 0.015% $\sigma =$ 0.00057 $- {}^3$H 12.26 y β^- 0.018
3	Li $\sigma_t = 71$	$^6Li - 7.5\% - \sigma$ (n,α) = 950 $\sigma = \leqslant 0.1 - {}^7Li$ 92.5% $\sigma = 0.033 - {}^8$Li 0.84 s β^- 13
4	Be $\sigma_t = 0.01$	^{7}Be 53 d K $\gamma - {}^9Be$ 100% $\sigma = 0.01 - {}^{10}$Be $2.7 \cdot 10^6$ y β^- 0.56 ^{11}Be 14 s $^-$
5	B $\sigma_t = 755$	^8B 0.5 s β^+ 14 $- {}^{10}B$ 18.8% σ(n,α) = 4020 $-$ ^{11}B 81.2% $\sigma < 0.05$
6	C $\sigma_t = 0.003$	^{10}C 19 s β^+ 1.9 $\gamma - {}^{11}$C 20.5 m β^+ 0.96 $-$ ^{12}C 98.89% $\sigma = 0.0032 - {}^{13}$C 1.11% $\sigma = 0.0009$ $- {}^{14}$C 5600 y β^- 0.158 $\sigma = < 10^{-6} - {}^{15}$C 2.3 s β^- 9.8 γ
7	N $\sigma_t = 1.9$	^{13}N 10 m β^+ 1.20 $- {}^{14}N$ 99.63% σ (n,p) = 1.8 $\sigma = 0.1 - {}^{15}N$ 0.37% $\sigma = 0.00002 - {}^{16}$N 7.4 s β^- 10.4 $\gamma - {}^{17}$N 4.14 s β^- 3.7
8	O $\sigma_t = 0.0002$	^{14}O 72 s β^+ 1.83 $\gamma - {}^{15}$O 2.1 m β^+ 1.7 $- {}^{16}O$ 99.75% $\sigma < 0.00002 - {}^{17}O$ 0.037% σ(n,α) = 0.5 $- {}^{18}O$ 0.204% $\sigma = 0.0002 - {}^{19}$O 29 s β^- 4.4 γ
9	F $\sigma_t = 0.009$	^{17}F 66 s β^+ 1.75 $- {}^{18}$F 1.87 h β^+ 0.65 $- {}^{19}F$ 100% $\sigma = 0.009 - {}^{20}$F 11 s β^- 5.42 $\gamma - {}^{21}$F 5 s β^-
11	Na $\sigma_t = 0.53$	^{21}Na 23 s β^+ 2.50 $- {}^{22}$Na 2.6 y β^+ 0.54 K $\gamma - {}^{23}Na$ 100% $\sigma = 0.53 - {}^{24}$Na 15.0 h β^- 1.39 $\gamma - {}^{25}$Na 60 s β^- 4.0 γ

Atomic N°	Element Total Absorption Cross Section (barn)	Stable Isotope (in italic): %, Thermal Neutrons Cross Section (barn). Radioisotope: Period, Emissions and Energy (MeV)
12	Mg $\sigma_t = 0.063$	^{23}Mg　12 s　β^+ 3.0 $-$ ^{24}Mg　78.8%　$\sigma = 0.03$ $-$ ^{25}Mg 10.1%　$\sigma = 0.27$ $-$ ^{26}Mg　11.1%　$\sigma = 0.03$ $-$ ^{27}Mg 9.5 m　β^- 1.75　γ $-$ ^{28}Mg　21.3 h　β^- 0.45　γ
13	Al $\sigma_t = 0.23$	^{24}Al　2.1 s　$\beta^+ < 8.5$　γ $-$ ^{25}Al　7.3 s　β^+ 3.24 $-$ ^{26}Al　6.5 s　β^+ 3.21 $-$ ^{26m}Al　($\simeq 10^6$ y)　β^+ 1.20　γ $-$ ^{27}Al　100%　$\sigma = 0.23$ $-$ ^{28}Al　2.30 m β^- 2.87　γ $-$ ^{29}Al　6.6 m　β^- 2.5　γ
14	Si $\sigma_t = 0.13$	^{26}Si　1.7 s　β^+ $-$ ^{27}Si　4.4 s　β^+ 3,8 $-$ ^{28}Si　92.17% $\sigma = 0.1$ $-$ ^{29}Si　4.71%　$\sigma = 0.3$ $-$ ^{30}Si　3.12%　$\sigma =$ 0.11 $-$ ^{31}Si　2.62 h　β^- 1.48　γ $-$ ^{32}Si　$\simeq 300$ y β^- 0.1
15	P $\sigma_t = 0.20$	^{29}P $-$ 4.5 s　β^+ 3.94　γ $-$ ^{30}P　2.5 m $-$ β^+ 3.3 $-$ ^{31}P　100%　$\sigma = 0.20$ $-$ ^{32}P　14.5 d　β^- 1.71 $-$ ^{33}P 25 d　β^- 0.25 $-$ ^{34}P　12.4 s　β^- 5.1　γ
16	S $\sigma_t = 0.49$	^{31}S　2.6 s　β^+ 4.4 $-$ ^{32}S　95.0% $-$ ^{33}S　0.75% $\sigma(n,p) = 0.002$ $-$ ^{34}S　4.2%　$\sigma = 0.26$ $-$ ^{35}S　87 d　β^- 0.167 $-$ ^{36}S　0.017　$\sigma = 0.14$ $-$ ^{37}S　5.0 m　β^- 4.7　γ
17	Cl $\sigma_t = 33$	^{33}Cl $-$ 2.8 s　β^+ 4.2　γ $-$ ^{34m}Cl　32.4 m　β^+ 2.5　I T γ $-$ ^{34}Cl　1.5 s　β^+ 4.5 $-$ ^{35}Cl　75.53%　$\sigma = 44$　$\sigma(n,p) =$ 30 $-$ ^{36}Cl　$3 \cdot 1 \cdot 10^5$ y　β^- 0.71　K $-$ ^{37}Cl　24.47%　$\sigma =$ (0.005 + 0.56) $-$ ^{38}Cl　37.3 m　β^- 4.8　γ $-$ ^{38m}Cl 1 s　I T　0.66 $-$ ^{39}Cl　55 m　β- 3.0　γ $-$ ^{40}Cl　1.4 m $\beta^- \sim 7$　γ
19	K $\sigma_t = 2.0$	^{38m}K　0.95 s　β^+ 5.1 $-$ ^{38}K　7.7 m　β^+ 2.7　γ $-$ ^{39}K 93.2%　$\sigma = 1.9$ $-$ ^{41}K　6.8%　$\sigma = 1.1$ $-$ ^{42}K　12.5 h β^- 3.5　γ $-$ ^{43}K　22 h　β^- 1.84　γ $-$ ^{44}K　22 m β^- 4.9　γ $-$ ^{45}K　34 m　β^-
20	Ca $\sigma_t = 0.43$	^{39}Ca　1.0 s　β^- 5.7 $-$ ^{40}Ca　96.9%　$\sigma = 0.2$ $-$ ^{41}Ca $1.1 \cdot 10^5$ y　K $-$ ^{42}Ca　0.64%　$\sigma = 40$ $-$ ^{43}Ca　0.14% $-$ ^{44}Ca　2.1%　$\sigma = 0.6$ $-$ ^{45}Ca　160 d　β^- 0.25 $-$ ^{47}Ca 4.7 d　β^- 2.0　γ $-$ ^{48}Ca　0.18%　$\sigma = 1.1$ $-$ ^{49}Ca　8.7 m β^- 2.0　γ
21	Sc $\sigma_t = 23$	^{41}Sc　0.87 s　β^+ 5 $-$ ^{42}Sc　0.66 s　$\beta^+ \sim 5$ $-$ ^{43}Sc　3.9 h　β^+ 1.19 γ $-$ ^{44m}Sc　2.4 d　I T　0.271 $-$ ^{44}Sc　4.0 h　β^+ 1.47 $-$ K γ $-$ ^{45}Sc　100%　$\sigma = (10 + 13)$ $-$ ^{46m}Sc　20 s　I T　0.14 e$^-$ $-$ ^{46}Sc　85 d　β^- 0.36　β $-$ ^{47}Sc　3.4 d　β^- 0.60　γ $-$ ^{48}Sc　44 h β^- 0.64　γ $-$ ^{49}Sc　57 m　β^- 2.0 $-$ ^{50}Sc　1.7 m　$\beta^- \sim 3.5$ γ

Atomic N°	Element Total Absorption Cross Section (barn)	Stable Isotope (in italic): %, Thermal Neutrons Cross Section (barn). Radioisotope: Period, Emissions and Energy (MeV)
22	Ti $\sigma_t = 6.0$	^{43}Ti 0.6 s $\beta^+ - ^{44}$Ti $>$ 20 y K $\gamma - ^{45}$Ti 3.08 h β^+ 1.02 K ^{46}Ti 8.0% $\sigma = 0.6 - ^{47}Ti$ 7.4% $\sigma = 1.6 - ^{48}$Ti 73.8% $\sigma = 7.8 - ^{49}Ti$ 5.5% $\sigma = 1.8 - ^{50}Ti$ 5.3% $- \sigma =$ $0.14 - ^{51}$Ti 5.80 m $\beta - 2.1$ γ
23	V $\sigma_t = 4.9$	^{45}V 1 s $\beta^+ - $ ^{47}V 31 m β^+ 1.89 $- ^{48}$V 16.2 d β^+ 0.69 K $\gamma - ^{49}$V $\sim$1 y K $- ^{50}V$ 0.25% $\sigma = \sim 100 - ^{51}V$ 99.75% $\sigma = 4.5 - ^{52}$V 3.77 m β 2.6 $\gamma - ^{53}$V 2.0 m β^- 2.50 γ ^{54}V 55 s β^- 3.3 γ
24	Cr $\sigma_t = 3.1$	^{46}Cr 1.1 s $\beta^+ - ^{48}$Cr 23 h K $\gamma - ^{49}$Cr 42 m β^+ 1.54 $\gamma - ^{50}Cr$ 4.4% $\sigma = 16 - ^{51}$Cr 27 d K $\gamma - ^{52}Cr$ 83.7% $\sigma = 0.8 - ^{53}Cr$ 9.5% $\sigma = 18 - $ ^{54}Cr 2.4% $\sigma = 0.37 - ^{55}$Cr 3.6 m β^- 2.8
25	Mn $\sigma_t = 13.3$	^{51}Mn 45 m β^+ 2.2 $- ^{52m}$Mn 21 m β^+ 2.7 $\gamma - $ ^{52}Mn 5.7 d K β^+ 0.6 $\gamma - ^{53}$Mn $\sim$140 y K $- $ ^{54m}Mn 2 m I T? β^- o e$^- - ^{54}$Mn 300 d K $\gamma - $ ^{55}Mn 100% $\sigma = 13.3 - ^{56}$Mn 2 h 58 β^- 2.8 $\gamma - $ ^{57}Mn 1.7 m β^- 2.6 γ
26	Fe $\sigma_t = 2.5$	^{52}Fe 8 h β^+ 0.80(2.7) K $\gamma - ^{53}$Fe 9 m β^+ 2.6 $\gamma - ^{54}Fe$ 5.9% $\sigma = 2.2 - $ ^{55}Fe 2.9 y K $- $ ^{56}Fe $- $ 91.6% $\sigma = 2.6 - ^{57}Fe$ 2.20% $\sigma = 2.4 - $ ^{58}Fe 0.33% $\sigma = 0.9 - ^{59}$Fe 45 d β^- 1.56 $\gamma - $ ^{60}Fe $3 \cdot 10^5$ y β^- $(\gamma)0.059) - ^{61}$Fe 5.5 m β^- γ
27	Co $\sigma_t = 37$	^{55}Co 18 h β^+ 1.50 K $\gamma - ^{56}$Co 77 d K β^+ 1.50 $\gamma - ^{57}$Co $- $ 267 d K, e$^-$ $\gamma - ^{58m}$Co 9 h I T 0.025 e$^- - ^{58}$Co 71 d K β^+ 0.48 $\gamma - ^{59}$Co 100% $\sigma = $ $(18 + 19) - ^{60m}$Co 10.5 m I T 0.059 β^- 1.5 γ $\sigma = $ $\sim 100 - ^{60}$Co 5.2 y β^- 0.31 γ $\sigma \sim 6$ ^{61}Co 1.65 h β^- 1.22 $\gamma - ^{62}$Co 1.6 m β^- $\gamma - ^{62m}$Co 14 m β^- 2.8 γ
28	Ni $\sigma_t = 4.6$	^{56}Ni 6.4 d K $\gamma - ^{57}$Ni 36 h K β^+ 0.84 $\gamma - ^{58}Ni$ 68.0% $\sigma = 4.3 - ^{59}$Ni $8 \cdot 10^4$ y K $- ^{60}Ni$ 26.2% $\sigma = 2.6 - ^{61}Ni$ 1.1% $\sigma = 2 - $ ^{62}Ni 3.7% $\sigma = 15 - ^{63}$Ni 80 y β^- 0.063 $- $ ^{64}Ni 1.0% $\sigma = 2 - ^{65}$Ni 2.56 h β^- 2.10 γ ^{66}Ni 56 h β^- 0.3
29	Cu $\sigma_t = 3.7$	^{58m}Cu 9.5 m $\beta^+ < 0.7 - ^{58}$Cu 3 s $\beta^+ \sim 8 - ^{59}$Cu 81 s β^+ 3.7 $\gamma - ^{60}$Cu 24 m β^+ 3.9 $\gamma - ^{61}$Cu 3.3 h β^+ 1.22 K $\gamma - ^{62}$Cu 9.9 m β^+ 2.9 $- ^{63}Cu$ 69.0% $\sigma = 4.4 - ^{64}$Cu $12 \cdot 8$ h K β^- 0.57 β^+ 0.66

Atomic N^o	Element Total Absorption Cross Section (barn)	Stable Isotope (in italic): %, Thermal Neutrons Cross Section (barn). Radioisotope: Period, Emissions and Energy (MeV)
		$\gamma - {}^{65}Cu$ 31.0% $\sigma = 2.2 - {}^{66}Cu$ 5.1 m β^- 2.63 γ $\sigma = 140 - {}^{67}Cu$ 61 h β^- 0.58 γ $e^- - {}^{68}Cu$ 32 s β^- 3.0 γ
30	Zn $\sigma_t = 1.10$	^{60}Zn 2.1 m $- {}^{61}$Zn 1.5 m β^+ 5 $- {}^{62}$Zn 9 h K β^+ 0.66 γ $e^- - {}^{63}$Zn 38 m β^+ 2.36 K γ ${}^{64}Zn$ 48.9% $\sigma = 0.5 - {}^{65}$Zn 245 d K β^+ 0.33 $\gamma - {}^{66}Zn$ 27.8% $- {}^{67}Zn$ 4.1% ${}^{68}Zn$ 18.6% $\sigma = (0.09 + 1.0) - {}^{69m}$Zn 14 h IT $- {}^{69}$Zn 52 m β^- 0.90 $- {}^{70}Zn$ 0.63% $\sigma = 0.09$ ^{71m}Zn 3 h β^- 1.5 $\gamma - {}^{71}$Zn 2.2 m β^- 2.4 $\gamma - {}^{72}$Zn 49 h β^- 1.6 γ
31	Ga $\sigma_t = 2.9$	^{64}Ga 2.5 m $\beta^+ \sim 5$ $\gamma - {}^{65m}$Ga 15 m I T 0.052 γ, β^+ 2.5 $- {}^{65}$Ga 8 m β^+ 2.2 $- {}^{66}$Ga 9.4 h β^+ 4.15 K $\gamma - {}^{67}$Ga 78 h K $\gamma - {}^{68}$Ga 68 m β^+ 1.88 γ K $- {}^{69}Ga$ 60.1% $\sigma = 1.9 - {}^{70}$Ga 21 m β^- 1.65 $\gamma - {}^{71}Ga$ 39.9% $\sigma = 4.6 - {}^{72}$Ga 14.1 h β^- 3.17 $\gamma - {}^{73}$Ga 5 h β^- 1.4 γ
32	Ge $\sigma_t = 2.3$	^{66}Ge 2.5 h K β^+ (?) $- {}^{67}$Ge 19 m β^+ 3.4 γ $- {}^{68}$Ge 250 d K$- {}^{69}$Ge 40 h K β^+ 1.21 $\gamma -$ ^{70}Ge 20.5% $\sigma = 3.4 - {}^{71}$Ge 12 d K$- {}^{72}Ge$ 27.4% $\sigma = 1.0 - {}^{73}Ge$ 7.8% $\sigma = 14$ ^{73m}Ge 0.53 s I T 0.054 $\gamma - {}^{74}Ge$ 36.5 $\sigma = (0.2 + 0.5)$ $- {}^{75m}$Ge 49 s I T 0.14 $- {}^{75}$Ge 82 m β^- 1.18 $\gamma - {}^{76}Ge$ 7.8% $\sigma = (0.015 + 0.30) - {}^{77m}$Ge 52 s β^- 2.9 I T 0.16 $\gamma - {}^{77}$Ge 12 h β^- 2.2 $\gamma -$ ^{78}Ge 86 m β^- 0.9 γ
33	As $\sigma_t = 4.3$	^{68}As ~ 7 m $\beta^+ - {}^{69}$As 15 m β^+ 2.9 $\gamma - {}^{70}$As 50 m β^+ 2.5 $\gamma - {}^{71}$As 62 h β^+ 0.81 $\gamma - {}^{72}$As 26 h K β^1 3.34 $\gamma - {}^{73}$As 76 d K (e^-) $\gamma -$ ^{74}As 17 d K β^- 1.36 β^+ 1.53 $\gamma - {}^{75}As$ 100% $\sigma = 4.3 - {}^{76}$As 26.7 h β^- 2.96 $\gamma - {}^{77}$As 39 h β^- 0.69 $\gamma - {}^{78}$As 90 m β^- 4.1 $\gamma - {}^{79}$As $-$ 9 m β^- 2.3 (γ 0.096) ^{80}As ~ 36 s β^-
34	Se $\sigma_t = 13$	^{70}Se 44 m $\beta^+ - {}^{72}$Se 9.7 d K $- {}^{73m}$Se 7.1 h β^+ 1.65 $\gamma - {}^{73}$Se 44 m β^+ 1.7 $- {}^{74}Se$ 0.93% $\sigma = 40 - {}^{75}$Se 127 d K $\gamma - {}^{76}Se$ 9.1% $\sigma = (7 + 78) - {}^{77m}$Se 17 s I T 0.16 $- {}^{77}Se = 7.5\%$ $\sigma = 41$ $- {}^{78}Se$ 23.6% $\sigma = 0.4$ ^{79m}Se 3.9 m I T 0.096 $e^- - {}^{79}$Se $7 \cdot 10^4$ y β^- 0.16 $- {}^{80}Se$ 49.9% $\sigma =$ $(0.03 + 0.5) - {}^{81m}$Se 57 m I T 0.103 $e^- - {}^{81}$Se

Atomic N°	Element Total Absorption Cross Section (barn)	Stable Isotope (in italic): %, Thermal Neutrons Cross Section (barn). Radioisotope: Period, Emissions and Energy (MeV)
		18 m β^- 1.38 − ^{82}Se 9.0% σ = (0.06 + 0.004) − ^{83m}Se 69 s β^- 3.4 − ^{83}Se 25 m β^-1.5 γ−^{84}Se ~2 m β^-
35	Br σ_t = 6.6	^{74}Br 36 m β^+ K −^{75}Br 1.6 h K β^+ 1.70 γ−^{76}Br 17 h β^+ 3.57 K γ−^{77}Br 57 h K β^+ 0.34 γ−^{78m}Br 6.4 m I T γ−^{78}Br <6 m β^+ 2.4 −^{79}Br 50.6% σ = (2.9 + 8.5) − ^{80m}Br 4.6 h I T 0.05 γ−^{80}Br 18 m β^- 2.0 K β^+ 0.86 γ− ^{81}Br 49.4% σ = 2.6 − ^{82}Br 35.9 h β^- 0.46 γ− ^{83}Br 2.3 h β^- 0.94 γ−^{84}Br 32 m β^- 4.68 γ−^{85}Br 3.0 m β^- 2.5 − ^{87}Br 56 s β^- 8.0 γ− ^{88}Br 16 s β^- − ^{89}Br 4.5 s β^-
37	Rb σ_t = 0.7	^{81}Rb 4.7 h K β^+ 1.0 γ−^{82m}Rb 6.3 h K β^+ 0.77 γ−^{83}Rb 75 s β^+ 3.2 − ^{83}Rb 83 d K γ−^{84m}Rb 21 m I T 0.46 K γ−^{84}Rb 33 d K β^+ 1.7 γ β^- 0.4 − ^{85}Rb 72.2% σ = (0.05 + 0.8) − ^{86m}Rb 1 m I T 0.56 − ^{86}Rb 18.6 d β^- 1.77 γ−^{87}Rb 27.8% 4.3 · 10^{10} y β^- 0.27 σ = 0.14 − ^{88}Rb 18 m β^- 5.2 γ σ = <200 − ^{89}Rb 15 m β^- 3.9 γ−^{90}Rb 2.7 m β^- 5.7 γ−^{91m}Rb 1.7 m β^- 4.6 β−^{91}Rb 14 m β^- 3.0 γ−^{92}Rb ~80 s β^-
38	Sr σ_t = 1.3	^{81}Sr 29 m β^+ − ^{82}Sr 26 d K (β^+ 3.2) − ^{83}Sr 33 h β^+ 1.2 K γ−^{84}Sr 0.55 % σ = 1 − ^{85m}Sr 70 m I T γ K γ−^{85}Sr 65 d K − γ−^{86}Sr 9.8% σ = (1.3 + ?) − ^{87m}Sr 2.8 h I T 0.39 −^{87}Sr 7.0% − ^{88}Sr 82.7% σ = 0.005 − ^{89m}Sr ~10 d I T γ− ^{89}Sr 54 d β^- 1.48 (γ 0.91) σ = <130 − ^{90}Sr 28 y β^- 0.54 σ = 1 − ^{91}Sr 9.7 h β^- 2.67 γ− ^{92}Sr 2.7 h β^- ~0.05 γ−^{93}Sr 7 m β^- −^{94}Sr ~2 m β^-
39	Y σ_t = 1.3	^{82}Y 70 m β^+ 2 − ^{83}Y 3.5 h −^{84}Y 3.7 h − β^+ 2.0 K γ−^{85}Y 5 h − ^{86}Y 15 h β^+ 1.80 γ−^{87m}Y 14 h I T ^{87}Y 80 h K β^+ 0.7 γ−^{88}Y 105 d K β^+ 0.83 γ−^{89m}Y 16 s I T − ^{89}Y 100% σ = 1.3 − ^{90}Y 64.0 h β^- 2.27 σ = 6 − ^{91m}Y 50 m I T − ^{91}Y 58 d β^- 1.54 γ−^{92}Y 3.5 h β^- 3.60 γ−^{93}Y 10 h β^- 3.1 γ−^{94}Y 17 m β^- 5.4 γ−^{95}Y 10 m β^-

Atomic N°	Element Total Absorption Cross Section (barn)	Stable Isotope (in italic): %, Thermal Neutrons Cross Section (barn). Radioisotope: Period, Emissions and Energy (MeV)
40	Zr $\sigma_t = 0.18$	^{86}Zr ~17 h K γ—^{87}Zr 1.6 h β⁺ 2.10 K γ—^{88}Zr 85 d K γ—^{89m}Zr 4.4 m IT β⁺ 2.4 γ—^{89}Zr 79 h K β⁺ 0.90 (γ) —^{90m}Zr 0.8 s IT—^{90}Zr 51.5% σ = 0.1 —^{91}Zr 11.2% σ = 1 —^{92}Zr 17.1% σ = 0.2 —^{93}Zr 9·10³ y β⁻ 0.063 (γ) σ = <5 —^{94}Zr 17.4% σ = 0.1 —^{95}Zr 65 d β⁻ 0.88 γ—^{96}Zr 2.8% σ = 0.1 —^{97}Zr 17 h β⁻ 1.91 γ
41	Nb $\sigma_t = 1$	^{89m}Nb ~ 2 h β⁺ (γ) —^{89}Nb 1.9 h β⁺ 2.9 —^{90m}Nb 24 s IT—^{90}Nb 14.6 h β⁺ 1.50 γ—^{91m}Nb 62 d IT K γ—^{91}Nb long K—^{92m}Nb 13 h K γ—^{92}Nb 10 d K γ—^{93m}Nb 3.7 y IT—^{93}Nb 100% σ = (1 + ?) —^{94m}Nb 6.6 m IT β⁻ 1.3 γ— ^{94}Nb 20.000 y β⁻ 0.5 γ σ = 15 —^{95m}Nb 84 h IT —^{95}Nb 35 d β⁻ 0.16 γ—^{96}Nb 23 h β⁻ 0.7 γ—^{97m}Nb 1 m IT—^{97}Nb 72 m β⁻ 1.27 γ —^{98}(?)Nb 30 m β⁻—^{99}Nb 2.5 m β⁻ 3.2
42	Mo $\sigma_t = 2.5$	^{90}Mo 5.7 h K β⁺ 1.2 γ—^{91m}Mo 66 s IT β⁺ 3.99 γ—^{91}Mo 15.6 m β⁺ 3.44 —^{92}Mo 15.7% σ = (<0.006 + ?) ^{93m}Mo 6.9 h IT γ—^{93}Mo >2 y K—^{94}Mo 9.3% —^{95}Mo 15.7% σ = 14 —^{96}Mo 16.5% σ = 1 ^{97}Mo 9.5% σ = 2 —^{98}Mo 23.8% σ = 0.13 —^{99}Mo 67 h β⁻ 1.23 γ—^{100}Mo 9.5% σ = 0.2 —^{101}Mo 15 m β⁻ 2.2 γ—^{102}Mo 11 m β⁻ 1 —^{105}Mo <2 m β⁻
44	Ru $\sigma_t = 2.5$	^{94}<Ru 0.9 m β⁺ γ—^{94}Ru 1 h K—^{95}Ru 98 m K β⁺ 1.2 γ—^{96}Ru 5.6% σ = 0.01 —^{97}Ru 2.9 d K γ—^{98}Ru 1.9% —^{99}Ru 12.7% —^{100}Ru 12.7% —^{101}Ru 17.0% —^{102}Ru 31.5% σ = 1.2 —^{103}Ru 40 d β⁻ 0.69 γ—^{104}Ru 18.6% σ = 0.7 —^{105}Ru 4.5 h β⁻ 1.15 γ—^{106}Ru 1.0 y β⁻ 0.04 (3.53) γ—^{107}Ru 4.5 m β⁻ 4.3 γ—^{108}Ru ~4 m β⁻
45	Rh $\sigma_t = 150$	^{97}Rh 35 m β⁺ —^{98}Rh 9 m β⁺ 3.3 γ—^{99m}Rh 15 β⁺ γ ^{99}Rh 4.5 h β⁺ 0.74 γ—^{100}Rh 21 h β⁺ 2.62 K γ—^{101m}Rh ~5 y IT γ—^{101}Rh 4.5 d K γ—^{102}Rh 220 d K β⁻ 1.15 β⁺ 1.24 γ—^{103}Rh 100% σ = (12 + 138) —^{103m}Rh 54 m IT ^{104m}Rh 4.4 m IT γ β⁻—^{104}Rh 42 s β⁻ 2.5 γ—^{105m}Rh 30 s IT—^{105}Rh 36 s β⁻ 0.56 γ—^{106m}Rh 2 h β⁻ ~1 γ—^{106}Rh 30 s β⁻ 3.53 γ—^{107}Rh 22 m β⁻ ~2 γ—^{108}Rh 18 s β⁻ ~4 γ—^{109}Rh <1 h

Atomic N°	Element Total Absorption Cross Section (barn)	Stable Isotope (in italic): %, Thermal Neutrons Cross Section (barn). Radioisotope: Period, Emissions and Energy (MeV)
46	Pd $\sigma_t = 8$	^{98}Pd 17 m $\beta^+ - ^{99}$Pd 24 m $\gamma - ^{100}$Pd 4.0 d K $\gamma - ^{101}$Pd 8 h K β^+ 2.3 ? $- ^{102}Pd$ 1.0% $\sigma = 4.8 - ^{103}$Pd 17 d K $\gamma - ^{104}Pd$ 11.0 % $- ^{105m}$Pd 23 s I T $- ^{105}Pd$ 22.2% $- ^{106}Pd$ 27.3% $- ^{107}$Pd $7 \cdot 10^6$ y β^- 0.04 $- ^{108}Pd$ 26.7% $\sigma =$ (0.07 + 11) $- ^{109m}$Pd 4.8 m I T $- ^{109}$Pd 13.6 h β^- 1.0 $(\gamma) - ^{110}Pd$ 11.8% $\sigma = (? + 0.4) - ^{111m}$Pd 5.5 h I T β^- $\gamma - ^{111}$Pd 22 m β^- 2.14 $\gamma -$ ^{112}Pd 21 h β^- 0.28 γ $- ^{113}$Pd 1.5 m β^-
47	Ag $\sigma_t = 60$	^{102}Ag 16 m $\beta^+ - ^{103}$Ag 1.1 h β^+ 1.3 K γ $- ^{104}$Ag 27 m β^+ 2.70 $\gamma - ^{105}$Ag 40 d K $\gamma - ^{106m}$Ag 24 m β^+ 1.96 K $\gamma - ^{106}$Ag 8.3 d K $\gamma - ^{107m}$Ag 44 s I T $- ^{107}Ag$ 51.4% $\sigma = 30$ $- ^{108}$Ag 2.3 m β^- 1.77 K β^+ 0.80 $\gamma - ^{109m}$Ag 40 s I T $- ^{109}Ag$ 48.6% $\sigma = (2 + 82) - ^{110m}$Ag 270 d β^- 0.53 I T $\gamma - ^{110}$Ag 24 s β^- 2.88 γ $- ^{111m}$Ag 75 s I T $- ^{111}$Ag 7.5 d β^- 1.04 $\gamma -$ ^{112}Ag 3.12 h β^- 4.1 $\gamma - ^{113}$Ag 5.3 h β^- 2.0 $\gamma - ^{114}$Ag 2 m $\beta^- - ^{115}$Ag 21 m β^- 3 γ
48	Cd $\sigma_t = 3,300$	^{104}Cd 59 m K $\gamma - ^{105}$Cd 55 m K β^+ 1.69 $\gamma - ^{106}Cd$ 1.22 % $\sigma = 1 - ^{107}$Cd 6.7 h K β^+ 0.32 $\gamma - ^{108}Cd$ 0.88% $- ^{109}$Cd 1.3 y K L $(\gamma) - ^{110}Cd$ 12.4% $\sigma = (0.2 + ?) - ^{111m}$Cd 49 m I T $\gamma - ^{111}Cd$ 12.8% $- ^{112}Cd$ 24.0% $\sigma = (0.03 + ?)$ $- ^{113m}$Cd 5 y β^- 0.58 $- ^{113}Cd$ 12.3% $\sigma = 27,000$ $- ^{114}Cd$ 28.8% $\sigma = (0.14 + 1.1) - ^{115m}$Cd 43 d β^- 1.63 $\gamma - ^{115}$Cd 54 h β^- 1.11 $\gamma - ^{116}Cd$ 7.6% $\sigma = 1.4 - ^{117m}$Cd 3.0 h I T $- ^{117}$Cd 50 m β^- 3.0 $\gamma - ^{118}$Cd ~30 m $(\beta^-$ 4)
49	In $\sigma_t = 190$	^{107}In 30 m β^+ 2 $\gamma - ^{108}$In 50 m β^+ 2.3 $-$ ^{109}In 4.3 h K β^+ 0.7 $\gamma - ^{110m}$In 5.0 h K I T $\gamma - ^{110}$In 66 m β^+ 2.25 K $\gamma - ^{111}$In 2.80 d K $\gamma - ^{112m}$In 21 m I T $(\gamma) - ^{112m}$In 2.5 s I T $- ^{112}$In 14 m β^- 0.66 K β^+ 1.52 $-$ ^{113}In 4.2% $\sigma = (61 + 2) - ^{113m}$In 1.73 h I T $-$ ^{114m}In 49 d I T K $\gamma - ^{114}$In 72 s β^- 1.98 K $\beta^+ \sim 1$ $\gamma - ^{115m}$In 4.5 h I T β^- 0.83 $-$ ^{115}In 95.8% $6 \cdot 10^{14}$ y β^- 0.6 $\sigma = (145 + 52) -$ ^{116m}In 54.0 m β^- 1.00 $\gamma - ^{116}$In 13 s β^- 3.3 $- ^{117m}$In 1.9 h β^- 1.77 I T $\gamma - ^{117}$In 1.1 h β^- 0.74 $\gamma - ^{118m}$In 4.5 m β^- 1.5 $\gamma - ^{118}$In <1 m β^- 4 $- ^{119}$In 18 m β^- 2.7

Atomic N°	Element Total Absorption Cross Section (barn)	Stable Isotope (in italic): %, Thermal Neutrons Cross Section (barn). Radioisotope: Period, Emissions and Energy (MeV)
50	Sn $\sigma_t = 0.6$	^{108}Sn 4 h K $-^{109}$Sn 18 m K β^+ $\gamma-^{110}$Sn 4 h K $\gamma-^{111}$Sn 35 m K β^+ 1.51 $-^{112}Sn$ 1.02 % $\sigma = 1.3 -^{113}$Sn 112 d K L $\gamma-^{114}Sn$ 0.69% $-^{115}Sn$ 0.38% $-^{116}Sn$ 14.3% $\sigma = (0.006 + ?)$ $-^{117m}$Sn 14 d I T $\gamma-^{117}Sn$ 7.6% $-^{118}$Sn 24.1% $\sigma = (0.01 + ?) -^{119m}$Sn 275 d I T $\gamma-$ ^{119}Sn 8.5% $-^{120}Sn$ 32.5% $\sigma = (\sim 0.001 + 0.1) -$ ^{121m}Sn > 1 y β^- 0.42 $-^{121}$Sn 27 h β^- 0.38 $-$ ^{122}Sn 4.8% $\sigma = (0.001 + 0.2) -^{123m}$Sn 130 d β^- 1.42 $-^{123}$Sn 40 m β^- 1.26 $\gamma-^{124}Sn$ 6.1% $\sigma = (0.2 + 0.004) -^{125m}$Sn 9.5 m $- \beta^-$ 2.1 $\gamma-^{125}$Sn 10 d β^- 2.4 $\gamma-^{126}$Sn 50 m $\beta^- -^{127}$Sn 1.5 h $\beta^- -^{130}$Sn 2.6 m $\beta^- -^{131}$Sn 3.4 m $\beta^- -^{132}$Sn 2.2 m β^-
51	Sb $\sigma_t = 5.5$	^{116m}Sb 15 m β^+ 2.4 $\gamma-^{116}$Sb 60 m β^+ 1.4 $\gamma-^{117}$Sb 2.8 h K $\gamma-^{118m}$Sn 3.5 m β^+ 3.1 I T $-^{118}$Sb 5.1 h K β^+ 0.7 $\gamma-^{119}$Sb 38 h K $\gamma-^{120m}$Sb 5.8 d K $\gamma-^{120}$Sb 17 m β^+ 1.70 $\gamma-^{121}Sb$ 57% $- \sigma = 7.0 -^{122m}$Sb 3.5 m I T $\gamma-^{122}$Sb 2.8 d β^- 1.98 K β^+ 0.5 $\gamma-$ ^{123}Sb 43% $\sigma = (0.03 + 0.03 + 3.4) -^{124m}$Sb 21 m I T $\beta^- -^{124m}$Sb 1.3 m β^- 3 I T $-^{124}$Sb 60 d β^- 2.31 $\gamma-^{125}$Sb 2.7 y β^- 0.62 $\gamma-^{126m}$Sb 9 h β^- 1 $\gamma-^{126}$Sb 28 d β^- 1.9 $-^{127}$Sb 93 h β^- 1.57 $\gamma-^{128}$Sb ~ 1 h $\beta^- -^{129}$Sb 4.6 h β^- 1.70 γ $-^{130m}$Sb 10 m β^- 2.9 $\gamma-^{130}$Sb 40 m β^- γ $-^{131}$Sb 22 m β^- 1.1 $-^{132}$Sb 2 m $\beta^- -^{133}$Sb 4.1 m $\beta^- -^{134}$Sb 0.8 m β^-
52	Te $\sigma_t = 4.6$	^{116}Te ~ 3 h $(\beta^+$ 2.4$) -^{117}$Te 2.5 h β^+ 2.5 $-$ ^{118}Te 6.0 d K $(\beta^+$ 3.1$)$ $(\gamma) -^{119m}$Te 16 h K $\gamma-^{119}$Te 4.5 d K $\gamma-^{120}Te$ 0.091% $\sigma =$ $< 140 -^{121m}$Te 150 d I T $\gamma-^{121}$Te 17 d K $\gamma-^{122}Te$ 2.5% $\sigma = (1 + 2) -^{123m}$Te 104 d I T $\gamma-^{123}Te$ 0.88% $\sigma = 400 -^{124}Te$ 4.6% $\sigma =$ $(5 + 2) -^{125m}$Te 58 d I T $\gamma-^{125}Te$ 7.0% $\sigma = 1.5 -^{126}Te$ 18.7% $\sigma = (0.09 + 0.8) -^{127m}$Te 110 d I T $-^{127}$Te 9.3 h β^- 0.68 $-^{128}Te$ 31.8% $\sigma = (0.016 + 0.14) -^{129m}$Te 33 d I T $-^{129}$Te 72 m β^- 1.46 $\gamma-^{130}Te$ 34.4% $\sigma = (0.01 + 0.2) -^{131m}$Te 30 h β^- 2.5 I T $\gamma-^{131}$Te 25 m β^- 2.1 $\gamma-$ ^{132}Te 77 h β^- 0.22 γ $(\beta^-$ 2.1$)$ $(\gamma) -^{133m}$Te 63 m I T $-^{133}$Te 2 m β^- 2.4 $\gamma-^{134}$Te 44 m β^-

Atomic N°	Element Total Absorption Cross Section (barn)	Stable Isotope (in italic): %, Thermal Neutrons Cross Section (barn). Radioisotope: Period, Emissions and Energy (MeV)
53	I $\sigma_t = 6.3$	^{119}I 18 m β^+ — ^{120}I > 1.3 h β^+ 4.0 — ^{121}I 1.4 h β^+ 1.13 γ — ^{122}I 3.5 m β^+ 3.12 ^{123}I – 13 h K γ — ^{124}I 4.5 d K β^+ 2.20 γ — ^{125}I 60 d K L γ — ^{126}I 13.3 d K β^- 0.87 β^+ 1.11 γ — ^{127}I 100% $\sigma = 6.3$ — ^{128}I 25.0 m β^- 2.12 K γ — ^{129}I $1.7 \cdot 10^7$ y β^- 0.15 γ $\sigma = 30$ — ^{130}I 12.6 h β^- 1.02 γ — ^{131}I 8.05 d β^- 0.81 γ $\sigma = {\sim}600$ — ^{132}I 2.3 h β^- 2.12 γ — ^{133}I 21 h β^- 1.3 γ — ^{134}I 52 m β^- 2.5 γ — ^{135}I 6.7 h β^- 1.4 γ — ^{136}I 86 s β^- 6.4 γ — ^{137}I 22 s β^- — ^{138}I 5.9 s β^- — ^{139}I 2.7 s β^-
55	Cs $\sigma_t = 31$	^{123}Cs 6 m β^+ — ^{125}Cs 45 m K β^+ 2.05 γ — ^{126}Cs 1.6 m β^+ 3.8 K γ — ^{127}Cs 6.2 h K β^+ 1.06 γ — ^{128}Cs 3.8 m β^+ 3.0 γ — ^{129}Cs 31 h K γ — ^{130}Cs 30 m β^+ 1.97 K β^- 0.44 — ^{131}Cs 9.7 d K L — ^{132}Cs 6.2 d K γ — ^{133}Cs 100% $\sigma = (0.016 + 31)$ — ^{134m}Cs 3.1 h – I T β^- 0.55 γ — ^{134}Cs 2.3 y β^- 0.65 γ — ^{135}Cs $2.0 \cdot 10^6$ y β^- 0.21 $\sigma = {\sim}15$ — ^{136}Cs 13 d β^- 0.66 γ — ^{137}Cs 30 y β^- 1.18 (γ) $\sigma = <2$ — ^{138}Cs 32 m β^- 3.40 γ — ^{139}Cs 9.5 m β^- ${\sim}4$ γ — ^{140}Cs 66 s β^- — ^{142}Cs ${\sim}1$ m – β^-
56	Ba $\sigma_t = 1.2$	^{126}Ba 97 m K $(\beta^+$ 3.8$)$ γ — ^{127}Ba ${\sim}12$ m — ^{128}Ba 2.4 d K $(\beta^+$ 3.0$)$ γ — ^{129}Ba 1.9 h β^+ 1.6 γ — ^{130}Ba 0.101% $\sigma = 6$ — ^{131}Ba 11.6 d K γ — ^{132}Ba 0.097% $\sigma = 3$ — ^{133m}Ba 39 h I T — ^{133}Ba 8 y K γ — ^{134}Ba 2.42% $\sigma = <4$ — ^{135m}Ba 29 h I T — ^{135}Ba 6.6% $\sigma = 5$ — ^{136}Ba 7.8% $\sigma = <1$ — ^{137m}Ba 2.60 m I T — ^{137}Ba 11.3% $\sigma = 4$ — ^{138}Ba 71.7% $\sigma = 0.55$ — ^{139}Ba 85 m β^- 2.38 γ $\sigma = 4$ — ^{140}Ba 12.8 d β^- 1.02 γ — ^{141}Ba 18 m β^- 2.8 γ — ^{142}Ba 6 m β^- — ^{143}Ba <0.5 m β^-
57	La $\sigma_t = 8.9$	^{131}La 58 m β^+ 1.6 — ^{132}La 4.5 h β^+ 3.5 γ — ^{133}La 4 h K β^+ 1.2 γ — ^{134}La 6.5 m K β^+ 2.7 — ^{135}La 19 h K γ — ^{136}La 9 m K β^+ 2.1 — ^{137}La > 10^5 y — ^{138}La 0.089% $2 \cdot 10^{11}$ y K β^- 1.0 γ — ^{139}La 99.911% $\sigma = 8.9$ — ^{140}La 40.2 h β^- 2.15 γ $\sigma = 3$ — ^{141}La 3.8 h β^- 2.43 γ — ^{142}La 77 m β^- > 2.5 γ — ^{143}La ${\sim}19$ m β^-

Atomic N°	Element Total Absorption Cross Section (barn)	Stable Isotope (in italic): %, Thermal Neutrons Cross Section (barn). Radioisotope: Period, Emissions and Energy (MeV)
58	Ce $\sigma_t = 0.7$	^{133}Ce 6.3 h K β^+ 1.3 $\gamma-^{134}Ce$ 72 h K — ^{135}Ce 22 h K β^+ 0.8 $-^{136}Ce$ 0.19% $\sigma =$ (~2 x ~20) $-^{137m}Ce$ 35 h I T $-^{137}Ce$ 9 h K — $\gamma-^{138}Ce$ 0.26% $\sigma = 1 - ^{139}Ce$ 140 d K $\gamma-^{140}Ce$ 88.47% $\sigma = 0.6 - ^{141}Ce$ 32 d β^- 0.57 $\gamma-^{142}Ce$ 11.08% $\sigma = 1 - ^{143}Ce$ 33 h β^- 1.38 γ $\sigma = 6 - ^{144}Ce$ 285 d β^- 0.30 $\gamma-^{145}Ce$ 3.0 m β^- 2.0 $\gamma-^{146}Ce$ 14 m β^- 0.7 γ
59	Pr $\sigma_t = 11$	^{135}Pr 22 m β^+ 2.5 $\gamma-^{136}Pr$ 70 m β^+ 2.0 γ $-^{138}Pr$ 2.0 h β^+ 1.4 $\gamma-^{139}Pr$ 4.5 h $- \beta^+$ 1.0 $\gamma-^{140}Pr$ 3.4 m β^+ 2.3 K $-^{141}Pr$ 100% $\sigma = 11 - ^{142}Pr$ 19.1 h β^- 2.16 $\gamma-^{143}Pr$ 13.8 d β^- 0.92 $-^{144}Pr$ 17 m β^- 2.98 $\gamma-^{145}Pr$ 5.9 h β^- 1.7 $\gamma-^{146}Pr$ 24 m β^- 3.7 γ
60	Nd $\sigma_t = 48$	^{138}Nd 22 m β^+ 2.4 $-^{139}Nd$ 5.5 h K β^+ 3.1 $\gamma-^{140}Nd$ 3.3 d K $-^{141}Nd$ 2.4 h K β^+ 0.7 $\gamma-^{142}Nd$ 27.1% $\sigma = 17 - ^{143}Nd$ 12.2% $\sigma =$ 320 $-^{144}Nd$ 23.9% $\sim 2 \cdot 10^{15}$ y α 1.8 $\sigma = 5 -$ ^{145}Nd 8.3% $\sigma = 44 - ^{146}Nd$ 17.2% $\sigma = 2 - ^{147}$ Nd 11.6 d β^- 0.81 $\gamma-^{148}Nd$ 5.7% $\sigma = 4 -$ ^{149}Nd 1.8 h β^- 1.5 $\gamma-^{150}Nd$ 5.6% $\sigma = 3 -$ ^{151}Nd 15 m β^- 1.9 γ
62	Sm $\sigma_t = 10,000$	^{143}Sm 8 m β^+ 2.3 $-^{144}Sm$ 3.1% $\sigma \sim 0.03 -$ ^{145}Sm 1.0 y K $\gamma-^{146}Sm$ $\sim 5 \cdot 10^7$ y α 2.5 $-^{147}Sm$ 1.3 $\cdot 10^{11}$ y α 2.18 15.0% $-^{148}Sm$ 11.2% $-^{149}Sm$ 13.8% $\sigma = 66,000 - ^{150}Sm$ 7.4% $-^{151}Sm$ 80 y β^- 0.076 γ $\sigma = 12,000 - ^{152}Sm$ 26.8% $\sigma = 140 - ^{153}Sm$ 47 h β^- 0.81 $\gamma-^{154}Sm$ 22.7% $\sigma = 5 - ^{155}Sm$ 23 m β^- 1.8 $\gamma-^{156}Sm$ ~ 10 h β^- 0.9
63	Eu $\sigma_t = 4,300$	^{144}Eu 18 m β^+ 2.4 $-^{145}Eu$ 5 d K $-^{146}Eu$ 38 h K $-^{147}Eu$ 24 d K γ α 2.9 $-^{148}Eu$ 58 d K $\gamma-^{148}Eu$ 120 d K $\gamma-^{150}Eu$ 14 h β^- 1.1 $-^{151}Eu$ 47.8% $\sigma = (1400 + 7200) - ^{153m}Eu$ 9.3 h β^- 1.88 K $\gamma-^{152}Eu$ 13 y $-$ K $- \beta^-$ 0.70 γ $\sigma = 5000 - ^{153}Eu$ 52.2% $\sigma = 400 - ^{154}Eu$ 16 y β^- 1.5 γ $\sigma = 1400 - ^{155}Eu$ 1.7 y β^- 0.25 γ $\sigma = 13,000 - ^{156}Eu$ 15 d β^- 2.4 $\gamma-^{157}Eu$ 15 h β^- ~ 1.7 γ $-^{158}Eu$ 60 m β^- 2.6 γ $-^{159}Eu$ 20 m β^-

Atomic N°	Element Total Absorption Cross Section (barn)	Stable Isotope (in italic): %, Thermal Neutrons Cross Section (barn). Radioisotope: Period, Emissions and Energy (MeV)
64	Gd $\sigma_t = 38{,}000$	^{148}Gd >35 $\alpha\,3.2 - ^{149}$Gd 9 d K γ $\alpha\,3.0$ $- ^{150}$Gd $>10^5$ y $\alpha\,2.7 - ^{151}$Gd ~ 150 d K γ $- ^{152}Gd$ 0.20% $\sigma = <180 - ^{153}$Gd 236 d K $\gamma - ^{154}Gd$ 2.15% $- ^{155}Gd$ 14.7% $\sigma = \sim 70{,}000 -$ ^{156}Gd 20.5% $- ^{157}Gd$ 15.7% $\sigma = \sim 180{,}000 -$ ^{158}Gd 24.9% $\sigma = 4 - ^{159}Gd$ 18 h $\beta^-\,0.95$ γ $- ^{160}Gd$ 21.9% $\sigma = 0.8$ ^{161}Gd 3.7 m $\beta^-\,1.6$ γ
65	Tb $\sigma_t = 45$	^{149}Tb 4.1 h $\alpha\,3.95$ K $- ^{150}$Tb 19 h K $\alpha\,3.4$ $- ^{153}$Tb 5.1 d K $\gamma - ^{154m}$Tb 7 h K β^-? $- ^{154}$Tb 17.2 h K $\beta^+\,2.75$ $\gamma - ^{156m}$Tb 5 h $\beta^-\,0.14 - ^{156}$Tb 5 d K β^-? 0.6 $\gamma - ^{159}Tb$ 100% $\sigma = 45 - ^{160}$Tb 72 d $\beta^-\,0.85$ γ $\sigma = \sim 600 -$ ^{161}Tb 7 d $\beta^-\,0.55$ $\gamma - ^{162}$Tb 14 m
66	Dy $\sigma_t = 1100$	^{156}Dy 0.052% $- ^{157}$Dy 8.2 h K $\gamma - ^{158}Dy$ 0.090% $- ^{159}$Dy 134 d K L $- ^{160}Dy$ 2.29% $-$ ^{161}Dy 18.9% $- ^{162}Dy$ 25.5% $- ^{163}Dy$ 25.0% $-$ ^{164}Dy 28.2% $\sigma = 2700 - ^{165m}$Dy 1.3 m I T $\beta^- \sim 1$ $\gamma - ^{165}$Dy 2.32 h $\beta^-\,1.25$ γ $\sigma = 4700$ $- ^{166}$Dy 82 h $\beta^-\,0.3$ γ
67	Ho $\sigma_t = 64$	^{160}Ho 5.0 h γ K $- ^{161}$Ho 2.5 h K $\gamma -$ ^{162}Ho 22 m K $\beta^+\,1.3$ $\gamma - ^{164}$Ho 37 m K $\beta^-\,0.09$ I T $\gamma - ^{165}Ho$ 100% $\sigma = (>0.007 + 64)$ $- ^{166m}$Ho >30 y $\beta^-\,0.2$ $\gamma - ^{166}$Ho 27.2 h $\beta^-\,1.85$ $\gamma - ^{167}$Ho 3.0 h $\beta^-\,1.0$ $\gamma - ^{169}$Ho 1.6 h β^-
68	Er $\sigma_t = 170$	^{160}Er 29 h K $- ^{161}$Er 3 h K $\beta^+\,1.2$ $\gamma -$ ^{162}Er 0.136% $- ^{163}$Er 75 m K $\gamma - ^{165}Er$ 1.56% $- ^{165}$Er 10 h K $- ^{166}Er$ 33.4% $-$ ^{167}Er 22.9% $- ^{168}Er$ 27.1% $\sigma = 2 - ^{169}$Er 9.4 d $\beta^-\,0.33 - ^{170}Er$ 14.9% $\sigma = 9 - ^{171}$Er 7.5 h $\beta^-\,1.48$ γ
69	Tu $\sigma_t = 125$	^{165}Tm 25 h K $\gamma - ^{166}$Tm 7.7 h K β^+ 2.1 $\gamma - ^{167}$Tm 9.6 d K $\gamma - ^{168}$Tm 87 d K $\gamma - \beta^-\,0.5$ (?) $- ^{169}Tm$ 100% $\sigma = 125 -$ ^{170}Tm 129 d $\beta^-\,0.97$ γ $\sigma = \sim 2000 - ^{171}$Tm 1.9 y $\beta^-\,0.10 - ^{172}$Tm 19 m $\beta^- - ^{174}$Tm ~ 2 d β^-

Atomic N°	Element Total Absorption Cross Section (barn)	Stable Isotope (in italic): %, Thermal Neutrons Cross Section (barn). Radioisotope: Period, Emissions and Energy (MeV)
70	Yb $\sigma_t = 37$	^{166}Yb 54 d K $\gamma - ^{167}$Yb 18 m K $\gamma - ^{168}Yb$ 0.14% $\sigma = 11{,}000 - ^{169}$Yb 32 d K $\gamma - ^{170}Yb$ 3.03% $- ^{171}Yb$ 14.3% $- ^{172}Yb$ 21.8% $- ^{173}Yb$ 16.2% $- ^{174}Yb$ 31.8% $\sigma = \sim 60 - ^{175}$Yb 4.2d $\beta^- 0.47$ $\gamma - ^{176}Yb$ 12.7% $\sigma = 7 - ^{177}$Yb 2.0 h $\beta^- 1.30$ γ
71	Lu $\sigma_t = 111$	^{170}Lu 1.7 d K $\gamma - ^{171m}$Lu 8.5 d K $\gamma -$ ^{171}Lu 1.6 y K $\gamma - ^{172m}$Lu 4.0 h $\beta^+ 1.2 -$ ^{172}Lu 6.7 d K $\gamma - ^{173}$Lu 1.4 y K $\gamma -$ ^{174}Lu 165 d K $\beta^- 0.6$ $\gamma - ^{175}Lu$ 97.40% $\sigma = (18 + ?) - ^{176m}$Lu 3.7 h $\beta^- 1.2$ $\gamma - ^{176}Lu$ 2.60% $3 \cdot 10^{10}$ y $\beta^+ 0.42$ K γ $\sigma = 3{,}200 - ^{177}$Lu 6.8 d $\beta^- 0.50$ γ ^{178}Lu 22 m $\beta^- - ^{179}$Lu ~ 5 h β^-
72	Hf $\sigma_t = 105$	^{170}Hf 1.8 h $\beta^+ 2.4 - ^{171}$Hf 16 h K $\gamma - ^{172}$Hf ~ 5 y K $\gamma - ^{173}$Hf 24 h K $\gamma - ^{174}Hf$ 0.18% $\sigma = \sim 1000 - ^{175}$Hf 70 d K $\gamma - ^{176}Hf - 5.2\% -$ $\sigma = <30 - ^{177}Hf$ 18.5% $\sigma = 370 - ^{178}Hf = 27.1\%$ $\sigma = 80 - ^{179m}$Hf 19 s I T $\gamma - ^{179}Hf$ 13.8% $\sigma = 65 - ^{180m}$Hf 5.5 h I T $\gamma - ^{180}Hf$ 35.2% $\sigma = 10 - ^{181}$Hf 46 d $\beta^- 0.41$ γ
73	Ta $\sigma_t = 22$	^{176}Ta 8.0 h K $\gamma - ^{177}$Ta 2.2 d K $\gamma - ^{178m}$Ta 9.3 m K $\beta^+ 1.1$ $\gamma - ^{178}$Ta 2.1 h $-$ K $\beta^+ \sim 1$ $\gamma - ^{179}$Ta ~ 600 d K $\gamma - ^{180m}$Ta 8.1 h K $\beta^- 0.70$ $\gamma - ^{180}Ta$ 0.012% $- ^{181}Ta$ 99.988 $\sigma = (0.07 + 22) - ^{182m}$Ta 16 m I T β^-? ^{182}Ta 112 d $\beta^- 0.51 - \gamma$ $\sigma = 20{,}000 - ^{183}$Ta 5.2 d $\beta^- 0.62$ $\gamma - ^{184}$Ta 8.7 h $\beta^- 1.26$ $\gamma - ^{185}$Ta 49 m $\beta^- 1.7$ $\gamma - ^{186}$Ta 10 m $\beta^- 2.2$ γ
74	W $\sigma_t = 18$	^{176}W 1.3 h K $\beta^+ \sim 2$ $\gamma - ^{177}$W 2.2 h K $\gamma - ^{178}$W 21 d K $\gamma - ^{179}$W 30 m K $- ^{180}W$ 0.14% $- 10^{14}$ y(?)$- \alpha$ 3 (?) $\sigma = <20 - ^{181}$W 140 d K L $\gamma - ^{182}W$ 26.2% $\sigma = (0.5 + 20) - ^{183m}$W 5.5 s I T $\gamma - ^{183}W = 14.3\%$ σ $11 - ^{184}W$ 30.7% $\sigma = 2.0 - ^{185m}$W 1.7 m I T $\gamma - ^{185}$W 74 d $\beta^- 0.43$ $\gamma - ^{186}W$ 28.7% $\sigma = 36 - ^{187}$W 24 h $\beta^- 1.31$ γ $\sigma = \sim 80 - ^{188}$W 65 d β^-

Atomic N°	Element Total Absorption Cross Section (barn)	Stable Isotope (in italic): %, Thermal Neutrons Cross Section (barn). Radioisotope: Period, Emissions and Energy (MeV)
75	Re $\sigma_t = 86$	^{180}Re 2.4 m β^+ 1.1 $\gamma - {}^{182m}$Re 13 h K $\gamma -$ ^{182}Re 64 h K $\gamma - {}^{183}$Re 150 d K $\gamma -$ ^{184m}Re 50 d K $\gamma - {}^{184}$Re 2.2 d K $\gamma -$ ^{185}Re 37.1% $\sigma = 105 - {}^{186}$Re 91 h β^- 1.07 K $\gamma - {}^{187}Re$ 62.9% $\sim 5 \cdot 10^{10}$ y $\beta^- < 0.008$ $\sigma = 75$ $- {}^{188m}$Re 20 m I T $\gamma - {}^{188}$Re 17 h β^- 2.12 γ $\sigma = < 3 - {}^{189}$Re ~ 200 d β^- 0.2 $\gamma - {}^{190}$Re 3 m β^- 1.7 $\gamma - {}^{191}$Re 10 m β^- 1.8
76	Os $\sigma_t = 15$	^{182}Os 24 h K $- {}^{183m}$Os 10 h K $\gamma - {}^{183}$Os 15 h K $\gamma - {}^{184}Os$ 0.018% $\sigma = \sim 20 - {}^{185}$Os 95 d K L $\gamma - {}^{186}Os$ 1.59% $- {}^{187m}$Os 35 h I T? $- {}^{187}Os$ 1.64% $- {}^{188}$Os 13.3% ^{189}Os 16.1% $- {}^{190m}$Os 9 m I T $\gamma - {}^{190m}$Os 6 h I T? $-$ ^{190}Os 26.4% $\sigma = 8 - {}^{191m}$Os 14 h I T $-$ ^{191}Os 16 d $- \beta^-$ 0.14 $\gamma - {}^{192}Os$ 41.0% $\sigma = 1.6$ $- {}^{193}$Os 31 h β^- 1.10 γ $\sigma = 200 - {}^{194}$Os ~ 2 y β^-
77	Ir $\sigma_t = 460$	^{187}Ir 12 h K β^+ 2.2 $\gamma - {}^{188}$Ir 41 h K β^+ 2 $\gamma - {}^{189}$Ir 11 d K $\gamma - {}^{190m}$Ir 3 h K β^+ 2.0 $\gamma - {}^{190}$Ir 11 d K $\gamma - {}^{191m}$Ir 5 s I T $\gamma - {}^{191}Ir$ 38.5% $\sigma = (250 + 750) - {}^{192m}$Ir 1.4 m I T $\beta^- - {}^{192}$Ir $- 74$ d β^- 0.67 K $\gamma - {}^{193}Ir$ 61.5% $\sigma = 120 - {}^{194}$Ir 19 h β^- 2.24 $\gamma - {}^{195}$Ir 2.3 h β^- 2.1 $\gamma - {}^{196}$Ir 9.7 d β^- 0.08 $\gamma -$ ^{197}Ir 7 m β^- 1.6 $\gamma - {}^{198}$Ir 50 s β^- 3.6 γ
78	Pt $\sigma_t = 10$	^{187}Pt 3 h $- {}^{188}$Pt 10 d K $\gamma - {}^{189}$Pt 11 h K $\gamma - {}^{190}Pt$ 0.012% $\sim 10^{12}$ y (?) α 3.3 (?) $\sigma = \sim 90$ $- {}^{191}$Pt 3.0 d K $\gamma - {}^{192}Pt$ 0.78% $\sigma = 8 -$ ^{193m}Pt 3.4 d I T $- {}^{193}$Pt long L $- {}^{194}Pt$ 32.8% $\sigma = 1.2 - {}^{195m}$Pt ~ 6 d I T $\gamma - {}^{195}Pt$ 33.7% $\sigma = 27 - {}^{196}Pt$ 25.4% $\sigma = (\sim 0 + 1.2) - {}^{197m}$Pt 1.4 h I T $- {}^{197}$Pt 19 h β^- 0.67 $\gamma - {}^{198}Pt$ 7.2% $\sigma = 4 - {}^{199}$Pt 30 m $\beta^- \sim 1.2$ γ
79	Au $\sigma_t = 98$	^{187}Au ~ 15 m $- {}^{188}$Au ~ 10 m $- {}^{189}$Au 42 m K $\gamma - {}^{191}$Au 3 h K $\gamma - {}^{192}$Au 4.8 d K β^+ 1.9 $\gamma - {}^{193m}$Au 4 s I T $\gamma - {}^{193}$Au 17 h K $\gamma - {}^{194}$Au 39 h K β^+ 1.55 $\gamma - {}^{195m}$Au 30 s I T $\gamma - {}^{195}$Au 180 d K $\gamma - {}^{196m}$Au 14 h K I T $- {}^{196}$Au 5.6 d K β^- 0.27 $\gamma -$ ^{197m}Au 7.4 s I T $\gamma - {}^{197}Au = 100\%$ $\sigma = 98 -$ ^{198}Au 2.70 d β^- 1.37 $\gamma - \sigma = 26,000 - {}^{199}$Au

Atomic N°	Element Total Absorption Cross Section (barn)	Stable Isotope (in italic): %, Thermal Neutrons Cross Section (barn). Radioisotope: Period, Emissions and Energy (MeV)
		3.15 d β^- 0.46 γ $\sigma = {\sim}30 - {}^{200}$Au 48 m β^- 2.2 $\gamma - {}^{201}$Au 26 m β^- 1.5 $\gamma - {}^{202}$Au ${\sim}25$ s $\beta^- - {}^{203}$Au 55 s β^- 1.9 γ
80	Hg $\sigma_t = 350$	^{189}Hg 25 m β^+ K $\gamma - {}^{190}$Hg 90 m K $- {}^{191}$Hg 57 m K $\gamma - {}^{192}$Hg 6 h K β^+ 1.2 $\gamma - {}^{193m}$Hg 12 h I T γ K $- {}^{193}$Hg 5 h K γ ^{194}Hg ${\sim}130$ d K $- {}^{195m}$Hg 40 h K I T $\gamma - {}^{195}$Hg 9.5 h K $\gamma - {}^{196}Hg$ 0.15% $\sigma = 2500 - {}^{197m}$Hg 25 h I T K $\gamma - {}^{197}$Hg 65 h K $\gamma - {}^{198}Hg$ 10.0% $- {}^{199m}$Hg 43 m I T $\gamma - {}^{199}Hg$ 16.9% $\sigma = 2000 - {}^{200}Hg$ 23.1% $\sigma = {<}50 - {}^{201}Hg$ 13.2% $\sigma = {<}50 - {}^{202}Hg$ 29.8% $\sigma = 3 - {}^{203}$Hg 48 d β^- 0.21 $\gamma - {}^{204}Hg = 6.8\%$ $- \sigma = 0.4 - {}^{205}$Hg $- 5.2$ m $- \beta^-$ 1.6 $- \gamma -$
81	Tl $\sigma_t = 3.3$	^{195}Tl 1.2 h $- {}^{196}$Tl ${\sim}4$ h K $\gamma - {}^{197}$Tl 2.8 h K $\gamma - {}^{198m}$Tl 1.9 h K I T $\gamma - {}^{198}$Tl 5 h K $\gamma - {}^{199}$Tl 7.4 h K $\gamma - {}^{200}$Tl 27 h K β^+ $\gamma - {}^{201}$Tl 3.0 d K $\gamma - {}^{202}$Tl 12 d K L $\gamma - {}^{203}Tl$ 29.5% $\sigma = 11 - {}^{204}$Tl 4.1 y β^- 0.76 K $- {}^{205}Tl$ 70.5% $\sigma = 0.11 - {}^{206}$Tl 4.20 m β^- 1.51 $- {}^{207}$Tl 4.78 m β^- 1.45 $\gamma - {}^{208}$Tl 3.1 m β^- 1.79 $\gamma - {}^{209}$Tl 2.2 m β^- 2.3 $\gamma - {}^{210}$Tl 1.32 m β^- 1.9 γ
82	Pb $\sigma_t = 0.17$	^{197}Pb 42 m K I T (?) $\gamma - {}^{198}$Pb 2.3 h K $\gamma - {}^{199m}$Pb 12 m I T $- {}^{199}$Pb 1.5 h K $\gamma - {}^{200}$Pb 21 h K $\gamma - {}^{201m}$Pb 1.0 m I T $- {}^{201}$Pb 9 h K $\gamma - {}^{202m}$Pb 3.5 h I T K $\gamma - {}^{202}$Pb ${\sim}105$ y L $- {}^{203m}$Pb 6 s I T $- {}^{203}$Pb 52 h K $\gamma - {}^{204m}$Pb 68 m I T $\gamma - {}^{204}Pb$ 1.3% $\sigma = 0.9 - {}^{205}$Pb $> 10^6$ y L ${}^{206}Pb$ 26% $\sigma = 0.03 - {}^{207m}$Pb 0.80 s I T $\gamma - {}^{207}Pb$ 21% $\sigma = 0.73 - {}^{208}Pb$ 52% $\sigma = 0.00045 - {}^{209}$Pb 3.3 h β^- 0.62 $- {}^{210}$Pb 20 y β^- 0.020 $\gamma - {}^{211}$Pb 36.1 m β^- 1.4 $\gamma - {}^{212}$Pb 10.64 h β^- 0.58 $\gamma - {}^{214}$Pb 26.8 m β^- 0.7 γ
83	Bi $\sigma_t = 0.033$	^{198}Bi 7 m K α 5.83 $- {}^{199}$Bi ${\sim}25$ m K α 5.47 $- {}^{200}$Bi 35 m K $- {}^{201m}$Bi 1.0 h K α 5.15 $- {}^{201}$Bi ${\sim}2$ h K $- {}^{202}$Bi 1.6 h K $- {}^{203}$Bi 12 h K α 4.85 $\gamma - {}^{204}$Bi 12 h K $\gamma - {}^{205}$Bi 14 d K $\gamma - {}^{206}$Bi 6.4 d K $\gamma - {}^{207}$Bi 8.0 y K L $\gamma - {}^{208}$Bi $- {}^{209}Bi$ 100% 3.10^{17} y (?)

Atomic N°	Element Total Absorption Cross Section (barn)	Stable Isotope (in italic): %, Thermal Neutrons Cross Section (barn). Radioisotope: Period, Emissions and Energy (MeV)
		α 3 (?) $\sigma = 0.019 + 0.014) - ^{210m}Bi$ 5.0 d β^- 1.17 $\alpha - ^{210}Bi$ 2.6 · 10^6 y α 4.94 $\beta^- - ^{211}Bi$ 2.15 m α 6.62 β^- $\gamma - ^{212}Bi$ 60.5 m β^- 2.25 α 6.09 $\gamma - ^{213}Bi$ 47 m β^- 1.39 γ α 5.9 $- ^{214}Bi$ 19.7 m β^- 3.17 α 5.5 $\gamma - ^{215}Bi$ 8 m β^-
90	Th $\sigma_t = 7.5$	^{223}Th $<$1 m α 7.5 $- ^{224}Th$ $<$9 m α 7.13 $-$ ^{225}Th 8 m α 6.57 K $- ^{226}Th$ 31 m α 6.34 $\gamma - ^{227}Th$ 18.2 d α 5.97 6.03 γ $\sigma = 1500 -$ ^{228}Th 1.90 y α 5.42 γ $\sigma = 120 - ^{229}Th$ 7300 y α 5.02 $\sigma = \sim 45 - ^{230}Th$ 80.000 y $- \alpha$ 4.68 γ $\sigma = 35 - ^{231}Th$ 25.6 h β^- 0.30 $\gamma - ^{232}Th$ 100% 1.39 x 10^{10} y $- \alpha$ 3.99 γ F S $\sigma = 7.5 - ^{233}Th$ 23.3 m β^- 1.23 γ $\sigma = 1400 - ^{234}Th$ 24.10 d β^- 0.19 γ $\sigma = 1.8 - ^{235}Th$ $<$5 m β^-
92	U $\sigma_t = 7.8$	^{227}U 1.3 m α 6.8 $- ^{228}U$ 9.3 m α 6.67 K $-$ ^{229}U 58 m K α 6.42 $- ^{230}U$ 21 d α 5.89 γ $\sigma = 25 - ^{231}U$ 4.3 d K α 5.45 γ $\sigma \sim 400$ ^{232}U 74yy α 5.32 γ $\sigma \sim 300 - ^{233}U$ 1.62 · 10^5 y α 4.82 γ $\sigma = 60 - ^{234}U$ 0.0055% 2.50 · 10^5 y α 4.76 γ F S $\sigma = 80 - ^{235}U$ 0.72% 7.1 · 10^8 y α 4.58 F S γ $\sigma = 108 - ^{236}U$ 2.39 · 10^7 y α 4.50 γ F S $\sigma = 8 - ^{237}U$ 6.75 d β^- 0.24 γ $- ^{238}U$ 99.27% 4.51 · 10^9 y α 4.18 γ F S $\sigma = 2.8 - ^{239}U$ 23.5 m β^- 1.21 γ $\sigma = 22 - ^{240}U$ 14 h β^- 0.36 (γ)

AUTHOR INDEX

Abbey, S. 334,336,347
Abdullaev, A. A. 384,406
Abdullaev, K. 384,386,406
Abrams, I. A. 398,406,408
Achmanova, M. V. 88
Adams, F. C. 345,347
Addink, N. W. H. 101,127
Adler, I. 334,347
Adler, L. 349
Adriaenssens, E. 251,254
Adrian, P. 255
Agarwal, M. 343,347
Ageeva, L. V. 129
Agemian, H. 211,216,254
Aggett, J. 235,240,254
Agnew, W. F. 255
Agranov, K. I. 10,25
Agterdenbos, J. 89,290,293
Aguilar, C. 410
Ahrens, L. H. 35,88
Aitken, D. W. 348
Akaiwa, H. 411
Akbarov, U. 383,406
Akselsson, R. 348
Al-Atyia, M. J. 406
Albert, P. 364,380,381,406,407,410
Alberts, J. J. 225,254
Albrecht, L. 347
Albu Yaron, A. 388,406
Alder, J. F. 252,254
Aldous, K. M. 147,174,259,277,293
Alexander, M. E. 338,347
Alfer, H. 408
Alger, D. 285,293
Alkemade, C. Th. J. 133,156,158,161,
    174,265,293
Alian, A. 382,383,406

Alimarin, I. P. 212,254,262
Allabergenov, B. R. 406
Allam, B. 406
Allen, W. J. F. 110,127
Allie, W. 349
Alonso-Pascual, J. J. 127
Al-Shahristani, H. 388,406
Alvarez-Herrero, C. 110,116,127
Amiel, C. 206
Amiel, S. 406
Amirshahi, L. 131
Amos, M. D. 15,25,138,167,174,246,254,
    279,291,292,293
Anand, V. D. 235,254
Anan'ev, V. S. 110,127
Anastase, S. 382,406
Andersen, C. A. 63,88
Anderson, D. H. 129,131
Anderson, J. 394,406
Anderson, M. R. 408
Anderson, R. G. 285,293,294
Ando, A. 258
Andreeva, T. P. 89,131
Andren, A. W. 290,294
Angell, G. R. 349
Anglin, J. H. 8,13,25
Angren, A. 295
Anni, C. K. 27
Anthony, N. R. 128,255
Antic, E. 127
Antonacopoulos, N. 234,254
Aoba, K. 216,254
Araki, M. 262
Araktingi, Y. E. 253,254,255,259
Aras, N. K. 410
Archer, M. C. 236,254
Argauer, R. J. 27

459

SUBJECT INDEX